*Experimental Methods in the Physical Sciences*

VOLUME 41

OPTICAL RADIOMETRY

# EXPERIMENTAL METHODS IN THE PHYSICAL SCIENCES

Robert Celotta and Thomas Lucatorto, *Editors in Chief*

*Founding Editors*

L. MARTON
C. MARTON

*Volume 41*

# Optical Radiometry

*Edited by*

Albert C. Parr
Raju U. Datla
*National Institute of Standards and Technology*
*Gaithersburg, Maryland, USA*

James L. Gardner
*CSIRO National Measurements Laboratory*
*West Lindfield, New South Wales, Australia*



2005

ELSEVIER
ACADEMIC
PRESS

Amsterdam Boston Heidelberg London New York Oxford Paris
San Diego San Francisco Singapore Sydney Tokyo

ELSEVIER B.V.
Radarweg 29
P.O. Box 211
1000 AE Amsterdam
The Netherlands

**ELSEVIER Inc.**
**525 B Street, Suite 1900**
**San Diego**
**CA 92101-4495**
**USA**

ELSEVIER Ltd
The Boulevard, Langford
Lane, Kidlington
Oxford OX5 1GB
UK

ELSEVIER Ltd
84 Theobalds Road
London
WC1X 8RR
UK



First edition 2005

Library of Congress Cataloging in Publication Data
A catalog record is available from the Library Congress.

British Library Cataloguing in Publication Data
A catalogue record is available from the British Library.

ISBN: 0 12 475988 2
ISSN: 1079-4042

∞ The paper used in this publication meets the requirements of ANSI/NISO Z39.48-1992 (Permanence of Paper).
Printed in the USA.

# CONTENTS

# CONTRIBUTORS

Numbers in parentheses indicate the pages on which authors' contributions begin.

ROBERT A. BARNES (453), *Science Applications International Corporation, Beltsville, Maryland, USA*

L. P. BOIVIN (97), *National Research Council of Canada, Ottawa, Canada*

STEVEN W. BROWN (155), *National Institute of Standards and Technology, Gaithersburg, Maryland, USA*

JAMES J. BUTLER (453), *NASA's Goddard Space Flight Center, Greenbelt, Maryland, USA*

RAJU U. DATLA (1, 535), *National Institute of Standards and Technology, Gaithersburg, Maryland, USA*

GORDON W. DAY (367), *National Institute of Standards and Technology (retired), Boulder, Colorado, USA*

GEORGE P. EPPELDAUER (155), *National Institute of Standards and Technology, Gaithersburg, Maryland, USA*

NIGEL P. FOX (35), *National Physical Laboratory, Teddington, UK*

J. L. GARDNER (291, 547), *National Measurement Institute, Lindfield, Australia*

JÖRG HOLLANDT (213), *High-Temperature and Vacuum Physics Department, Physikalisch-Technische Bundesanstalt, Berlin, Germany*

B. CAROL JOHNSON (453), *National Institute of Standards and Technology, Gaithersburg, Maryland, USA*

ROMAN KLEIN (213), *Photon Radiometry Department, Physikalisch-Technische Bundesanstalt, Berlin, Germany*

KEITH R. LYKKE (155), *National Institute of Standards and Technology, Gaithersburg, Maryland, USA*

ALAN MIGDALL (213), *Optical Technology Division, National Institute of Standards and Technology, Gaithersburg, Maryland, USA*

YOSHI OHNO (327), *National Institute of Standards and Technology, Gaithersburg, Maryland, USA*

ALBERT C. PARR (1, 535), *National Institute of Standards and Technology, Gaithersburg, Maryland, USA*

JOSEPH P. RICE (35), *National Institute of Standards and Technology, Gaithersburg, Maryland, USA*

JOACHIM SEIDEL (213), *High-Temperature and Vacuum Physics Department, Physikalisch-Technische Bundesanstalt, Berlin, Germany*

ERIC L. SHIRLEY (409, 535), *National Institute of Standards and Technology, Gaithersburg, Maryland, USA*

GERHARD ULM (213), *Photon Radiometry Department, Physikalisch-Technische Bundesanstalt, Berlin, Germany*
MICHAEL WARE (213), *Department of Physics and Astronomy, Brigham Young University, Provo, Utah*

# VOLUMES IN SERIES

# EXPERIMENTAL METHODS IN THE PHYSICAL SCIENCES

(formerly Methods of Experimental Physics)



Volume 1. Classical Methods
*Edited by* Immanuel Estermann

Volume 2. Electronic Methods, Second Edition (in two parts)
*Edited by* E. Bleuler and R. O. Haxby

Volume 3. Molecular Physics, Second Edition (in two parts)
*Edited by* Dudley Williams

Volume 4. Atomic and Electron Physics—Part A: Atomic Sources and Detectors; Part B: Free Atoms
*Edited by* Vernon W. Hughes and Howard L. Schultz

Volume 5. Nuclear Physics (in two parts)
*Edited by* Luke C. L. Yuan and Chien-Shiung Wu

Volume 6. Solid State Physics—Part A: Preparation, Structure, Mechanical and Thermal Properties; Part B: Electrical, Magnetic and Optical Properties
*Edited by* K. Lark-Horovitz and Vivian A. Johnson

Volume 7. Atomic and Electron Physics—Atomic Interactions (in two parts)
*Edited by* Benjamin Bederson and Wade L. Fite

Volume 8. Problems and Solutions for Students
*Edited by* L. Marton and W. F. Hornyak

Volume 9. Plasma Physics (in two parts)
*Edited by* Hans R. Griem and Ralph H. Lovberg

Volume 10. Physical Principles of Far-Infrared Radiation
*Edited by* L. C. Robinson

Volume 11. Solid State Physics
*Edited by* R. V. Coleman

Volume 12. Astrophysics—Part A: Optical and Infrared Astronomy
*Edited by* N. Carleton

Part B: Radio Telescopes; Part C: Radio Observations
*Edited by* M. L. Meeks

Volume 13. Spectroscopy (in two parts)
*Edited by* Dudley Williams

Volume 14. Vacuum Physics and Technology
*Edited by* G. L. Weissler and R. W. Carlson

Volume 15. Quantum Electronics (in two parts)
*Edited by* C. L. Tang

Volume 16. Polymers—Part A: Molecular Structure and Dynamics; Part B: Crystal Structure and Morphology; Part C: Physical Properties
*Edited by* R. A. Fava

Volume 17. Accelerators in Atomic Physics
*Edited by* P. Richard

Volume 18. Fluid Dynamics (in two parts)
*Edited by* R. J. Emrich

Volume 19. Ultrasonics
*Edited by* Peter D. Edmonds

Volume 20. Biophysics
*Edited by* Gerald Ehrenstein and Harold Lecar

Volume 21. Solid State Physics: Nuclear Methods
*Edited by* J. N. Mundy, S. J. Rothman, M. J. Fluss, and L. C. Smedskjaer

Volume 22. Solid State Physics: Surfaces
*Edited by* Robert L. Park and Max G. Lagally

Volume 23. Neutron Scattering (in three parts)
*Edited by* K. Skold and D. L. Price

Volume 24. Geophysics—Part A: Laboratory Measurements; Part B: Field Measurements
*Edited by* C. G. Sammis and T. L. Henyey

Volume 25. Geometrical and Instrumental Optics
*Edited by* Daniel Malacara

Volume 26. Physical Optics and Light Measurements
*Edited by* Daniel Malacara

Volume 27. Scanning Tunneling Microscopy
*Edited by* Joseph Stroscio and William Kaiser

Volume 28. Statistical Methods for Physical Science
*Edited by* John L. Stanford and Stephen B. Vardaman

Volume 29. Atomic, Molecular, and Optical Physics—Part A: Charged Particles; Part B: Atoms and Molecules; Part C: Electromagnetic Radiation
*Edited by* F. B. Dunning and Randall G. Hulet

Volume 30. Laser Ablation and Desorption
*Edited by* John C. Miller and Richard F. Haglund, Jr.

Volume 31. Vacuum Ultraviolet Spectroscopy I
*Edited by* J. A. R. Samson and D. L. Ederer

Volume 32. Vacuum Ultraviolet Spectroscopy II
*Edited by* J. A. R. Samson and D. L. Ederer

Volume 33. Cumulative Author Index and Tables of Contents, Volumes 1–32

Volume 34. Cumulative Subject Index

Volume 35. Methods in the Physics of Porous Media
*Edited by* Po-zen Wong

Volume 36. Magnetic Imaging and its Applications to Materials
*Edited by* Marc De Graef and Yimei Zhu

Volume 37. Characterization of Amorphous and Crystalline Rough Surface: Principles and Applications
*Edited by* Yi Ping Zhao, Gwo-Ching Wang, and Toh-Ming Lu

Volume 38. Advances in Surface Science
*Edited by* Hari Singh Nalwa

Volume 39. Modern Acoustical Techniques for the Measurement of Mechanical Properties
*Edited by* Moises Levy, Henry E. Bass, and Richard Stern

Volume 40. Cavity-Enhanced Spectroscopies
*Edited by* Roger D. van Zee and J. Patrick Looney

Volume 41. Optical Radiometry
*Edited by* A. C. Parr, R. U. Datla, and J. L. Gardner

# PREFACE

*Genesis, Chapter 1*
*And God said, let there be light; and there was light.*
*And God saw the light, that it was good: and God divided the light from the darkness*

Since the beginning of man's quest to understand the nature of his surroundings, the drive to understand light and visual sensation has been a prime activity because of the importance of vision in man's ability to survive. As Aristotle said in Metaphysics, "... above all the other senses, sight helps us to know things and reveals many distinctions". Early civilizations often worshipped the Sun or otherwise held it in high regard because of the manifest importance of the radiant energy that it supplies to the earth.

At its core, radiometry concerns the measurement of all aspects of light or optical radiant power, including its polarization and spectral properties and hence is one aspect of the modern scientific pursuit of Aristotle's categorization of the senses. Modern applications of light range from remote sensing to optical communication. The latter is the backbone of the internet, which in Aristotle's words "reveals many distinctions". Historically, the term radiometry has been applied to such measurements in the wavelength range from about 5 nm to 100 μm, a range of over four decades in photon energy. Such a broad span of energy require the use of many different types of detectors and sources to provide measurements over the entire range.

The advent of cryogenic radiometers in the last two decades has revolutionized the realization of radiometric scales. Cryogenic radiometers can be utilized as very sensitive detectors over the entire radiometric spectral range. The fact that they lend themselves readily to accurate power measurements by electrical substitution has made them the most accurate radiometers in existence, often achieving uncertainties of less than 0.01%. This fact, together with the significant advances in solid-state detectors and laser technology, has opened a new era in high accuracy, broadly applicable in many areas of radiometry. These changes have been spurred in part by a redefinition of the fundamental International System of Units (SI) unit, the candela, as well as the demand for improved radiometric measurements by applications in science and industry. The SI system is maintained with international accord and a program of intercomparison and verification carried out under the auspices of the Bureau International des Poids et Mesures (BIPM) in Sevres, France (www.bipm.org).

Prior to 1948, the SI units for photometric quantities were maintained by comparisons of various types of candles, lamps, and other types of mutually agreed-upon light sources. In 1948, the Conférence Générale des Poids et Mesures (CGPM), the General Conference on Weights and Measures adopted a definition of the unit of luminous intensity, called the "new candle", that depended upon the radiation properties of platinum at the temperature of its phase transition from a solid to a liquid. Most national laboratories found it difficult and expensive to maintain a platinum freezing point blackbody and hence a consensus developed to change the definition of the candela to one based solely upon the measurement of optical power and remove the reliance for the definition of the candela upon the performance of high-temperature sources. In 1979, the CGPM adopted a recommendation from the Comité International des Poids et Mesures (CIPM), the International Committee on Weights and Measures and redefined the candela in terms of a specific amount of optical power from a source which could be measured by an appropriate detector.

The 1979 candela is defined as follows:

> The candela is the luminous intensity, in a given direction, of a source that emits monochromatic radiation of frequency $540 \times 10^{12}$ hertz and that has a radiant intensity in that direction of 1/683 watts per steradian. (16th CPGM, 1979, Resolution 3)

The factor 1/683 W per steradian was chosen based upon the best estimate at that time of the freezing point of platinum in order to have the new definition of the candela agree with the old one. This definition of the candela, and the definition of the standard observer's spectral response by the CIE, decoupled the photometric units from the radiation properties of a platinum blackbody and allowed the photometric units to be determined by the measurement of optical power with appropriate photodetectors.

This conversion to a detector-based photometric scale was a major driving force that spurred the development of improved optical radiation detectors in the visible wavelength region and fostered the migration of the traditional source-based radiometry and photometry to procedures that were based upon the use of accurately calibrated detectors. By the 1970s, improvements in solid state detectors had progressed such that silicon and other types of materials could be used to manufacture stable solid state photodetectors that could be routinely used for accurate radiometric and photometric measurements.

At about the same time, silicon and other solid state devices were being perfected for use in radiometry and photometry, the technology to make better electrical substitution radiometers was developed. Electrical substitution radiometers are constructed by devising an absorbing receiver that collects the optical power, and as a result, undergoes a temperature rise with respect to a fixed-temperature heat reservoir. The optical power is then determined by comparison with the electrical power used to accomplish the same temperature change in the system and hence relates optical and electrical power in appropriate SI units. As mentioned above, electrical substitution radiometers operating at liquid helium temperatures and called cryogenic radiometers are the most accurate rendition of the electrical substitution radiometers.

This book describes the changes that have occurred in radiometry over the last decade or so and the way that these changes have improved accuracy in a host of applications. The technical community that uses radiometric measurements in the pursuit of their goals should be able to gain new insight from this book on how to improve the quality of their measurements. The development of high-quality transfer standard detectors that can be calibrated using the cryogenic devices has improved areas as diverse as photometry and remote sensing using earth-orbiting satellites. Wavelength-filtered radiometers have been developed that can be used to assign spectral source scales for irradiance and radiance as well as improve the uncertainties in using a synchrotron storage ring as a radiometric source. Well-characterized filter detector instruments have been developed that are used to routinely measure temperatures of blackbody sources to within a few tenths of a degree Kelvin. The advent of optical communication using laser sources and fiber optics has generated a new set of demands for radiometric standards and measurement techniques that are unique to this industry.

The major accomplishments in radiometry over the last two decades have been presented at periodic conferences that are called International Conference on New Developments in Radiometry, or NEWRAD for short. The proceedings of the more-recent conferences have been published in the journal *Metrologia* (www.iop.org/journals/metrologia). The editors of this volume hope that the contributions from the experts who have volunteered to make them will be of considerable use to the wide range of scientists and engineers who routinely use radiometry and photometry to accomplish their technical objectives. The authors have all participated in the NEWRAD conferences over the years and it is our hope that the topics we have selected are of the most general use to readers wanting to become acquainted with current radiometric practices.

Throughout this book, the various authors may make reference to commercially available devices and refer to the item's origin of manufacture.

This should not be construed to be an endorsement of any commercial entity or product as there may be other items that may provide the equivalent functionality. These commercial descriptions are introduced solely for clarifying and describing a particular apparatus for the readers of this volume.

*National Institute of Standards and Technology* Albert C. Parr
*Gaithersburg, Maryland, USA* Raju U. Datla
*July 2005* James L. Gardner

# 1. INTRODUCTION TO OPTICAL RADIOMETRY*

Raju U. Datla, Albert C. Parr
National Institute of Standards and Technology, Gaithersburg, Maryland, USA

## 1.1 Background

Radiometry is the science of measuring electromagnetic radiation in terms of its power, polarization, spectral content, and other parameters relevant to a particular source or detector configuration. An instrument which measures optical radiation is called a radiometer. While in many parts of the world, the term radiometer is exclusively applied to devices which monitor radiance, we will use the word in a more general sense to mean a device which measures one of several optical-power-dependent quantities. Radiance, the optical power from or through an area within some solid angle, is one of several optical terms that will be defined and discussed in this chapter. It has been the authors' experience that the meaning of terms like radiance and other radiometric expressions is one of the more vexing problems faced by scientist new to the field. It is our hope that this issue, among others, will be clarified by this chapter.

A radiometer will have as its essential component a detector or sensor of the optical radiation and, with it, associated optical and electronic elements to generate a signal that is representative of the quantity being monitored. A major technical challenge in radiometry involves the characterization of a source of radiation which in turn requires the characterization of a detector system that will measure the source optical power and from which the characteristics of the source can be determined. For example, it is necessary to characterize light sources used for illumination of buildings so as to most efficiently use electricity, and at the same time provide adequate illumination for the inhabitants of a building. This in turn requires that specialized optical detectors be used that allow the illumination engineer to measure the light in a manner that is relevant to human vision by properly accounting for the visual response of the human eye. There are many uses of radiometry in industrial application to monitor manufacturing processes and in scientific and technical activities that utilize the sensing of optical radiation to deduce information about a wide range of physical, chemical, and biological processes.

---

*Contribution of the National Institute of Standards and Technology.



EXPERIMENTAL METHODS IN THE PHYSICAL SCIENCES, vol. 41
ISSN 1079-4042 DOI: 10.1016/S1079-4042(05)41001-2
Published by Elsevier Inc.

The earth remote-sensing community relies upon complex radiometric systems to explore the earth's radiation budget, to monitor land and ocean environmental health, and to explore global climate change issues [1]. The science of radiometry encompasses all these varied needs for sensing and measuring light, and as a result it is a multifaceted discipline with many different techniques to meet varied technical needs. In cases where a radiometer is developed to sense some particular physical phenomena or process, the radiometer and its associated optical system is often called a sensor or optical sensor, or even given a specific name that denotes a purpose, such as pyrometer, which is an optical sensor for measuring temperature. The exact meanings of the terms detector, sensor, or radiometer in general needs to be decided from their context. This book is intended to be an introduction for the reader to the recent innovations in radiometry that have been developed to take advantage of the technical advances of the past several decades.

Early efforts in radiometry were associated with the desire to understand visual sensations. Many of the early scientific writings from the time of the Greeks and Romans through the middle ages involved attempts to understand the eye and its relation to visual phenomena. Scientists as distinguished as Newton, Kepler, and Descartes spent considerable effort in attempts to understand vision, and in addition they contributed significantly to understanding the function of the eye's lens and formulated theories to explain color perception [2, 3]. The history of the study of vision is a central theme in the evolution of scientific thought, and the reader is encouraged to pursue this fascinating topic in some of the references cited here [2].

Most of the early works on radiometry were in an area we now call photometry. Photometry is the science of measuring light taking into account the wavelength response or sensitivity of the human eye. Hence, photometry is one of a number of radiometric techniques that use a wavelength selective detector system to measure a quantity of interest. Instruments that are designed to measure light as the human eye does are called photometers. The reader should be aware that in various parts of the world and in other scientific disciplines, the term photometer can refer to some other sort of instrument for measuring light that is not directly related to human vision. Photometry will be covered extensively in Chapter 7.

The first efforts in quantitative radiometry are attributed to Bouguer and Lambert, who developed photometers and attempted to quantify the measurement of the visual effects of visible radiation in the period, 1725–1760 [3]. Lambert laid out the theoretical foundations of photometry, the principles of which remain in modern practice. He established mathematical relationships that include the law of addition of illuminations, the inverse square law, the cosine law of illumination, the cosine law of emission, etc. The concepts laid out by Lambert for photometry have been extended and generalized to measurements involving the infrared and ultraviolet parts of the electromagnetic spectrum and form an integral part of modern radiometric practice.

This chapter will review the history of radiometry and define the terminology and basic methodology that underpin modern radiometric practices. The remaining chapters will develop in greater detail the important critical elements of modern radiometry.

## 1.2 Basics of Radiometry and Important Milestones

The initial impetus for radiometry in the 18th and 19th centuries had been the development of quantitative measurements in the physical sciences and, in particular, the efforts by astronomers to quantify the varying intensity of the observed stars. The astronomer Sir William Herschel (1800) discovered infrared radiation by comparing the temperature rise of liquid in glass thermometers placed in different spectral parts of the dispersed solar radiation. He showed that heating occurred in the red portion of the spectrum and also in the invisible portion at longer wavelengths as part of his efforts to develop means to quantify stellar intensity measurements [3]. Similarly, in 1802, ultraviolet light was discovered by Johann Ritter, who used chemical activity of light in analyzing the spectrum of white light and noted that the activity extended to shorter wavelengths than the visible blue light [4]. There were many contributions to the understanding of electricity and magnetism in the 19th century that culminated in the developing of a comprehensive theory of electromagnetism by Maxwell in 1864 [5]. Maxwell's equations predicted the existence of electromagnetic waves traveling at the speed of light. The existence of these waves was confirmed by Hertz in 1887 [6]. By the end of the 19th century, there were measurements of the speed of light by a variety of laboratory, terrestrial, and astronomical techniques that gave an experimental value for the speed of light and the waves that Hertz discovered. The presently accepted value of the speed of light is 299,792,458 m/s and is an exact number by international convention [7]. Maxwell's work provided an underpinning for the explanation of all forms of electromagnetic radiation including that of light from the ultraviolet through the infrared. Table 1.1 shows the internationally recognized designations for the various wavelength regions [8] commonly used in radiometric measurements discussed in this book.

Toward the end of the 19th century, there was intense interest in the scientific community to correctly explain the observations being made concerning the spectrum of radiation from high-temperature sources. This interest helped precipitate the development of new kinds of optical radiation

TABLE 1.1. Commonly Named Wavelength Regions

| Region | Wavelength interval |
|---|---|
| UV-C | 100–280 nm |
| UV-B | 280–315 nm |
| UV-A | 315–400 nm |
| Visible | 380–780 nm |
| IR-A | 780–1400 nm |
| IR-B | 1.4–3 μm |
| IR-C | 3–1000 μm |

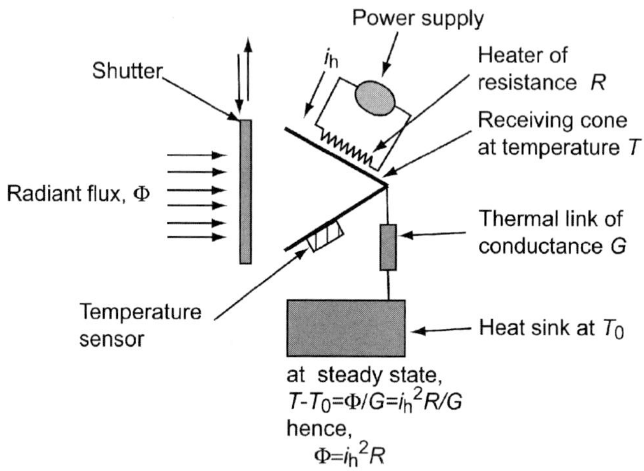


FIG. 1.1. Schematic diagram of an electrical substitution radiometer.

detectors that were designed to make direct measurements of the amount of radiation in portions of dispersed spectra. One of the first radiometers developed was a device that compared optical power to electrical power and hence became known as an electrical substitution radiometer. A schematic representation of such a device is shown in Figure 1.1. The device has a shutter that can be opened to allow light to fall upon the receiving cone which is usually coated with an absorbing material. This will cause the temperature of the cone to rise to some equilibrium value $T$ that depends on the conductance of the thermal link and the heat sink that is maintained at $T_0$ and other parameters of the system. When the shutter is closed, an electrical current $i_h$ is passed through the heater that then maintains the temperature $T$ of the cone. Neglecting correction due to various losses, this equivalence of temperature implies that the optical power $\Phi$ is equal to the electrical power $i_h^2R$. Devices using this principle today are of fundamental importance in radiometry and their modern application is discussed in Chapter 2.

Kurlbaum and Ångstrom, working separately in the 1890s, are credited with developing the first electrical substitution radiometers for measuring a physical process [9]. They performed measurements of the spectral distribution of the radiation from blackbody sources using these radiometers. The attempt to understand blackbody radiation was an important scientific topic at the end of the 19th century. The spectrum from a blackbody was shown to rise in radiance from low levels at short wavelengths, reach a maximum, and then decrease in radiance at longer wavelengths. As we mentioned earlier in this chapter, radiance is the amount of optical power from a surface area that is emitted into a solid angle. The wavelength position of the maximum radiance shifted toward shorter wavelengths at higher temperatures. A perfect blackbody source is a radiator that would absorb and emit with unit efficiency, while any realistic source will have an efficiency that is wavelength-dependent and is less than unity. In the early 20th century, Coblentz at the Bureau of Standards in the US developed a thermopile detector system that relied upon electrical substitution to measure the Stefan–Boltzmann constant and the constants in Planck's radiation law [10–12]. Coblentz's radiometer was conceptually similar to that shown in Figure 1.1, except the receiving surface was an absorbing thermocouple array which directly gave a signal. Similar work, too extensive to review here, was carried on in a number of laboratories worldwide. A later section of this book will deal with the details of blackbody sources, and an excellent description of the development of the various early electrical substitution radiometers can be found in Hengstberger's book [9]. Electrical substitution radiometers are sometimes called absolute radiometers because they measure the optical power directly using fundamental physical relationships and do not rely upon some other type of optical device for their calibration.

Planck, after attempts by Rayleigh, Boltzmann, and others, developed a theory that correctly accounted for the spectral distribution of blackbody radiation. Planck's theory necessitated the hypothesis that the radiators emitting energy in the blackbody source emit energy only in multiples of a quantity that is proportional to the frequency of the radiation [13]. This insight by Planck is credited with the development of modern ideas on the quantum nature of physical phenomena. The proportionality constant between the energy $h\nu$ the quantized unit of light, later named the photon, and its frequency $\nu$ is called Planck's constant $h$.

One outgrowth of Planck's formulation of the photon nature of light is the seeming variance with the wave nature of light as predicted by Maxwell's equation. Most of the issues in radiometry can be understood using geometrical optics or wave notions, but in some parts of this book it will be appropriate to introduce descriptions based upon the quantum theory of radiation in which the photon or particle nature of light is necessary for understanding [14]. For example, the photoelectric effect was first properly described by Einstein by invoking the quantum nature of light [14, 15]. Classical wave theory of light suggested that the energy of the photoelectrons emitted from surfaces when light is incident should increase with the intensity of the light. Instead it was found that the energy distribution of the electrons only depended upon the frequency of the light which led Einstein to suggest the explanation that the energy of the light was proportional to its frequency in the same manner that Planck had hypothesized to explain blackbody radiation. These two experiments, blackbody radiation and the photoelectric effect, and their explanations, are a major underpinning of modern quantum theory.

An important milestone in improving the accuracy in radiometry that demands special note was the development of a cryogenic electrical substitution radiometer by Quinn and Martin at the National Physical Laboratory in the UK. This instrument enabled the measurement of the Stefan–Boltzmann constant to an uncertainty of 100 parts in a million [16]. The advantage of cryogenic operation, usually at liquid helium temperature (4 K), is that various sources of error that affect the establishment of equivalence between the electrical and optical heating of the cavity are eliminated or greatly reduced. These include errors caused by radiative loss, conductive loss in electrical leads, convection losses, and others. Cryogenic radiometers are now the standards by which most national metrology institutes maintain their radiometric quantities and they have had a profound impact on lowering the measurement uncertainties associated with radiometric measurements. Chapter 2 is devoted to the important topic of cryogenic radiometers and their use.

The next section describes the nomenclature associated with the practice of radiometry and photometry and discusses the essential geometrical aspects of radiometry that are essential to its understanding and use. In Section 1.4, the basic problem of radiometry is introduced in terms of the measurement equation. The measurement equation is a method of analyzing radiometric measurement arrangements and, for example, allows for the output of a radiometer to be expressed in terms of the source quantities and the geometrical, electrical, optical, and other relevant properties of the optical sensor system. This is the key to understanding the sources of uncertainty and deducing the expected quality of a measurement. In general, the measurement equation allows one to extract physically meaningful data from a measurement of optical radiation with some detector system. The use of measurement equations is an integral part of most of this book, and their use in deducing uncertainty estimates is explored as appropriate.

An example of calculating the uncertainty in a radiometric measurement using the measurement equation is presented in Section 1.5.2, and Chapter 6 deals with uncertainty according to internationally accepted practice [17].

## 1.3 Radiometric Terminology

The radiometric terminology in this book conforms to the definitions accepted by the International Standards Organizations (ISO) and the International Commission on Illumination (CIE) [8, 18]. Table 1.2 summarizes few of the commonly used radiometric quantities and their corresponding photometric quantities. The symbols representing spectral radiometric quantities, for example, spectral irradiance, are formed by adding a subscript appropriate to the spectral quantity, for example, wavelength $\lambda$, to symbolize the spectral irradiance $E_\lambda$. The denominators of spectral units have an additional unit of length and, in the case of spectral irradiance, the dimension is $W/m^3$. More commonly the wavelength is measured in nanometers (nm), and the dimension becomes $W/(m^2\,nm)$.

The radiometric quantities listed in Table 1.2 can be visualized with reference to Figure 1.2a, in which an emitting surface designated by $dA_1$ acts as a source of radiation that impinges upon a receiving surface designated by $dA_2$. For the purposes of this discussion, $dA_1$ emits uniformly in all directions and the two surfaces are both centered on and perpendicular to the centerline. A source of radiation emitting equally in all directions is called a Lambertian emitter. The radiant intensity is conceptualized as the power from a point on the surface emitted into the solid angle shown by the cone of light starting at the origin of $dA_1$ and intersecting $dA_2$, and hence is the power per steradian. More practically, the radiant intensity is the amount of

TABLE 1.2. Radiometric and Photometric Quantities and their Units

| Radiometric quantity | Symbol | Units | Units | Symbol | Photometric quantity |
|---|---|---|---|---|---|
| Radiant energy | $Q$ | J | lm s | $Q_v$ | Luminous energy |
| Radiant flux (power) | $P$, $\Phi$ | W | lm | $\Phi_v$ | Luminous flux |
| Irradiance | $E$ | $W/m^2$ | $(lm/m^2) = lx$ | $E_v$ | Illuminance |
| Radiance | $L$ | $W/(m^2\,sr)$ | $lm/(m^2\,sr)$ | $L_v$ | Luminance |
| Radiant intensity | $I$ | W/sr | (lm/sr) = cd | $I_v$ | Luminous intensity |
| Radiant exitance | $M$ | $W/m^2$ | $lm/m^2$ | $M_v$ | Luminous exitance |
| Radiant exposure | $H$ | $W\,s/m^2$ | lx s | $H_v$ | Luminous exposure |
| Radiance temperature | $T$ | K | K | $T_c$ | Color temperature |

*Note*: J = joule, W = watt, lm = lumen, lx = lux, m = meter, sr = steradian, s = second, cd = candela, K = kelvin.

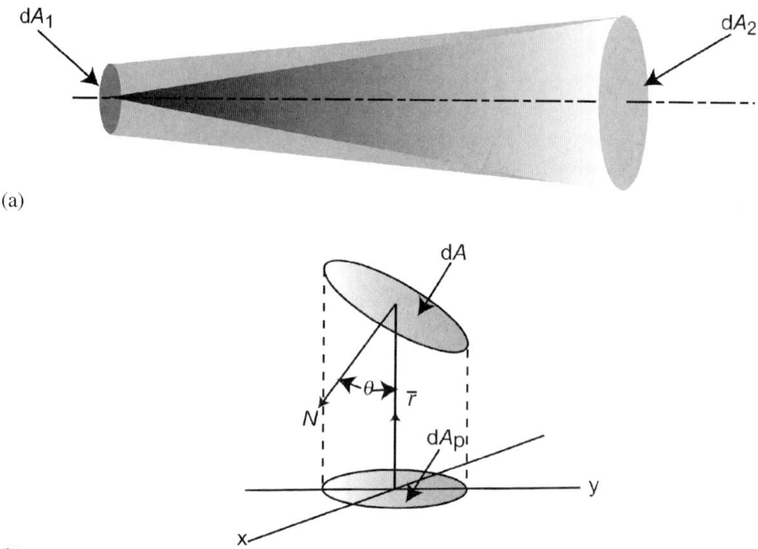


(a)

(b)

FIG. 1.2. (a) Schematic of a source of radiation at $dA_1$ which illuminates a second surface $dA_2$. (b) Schematic which shows the projected area $dA_p$ of an area $dA$.

power per steradian passing through a surface subtending a given solid angle which can be realized in a situation when the observer is far from a small source that can be considered a point source. The term intensity in optics is often used in differing ways and can cause great confusion [19, 20]. The *radiance* of the source area $dA_1$ is defined by the amount of optical power from $dA_1$ within the angular space defined by the truncated cone of radiation emitted from the entire surface element $dA_1$ and which is incident upon the area $dA_2$. Radiance is the optical power per area of the source per steradian of solid angle defined in some direction of the propagation and, in this case, the solid angle is defined by the distance between the two surfaces and the size of the areas. Assuming the total optical power passing through the surface $dA_2$ is evenly distributed over the area $dA_2$, the *irradiance* is defined by dividing the radiant flux by the area $dA_2$. Irradiance is then the power per unit area in some region of space and is a very useful quantity for describing the energy obtainable from an optical source at a given position. For example, the energy from the Sun is usually given in irradiance, $W/m^2$, for use in estimating the amount of solar energy available in some configuration. These radiometric quantities are developed more fully in the following sections.

The quantity, $dA \cos \theta$, often in radiometry and is called the projected area $dA_p$. This concept can be seen from the geometry shown in Figure 1.2b where an area $dA$, whose normal vector $N$ is oriented at an angle $\theta$ with respect to a plane defined by coordinates $x$, $y$ and which has a normal direction shown by $\bar{\mathbf{r}}$. The inclination produces a projected area $dA_p = dA \cos\theta$ in the $xy$ plane and represents the area of $dA$ as viewed from the $xy$ plane. This concept is useful in describing the amount of flux passing through a plane due to some external source such as might be represented by an emitting surface element $dA$.

Radiometric quantities can be functions of wavelength $\lambda$, frequency, $\nu$, or wavenumber, $\sigma$. The quantities $\lambda$, $\nu$, and $\sigma$ are related by

$$\lambda = \frac{c}{n\nu} = \frac{1}{\sigma} \tag{1.1}$$

In Eq. (1.1), $c$ is the velocity of light, and $n$ the index of refraction of the medium in which the light is propagating. If radiometric quantities are functions of $\lambda$, $\nu$, and $\sigma$, they are designated by the same term preceded by the adjective *spectral* and by the same symbol followed by $\lambda$, $\nu$, or $\sigma$, in parentheses to indicate the functional dependence; for example, spectral emissivity $\varepsilon(\lambda)$. Some of the quantities, for example, radiance, can be functions of wavelength (or frequency or wavenumber) and it is then called the *spectral radiance* and is represented by the symbol for the quantity with the subscript $\lambda$ ($\nu$ or $\sigma$), depending on the quantity chosen for the independent variable. Using this notation, the spectral radiance would be represented symbolically by $L_\lambda$ with the functional dependence on $\lambda$ implicit, as indicated here, or included the explicitly by formally including the variable in parentheses, i.e., $L_\lambda(\lambda)$. The subscript indicates here, as in calculus notation, that the quantity $L_\lambda$ is differential with respect to $\lambda$ and, hence, in this case, the spectral radiance is the radiance per wavelength interval. The equations governing the spectral radiometric quantities can be converted into one or the other of the possible independent variables by using the ordinary rules of algebra and calculus for substitution of variables in equations.

The photometric quantities on the right side of Table 1.2 are obtained from the corresponding spectral radiometric quantities by integrating the spectral radiometric quantity weighted with the function called the *spectral luminous efficiency*, $V(\lambda)$, over the visible wavelength region [21]. The $V(\lambda)$ function is used for characterizing the human visual response under good lighting conditions, and there are other functions defined for the human visual response under lighting conditions that are less than optimal.

### 1.3.1 Radiance

The primary quantity measured by radiometers is optical power $\Phi$ incident upon the detector. While in some cases the power is the quantity of interest, in most radiometric measurements, one is trying to deduce some other quantity such as the radiance of a source or the irradiance incident upon some surface. As mentioned in Section 1.2, we develop the detailed concepts of radiance first and then show how the irradiance is related to the radiance and geometrical factors. In the previous section, these quantities were defined in general terms as shown in Figure 1.2 and in this section, we develop the ideas of radiance and the other quantities in the detail necessary to fully define radiometric measurement arrangements and provide the framework necessary to estimate appropriate uncertainties.

The relationship between quantities such as power, radiance, and irradiance can be demonstrated by considering a general type of radiometric measurement situation that is shown in Figure 1.3. In Figure 1.3, $x_1$ and $y_1$ describe a coordinate system centered on a source that emits radiation from a differential element of area $dA_1$ positioned on the larger area shown as $A_1$. This source is characterized by its radiance, which is the amount of optical power per unit area of the source emitted per unit of solid angle. In order to visualize these quantities, it is useful to describe a bundle radiation from $dA_1$ that is incident upon a second surface in a differential element $dA_2$ of a larger area $A_2$ centered on coordinate system $x_2$ and $y_2$. The lines connecting

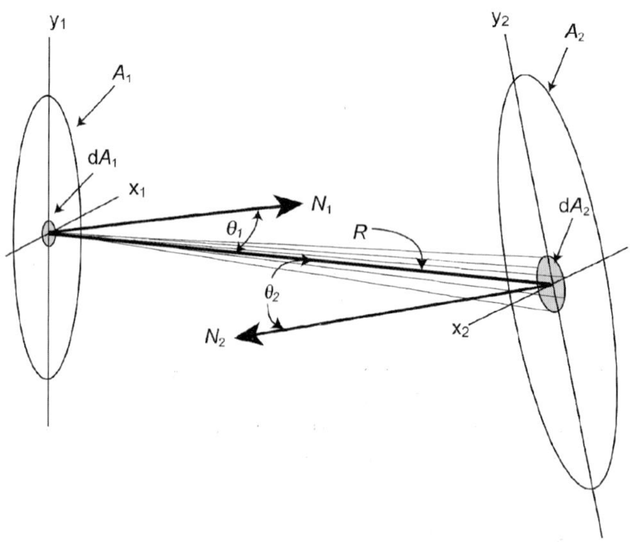


FIG. 1.3. Generalized configuration of optical radiation passing from one surface to another. The surface $A_1$ is considered a source region that originates rays of optical radiation passing onto surface $A_2$.

surface elements $dA_1$ and $dA_2$ in Figure 1.3 indicate some, but certainly not all, of the possible path rays of light traverse between the surfaces.

The radiation incident upon $A_2$ can be characterized in terms of the radiance, but often the irradiance, the amount of optical power per unit area, is a more useful quantity for characterizing radiation incident upon a surface. A discussion of Figure 1.3 shows the relationship between these fundamental quantities. $A_1$ and $A_2$ are shown in Figure 1.3 as being circular for schematic reasons, but they can be of any shape that describes a source of radiation and a surface of interest through which it propagates. For a typical example in a radiometric measurement, $A_2$ is the entrance aperture of a detector system and $A_1$ describes the aperture of the source providing the radiation.

$R$ is a line whose length is the distance between the origins of the two surface area elements, and $N_1$ and $N_2$ are the normal vectors to the surfaces at angles $\theta_1$ and $\theta_2$ with respect to $R$. The coordinate systems are centered on the apertures for convenience. It is useful to discuss this general type of radiometric configuration in order to define the radiometric quantities involved and to become aware of the consequences of approximations made when simplifying for actual measurement arrangements.

If $L_1$ is the radiance of the source at $dA_1$, the amount of flux $\Delta\Phi_1$ in the beam that leaves the element of area $dA_1$ and that passes through element of area $dA_2$ is

$$\Delta\Phi_1 = \frac{L_1 dA_1 \cos\theta_1 dA_2 \cos\theta_2}{R^2} \tag{1.2}$$

This equation defines radiance and underscores its fundamental properties in describing the propagation of fluxes of optical radiation. Equation (1.2) relates the optical power passing through a region of space to a property of the source, the radiance $L$, and purely geometric considerations. In general, the radiance is a function of the coordinates defining $dA_1$ as well as the angles that define the direction of propagation of the light leaving surface $dA_1$, and thus evaluating Eq. (1.2), in real circumstances, can be difficult. In many cases, simplifying assumptions must be made to obtain a result. It is important, however, to start with this complex definition of radiance in order to understand the implications of the approximations that are made to evaluate the flux in configurations used in practical radiometric measurements. Terms on the left portion of the right-hand side of Eq. (1.2) can be grouped and the expression written in a different manner. Using the following definition

$$d\omega_2 = \frac{dA_2 \cos\theta_2}{R^2} \tag{1.3}$$

where $d\omega_2$ is the solid angle subtended at $dA_1$ by area $dA_2$, we can rearrange Eq. (1.2) in the following form to explicitly define radiance

$$L_1 = \frac{\Delta\Phi_1}{d\omega_2\, dA_1 \cos\theta_1} \tag{1.4}$$

This equation defines the radiance in terms of the optical flux and geometry of the source and some area through which the flux passes and is useful, along with Figure 1.3, in conceptualizing the meaning of the quantity radiance. We see in Table 1.2 that radiance has units of $W/(m^2\,sr)$ and from Eq. (1.4) that radiance is the optical power of a source emitted into a solid angle defined by the region of space in which the power is directed. It is important to keep track of the directions and angles included between source and observer in determining radiance due to the angular component, $\cos\theta_1$, in Eq. (1.4).

In the absence of any dissipative mechanisms in the space between $A_1$ and $A_2$, the flux in the beam leaving $dA_1$ within the solid angle shown in the construction of our example, is equal to that which passes through $dA_2$. One can create an equation much like that of Eq. (1.2) that describes the relationship between the radiance and flux at $dA_2$ and show that the radiance is a conserved quantity in the beam [22–24]. This can easily be seen if we introduce a radiance $L_2$ that represents the radiance of the beam at surface element $dA_2$ and realize that Eq. (1.2) can express the flux $\Delta\Phi_2$ at that surface by permuting the variable subscripts, $1 \leftrightarrow 2$, and generate the same equation as Eq. (1.2) due to the symmetry in the variables. This argument which assumes that the radiance is not varying over the elements of area $dA_1$ and $dA_2$ leads to Eq. (1.5a) which states the conservation of radiance in a beam which is propagating in a non-dissipative medium whose index of refraction is unity. This fundamental relationship underscores the importance of the concept of radiance in radiometry. This relationship undergoes a slight modification if the index of refraction of the medium in which the optical radiation is traveling differs from unity as assumed so far in our discussion. Application of Snell's law to the angles defining the angular quantities in the definitions of radiance leads to a factor of $n^2$ normalizing the radiance, where $n$ is the index of refraction at the defining surfaces for the determination of the radiance, $dA_1$ and $dA_2$ in this example. Using $n_1$ and $n_2$ to represent the index of refraction of the medium at the surfaces $dA_1$ and $dA_2$ respectively, we can write Eq. (1.5b), which defines the conserved quantity in circumstances when the index of refraction is different from unity [25]. The quantity $L/n^2$ is sometimes referred to as the *reduced radiance* and becomes the generalized conserved quantity in the presence of media with varying indices of refraction. This relationship depends upon the conservation of flux in a beam and hence factors due to any scattering because of index variation or other factors are not accounted for in this formulation and would have to be dealt with separately.

$$L_1 = L_2 = \text{constant} \tag{1.5a}$$

$$\frac{L_1}{n_1^2} = \frac{L_2}{n_2^2} = \text{constant} \tag{1.5b}$$

### 1.3.2 Radiance in Terms of Projected Area and Projected Solid Angle

Dropping subscripts on the solid angle and letting the elemental areas and solid angles become differentials by taking limits, and realizing that the flux depends upon two quantities, Eq. (1.4) can be rewritten in terms of the projected area as

$$L = \frac{\mathrm{d}^2\Phi}{\mathrm{d}\omega\,\mathrm{d}A_\mathrm{p}} \tag{1.6}$$

Written in this way, the radiance is seen as the amount of flux per unit projected area of the source per unit solid angle subtended by the area to which the flux is headed. The reader is referred to the literature for details of the mathematical limiting procedures to arrive at the mathematical correct differential form of radiance [23, 25, 26].

An alternative and sometimes useful variant of the development of the projected area is the introduction of the concept of projected differential solid angle $\mathrm{d}\Omega = \cos\theta\,\mathrm{d}\omega$. The projected solid angle $\Omega$ is then given by

$$\Omega = \int_\omega \mathrm{d}\Omega = \int_\omega \cos\theta\,\mathrm{d}\omega \tag{1.7}$$

In this notation the expression for the radiance in Eq. (1.6) is written as

$$L = \frac{\mathrm{d}^2\Phi}{\mathrm{d}\Omega\,\mathrm{d}A} \tag{1.8}$$

### 1.3.3 Radiant Flux and Irradiance

Inspection of Eq. (1.2) indicates that the total radiant flux $\Phi_1$ passing through the surface $A_2$ which originates from surface $A_1$ can be expressed with an integral relation over the surface variables of the two surfaces of interest, i.e., $A_1$ and $A_2$ as shown in Figure 1.3. Therefore,

$$\Phi_1 = \int_{A_1}\int_{A_2} \frac{L_1 \mathrm{d}A_1 \cos\theta_1\,\mathrm{d}A_2 \cos\theta_2}{R^2} \tag{1.9}$$

An additional implicit consideration in Eq. (1.9) is the fact that $L_1$, in addition to being a function of the spatial coordinates of the emitting surface, is also in general a function of the wavelength. In this situation, the relationship would describe the *spectral flux* leaving the surface that is a result of the *spectral radiance* of the surface.

In general, the integral in Eq. (1.9) is difficult to perform, particularly if $L_1$ is a function of the spatial and angular coordinates. Additionally if the apertures are large, the angles $\theta_1$ and $\theta_2$ and the distance $R$ all have complicated relationships across the areas that make the solution of Eq. (1.9) very difficult. The propagation of optical flux represented by Eq. (1.9) is related to problems in other areas of physics such as heat transfer and some of the techniques developed for those problems can be utilized for calculations of radiometric flux transfer [27, 28].

Although Eq. (1.9) is in general difficult to solve exactly, in many situations in radiometry it is fortunate that circumstances exist or assumptions can be made which allow for simplification and solution of the complex integral relationship. Often sources of radiometric interest can be considered Lambertian; they radiate a constant radiance in all directions into a hemisphere, and furthermore they emit the same at every point in the source plane $A_1$. If this is the case, we can rewrite Eq. (1.9) in the following manner by removing the radiance from the functional considerations,

$$\Phi_1 = L_1 \int_{A_1}\int_{A_2} \frac{\mathrm{d}A_1 \cos\theta_1 \mathrm{d}A_2 \cos\theta_2}{R^2} = L_1 A_1 \pi F_{12} = L_1 T_{12} \tag{1.10}$$

where $F_{12}$ only depends on the geometry and is called the configuration factor, which is obtained by evaluation of the double integral. Other terms for the configuration factor include view, shape, or exchange factor. For many cases in ordinary radiometry, the configuration factor can be looked up in literature [27, 28] or evaluated using numerical techniques and computers. The quantity $T_{12} = A_1 \pi F_{12}$ occurs often in radiometry and is called the *throughput* of the particular optical arrangement. In the absence of dissipative effects, when the flux is conserved in an optical system, the invariance of the radiance implies the invariance of the throughput of the system. Using the same arguments as above, one can readily show the reciprocity relations for radiometric systems, $A_1 F_{12} = A_2 F_{21}$ or $T_{12} = T_{21}$, which indicates the reversibility of the propagation of the optical beam.

Additional insight and quantities can be seen by considering a bundle of rays leaving surface $A_1$ in Figure 1.3 and applying Eq. (1.8) to calculate the flux in a bundle of rays leaving the surface with some simplifications. In this approximation, we assume that the projected solid angle of the beam is independent of the position on $A_1$. This means, in practice, that the source size and aperture size of $A_2$ are small compared to the distance $R$.

Reinserting subscripts into Eq. (1.8) and integrating since we can separate the integrands by assumption, we have

$$\Phi_1 = L \int_{A_1} \mathrm{d}A_1 \int_{\omega_2} \cos\theta_1 \, \mathrm{d}\omega_2 = LA_1\Omega_{12} \tag{1.11}$$

In this approximation, often useful in radiometry, the throughput $T_1 = A_1\Omega_{12}$, can often be simply approximated and the flux and radiance directly and simply related. In more complicated situations where the dimensions are large, the more exact expressions utilizing the configuration factor must be used. For example, in a simple case where the source is a uniform small circular aperture $A_1$ separated by a distance $R$ that is large ($R > 20$ radius of $A_1$) from a receiving aperture of $A_2$ of similar small size and the two apertures are perpendicular to the line connecting their centers, we have the flux incident upon $A_2$, $\Phi_2$, equal by construction to the flux leaving $A_1$, $\Phi_1$,

$$\Phi_2 = \Phi_1 \cong L_1 \frac{A_1 A_2}{R^2}, \qquad E_2 = \frac{\Phi_2}{A_2} = \frac{L_1 A_1}{R^2} \tag{1.12}$$

In this case, the irradiance $E_2$ at the surface $A_2$ is simply related to the radiance of the beam and geometric factors. The irradiance and its spectral analog, the *spectral irradiance*, are important radiometric quantities because they specify the optical power in a cross-sectional area of an optical beam. The irradiance is an important quantity because most detectors measure optical power, and by knowing the aperture area defining what the detector intercepts, the irradiance can be determined and related to radiance by equations like Eq. (1.12).

The more general case of two circular apertures whose centers are on a common centerline and which are oriented such that their aperture planes are normal to the centerline is an important arrangement in radiometry and will be discussed here as a further example. Letting $r_1$ be the radius of $A_1$ and $r_2$ the radius of $A_2$, the configuration factor determined by the integral in Eq. (1.10) becomes [27, 29]

$$F_{12} = \frac{1}{2}\left[\left(\frac{r_1^2 + R^2 + r_2^2}{r_1^2}\right) - \left[\left(\frac{r_1^2 + R^2 + r_2^2}{r_1^2}\right)^2 - 4\frac{r_2^2}{r_1^2}\right]^{1/2}\right] \tag{1.13}$$

From Eq. (1.10) and writing $A_1 = \pi r_1^2$ and assuming the radiance is a constant over the surface, we can write an exact expression for the optical power at the second aperture due to the radiance from $A_1$ as

$$\Phi = L\frac{\pi^2}{2}[(r_1^2 + r_2^2 + R^2) - [(r_1^2 + r_2^2 + R^2)^2 - 4r_1^2 r_2^2]^{1/2}] \tag{1.14}$$

In many cases $R$ is sufficiently larger than either of the radii of the apertures and the term can be factored out and the expression expanded and simplified. We have then,

$$\begin{aligned}\Phi &= \frac{L\pi r_1^2 \pi r_2^2}{(r_1^2 + r_2^2 + R^2)}\left(1 + \frac{r_1^2 r_2^2}{(r_1^2 + r_2^2 + R^2)^2} + \text{higher terms}\right) \\ &= \frac{LA_1 A_2}{D^2}[1 + \delta + \text{higher terms}] \qquad (1.15)\end{aligned}$$

where we have $D^2 = r_1^2 + r_2^2 + R^2$ and $\delta = r_1^2 r_2^2 / D^4$.

This result gives a correction to the approximations used in Eq. (1.12) and is useful for many radiometric applications and for uncertainty analysis, although with the ready availability of modern computers, the exact expression can often be easily derived. As long as the source is Lambertian, this equation is useful for calculating the radiance from a measured flux and for estimating the contributions to the uncertainty by evaluating the significance of higher-order terms in the above expansion. In this example, if an optical detector were behind the aperture $A_2$, its measurement of the optical power $\Phi$ would allow the determination of the radiance of the source at $A_1$. It can be seen by comparison of Eq. (1.12) to the results in Eq. (1.15) that the more exact calculation using the configuration factor results in an equation of the same general form but with corrections to the geometrical factors. The irradiance at $A_2$ in the approximations implied by Eq. (1.15) is simply related to the radiance of the source and a geometric factor and is found by dividing both sides of Eq. (1.15) by $A_2$.

The reader is referred to the extensive literature on specific optical systems to find approximations used for the configuration factor to calculate the throughput for various specific optical arrangements, including those with lenses and other beams forming and steering devices [22–24, 30, 31]. In any radiometric system where these equations and approximations are used, it is necessary to keep in mind that the degree of approximation used also introduces uncertainties into the measurement process that must be accounted for in the uncertainty budget for the measurement. The simplifications shown above, such as by Eq. (1.15), rely upon the uniformity of the optical beam and the Lambertian nature of the source as well as the lack of dissipation of the beam by scattering or absorption of optical radiation. If a source is not Lambertian and uniform, extensive measurements and characterizations are necessary to understand the relationship between the radiance of the source and any measured optical power.

### 1.3.4 Radiant Intensity

Another quantity that is useful in radiometry for certain applications is called the *radiant intensity* and is usually denoted by the symbol $I$. This quantity usually is associated with point sources or those of negligible dimensions and is the amount of flux per solid angle. In terms of the variables used above, the radiant intensity is

$$I = \frac{\mathrm{d}\Phi}{\mathrm{d}\omega} \tag{1.16}$$

This quantity corresponds to the flux in the bundle of radiation shown in Figure 1.3 if the source element of area $\mathrm{d}A_1$ is collapsed to a point. The corresponding photometric quantity is the *luminous intensity*, which is measured in the SI base unit of the candela.

## 1.4 Radiometric Measurements

The main measurement problems posed in radiometry are the characterization of a source of optical radiation for its radiance, spectral radiance, or photometric quantities, and the development, characterization, and calibration of a detector system to make such measurements. For example, by measuring the spectral radiance of a blackbody source, its temperature can be inferred. In other applications, optical sensor systems that operate in narrow wavelength regions are used to deduce properties of celestial bodies or to monitor earth resources from orbiting spacecraft. These specific applications, as well as others, are discussed in the chapters 4, 5, 7, and 10 of this book.

The problems of correctly characterizing sources and detectors have been the traditional driving forces for improvements in radiometry. These issues continue to attract attention in order to meet current demands for increased accuracy of measurement for remote sensing, industrial applications, and scientific studies. Separate chapters of this volume will deal with the calibration and characterization of modern photodetector systems, and others will deal with the characterization of optical radiation sources that can be used as calibration sources. These once poorly connected efforts are merging as the technology for building and characterizing stable and accurate photodetector systems, which allows for very accurate determination of source characteristics, and hence the traditional separate technologies of characterizing sources and detectors is merging into one of characterizing detector systems [32, 33]. While sophisticated optical detector systems often form the basis of fundamental standards maintenance for both sources and detectors, well understood and characterized optical radiation sources are often very useful for the calibration of the spectral characteristics of optical sensor systems.

### 1.4.1 Detector Responsivity

A detector of optical radiation generates an output, usually electrical, which can be related to the amount of optical flux incident upon the detector. For example, in solid-state detectors, the absorption of a photon results in the promotion of a charge carrier to the conduction band and a resultant current in an electrical circuit. At a particular wavelength, this current is found to be proportional to the optical flux over a large dynamic range [25, 34]. Various types of solid-state devices are used to detect optical radiation from the X-ray to the infrared wavelength region and are the backbone of many complicated radiometric sensor instruments [34, 35].

Figure 1.4 shows a typical configuration of how a solid state or other type of detector might be used to make a measurement of the properties of a source of radiation. A source of optical radiation that has a defining aperture of area $A_S$ and radius $r_S$ is a distance $d$ from an aperture $A_D$ with a radius $r_D$. $A_D$ defines the flux boundary of the radiation that passes through a filter F and is incident upon a detector D. For solid-state detectors, the output signal is usually a current, and hence a signal response $r$ which is proportional to the current produced by the detector is generated by the signal amplifier. In electrical substitution radiometers the signal would be proportional to the electrical power needed to generate a response equivalent to the optical power. This relationship between optical power and the output signal response $r$ is called $R$, the responsivity of the detector; hence we have

$$r(\lambda) = R(\lambda)\Phi(\lambda) \tag{1.17}$$

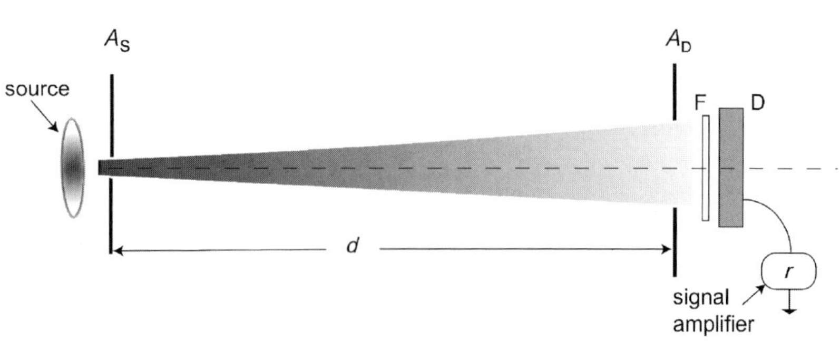


FIG. 1.4. A Lambertian source behind aperture $A_S$ illuminates a defining aperture $A_D$. The radiation passes through an optical filter F and is measured by a detector D that generates a signal response $r$.

As indicated, the signal, responsivity, and flux in general depend upon the wavelength of the radiation. This is indicated in Eq. (1.17) by explicitly indicating the functional dependence upon the wavelength $\lambda$. In general, the responsivity $R$ of a detector can depend upon a host of parameters including temperature and other environmental quantities, polarization of the light, spatial effects on its own receiving surface, angle of incidence, and others. The complete characterization of the various factors in the response of a detector is essential to the accurate operation of the system in which it is employed and to properly generate an error budget for the system. A system composed of a detector, electronics, and optics including wavelength selection, is often referred to as a *sensor*. The terms detector and sensor as well as radiometer are sometime used interchangeably and, as mentioned earlier in this chapter, the meaning must be inferred from the context of the discussion. Subsequent chapters of this book describe the characterization and calibration of various detectors and their use in measurement systems.

### 1.4.2 The Measurement Equation

Henry Kostkowski and Fred Nicodemus of the National Bureau of Standards (now NIST) introduced the concept of a "measurement equation" in radiometry [23, 26]. The measurement equation describes receiver output due to the optical radiation received from a specific source configuration. It is a *system equation*; i.e., it models the system performance in terms of the subsystem and component specifications and provides not only the measurement quantities required, but also serves as the basis for estimating the uncertainties of the measurement.

#### *1.4.2.1 Measurement equation for a filter radiometer*

The situation depicted in Figure 1.4 is a simple, yet common example encountered in radiometry. If we assume the source is a Lambertian emitter with a spectral radiance $L_\lambda(\lambda)$, the source and detector apertures, $A_S$ and $A_D$ with radii $r_S$ and $r_D$, respectively, are circular and perpendicular to the line connecting their centers and are a distance $d$ apart, and the detector has a uniform spatial response, then we can use the results of the previous section to generate the measurement equation. The transmittance of the filter is represented by $\tau(\lambda)$ and the spectral responsivity of the detector to optical power will be designated $R(\lambda)$. Using the results of Eqs. (1.15) and (1.17), we can write the signal response $r$ in terms of an integral over wavelength

$$r = \frac{A_S A_D (1 + \text{correction})}{(r_S^2 + r_D^2 + d^2)} \int_\lambda L_\lambda(\lambda)\tau(\lambda)R(\lambda)\,\mathrm{d}\lambda \tag{1.18}$$

This result is obtained by using the expression for the optical flux as shown in Eq. (1.15), modifying it by the transmittance of a filter and integrating over the wavelength region where the filter has significant transmittance. To determine the spectral radiance of the source from this equation, further assumptions need to be made in order to evaluate the integral. If the functional form of the spectral radiance is known, such as if it is a blackbody, as the transmittance and detector responsivity are known, the equation can often be iteratively solved for the spectral radiance and hence the temperature of the source. In other situations, the bandpass of the filter is narrow compared to the variations in the incident *radiance* distribution. This allows the values in the integral to be calculated by taking appropriate average values of the spectral quantities over the bandpass $\Delta\lambda$ of the instrument. Defining the quantities in front of the integral sign in Eq. (1.18) as $C$, we can rewrite the equation in a more useful form

$$\begin{aligned} r &= CL_{\lambda_0}(\lambda_0)\tau(\lambda_0)R(\lambda_0)\Delta\lambda \\ &= R_{\mathrm{L}}L_{\lambda_0} \end{aligned} \tag{1.19}$$

where $R_{\mathrm{L}}$ is called the *total radiance responsivity* of the radiometer. In this approximation, it is assumed that the spectral radiance is a constant in the wavelength interval determined by $\Delta\lambda$, and $\Delta\lambda$ is chosen so that the product of the transmittance, responsivity and $\Delta\lambda$ gives the value that the integral over these quantities would yield. In other words, the integral over these quantities is replaced by a width and average value of the integrands. The value of $\lambda_0$ in the equation is chosen such that the value of the functions evaluated at this point and the value for the bandpass give the best estimate of the integral approximation. This equation is often used in radiometry and it is important to remember the assumptions that have been made and how they may contribute to the uncertainty of the measurement. For example, one must evaluate how well this approximation for the integral relation actually reproduces the more exact expression and add a component in the error budget for the added uncertainties due to the approximations. In Section 1.5, we discuss how the radiometer could be alternatively calibrated with a known spectral radiance source.

#### *1.4.2.2 Measurement equation for a spectral radiometer*

A spectral radiometer is a device which performs wavelength selectivity in its measurement of optical radiation. While in one sense the device shown in Figure 1.4 has wavelength selectivity, it is not normally referred to as a spectral radiometer since its wavelength selectivity is fixed. Figure 1.5 shows schematically the principal components of a spectral radiometer that is designed to measure spectral radiance. Substitution of different collection

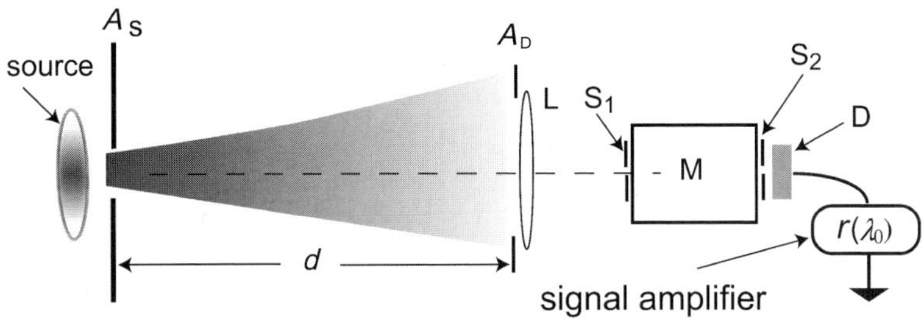


FIG. 1.5. A source illuminates an aperture $A_S$ that transmits radiation to a defining aperture $A_D$. The radiation is focused by a lens L onto a monochromator system M that disperses the radiation onto a detector D. The output from D is conditioned by a signal amplifier that generates a wavelength-dependent signal $r(\lambda_0)$ at each wavelength setting $\lambda_0$.

optics could easily make this device a spectral irradiance or spectral flux radiometer with minimal fundamental changes in the formulation we present here.

In Figure 1.5, a source illuminates an aperture $A_S$ which transmits light that fills the detector aperture $A_D$ that is a distance $d$ from the source aperture. As in the previous discussion of the filter radiometer, the two apertures are perpendicular to the axis joining their centers. The optical radiation passing through $A_D$ is imaged by a lens L onto a monochromator with an entrance slit $S_1$ and an exit slit $S_2$. The dispersed radiation exiting the monochromator is incident upon a detector D which has its output connected to a signal amplifier that conditions the detector output and produces a response $r(\lambda)$ (that is dependent upon the monochromator's wavelength setting specified by $\lambda_0$. The overall spectral responsivity $R(\lambda_0, \lambda)$ of the imaging system, the monochromator, and the detector is a function of wavelength where the monochromator is set, $\lambda_0$, and the wavelength range $\lambda$ over which the instrument has sensitivity. Additionally, the responsivity is a function of the many variables that characterize the various elements of the system including the transmittance and reflectance of the various optical elements, the responsivity of the detection element, and dissipative effects such as scattering and diffraction.

It is often convenient to factor the overall system spectral responsivity $R(\lambda_0, \lambda)$ into a term that represents the wavelength selectivity called the slit scattering function $\rho(\lambda_0, \lambda)$ which has a finite amplitude only in the region around the set wavelength $\lambda_0$ and a term $R^{\mathrm{f}}(\lambda)$ that represents the overall responsivity as a function of the wavelength [23, 26].

$$R(\lambda_0, \lambda) = \rho(\lambda_0, \lambda) R^{\mathrm{f}}(\lambda) \tag{1.20}$$

In a case like we have described, the function $R^{\mathrm{f}}(\lambda)$ can be further factored into terms representing transmittance, reflection, detector responsivity, and other terms that characterize the optical system. Each situation that a particular instrument poses will likely be amenable to various ways of accounting for the total responsivity of the instrument and, hence, we will keep to the general case for our discussion and leave the specific cases to other chapters in this volume as well as the literature [23]. The responsivity of a spectral instrument normally has a nonzero amplitude in a region $\Delta\lambda$ around the set wavelength of $\lambda_0$ as shown in Figure 1.6 by the slit scattering function $\rho(\lambda_0, \lambda)$. For our example, we will assume the shape $\rho(\lambda_0, \lambda)$ will not vary with the wavelength setting of the monochromator and that the wavelength variation of the magnitude of the responsivity can be accounted for by responsivity factor $R^{\mathrm{f}}(\lambda)$ which in this context is the amplitude variation of the system response as the wavelength is varied.

The slit scattering function represents the wavelength selectivity of the monochromator and is due to the imaging of the entrance aperture on the exit aperture. It also contains information about scattering and diffraction in the monochromator optics such as its grating and mirrors, and accounts for transmittance of the monochromator away from the central wavelength $\lambda_0$. In an idealized case of equal entrance and exit apertures, the slit scattering function is a triangular shape with a width representing the resolution of the instrument. This function must be determined experimentally by using appropriate narrow wavelength light sources such as lasers. Details of these techniques can be found in the literature [23, 26].

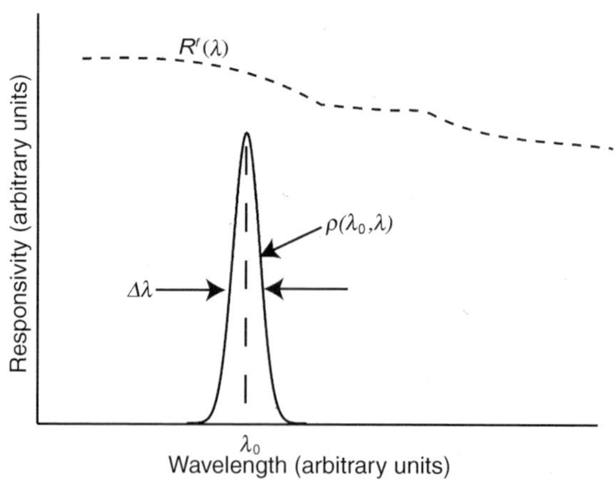


FIG. 1.6. Schematic of a responsivity relationship with an instrument with a slit scattering function $\rho(\lambda_0, \lambda)$ and a responsivity factor $R^{\mathrm{f}}(\lambda)$.

Inserting these expressions for the responsivity in Eq. (1.20) into Eqs. (1.18) and (1.19), and ignoring the small corrections, we can generate the measurement equation for the spectral radiometer.

$$\begin{aligned} r(\lambda_0) &= \frac{A_S A_D}{(r_S^2 + r_D^2 + d^2)} \int_\lambda L_\lambda(\lambda)\rho(\lambda_0, \lambda)R^f(\lambda)\,d\lambda \\ &= C \int_\lambda L_\lambda(\lambda)\rho(\lambda_0, \lambda)R^f(\lambda)\,d\lambda \end{aligned} \tag{1.21}$$

We have incorporated the geometric terms in the constant $C$. This constant will be different for different configurations caused by changes in apertures or the distances involved in the measurements. If rectangular slits were used instead of circular apertures, the configuration factor appropriate for slits could be obtained from the literature and the appropriate constant factor determined. The slit scattering function is usually very narrow compared to structure and variation in the spectral radiance and, as in the previous example, the term $L_\lambda(\lambda)$ can be removed from the integral and replaced by its suitably averaged value $L_\lambda(\lambda_0)$ to arrive at Eq. (1.22).

$$r(\lambda_0) = CL_\lambda(\lambda_0) \int_\lambda \rho(\lambda_0, \lambda)R^f(\lambda)\,d\lambda \tag{1.22}$$

## 1.5 Radiometric Calibration and Uncertainties

The measurement equations, such as Eqs. (1.19) and (1.22), form the basis for the uncertainty analysis in determining the radiometric quantities. In general, radiometric calibration of the sensor is performed by using the measurement equation to deduce the unknown radiometric quantity by *in situ* comparison with that of a standard under an identical geometrical setup. In that case, the associated geometrical factors cancel, leaving the solution for the unknown radiometric quantity in terms of just the two measured output signals (the unknown and the standard) and the known value for the standard. Alternatively, the standard could be used to evaluate the responsivity of the sensor first, and then the calibrated responsivity is used in the solution of the measurement equation to measure the unknown quantity from signals measured under the same or known geometrical conditions. In either case, the solutions are expressed as equations that are often referred to as calibration equations.

For example, in the case of a spectral radiometer, the measurement equation, Eq. (1.22), can be used to determine an unknown spectral radiance if the slit scattering function and the responsivity factor can be determined using appropriate approximations and accounting for uncertainties introduced. This is most often accomplished by employing a known spectral source, in this case a spectral radiance source, to determine the value of the integral at each wavelength of interest. Using the superscript $c$ to denote the values of the signal response and other components of Eq. (1.22) when a calibration source is used, we can write the relationship shown in Eq. (1.23). The signal output at each wavelength can be measured, and since the spectral radiance of the calibration source is known, the value of the integral in Eq. (1.22) can be determined using

$$\frac{r^c(\lambda_0)}{C^c L^c_\lambda(\lambda_0)} = \int_\lambda \rho(\lambda_0, \lambda) R^{\mathrm{f}}(\lambda)\, \mathrm{d}\lambda \tag{1.23}$$

By inverting Eq. (1.22) to determine the unknown spectral radiance using Eq. (1.23), we get the equation referred to as the calibration equation as shown in Eq. (1.24). It shows the unknown spectral radiance in terms of the signal response and the calibration quantities.

$$L_\lambda(\lambda_0) = \frac{C^c}{C} \frac{L^c_\lambda(\lambda_0)}{r^c(\lambda_0)} r(\lambda_0) \tag{1.24}$$

If the configuration factors are the same in both the calibration and the use of the instrument to perform measurements, the ratio of the configuration factors cancel, and if not, the geometrical terms involved in $C$ can sometimes indicate their ratio. Care must be exercised in a system with imaging optics to ensure that the entrance slit on the monochromators is illuminated in the same way in the calibration as in the use of the instrument. Another potential error is introduced by the fact that the slit scattering function does not go to zero outside the region of $\Delta\lambda$ but in fact has some finite value. If the calibration source and the source to be measured have different functional forms, then further uncertainties can occur due to the differences in accounting for the contributions in the wings of the slit scattering function's transmittance. Koskowski treats corrections to the measurement equation due to proper accounting of the transmittance in the wings away from the set wavelength $\lambda_0$ of the slit scattering function [26].

If the spectral radiometer is designed to be a spectral irradiance sensor system, the arguments follow similar paths, except one uses a calibration source of known spectral irradiance. Irradiance sensor systems often employ an integrating sphere with a known aperture as a collection device that is placed in front of the monochromator. Another technique to calibrate spectral instruments involves using tunable laser systems that calibrate spectral instruments by scanning a narrow wavelength laser line across the portion of the spectrum covered by the spectral instrument [36]. These techniques are further illustrated in Chapters 3 and 4. Another example is the measurement of the spectral radiance of an unknown source using the filter radiometer shown in Figure 1.4. It requires that the geometric factors, the filter transmittance, and detector responsivity be known in order to complete a measurement. Equivalently, if a calibration source of known spectral radiance is available, one can determine the total radiance responsivity $R_L$ and measure the unknown radiance as a ratio of the known radiance. In using this technique, it is important to note that the unknown spectral radiance source should have similar relative spectral radiance distributions as the calibration source, or extra uncertainties in the measurement will result from the difference. This comes about because of the averaging process used to reduce Eq. (1.18) to that of Eq. (1.19), which relies on being able to average the values in the integrand of Eq. (1.19). If a calibration source and an unknown source both do not have either roughly constant values over the bandpass or the same general functional form, then errors in applying Eq. (1.19) will result. A detailed examination of the uncertainty analysis for filter radiometers can be found in the literature [37].

In the measurement equation examples so far discussed, we have assumed that the detector is uniform and that there are no spatial, polarization, or temporal dependencies. In the more general case, this is not always realized, and a more detailed consideration of the measurement equation is necessary. For the calibration and uncertainty analysis of radiometers or complex electro-optical sensors for all the recognized dependencies, the goal is first to design calibration experiments using a standard source if necessary, and independently characterize the radiometer or sensor's overall system responsivity $R^T$ in the spectral, spatial, temporal, and polarization domains according to

$$R^T(\lambda_0, \lambda, \theta, \phi, t, P) = R(\lambda_0, \lambda)R(\theta, \phi)R(t)R(P) \tag{1.25}$$

where $R(\lambda_0, \lambda)$ is the overall system spectral responsivity, $R(\theta, \phi)$ is the *spatial responsivity*, also called the field of view responsivity, $R(t)$ is the *temporal responsivity* and $R(P)$ is the *polarization responsivity*. The measurement equations such as Eqs. (1.19) and (1.22) are generally derived for the major domain, that is, the spectral part, with certain assumptions made regarding the spatial and other domains. Therefore, the quantity that is most important to measure independently is the overall system spectral responsivity, $R(\lambda_0, \lambda)$ of the radiometer or sensor system. For spatial and other domains, deviations from the assumptions are assessed and applied as corrections to the measurement equations. Solutions to the modified measurement equations are obtained from results of the calibration experiment at the system level and are compared with predictions from component level specifications and measurements. This procedure allows for accurate calibration of the radiometer or the sensor and determination of the overall uncertainty budget.

We see that the spectral characterization is the major part of the calibration experiment. For broad-band filter radiometers, the shape of the spectral bandpass function for the radiometer system must be measured and compared to the calculated one to assess the uncertainties as the product of optical transmittance (reflectance) of apertures, filters, mirrors, etc. The calculated transmittance may not reflect the reality because multiple reflections and diffraction effects may alter the spectral characteristics of the throughput of the system. The spectral characterization is usually accomplished by scanning with a calibrated monochromator output with sufficient resolution and recording the system response across the wavelength band. This method of using monochromator output often suffers from the problems of not having sufficient spectral resolution or sufficient intensity in the output. However, modern developments in using tunable laser sources such as the SIRCUS facility at NIST solved these problems not only for the broadband filter radiometers but also for the complex spectroradiometers discussed in Chapter 4 [36, 38].

Also, in the case of filter radiometers, there is the problem of out-of-band leakage. In some applications, adequate long wavelength blocking is a major problem because of lack of availability of suitable materials and the difficulty of designing interference filters to serve the purpose. Short wavelengths can create fluorescence in the optical components and contribute to the out-of-band leakage. These factors, as well as contributions due to scattering, must be carefully assessed at the component level and at the system level to assess the overall uncertainty of the spectral responsivity of the sensor system. One method commonly used is to vary the temperature of a blackbody source to systematically change the peak of the Planckian radiation and measure the responsivity of the system. Long wavelength leakage can be detected and corrected by this method especially for mid or long wave infrared filter radiometers [22]. Again, the tunable laser facility such as SIRCUS is found to be very useful to assess this leakage in the case of the spectroradiometers used by NASA for sea surface temperature measurements and thus enables a correction to be applied to the overall system spectral responsivity. As mentioned earlier, comprehensive discussions of various methods to determine $R(\lambda_0, \lambda)$ can be found in the literature [22, 23]. It should be noted that the term *relative spectral responsivity* is introduced by some authors and refers to quantities such as we have defined but perhaps normalized to an integral or peak value. It is important to ascertain the precise definitions of terms in a particular discussion from the context.

The spatial characterization of a detector or a sensor is very important, because these devices often exhibit spatial non-uniformity. In the case of a sensor based on array detector, the spatial non-uniformity is evaluated by a calibration experiment where the array is flooded by a spatially uniform source at one irradiance level. Such sources are discussed in Chapter 5. In the case of a radiometer based on a single detector element, it is often evaluated by scanning the surface area of the detector with a small spot of radiation from a stable laser or a stable incoherent monochromatic source. The uncertainty evaluation would be different based on the application, whether the radiance, irradiance, or radiant power is measured by the system. These issues are discussed in Chapter 3.

The angular field-of-view characterization is also very important to assess the system performance for the desired linear field of view. It is necessary to know the spatial field of view responsivity $R(\theta, \phi)$ for the system, because errors can be made if a non-uniform source is measured with a non-uniform spatial field of view responsivity. As discussed by Wyatt the errors are due to on-axis performance and/or the measure of off-axis (out-of-field) rejection [22]. The on-axis performance is assessed by the response to a point source at angles close to the optical axis in comparison with the ideal designed field of view of the radiometer system. The off-axis performance is assessed by the response to a point source at angles far from the optical axis and evaluated to many orders of magnitude below the on-axis performance depending on the requirements of the system performance.

The *modulation transfer function* (MTF) is a parameter that describes the optical system response to spatial frequencies and is especially important for imaging systems with array detectors. Imaging radiometers are not dealt with extensively in this volume and the reader is referred to the references for discussion of this topic [26, 39].

It is important to characterize the temporal responsivity, $R(t)$, of a radiometer system, because the flux from the source being observed may change with time, or because intentional chopping of the radiation is employed to discriminate against background. All detectors have a characteristic response time before a signal is detected and an integration time for the signal to reach a stable value for measurement. Therefore, it is important to characterize the frequency response of the system. The frequency response is measured by observing the output response to a modulated light source [22]. This topic is discussed in Chapters 3 and 4.

It is also important to characterize the polarization responsivity, $R(P)$, of the radiometer sensor because mirrors and other possible materials in the optical beam path may introduce polarization or have polarization-dependent properties. For example, scattered optical radiation is frequently polarized, and hence the polarization sensitivity of a sensor measuring it must be characterized by employing polarizer and retarder combinations in various ways [22, 23]. This topic is not pursued in depth in this book.

Further more, *noise and drift* are important, because they affect the repeatability and reproducibility of the data. In general, noise and drift are characterized by employing a stable source of radiation, such as a blackbody, and collecting data at intervals throughout the dynamic range of the system for inclusion as a part of the calibration. Most optical sensor systems exhibit some degree of nonlinearity. The evaluation of the non-linearity of radiometers, associated corrections and uncertainties is very important and is discussed in Chapters 3 and 4.

The calibration of a radiometer system requires an experiment which includes characterization and determination of the corrections and sources of uncertainties listed above. This effort will yield the radiometer response as a function of the radiant, spectral, spatial, temporal, and polarization properties of an appropriate transfer standard. The transfer standard could be a well characterized source or another radiometer which has been calibrated and is traceable to international standards. The resulting equations, such as Eq. (1.24), are often called calibration equations and are derived from approximations in order to perform an inversion of the measurement equation. These approximations introduce elements of uncertainty which must be accounted for.

### 1.5.1 Uncertainty Nomenclature According to the ISO Guide

No measurement is complete unless it is associated with an uncertainty statement. The acceptance of the ISO guide to the Expression of Uncertainty in Measurement by the National Metrology Laboratories of various countries around the world paved the way for a uniform approach for the expression of uncertainty [17]. The basic concepts and nomenclature are discussed in Chapter 6.

The uncertainty in the result of a measurement generally consists of several components of uncertainty based on the measurement equation for the measurand $Y$. If we denote the estimated value for $Y$ as $y$ and the best estimates for various other component variables in the measurement equation as $x_1, x_2, x_3, \ldots, x_M$, the measurement equation can be written as

$$y = f(x_1, x_2, x_3, \ldots, x_M) \tag{1.26}$$

The total uncertainty, called the *combined standard uncertainty* $u_c(y)$, for $M$ statistically independent mean values of components, with uncertainties $u(x_1), u(x_2), u(x_3), \ldots, u(x_M)$, is calculated using Eq. (1.26) and the law of propagation of uncertainties

$$u_c(y) = \left[ \sum_{j=1}^{M} \left[ \frac{\partial f}{\partial x_j} \right]^2 u^2(x_j) \right]^{1/2} \tag{1.27}$$

However, if correlations are present between the components, then the covariances of the variables must be estimated and Eq. (1.27) must be modified with additional terms reflective of the covariance between the variables. This subject is further treated extensively in Chapter 6. The *relative combined standard uncertainty* is defined as $u_{c,r}(y) = u_c(y)/y_a$ and is frequently used because it expresses an uncertainty as a fraction of the measurand, which is a useful way to compare results.

The uncertainty determined by statistical techniques on the basis of direct measurements is referred to in the *ISO guide* as Type A while those which are evaluated by other means (e.g., on the basis of scientific judgment) as Type B. Finally, the *expanded standard uncertainty* is denoted in the *ISO guide* as $U$ and is obtained for an approximate level of confidence (the interval that will cover the true value of the estimated parameter with a given confidence) using the *coverage* factor $k$. Thus we have $U = ku_c(y)$ and the measurand $Y = y_a \pm U$, where $y_a$ is the measurement result. For example, approximately 95% of the measurements will fall within $\pm\, 2u_c(y)$, of the mean which corresponds to the case $k \approx 2$ if the distribution represented by $y_a$ and $u_c(y)$ is approximately normal and the sample is large. A 99% level of confidence corresponds to $k \approx 3$. The *ISO guide* also recommends the use of a *coverage factor* of 2 as a default multiplier in which case the level of confidence is approximately 95%.

However, the earth remote sensing community has been dealing with measurement of small changes in signals over extended time periods and has introduced the concepts of *accuracy* and *stability* in a quantitative fashion for time-series analysis of data. Accuracy is defined by the ISO guide as the "closeness of the agreement between the result of the measurement and the true value of the measurand" [17]. So, the term accuracy is measured by the bias or systematic error of the data, that is, the difference between the short-term average of the measured value of a variable and the truth. The short-term average value is the average of a sufficient number of successive measurements of the variable under identical conditions such that the random error is negligible relative to the systematic error. The term stability may be thought of as the extent to which the accuracy remains constant with time. Stability is measured by the maximum excursion of the short-term average measured value of a variable under essentially identical conditions over a decade. The smaller the maximum excursion, the greater the stability of the data set. Chapter 10 uses this terminology and further discussion on this topic can be found in Reference [40].

### 1.5.2 Application to Radiometric Uncertainty Analysis

As a simple example to illustrate the evaluation of combined standard uncertainty, $u_c(y)$, let us consider Eq. (1.12) for measuring the flux $\Phi$ reaching the aperture of area $A_2$ of an absolute cryogenic radiometer (ACR) in

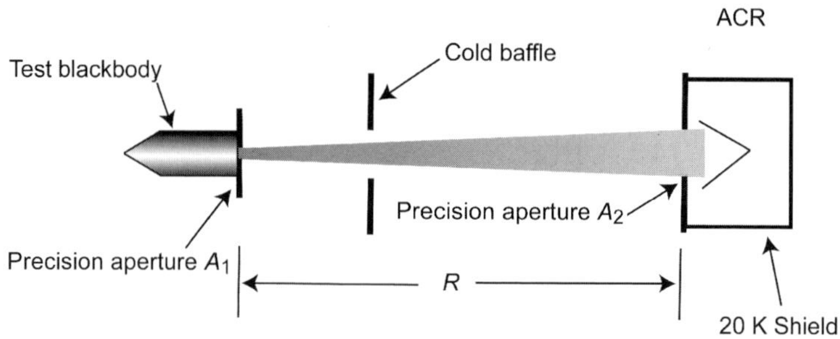


FIG. 1.7. Schematic of a blackbody calibration using an ACR.

Figure 1.7. The acronym ACR used here should not be confused with the same acronym commonly used in the literature for Active Cavity Radiometers, which are of non-cryogenic type [17]. This equation is useful as a close approximation for the calibration of the radiance temperature of a point source cryogenic blackbody that is equipped with pinhole apertures. Such blackbodies are used to calibrate infrared sensors in chambers that simulate the cold space background like what sensors in space operate in.

The setup shown in Figure 1.7 applies to the calibration of such blackbodies. The blackbody illuminates a precision aperture $A_1$ through which radiation passes to the limiting aperture $A_2$ on the ACR input. The ACR in Figure 1.7 is an electrical substitution radiometer that has the same response for all wavelengths of light and is used to measure the total optical power in absolute units of watts. These radiometers are discussed in detail in Chapters 2 and 3 and function like the radiometer shown in Figure 1.1. In order to calibrate the radiance temperature of the blackbody for its various temperature settings as read by the contact thermometers on the blackbody core, the Stefan–Boltzmann law for the *radiant exitance* $M$ is used. The *radiant exitance* is the flux per unit area of the source emitted into a hemisphere around the source. For a blackbody as a Lambertian source, integration of Eq. (1.4) for the whole hemisphere yields, $M = \pi L$. Therefore, Stefan–Boltzmann law for the set up shown in Figure 1.7 gives the flux $\Phi$ in terms of temperature $T$ and geometric factors as

$$\Phi \cong \frac{M}{\pi} \frac{A_1 A_2}{R^2}, \qquad M = \sigma T^4 \tag{1.28}$$

where $\sigma$ is the Stefan–Boltzmann constant whose value is $5.6704 \times 10^{-8}\,(\mathrm{W\,m^{-2}\,K^{-4}})$. If $r_1$ and $r_2$ are the radii of circular precision apertures $A_1$ and $A_2$, $R$ is distance between those apertures in Figure 1.7, the flux measured by the ACR is

$$\Phi \cong \frac{A_1 A_2}{R^2} \frac{\sigma}{\pi} T^4 \tag{1.29}$$

Equation (1.29) is the measurement equation to deduce the *radiance temperature* $T$ of the blackbody as

$$T = \left[ \frac{\Phi R^2}{\sigma r_1^2 \pi r_2^2} \right]^{1/4} \tag{1.30}$$

Using the procedures defined in the ISO guide to the Expression of Uncertainty in Measurement, the *combined relative standard uncertainty* $u_{c,r}(T)$ will have components of uncertainty consisting of four components arising from $r_1$, $r_2$, $R$, and $\Phi$ as

$$u_{c,r}(T) = \frac{1}{4}[u_r^2(\Phi) + 4u_r^2(r_1) + 4u_r^2(r_2) + 4u_r^2(R)]^{1/2} \tag{1.31}$$

The components $u_r(r_1)$, $u_r(r_2)$, and $u_r(R)$ are the relative uncertainties of the measurements of the geometrical terms and contains Type A uncertainties due to measurement statistics as well as uncertainties associated with temperature effects and other corrections to the values used in the calculation. The relative uncertainty in the flux measurement $u_r(\Phi)$ has both Type A and Type B contributions. The uncertainties in the flux measurement have contributions from the ACR itself and diffraction corrections, which are usually wavelength dependent. These various factors need to be evaluated and combined using the square root of the sum of the squares of the individual contributions provided the parameters are uncorrelated. Should correlation among the parameters be important, more detailed treatment like that discussed in Chapter 6 may be necessary. Any correction made to the flux due to diffraction will have an uncertainty associated with it and hence will contribute to the overall uncertainty in the flux in Eq. (1.31). The corrections for diffraction in radiometry are discussed in Chapter 9.

The treatment given here is a simplified analysis for a single setting of the blackbody temperature. A comprehensive analysis that includes data at different temperatures of the blackbody, a regression analysis to establish a calibration equation to predict radiance temperature in terms of the contact temperature reading of the blackbody, and a detailed analysis of the uncertainties involved in all the contributing components are presented in Appendix A.

## References


1. D. J. Dokken, Ed., "Strategic Plan for the U. S. Climate Change Science Program," p. 202. NOAA, Washington, DC, 2003.
2. D. Parks, "The Fire Within the Eye." Princeton University Press, Princeton, NJ, 1997.
3. S. F. Johnston, "A History of Light and Colour Measurement." Institute of Physics, Bristol and Philadelphia, 2001.
4. S. Perkowitz, "Empire of Light," 1st edition. Joseph Henry Press, Washington, DC, 1996.
5. J. C. Maxwell, "Treatise on Electricity and Magnetism," 3rd edition. Clarendon Press, Oxford, UK, 1892.
6. H. Hertz, "Electric Waves." Macmillan, New York and London, 1893.
7. P. J. Mohr and B. N. Taylor, CODATA recommended values of the fundamental constants: 1998, *J. Phys. Chem. Ref. Data* 28(6), 1713 (1999).
8. *International Lighting Vocabulary, CIE Publication 17.4*. Vol. CIE Publication 17.4. 1987, Vienna: Commission Internationale de L'Eclairage (CIE).
9. F. Hengstberger, "Absolute Radiometry." Academic Press, Boston, MA, 1989.
10. W. W. Coblentz, Present status of the determination of the constant of total radiation from a black body, *Bull. Bureau Stand.* 12, 553 (1915).
11. W. W. Coblentz, Studies of instruments for measuring radiant energy in absolute value: An absolute thermopile, *Bull. Bureau Stand.* 12, 503 (1915).
12. W. W. Coblentz, Various modifications fo Bismuth–Silver thermopiles having a continuous absorbing surface, *Bull. Bureau Stand.* 11, 131 (1914).
13. M. Planck, The theory of heat radiation, in "The History of Modern Physics, 1800–1950" (G. Holton, Ed.), Vol. 11, p. 470. Tomash/American Institute of Physics, New York, 1989.
14. R. Loudon, "The Quantum Theory of Light," 2nd edition. Clarendon Press, Oxford, UK, 1983.
15. A. Einstein, *Annl. Phys.* 17, 132 (1905).
16. T. J. Quinn and J. E. Martin, A radiometric determination of the Stefan–Boltzmann constant and thermodynamic temperatures between −408 °C and +1008 °C, *Phil. Trans. Roy. Soc. London* A316, 147 (1985).
17. *Guide to the Expression of Uncertainty in Measurement*, 1st edition, p. 101. International Organization for Standardization, Switzerland, Geneva, 1993.
18. *Quantities and Units*. "ISO Standards Handbook." International Organization for Standards, Geneva, 1993.

19. J. M. Palmer, Getting intense on intensity, *Metrologia* 30, 371 (1993).
20. W. L. Wolfe, Introduction to Radiometry, in "Tutorial Texts in Optical Engineering" (D. O'Shea, Ed.), p. 184. International Society for Optical Engineering, Bellingham, WA, 1998.
21. *The Basis of Physical Photometry*, Vol. CIE publication number 18.2. CIE, Vienna, 1983.
22. C. L. Wyatt, "Radiometric Calibration: Theory and Methods." Academic Press, Orlando, FL, 1978.
23. F. E. Nicodemus, *Self-Study Manual on Optical Radiation Measurements*. NBS Technical Note. Vol. TN910-1 through TN910-8. 1976–1985, U.S. Government Printing Office, Washington, DC. Available on CD from Optical Technology Division, NIST, Gaithersburg, MD 20899-8440.
24. F. Grum and R. Becherer, Radiometry, in "Optical Radiation Measurements," 1st edition (F. Grum, Ed.), Vol. 1, p. 335. Academic Press, San Diego, 1979.
25. R. W. Boyd, "Radiometry and the Detection of Optical Radiation." Wiley, New York, 1983.
26. H. J. Kostkowski, "Reliable Spectroradiometry," 1st edition, p. 609. Specroradiometric Consulting, La Plata, MD, 1997.
27. R. Siegel and J. R. Howell, "Thermal Radiation Heat Transfer," 3rd edition. Hemisphere Pubishing Co, Washington, DC, 1992.
28. F. P. Incropera and D. P. DeWitt, "Fundamentals of Heat and Mass Transfer," 4th edition, p. 886. Wiley, New York, 1996.
29. A. C. Parr, "A National Measurement System for Radiometry, Photometry, and Pyrometry Based upon Absolute Detectors." NIST Technical Note. Vol. TN 1421. U.S. Government Printing Office, Washington, DC, 1996.
30. C. L. Wyatt, "Radiometric System Design." Macmillan, New York, 1987.
31. W. R. McCluney, "Introduction to Radiometry and Photometry." Artech House, Boston, 1994.
32. H. W. Yoon, C. E. Gibson, and P. Y. Barnes, The realization of the NIST detector-based spectral irradiance scale, *Metrologia* 40(1), S172 (2003).
33. E. W. M. van der Ham, H. C. D. Bos, and C. A. Schrama, Primary realization of a spectral irradiance scale employing a monochromator based cryogenic radiometer, *Metrologia* 40(1), S181 (2003).
34. E. L. Derniak and D. G. Crowe, Optical Radiation Detectors, in "Wiley Series in Pure and Applied Optics" (S. S. Ballard, Ed.), p. 300. Wiley, New York, 1984.
35. G. H. Rieke, "Detection of Light: From the Ultraviolet to the Submillimeter." Cambridge University Press, Cambridge, UK, 1994.

*36.* S. W. Brown, G. P. Eppeldauer, and K. R. Lykke, NIST Facility for Spectral Irradiance and Radiance Responsivity Calibrations with Uniform Sources, *Metrologia* 37, 579–582 (2000).

*37.* B. K. Tsai and B. C. Johnson, Evaluation of uncertainties in fundamental radiometric measurements, *Metrologia* 35, 587–593 (1998).

*38.* K. R. Lykke, P.-S. Shaw, L. M. Hanssen, and G. P. Eppeldauer, Development of a monochromatic, uniform source facility for calibration of radiance and irradiance detectors from 0.2 μm to 12 μm, *Metrologia* 35(4), 479–484 (1998).

*39.* D. O'Shea, "Elements of Modern Optical Design," 1st edition, p. 402. Wiley, New York, 1985.

*40.* R. Datla, W. Emery, G. Ohring, R. Spencer, and B. Wielicki, Stability and accuracy requirements for passive satellite sensing instruments for global climate change monitoring, in "Post-Launch Calibration of Satellite Sensors" (S. A. Morain and A. M. Budge, Eds.), p. 193. A. A. Balkema, New York, 2004.

# 2. ABSOLUTE RADIOMETERS*,†

Nigel P. Fox
National Physical Laboratory, Teddington, UK

Joseph P. Rice
National Institute of Standards and Technology, Gaithersburg, Maryland, USA

## 2.1 Introduction

Many of the advances made in optical radiometric measurements in the last two or three decades (highlighted in the chapters of this book) are a direct consequence of the use of absolute optical radiation detectors (absolute radiometers). In this case, an absolute radiometer is an instrument that cannot only detect and quantify the level of incident radiation, and for which the means of quantification is by direct reference to another measurable physical phenomena (with a route of traceability to SI units of lower uncertainty than intended by the radiometer), but also is self-contained within the radiometer, i.e. the radiometer is not simply calibrated against another instrument or reference source.

Figure 2.1 illustrates this hierarchy and interdependency, with the absolute radiometer linked (in terms of traceability to SI) through fundamental constants (often via the electrical units). The linkage to spectrophotometry is less direct, but it is clear that many of the improvements in this field are a direct result of the implementation of improved detectors resulting from improved spectral responsivity measurements.

There are two principal underpinning concepts/methodologies in current usage within modern absolute radiometers: the use of electrical substitution (thermal) and that of quantum physics (photon). In both cases these basic concepts can be further subdivided in their operational detail but are largely generic in their overall implementation.

This chapter gives a brief introduction to the basic principles behind both these approaches but largely highlight more detailed literature which will allow the reader to explore in greater depth. The chapter concentrates on a particular type of electrical substitution radiometer (ESR) called a cryogenic radiometer, which has become the reference instrument of choice for all



†Portions of this paper written by J. P. Rice are a contribution of the National Institute of Standards and Technology and are in the public domain.



EXPERIMENTAL METHODS IN THE PHYSICAL SCIENCES, vol. 41
ISSN 1079-4042 DOI: 10.1016/S1079-4042(05)41002-4
Published by Elsevier Inc.

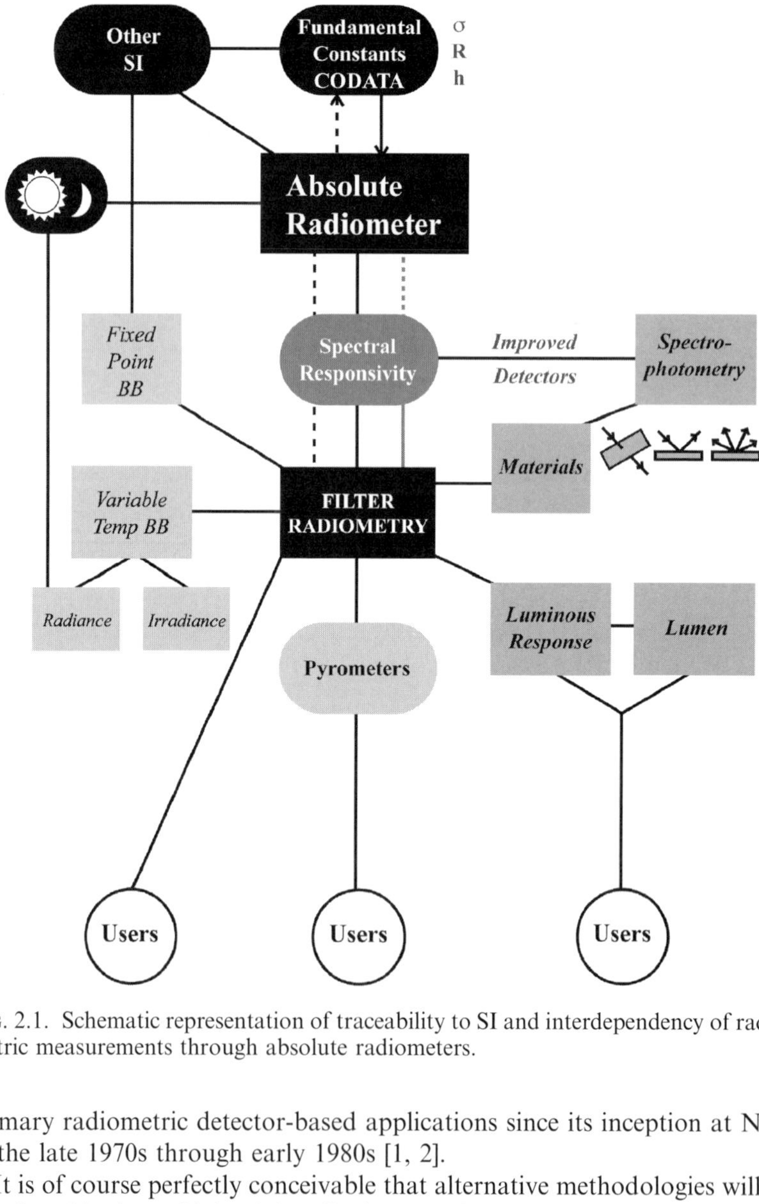


FIG. 2.1. Schematic representation of traceability to SI and interdependency of radiometric measurements through absolute radiometers.

primary radiometric detector-based applications since its inception at NPL in the late 1970s through early 1980s [1, 2].

It is of course perfectly conceivable that alternative methodologies will be developed in time which might offer improved or different routes to SI units, e.g., photon correlation described in Section 5.4, but at present the highest accuracy is attainable through the instruments/techniques described in this chapter.

## 2.2 Predictable Quantum Efficiency Detectors

### 2.2.1 Scales Based on Photon Detectors

These utilise solid-state physics modelling techniques to predict the response of photodiodes, commonly called the self-calibration technique. In this technique, a number of measurements can be performed to calculate the quantum efficiency of the device, i.e. the number of charge carriers reaching the sensing electrical circuit per incident photon, and consequently predict its spectral response. From our definition in Section 2.1, such a method can definitely be considered "absolute" and thus such characterised photodiodes, as "absolute radiometers". The details of this technique are best described by its originators, Geist and co-workers [3, 4], but for completeness its principles are summarised here.

### 2.2.2 Determination of Self-Calibration Technique Parameters

The responsivity $R_\lambda$ of a photodiode at a wavelength $\lambda$ can be described by

$$R_\lambda = \frac{\varepsilon \lambda e}{hc} \tag{2.1}$$

where $e$ is the electronic charge of an electron, $c$ the speed of light, $h$ Plancks constant and $\varepsilon$ the quantum efficiency. Since $e$, $h$ and $c$ are all constants, Eq. (2.1) can be simplified to

$$R_\lambda = \varepsilon \lambda K \tag{2.2}$$

where $K = e/hc$ is a constant.

The main hypothesis in the self-calibration technique is that the quantum efficiency $\varepsilon$, or perhaps more intuitively, the quantum deficiency $(1 - \varepsilon)$, can be determined from a number of simple experiments.

The first, and largest in value, parameter is the loss due to reflection from the front surface of the photodiode. It is assumed by the nature of the optical surface of the photodiode that this reflectance is dominated by its specular component; in fact, the light spot is hardly visible on the surface if viewed in a non-specular direction, indicating the lack of any diffuse component. The specular reflectance, $\rho$, can easily be measured by using a collimated intensity-stabilised laser beam (see Chapter 3 for more details of laser stabilisation) and positioning the photodiode at a slight angle to the normal (approximately 5°) using the following measurement sequence. The power of the incident beam can first be measured by substituting another, "reference photodiode" at the location of the photodiode being measured. Then after replacing back the original photodiode (on which the measurement is being made) the "reference photodiode" measures the power of the reflected beam. This process removes the need to calibrate the reference, as a ratio of the two signals gives the reflectance of the photodiode under test. The only additional parameter that has to be checked during this measurement is the angular sensitivity of this reflectance and any change in the photodiode's responsivity due to angle. This sensitivity is critical because the aim of the self-calibration technique is to determine all the loss mechanisms for a photodiode under a set of standard conditions, i.e. near-normal incidence. The former has been found to be insensitive at the uncertainty levels achievable here to angles less than about 10°. The latter can easily be measured again as a ratio of its response at an angle of 5°, to that of normal incidence under a constant irradiance. This correction is given by the ratio $R_0/R_5$ in the results in Table 2.1.

The remaining losses are those internal to the photodiode which effectively reduce the internal quantum efficiency, $\varepsilon_i$, from unity, resulting in Eq. (2.3), where $\varepsilon$ from Eq. (2.2) has been replaced by $(1-\rho)\varepsilon_i$ from the above discussion.

$$R_\lambda = (1-\rho)\varepsilon_\lambda \lambda K \tag{2.3}$$

In a $p^+nn^+$ photodiode, the major internal losses are due to recombination effects and are:

(1) in the rear of the photodiode beyond the depletion region surrounding the junction, particularly at long (red) wavelengths where the absorption coefficient is lower, and
(2) in the front region of the photodiode, close to the interface between the silicon and the silicon dioxide anti-reflection coating. Trapped positive

TABLE 2.1. Results of the Comparison of Spectral Responsivities of Photodiodes C80 and C81 at a Wavelength of 676.4 nm

| Parameter | Photodiode reference | |
|---|---|---|
| | C80 | C81 |
| $\varepsilon_F$ | 0.9989 | 0.9928 |
| $\varepsilon_R$ | 0.9982 | 0.9959 |
| $\varepsilon_A$ | 0.99987 | 0.99987 |
| $(1-\rho)$ | 0.9209 | 0.9216 |
| $\varepsilon_i$ | 0.9970 | 0.9886 |
| $R_0/R_5$ | 0.99968 | 0.99970 |
| $R_{SCT}$ | 0.50086 A/W | 0.49706 A/W |
| $R_{rad}$ | 0.50093 A/W | 0.49696 A/W |
| $R_{rad}/R_{SCT}$ | 1.00014 | 0.99980 |

charge at this interface attracts the minority carriers (electrons) and lengthens the time, compared to the carrier lifetime, taken to reach the depletion region and this increases the probability of recombination.

The first effect can be evaluated by applying a reverse bias to the photodiode, which extends the depth of the depletion region such that the collection efficiency in the rear of the photodiode can be brought up to unity. Measurements are generally straightforward and the effect is usually small. Figure 2.2 shows typical results for an EGG UV444 BQ photodiode and the associated correction $\varepsilon_R$, being the ratio of the photodiodes response at the plateau to that in the unbiased state.

A more difficult problem is that of counteracting the effect of the trapped positive charge at the $SiO_2$–Si interface. A possible solution is to measure the responsivity change when changing the associated recombination loss by storing a sufficiently large negative charge on the $SiO_2$ surface. This is called the "oxide bias" measurement. Success has been widely achieved through the use of a drop of water or dilute boric acid solution deposited on the oxide surface, as a removable transparent electrode (Fig. 2.3).

When a negative voltage is applied to the water drop, the response of a photodiode to constant monochromatic irradiation is seen to increase with increasing negative voltage until saturation occurs at a point where it is assumed that recombination losses have been eliminated (Fig. 2.4). However, as pointed out by Geist et al. [5], one consequence of the repeated application and removal of a water drop is a progressively deteriorating surface quality. Zalewski [6] also reported instability problems associated with the application of a negative bias to the surface, and during the course of critical studies of the self-calibration technique at NPL there were

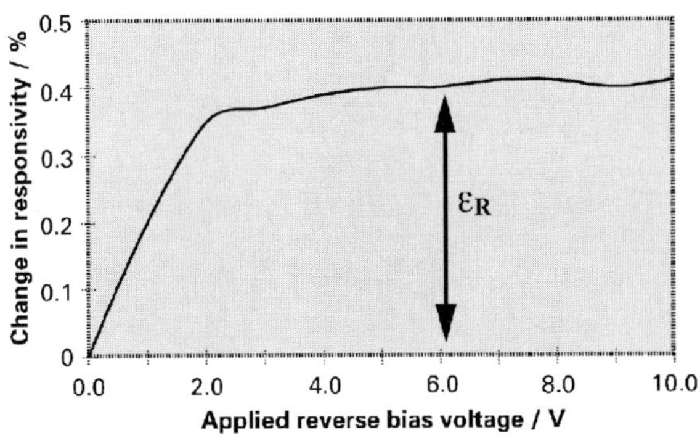


FIG. 2.2. Plot showing change in spectral responsivity for photodiode reference C81 as a function of reverse bias voltage at a wavelength of 676.4 nm.

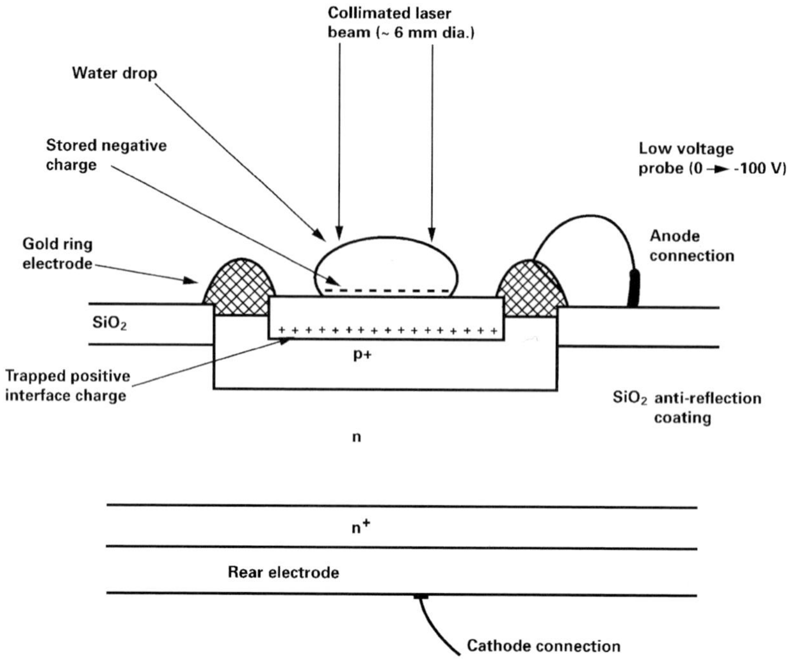


FIG. 2.3. Basic structure of a silicon photodiode and the experimental arrangement used for the water drop bias experiments.

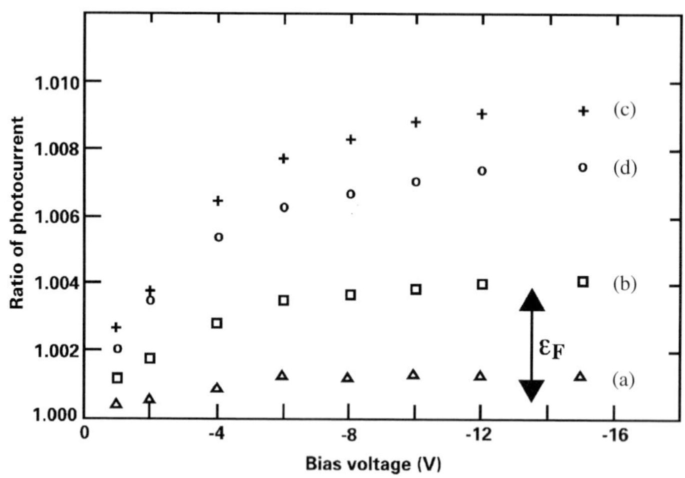


FIG. 2.4. Ratio of photocurrent for a typical silicon photodiode, relative to zero bias, as a function of oxide-bias voltage at (a) 568.2 nm, (b) 468.0 nm, (c) 406.7 nm. Four months later measurements were repeated at 468.0 nm (d).

indications both of instability linked with the magnitude of the bias voltage applied and of longer-term instability of internal quantum efficiency, possibly associated with the application of the self-calibration technique [7]. An example is apparent in Figure 2.4d, which shows that the size of the oxide bias quantum efficiency correction changed with time. However, with appropriate care, the ratio of biased to unbiased gives the correction for front region loss $\varepsilon_F$.

Geist et al. [4] describe one other loss mechanism which is currently based on an estimated loss due to bulk absorption in the heavily doped bulk region through an Auger recombination process, $\varepsilon_A$. However, as can be seen from Table 2.1, this is relatively small compared to the others.

Combining these loss mechanisms allows $\varepsilon_i$ to be determined as

$$\varepsilon_i = \varepsilon_F \varepsilon_R \varepsilon_A \tag{2.4}$$

An example set of results of the determination of the parameters making up the self-calibration technique for two typical photodiodes, together with their resultant calculated responsivity compared with that determined by a cryogenic radiometer (see Section 2.3.1) at a wavelength of 676.4 nm, is presented in Table 2.1.

### 2.2.3 Current Status of Photon-Based Techniques

It soon became apparent that the self-calibration technique had many advantages over room temperature ESRs, not least because of its simplicity and low cost. However, the technique was initially largely limited to the relatively narrow, although important, spectral region between around 400 and 900 nm with an uncertainty around 0.1%, which was comparable with the best ESRs of the time [8]. The technique lost favour as a primary standard in the latter half of the 1980s as cryogenic radiometers were developed which could achieve uncertainties of <0.01% [2] while uncertainties for the self-calibration technique was limited to at best about 0.05% [9, 10]. As will be discussed in Chapter 3, the concepts underpinning the "self-calibration" technique led to the development of "trap detectors" [11, 12] and has helped considerably in the establishment of models to interpolate spectral responsivity scales [13]. Such models have been further developed to extend the application into the UV spectral region [14, 15] and have allowed spectral responsivity scales to be established with uncertainties approaching 0.5% through extrapolation of visible measurements.

In summary, it can be seen that if appropriate care is taken when carrying out the self-calibration technique, then it is at least as accurate as any other technique for determining a spectral radiant power responsivity scale with the exception of cryogenic radiometry. Given its implicit simplicity and relatively low cost it is perhaps surprising that it is rarely used. It should be noted that without this technique and the stimulus it gave to the science of radiometry, cryogenic radiometry would probably not have been developed to the extent that it has, and consequently many of the advances discussed in this book may not have reached the maturity they have in the same timescale.

Since the inception of this concept some 25 years ago, there has been considerable progress in photodiode manufacture, and knowledge of properties of silicon. This has allowed Geist and others [10, 16] to continue the work, utilising software models developed for solar cell development, and have demonstrated that, subject to more detailed characterisation of key materials input parameters, the technique could be improved significantly, possibly to a level approaching 1 part in $10^8$ [17]. This is obviously a significant challenge to the community and, if achieved, leaves open the possibility of new concepts and approaches. One such method is the use of photon correlation techniques to establish scales directly in terms of "counted photons". These techniques are discussed in detail in Section 5.4 and will not be discussed further here.

## 2.3 Absolute Thermal Detectors

### 2.3.1 Principles of Electrical Substitution Radiometers

The origins of modern radiometry can be considered to date back to the late 19th century when the first absolute thermal radiometers were independently developed by Ångström [18] and Kurlbaum [19]. Ångström's radiometer was designed to measure incident solar radiation for meteorological purposes and Kurlbaum's was the first to be designed as an absolute standard for light measurement (the first attempt to replace a source as the primary standard).

The operating principle of these absolute radiometers was the same: *the comparison of the heating effect of optical radiation with that of a substituted amount of electrical Joule heating* and is now commonly called electrical substitution (ES) and the radiometers electrical substitution radiometers (ESRs). Figure 2.5 outlines the operating principle, and some of its limitations, in a more illustrative fashion.

A typical radiometer, operating in ambient conditions, consists of an optical absorbing element, which can in its simplest form be a metal disk, linked via a poorly conducting heat link, to a reference heat sink maintained at a constant temperature. The constancy of temperature can be maintained actively by a temperature control loop, or passively through its thermal mass.

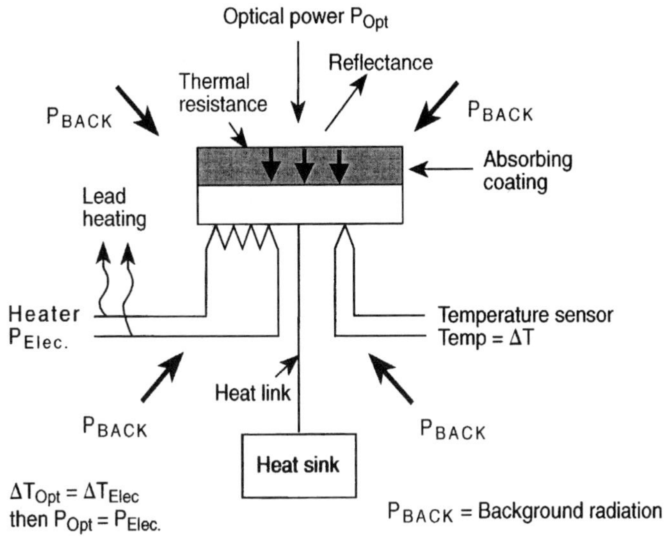


FIG. 2.5. Schematic representation of the electrical substitution principle. $P_{\mathrm{Back}}$ = background radiation input, $P_{\mathrm{Elec}}$ = electrical power input, $P_{\mathrm{Opt}}$ = optical power input. $P_{\mathrm{Elec}}$ is adjusted so that $\Delta T_{\mathrm{Elec}} = \Delta T_{\mathrm{Opt}}$. $\Delta T = T(P_{\mathrm{Elec+Back}})$ or $(P_{\mathrm{Opt+Back}})$-$T(P_{\mathrm{Back}})$.

In the latter case this might simply be a metal block with good contact to ambient air. When optical radiation is incident on the absorbing element, a proportion will be absorbed (dependent largely on the material properties of the surface) and lead to the establishment of a temperature gradient between the absorber and the reference heat sink, proportional to the amount of optical radiation incident upon it. This temperature gradient can be detected and quantified by some form of heat sensor, e.g. thermometer, or material attribute, e.g. thermo-resistive or pyroelectric effect. *It is important to recognise that in an absolute radiometer this temperature-sensing device is usually only used in a relative sense and more importantly as a "null-detector" and thus only its measurement resolution and short-term stability are key parameters.* To quantify the optical radiation in terms of radiant power with the SI units of Watts, requires the substitution at the absorbing element of a measurable and equivalent amount of heat energy generated by electrical Joule heating. This can of course most easily be achieved through the use of a resistive heater attached to the optical absorber and a measure of the potential drop across it and current passed through it. By adjusting the current flowing through the heater until the temperature-sensing element measures the same as when optical radiation was incident upon the radiometer (null-change), equivalence between the two forms of heating can be established and consequently the optical power can be determined.

Although the ES principle described above is very simple, it is also relatively easy to identify its limitations and dominant sources of uncertainty and error. The underpinning concept of ES is that the heating effect of the measured electrical power is directly equivalent to that of the incident optical power, i.e. the heat flow paths generated by each source of energy are directly equivalent. Unfortunately it is immediately obvious that in the general situation, and at ambient temperatures, this is not the case and a number of subsequent experiments need to be carried out to evaluate these sources of correction and in particular their consequential uncertainty.

Over the last century a wide range of techniques and methods have been devised to characterise such absolute radiometers and a good review of these can be found in Reference [20] and references therein. However, in the late 1970s, it was realised that many of these sources of error can be significantly reduced by cooling the radiometer to cryogenic temperatures, in particular the ~4 K temperature of liquid helium. The initial concept can be traced back to Ginnings and Reilly of NIST [21], who sought to build an instrument to measure the total radiation from a blackbody to measure thermodynamic temperature of the boiling point of water. However, while their attempt failed, largely due to diffraction and scattering effects, it served to inspire the developments of Quinn and Martin of NPL who succeeded in building the first absolute radiometer operating at cryogenic temperatures [1]. Their instrument again was designed to measure thermodynamic temperature, but in this case also to make a radiometric determination of the Stefan–Boltzmann constant, through a measure of the total radiation emitted from a blackbody operated at the triple point of water.

The following sections will discuss, in turn, the key conditions, listed below, that need to be satisfied by an absolute radiometer and in particular how these can be optimised by low-temperature operation.

1. The radiometer detector/absorber should have a high absorptance to ensure that all incident radiant flux is absorbed and contributes to a rise in temperature of the radiometer detector.
2. All supplied (measured) electrical power should be dissipated as heat in the detector, no energy should be dissipated in the connecting leads, and the amount of heat conducted away via the connecting leads should be minimised.
3. The heat flow path from the detector to the reference temperature heat sink should be identical for electrical and radiant heating, and should not be influenced by any difference in temperature gradients in the detector created by the two separate heating modes.

4. The thermometer which senses the temperature rise of the detector should have a small thermal capacity and be in thermal equilibrium with the detector; it should also have adequate resolution, sensitivity and short-term stability.
5. The temperature of the reference heat sink should remain constant during the period of measurement.
6. The detector should be shielded so that its operation is not influenced by any other sources of radiation, or heat generation source, and similarly its field of view limited to reduce scattered radiation falling on the detector.

#### *2.3.1.1 Absorption of optical radiation*

The use of a black (spectrally non-selective) low-reflectance coating, e.g. paint, metal black, carbon etc., can increase the absorptance of the optical absorber and remove any variability due to spectral content of the incident radiation. This can be further enhanced through changing the geometric shape of the absorber from, for example, a flat disc to a cavity such as a cone or enclosed cylinder (analogous to a black body) as shown in Figure 2.6. The larger the aspect ratio of exit aperture to plane of incidence of absorber (base of cavity where the optical beam is incident), the higher the absorptance, since the dominant source of reflection is diffuse from the first reflecting surface. This is an important factor in terms of cavity design. It means that the optimum black coating is the one with the lowest component

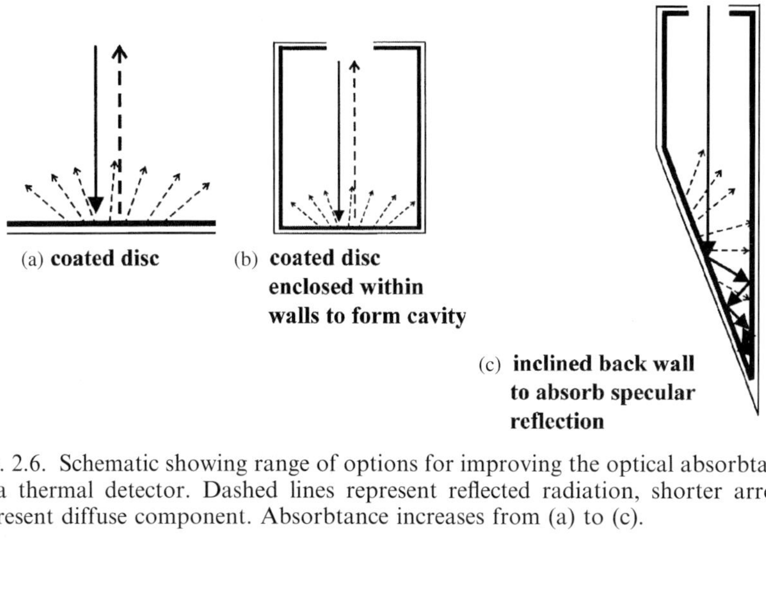


FIG. 2.6. Schematic showing range of options for improving the optical absorbtance of a thermal detector. Dashed lines represent reflected radiation, shorter arrows represent diffuse component. Absorbtance increases from (a) to (c).

of diffuse reflectance, not necessarily the lowest reflectance. For example, by careful cavity design, it can be seen that a specular reflecting surface can preferentially direct radiation, resulting in multiple reflections and thus effectively ensuring that all the radiation is buried and absorbed (Fig. 2.6c), with the exception of any diffuse component of reflection. Many specular reflecting surfaces by their very nature have low diffuse components of reflection and thus are often better than a nominally lower "total reflectance", diffuse black surface. In fact, even with a low reflecting diffuse black surface some advantage can be gained by using similar design principles, as these reduce the solid angle of the emitted surface with respect to the entrance aperture of the cavity by the cosine of the angle of incidence. Although increasing this angle often increases the specular reflectance of the surface (particularly for longer wavelengths), this reflected radiation is buried by multiple reflections.

The absorptivity of a diffusely reflecting cavity, $\alpha$, can be calculated by using either the integral-equation method [22] or the series reflectivity method [23, 24]; both methods are a function of the physical dimensions of the cavity as well as the surface emissivity. The absorptivity of a specular reflecting cavity can be calculated from the series reflectivity method and is a much simpler calculation involving ray tracing. In general, it is easier to design a highly absorbing cavity that is physically large compared to the diameter of the cavity aperture. If we consider a cylindrical cavity of radius $R$ with an aperture of radius $r$ at one end separated from a back wall at the other end by a distance $L$ and $\varepsilon$ is the emissivity of the interior surface, then for $\alpha > 0.99$ we require $r/R < 0.2$, $L/r > 5$ and $\varepsilon > 0.8$.

For an example we will use the cavity that was designed for the second NPL cryogenic radiometer [2], i.e. that designed specifically for radiometry (the primary standard (PS) radiometer). This is also the same cavity used in the HACR of NIST [25] and at PTB in Berlin [26]. This cavity is shown in Figure 2.7.

The radiometer cavity is made of 0.25-mm-thick electro-formed copper and is cylindrical in shape. It is 50 mm in diameter and 150 mm long. The lower end of the cavity is closed with a plate containing a 12-mm-diameter entrance aperture; the other end of the cavity is closed by a back wall inclined at 30° to the cylinder axis. It is coated internally with Chemglaze Z302, a specularly reflecting black paint; this paint has a diffuse reflectance of about 1% and a specular reflectance of 5% in the visible region of the spectrum. The absorptivity of the cavity has been calculated for a beam of visible radiation entering the cavity through the aperture along the axis of the cylinder. The beam is incident on the back wall and the cavity geometry is designed so that the specular component of the reflection is nearly totally absorbed (99.998% after the fifth reflection). Consideration need only be

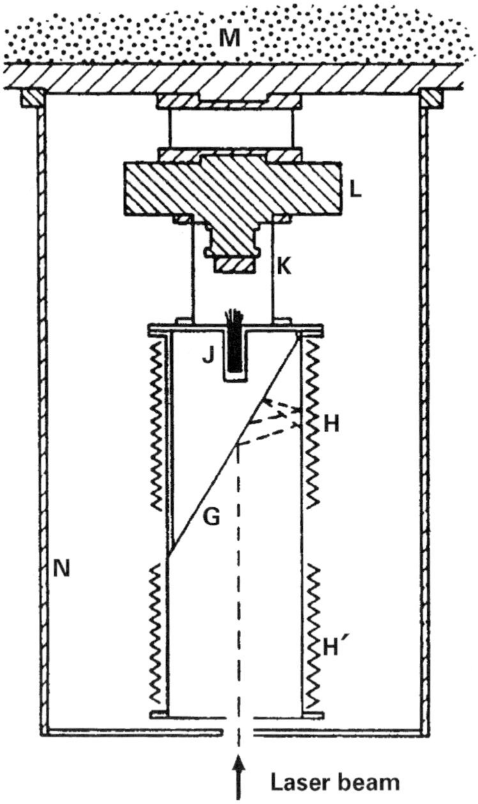


FIG. 2.7. Schematic diagram of the cavity designed for the second NPL cryogenic radiometer (Primary Standard PS radiometer). G radiometer cavity, H and H′ electrical heaters, J thermometer well, K heat link, L heat sink, M helium reservoir, N radiation shield.

given to the diffuse component of the reflection and this has been calculated using the series reflectivity method from the equation [24]

$$\alpha = 1 - I_1(L)\rho - I_2(L)\rho^2 - \cdots - I_n(L)\rho^n \tag{2.5}$$

where $\rho$ is the hemispherical diffuse reflectance of the wall and $I_1(L)$, $I_2(L), \ldots, I_n(L)$ are geometrical factors. As the diffuse reflectance is so small in this case, the $\rho^2$ and higher-order terms can be neglected. Also because $L \gg r$, $I_1(L)$ approximates to $\cos\theta\ (r^2/(r^2+L^2))$ and so for this cavity,

$$\alpha = 1 - \rho\cos\theta\frac{r^2}{(r^2+L^2)} \tag{2.6}$$

where $\theta$ is the angle of incidence of the beam at the rear wall.

From Eq. (2.6) and using $r = 6.0\,\text{mm}$, $L = 100\,\text{mm}$, $\rho = 0.01$ and $\theta = 60°$, the diffuse absorptance of the cavity is calculated to be 0.99998 in any part of the visible spectrum. This value of the cavity absorptance has also been confirmed experimentally by measuring the total reflectance of a model of the cavity using an integrating sphere and a helium–neon laser.

A full description of such a measurement technique can be found in Fox et al. [27]. Here they make use of a large integrating sphere to collect the radiation reflected by a cavity absorber (Fig. 2.8). In this case the reflectance is determined by comparison to a reference black itself calibrated against a standard white diffuser. As the cavity reflectance is very small, $<0.002\%$, uncertainty in the reference black reflectance value ($\sim 2.5\%$) becomes less significant.

Cavities of this size can obviously be constructed for both ambient temperature ESRs and cryogenic radiometers. However, another factor that

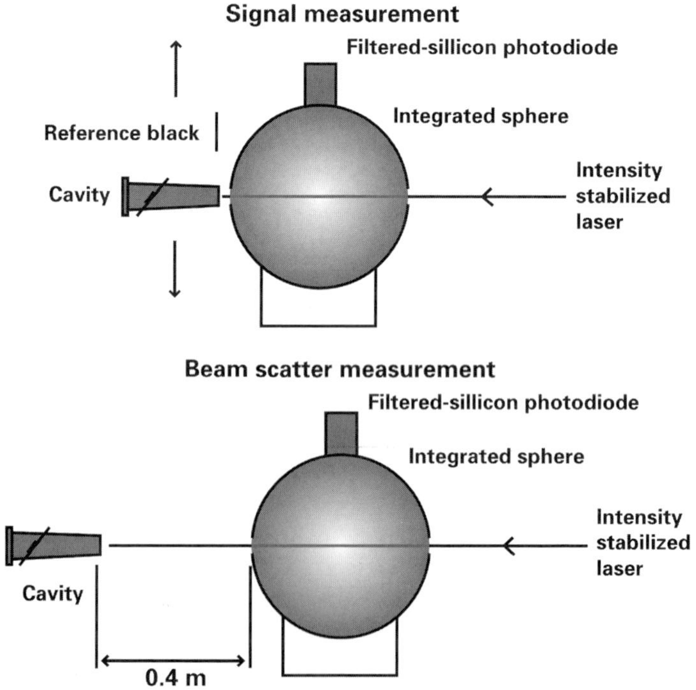


FIG. 2.8. Schematic diagram describing the measurement of the reflectance of a low reflectance cavity absorber. The lower schematic shows the use of the cavity itself as a "beam dump" to allow scatter from the laser beam within the integrating sphere to be determined.

must be taken into consideration when designing a cavity is the time constant, the time taken for thermal equilibrium to become established in the cavity after a power change, and its sensitivity.

Consider a constant radiant power $Q$ absorbed in a cavity at temperature $T_a$ and connected by a heat link of thermal resistance $R$ to a heat sink at temperature $T_r$. For a small temperature difference $dT = T_r - T_a$ and an incremental time period $dt$, the heat flow equation can be written as [28]

$$Q\,dt - (dT/R)\,dt = C\,dT \tag{2.7}$$

which can be solved to give the well-known equation

$$dT = RQ(1 - \exp[-t/RC]) \tag{2.8}$$

where $C$ is the thermal capacity of the cavity and is equal to the mass of the cavity times its specific heat. Thus, the temperature of the receiver will rise exponentially with a time constant $\tau = RC$.

If the cavity is made of copper, then by cooling the cavity from 293 to 4 K the specific heat of the cavity decreases by a factor of about 1000 and $\tau$ is correspondingly reduced. $\tau$ for the cavity shown in Figure 2.7 (mass = 260 g) at 4 K has been measured experimentally to be about 3 min, whereas at 293 K it would be of the order of many hours. Therefore, large cavities are impractical for ambient temperature ESRs and smaller cavities are normally used, which cannot be made so highly absorbing and hence require the application of large corrections (with their associated uncertainties) for the incomplete absorption of the incident radiation in the cavity.

One clear advantage of cooling the radiometer to 4 K is that a large cavity can be constructed with a high absorptance while maintaining an acceptable time constant. However, while the use of even relatively simple geometric shapes can greatly increase the absorptance, they also consequently add to the thermal mass of the absorber and thus because of the specific heat capacity of the substrate material, reduce the sensitivity of the radiometer. For the example given above a time constant of 3 min was obtained at cryogenic temperatures; while acceptable in its particular application, it is limiting in terms of convenience and in the measurement of low power levels or more time-variant input signals. Considerable effort has therefore been expended on the manufacture of very light-weight structures (often using electroforming techniques) and the development and use of low-reflectance coatings. In some cases, absorptance enhancement has been achieved using reflective elements thermally isolated from the black substrate. In such cases, the mirror constructs a pseudo-cavity by extending the position of the exit aperture away from the reflecting surface while not contributing to the mass [29] (Fig. 2.9).

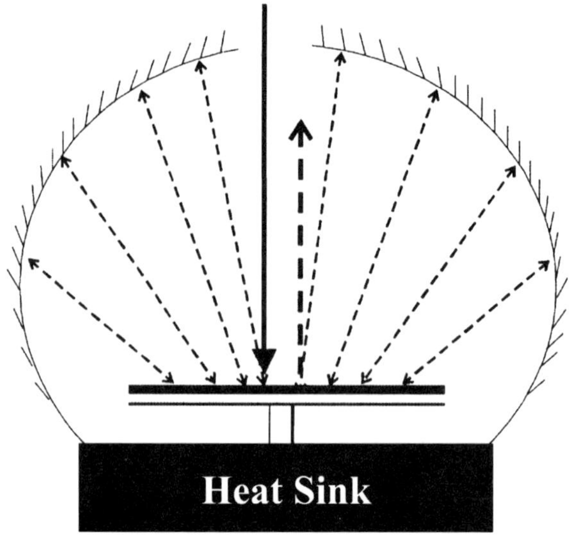


FIG. 2.9. Schematic showing absorbtance enhancement of a blackened disc through use of a reflecting hemisphere.

In addition to simply considering the absorption properties of a black coating it is also essential to consider its thermal resistance. For example, the absorption of radiation near the surface of an absorbing coating will establish a temperature gradient between it and the substrate. This surface can then dissipate heat energy to its surrounding by radiation (proportional to $T^4$), and if not in a vacuum, by conduction and convection to the surrounding air. Since most black coatings are relatively poor thermal conductors and the lowest reflectors, e.g. metal blacks and NPL super-black [30], have a very large effective surface area since their absorptive properties are enhanced by the fine structure of the surface (Fig. 2.10), any loss of energy in this way will not be accounted for by the radiometers temperature sensor and thus will not contribute to the energy balance.

### *2.3.1.2 Electrical lead heating*

The exploitation of electrical Joule heating, i.e. *the passing of an electrical current through a material with a finite electrical resistance will generate heat proportional to the square of the current flow and proportional to the resistance of the material*, is one of the underpinning principles behind the ESR. However, it also presents one of the largest sources of error. To obtain equivalence in the electrical and optical heat flow paths requires all the measured electrical input power to be converted to heat energy and fully

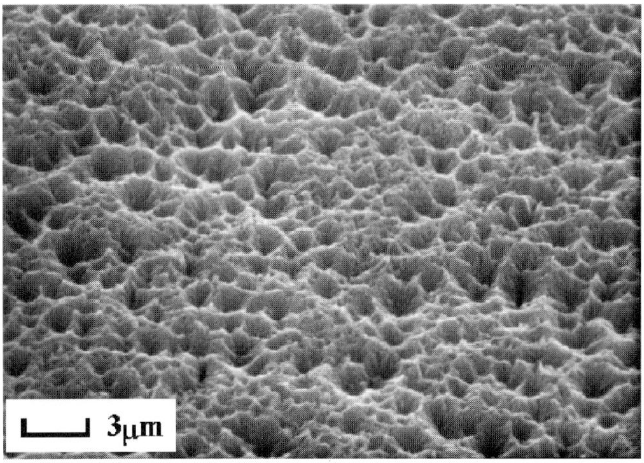


FIG. 2.10. Electron-micrograph of NPL "superblack" showing fine structure and consequential large surface area of black coating.

contribute to the heating of the radiometer absorber. This requires that the heating element is in perfect thermal contact with the absorber and has no internal thermal gradients to dissipate "unmeasured" heat (similar to the absorbing black coating described above). This situation can be closely approximated through careful positioning and design of the heater element. For example, often it is sandwiched between metal plates so that any lost heat is still collected by the absorber. In some cases the black absorbing coating itself is used directly, such as in the electrically calibrated pyroelectric radiometer (ECPR) [31, 32] now marketed by Laser Precision. In this instrument, the gold black coating is used as both optical absorber and electrical resistor. Such a process should in principle ensure good equivalence, although given the fine structure of the material it is possible that some thermal non-uniformity or temperature gradients may be different between optical and electrical inputs. The significant improvement in thermal conductivity obtainable through cooling to cryogenic temperatures together with the removal of air conductance and convection make this potential source of non-equivalence even less significant for cryogenic radiometers.

However, the biggest source of error arises from the electrical connection leads. These are the leads carrying the current to the heating element, which of course, in general, have a finite resistance and consequently will generate heat. At some point this heat cannot be collected by the absorber and so will represent a loss and source of non-equivalence. The effect can be minimised by ensuring that the potential difference across the heater element is

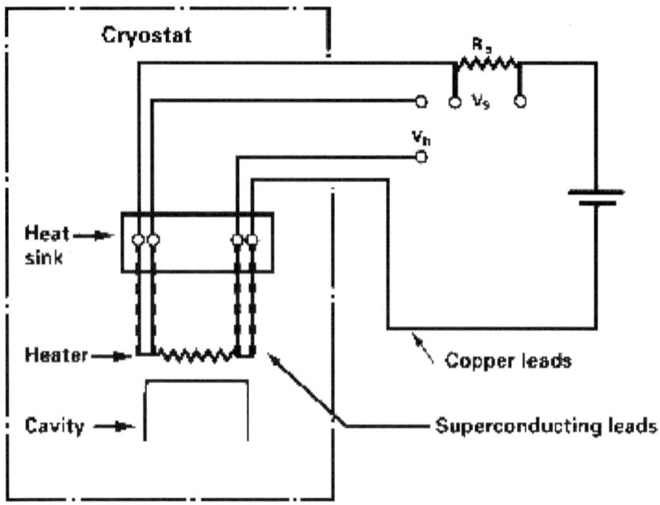


FIG. 2.11. A typical electrical heater circuit for a cryogenic radiometer.

measured as closely to the current leads as possible and that both are in good thermal contact with the absorber. Similarly, heat will be generated in the leads, which can be another source of thermal energy sensed by the detector.

A typical electrical measurement circuit can be seen in Figure 2.11. Here in its simplest form it can be seen that the electrical power can be determined through a measure of $V_h V_s / R_s$, where $V_h$ and $V_s$ are the voltages across the heater and standard resistor $R_s$, respectively. By careful selection of resistances this process can allow relatively high accuracy measurements to be made with low-cost instrumentation and standard methods, e.g. in measuring a power of 1 mW, a 1000 Ω standard resistor and heater together with a current of 1 mA would generate 1 V across each of the heater and resistor, all readily available as standard calibration reference values.

Cooling to cryogenic temperatures allows the final electrical linkage to be made using superconducting wire (e.g., niobium or a niobium/50% titanium alloy). In this way, there is no Joule heating in the connecting leads for temperatures below ~9 K, and all residual thermal losses in copper are removed by the heat sink. Similarly, superconductors also provide a reasonably good thermal isolation of the detector themselves, having a low thermal conductance. The availability of higher critical-temperature superconducting ceramics allows the same advantage to be obtained for higher operating temperatures, as exploited in the mechanically cooled radiometer of Fox et al. [27], which operates close to 15 K.

Some care needs to be taken in this process to avoid raising the temperature of the detector above the critical temperature $T_c$ of the superconductor. Although obvious, it is easy to forget that the radiometer works by raising the temperature of the detector several degrees above that of the reference temperature heat sink, and since, in general, the larger the temperature rise the more sensitive the detector, this can put an upper limit on the power measurable by the radiometer for a given heat link. It is also prudent to test the effective $T_c$ of the leads in the configuration actually used, as the effective critical current of the current leads may be derated either in the case of damaged leads or in the case of thermal runaway if the leads are not properly heat sunk.

The implementation of the circuit illustrated in Figure 2.11 above can vary significantly between different cryogenic radiometer designs, not only different heater technologies, e.g. wound wires or thin film layers, but also in operation. For example, a constant current can be passed through the heater and reduced as optical power is added or the exact opposite.

#### *2.3.1.3 Heat flow path*

The cavity detector and the reference heat sink are, by design, connected by a thermally conducting heat path. To show that the same fraction of heat generated in the cavity, either by the absorption of the radiant flux or by passing a current through the heater, will flow along this heat path by conduction to the heat sink in a similar manner, the following conditions must be fulfilled:

(1) there is no change in the radiative loss from the cavity when the electrical power is substituted for the radiant power;
(2) all the leads to the heater are thermally anchored to the heat link, thereby preventing an alternative heat conduction path from the cavity;
(3) the heater is securely anchored to the cavity to prevent overheating when a current is passed;
(4) there are no energy losses or gains by gas conduction or convection during the measurement sequence.

Conditions (2) and (3) can be satisfied by most types of carefully designed ESR. Condition (4) can be satisfied by placing the ESR in vacuum. Apart from pyroheliometers designed to measure the total solar irradiance in space, ambient temperature ESRs have rarely been designed to operate in vacuum. By contrast, cryogenic radiometers can only operate inside an evacuated chamber, where pressures of the order $2 \times 10^{-6}$ Pa are obtained and remain constant throughout the measurement sequence.

It is in satisfying condition (1) that ESRs have the greatest difficulty. A change in the radiative loss will arise if the temperature gradients, created in the cavity by the two heating modes, are significantly different. The loss is made up of two parts, the first from the exterior surface of the cavity to the surroundings, and the second from the interior surface through the cavity aperture. The ideal case is for the radiant power and the electrical power to be absorbed at identical locations within the cavity. However, in practice this is virtually impossible to achieve.

To illustrate the problem we shall take the cavity shown in Figure 2.7 and consider all the electrical power (1 mW) being applied at one end. The resulting temperature difference between the ends of the cavity will be about 10 mK and will be approximately the same for a cavity at a temperature of 5 and 293 K (the thermal conductivity of copper being approximately 0.04 W/m for both temperatures). Taking a value of the emissivity of the outside copper surface to be 0.05 and interior black-coated surface to be 1.0, then Table 2.2 shows the radiative losses from the cavity at 5 and 293 K (it is assumed that the cavity surroundings are also at these two temperatures, respectively).

For the cavity at 293 K the loss approaches 10% of the input power. Of course, such a large cavity would not be used for an ambient-temperature ESR and careful design would certainly lead to a significant reduction in the loss. However, this example does highlight the problems associated with temperature gradients for this type of ESR, whereas, for a cryogenic radiometer, the loss is less than 1 part in $10^6$ of the input power and it is virtually impossible to envisage temperature gradients in a cooled cavity that could lead to significant radiant losses.

Furthermore, it has been possible to check experimentally conditions (2) and (3) for a cryogenic radiometer. Two heaters, of similar specification, have been wound on the cavity at different locations as shown in Figure 2.7 ($H$ and $H'$). The same input power (1 mW) was applied to each heater in turn and the temperature rise of the cavity measured, with the results shown in Figure 2.12. It can be seen that the cavity responses are identical within the experimental noise (less than 2 parts in $10^5$, peak to peak, of the input power).

TABLE 2.2. Radiative Losses from the Cavity

| Cavity temperature | 5 K | 293 K |
|---|---|---|
| Radiative loss from external surfaces | $1.7 \times 10^{-10}$ W | $3.4 \times 10^{-5}$ W |
| Radiative loss from internal surface | $2.8 \times 10^{-10}$ W | $5.6 \times 10^{-5}$ W |
| Total loss | $4.5 \times 10^{-10}$ W | $9.0 \times 10^{-5}$ W |

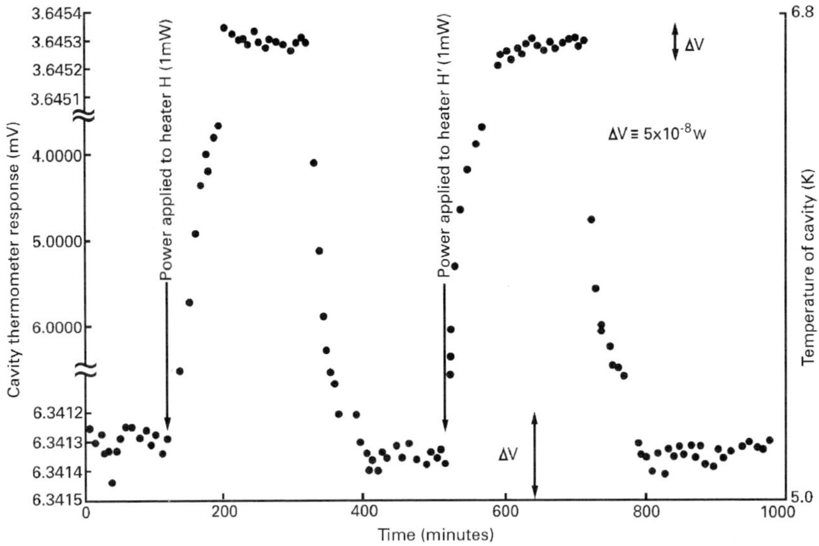


FIG. 2.12. The response of the cavity thermometer when equal amounts of electrical power (1 mW) are applied to heaters H and H′ in turn. The thermometer response is expressed in terms of the potential difference developed across the thermometer for a constant current of 10 μA. $\Delta V$ is the change in the voltage for a power change of $5 \times 10^{-8}$ W.

### *2.3.1.4 The temperature sensor*

The most common type of temperature sensor used with high-accuracy ambient temperature ESRs is a thermopile, i.e. an array of thermocouples joined in series. Their operation is based on the Seebeck effect, where a thermo-electric e.m.f. is built up in pairs of two dissimilar conductors when they experience a temperature gradient. An ideal thermopile should use materials with large Seebeck coefficients, it should have a large number of junctions to increase the output voltage, low electrical resistance to reduce the Johnson noise, a low thermal capacity to give a rapid response, and a low thermal conductance to maximise the temperature rise on irradiation. These requirements are often conflicting and the design calls for compromises. The devices are usually operated with hot (irradiated) and cold (heat sink) junctions; this arrangement serves to compensate for any drift in the temperature of the heat sink. The active elements may be metallic, for example, bismuth–silver, gold–nickel and copper–constantan or *p*- and *n*-type semiconductors, for example tellurium bismuth. A good example is that of the ESR developed by Boivin at the National Research Council, Canada (NRC), which incorporates a 30-junction silver–bismuth thin-film thermopile; the ESR has a sensitivity of 0.3 V/W, a noise equivalent power (NEP) of 20 nW, and an overall uncertainty of ~0.1% [33].

The temperature sensor most commonly used with cryogenic systems and some radiometers is a germanium resistance thermometer, GRT [34]. This type of thermometer consists of a small chip of *n*-type doped germanium supported in a strain-free manner with side arms for attaching potential and current leads. The chip is enclosed in a capsule, typically 3 mm diameter and 8.5 mm long, filled with helium gas. GRTs are extremely sensitive and reproducible (0.003%) and hence are very suitable for use in cryogenic radiometers. The GRT used with the cavity in Figure 2.7, *J*, is inserted into a central well as shown. The four thermometer leads are thermally anchored to the cavity; great care should be taken in this operation as the temperature of the chip is greatly influenced by the lead temperature. The thermometer completes a simple DC circuit in which the potential difference across the chip is measured with a nanovoltmeter when a constant current of 10 μA is passed. The sensitivity of the thermometer in this radiometer varies from 4.2 V/W (NEP 12 nW) to 1.2 V/W (NEP 42 nW) for power inputs of 0.2 and 1.4 mW, respectively. One problem associated with a GRT is the constancy of the self-heating of the Ge chip and 10 μA is approaching the maximum acceptable measuring current.

An alternative approach has arisen following the introduction of thin-film rhodium–iron thermometers. Unlike most metals and alloys, dilute rhodium–iron alloys (typically rhodium + 0.5% iron) still have an appreciable change in their resistivity at very low temperatures and, therefore, make good sensing elements for resistance thermometers. Wire-wound Rh–Fe thermometers have been available for the past 25 years [35], but these are physically large devices and not suitable for this application. However, it is now routine to deposit thin films of Rh–Fe (with similar resistivity characteristics to those of the wire) on slabs of sapphire, 6 mm × 5 mm × 0.5 mm [36]. A line pattern is then etched on the film and trimmed to give a resistance of approximately 450 Ω at 293 K falling to 40 Ω at 5 K. There are a number of advantages to using this type of device instead of a GRT as the cavity thermometer and they are the thermometer of choice for many of NPLs cryogenic radiometers [27]. These include:

(a) Better thermal contact to the cavity as the thermal anchoring of the leads is less critical.
(b) A higher measuring current can be used (100 μA) before the self-heating of the thermometer becomes unacceptable. This can sometimes be important when measuring relatively small powers and where the size of and consequential fluctuations in this additional joule heating can impact on the noise of the system.

(c) The response $\delta\Omega/\delta T$, for any thin-film Rh–Fe thermometer of the same type is virtually identical over the temperature range 5–10 K (this is not the case for a GRT). Hence, if a thin-film thermometer is also used to measure the temperature of the reference heat sink, then the resistance ratio of the heat sink thermometer vs. the cavity thermometer (which can be measured on one bridge circuit) will remain constant for small fluctuations of the heat sink temperature. This can sometimes have benefits when measuring relatively small signals as it is often easier to measure ratios than absolute values.

For high-sensitivity applications, use can also be made of the transition-edge superconductor thermometers. These, as the name suggests, make use of the rapid transition to zero resistance near the critical temperature of superconductors. Such a change in resistance is highly sensitive to temperature, but has little temperature range and so needs to be optimised for particular temperature/power ranges of operation (see Section 2.6.1).

In conclusion, except for low-power-measurement applications as discussed in Section 2.6.1, the temperature sensor is not a limiting factor in the accuracy of either an ambient temperature or cryogenic radiometer.

#### *2.3.1.5 The reference temperature heat sink*

The accuracy with which it is necessary to control the temperature of the heat sink depends upon two factors; first, the accuracy sought in the measurement of the radiant power and, second, the variation of the thermal conductivity of the heat link with temperature. For the present we shall ignore the second factor, but its relevance will be discussed in Section 2.3.2. Hence, if we wish to measure 1 mW of radiant power with an accuracy of 1 part in $10^4$ and arrange that this power input will cause a temperature rise of the cavity of 2 K, then we require the heat sink to remain stable to 0.2 mK.

Because of the high thermal diffusivity (thermal conductivity/specific-heat capacity) of metals at low temperatures, it is relatively straightforward to control the temperature of a block of copper to a few tenths of 1 mK using commercially available equipment. The heat sink system employed for the second NPL cryogenic radiometer, shown as L in Figure 2.7, is a block of copper at a temperature of about 5 K controlled by a commercial electronic temperature controller to within a stability of $\pm 0.2$ mK.

However, for the first NPL cryogenic radiometer (to measure the Stefan–Boltzmann constant), discussed in more detail in Section 2.5, we required the power to be measured with an uncertainty of 1 part in $10^5$ and for this radiometer we used a bath of superfluid helium (liquid helium cooled below its $\lambda$ point, 2.17 K). Superfluid helium makes an ideal reference temperature heat sink because of its unique physical properties, such as film flow and extremely high thermal conductivity. The heat sink in this case consisted of a 2 l copper reservoir filled with liquid helium which was continuously pumped through a precision room temperature needle valve to about 2 K, an AC Wheatstone bridge arrangement with the critical components immersed directly in the liquid helium providing the fine temperature control. Over a period of 2 h the total drift and short-term fluctuations of the temperature of the reservoir did not exceed 4 μK.

This AC bridge control system was later replaced with a digital system using a computer to provide on-line control of a DC thermometer resistance sensing system and a digital power supply. This combination proved even more successful and provided greater flexibility in selecting the reference temperature, while still maintaining the stability of the reference heat sink to ±2 μK.

The use of a differential temperature sensor (e.g., thermopiles as discussed in Section 2.3.1.4) has been common practice for ambient temperature ESRs, as precise independent temperature control of the heat sink is more difficult at these temperatures. However, the accuracy of these instruments is not usually limited by the temperature stability of the heat sink.

#### *2.3.1.6 Radiation shielding*

Similar arguments to those used in Section 2.3.1.3, relating to changes in the radiant loss from the cavity to the surroundings as a result of temperature changes in the cavity, apply equally to the shielding surrounding the cavity. By cooling the surrounding shields, the changes in the radiation exchange between these shields and the cavity are negligible compared to the radiant power to be measured. Quite simply, by cooling, the cavity can be effectively isolated from small temperature perturbations in its surroundings.

One further advantage of cooling the radiometer is that a cooled radiation trap can be placed in front of the cavity aperture. This not only determines the field of view but minimizes any background and scattered radiation from passing directly through the aperture into the cavity.

This subsection completes the discussion on the six conditions that were listed in Section 2.3.1 to show that the thermometer sensing the temperature rise of the absorber responds equally to equivalent radiant and electrical power. Before we continue with descriptions of some example cryogenic radiometers and their applications, it is appropriate at this stage to include a section on the thermal behaviour of the whole system. The equations developed in this section can be used to design the power range and the measurement sensitivity for a cryogenic radiometer.

### 2.3.2 Thermal Behaviour of the Whole System

Consider a cavity connected by a heat link (a stainless-steel tube) to a reference temperature heat sink at 2 K. The thermal conductivity, $\kappa(T)$, of stainless steel between 2 and 10 K is given by the relation [37]

$$\kappa(T) = k_{ss} T^{1.35}\,\mathrm{W\ cm^{-1}\ K^{-1}} \tag{2.9}$$

where $k_{ss} = 3.5 \times 10^{-4}$ and $T$ the absolute temperature in Kelvin. Hence, if the heat link has a cross-sectional area $A$ and length $L$, then the steady-state heat flow $Q$ across an element of length $\mathrm{d}l$ is

$$Q = k_{ss} T^{1.35} A\ \mathrm{d}T/\mathrm{d}l \tag{2.10}$$

which we can write as

$$Q = k_{ss} A T^{1.35}\ \mathrm{d}T/\mathrm{d}l \tag{2.11}$$

so that

$$Q \int_0^L \mathrm{d}l = k_{ss} A \int_{T_r}^{T_a} T^{1.35}\ \mathrm{d}T \tag{2.12}$$

where $T_r$ is the temperature of the 2 K heat sink and $T_a$ is the temperature of the cavity; hence,

$$Q = k_{ss} A (T_a^{2.35} - T_r^{2.35})/2.35L \tag{2.13}$$

The change in the heat flow, $\delta Q$, for the incremental changes in temperature $\delta T_r$ and $\delta T_a$ of the ends of the heat sink is

$$\delta Q = k_{ss} A L^{-1} T_a^{1.35} \delta T_a - k_{ss} A L^{-1} T_r^{1.35} \delta T_r \tag{2.14}$$

so that

$$\delta Q/Q = 2.35((T_a^{1.35}\delta T_a - T_r^{1.35}\delta T_r)/(T_a^{2.35} - T_r^{2.35})) \tag{2.15}$$

From Eq. (2.15) we can calculate the maximum permitted temperature changes that can occur in the heat sink, $\delta T_r$, and the cavity, $\delta T_a$, for a given accuracy of power measurement. Figure 2.13 shows the magnitude of $\delta T_r$ and $\delta T_a$ vs. $T_a$ in the range 2–8 K for $\delta Q/Q = 1 \times 10^{-5}$. From this figure it can be seen that the stability of the heat sink need not be so stringent as was suggested in Section 2.3.1.5. For example, for an accuracy in the power measurement of 1 part in $10^5$ at $T_a = 6$ K, we only require the 2 K heat sink to remain stable to 5 parts in $10^5$, whereas it is necessary for $T_a$ to remain stable to 4 parts in $10^6$.

If the temperature of the cavity is measured with a GRT having temperature-dependent resistance $R_a$, then by using Eqs. (2.13), (2.14) and (2.15) in conjunction with the responsivity of the thermometer, $\delta R_a/\delta T_a$, the sensitivity of the radiometer, $S = \delta R_a/\delta Q$, can be deduced. Figure 2.14 shows the

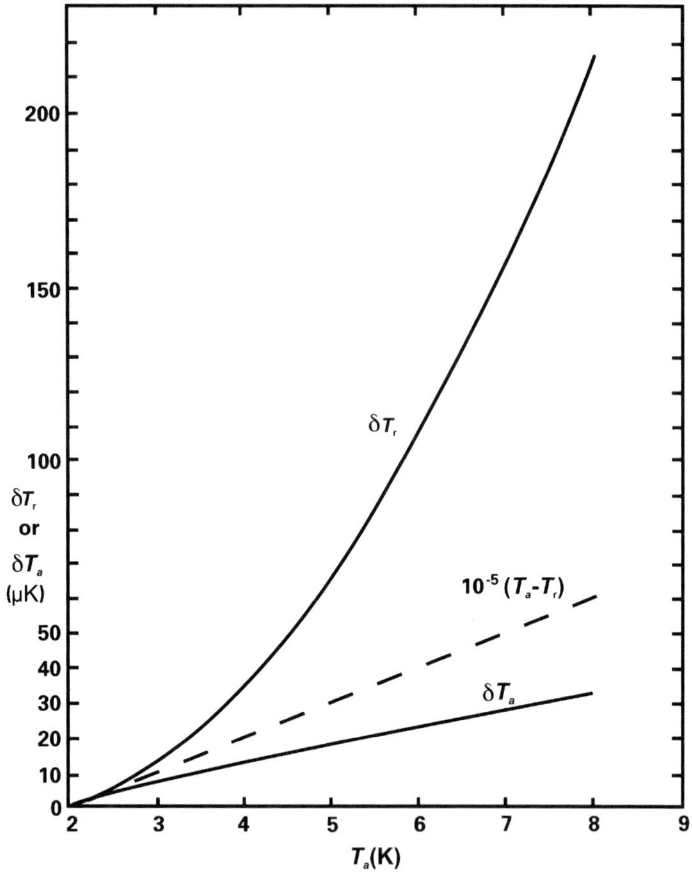


FIG. 2.13. Permitted temperature changes that can occur in the heat sink $\delta T_r$ and the cavity $\delta T_c$ for a measurement of power with an accuracy of 1 part in $10^5$.

change in the resistance of the GRT for a given increment of radiant power absorbed by the cavity, expressed in ohm per milliwatt, as a function of $T_a$ and $Q$. Figure 2.13 also shows the resistance of the GRT and the cavity temperature as a function of the power absorbed in the cavity.

By using the DC circuit described in Section 2.3.1.4, it is possible to measure the resistance of the GRT over the range 20–1000 Ω with a resolution of about 2 parts in $10^6$. Hence, by combining this resolution and the data shown in Figure 2.13, we can calculate the minimum detectable change of the radiant power, $\delta Q$, that can be measured by the radiometer. This is shown in Table 2.3 for various powers and for the reference temperature at 2 K.

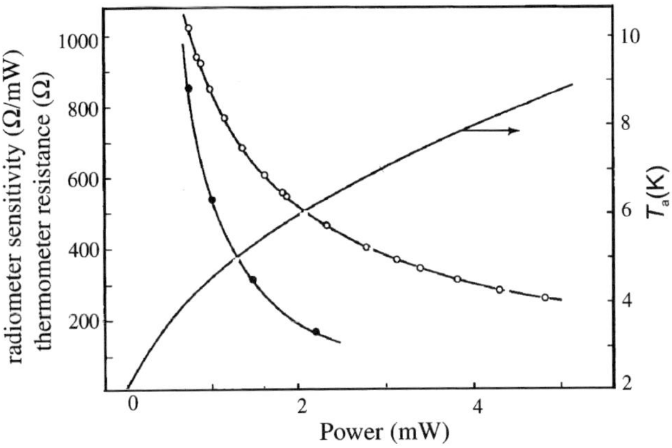


FIG. 2.14. The resistance of the cavity thermometer, o; its temperature, ——; and the radiometer sensitivity in Ω/mW using a stainless-steel heat link.

TABLE 2.3. The Sensitivity of the Radiometer

| $Q$ (mW) | $T_a$ (K) | $R_a$ (Ω) | $S$ (Ω m W$^{-1}$) | $\delta Q$ (nW) | $\delta Q/Q$ (parts in $10^6$) |
|---|---|---|---|---|---|
| 0.8 | 4.0 | 1017 | 830 | 1.2 | 1.5 |
| 1.3 | 5.0 | 700 | 330 | 3.0 | 2.3 |
| 2.7 | 6.8 | 400 | 110 | 9.1 | 3.4 |
| 3.5 | 7.5 | 340 | 80 | 12.0 | 3.4 |
| 4.8 | 8.6 | 256 | 50 | 20.0 | 4.2 |

It can be seen from the values in Table 2.3 that the sensitivity of this radiometer is greatest at the lowest temperatures, mainly because the sensitivity of the GRT is greatest there.

This example demonstrates that the radiometer sensitivity is determined by two parameters. The first is the temperature rise of the cavity for a given power input; the greater the temperature rise, the greater the sensitivity. The temperature rise is calculated from Eq. (2.13) and is a function of the thermal conductivity of the heat link and its physical dimensions. The second parameter is the responsivity of the temperature sensor. Therefore, by careful design of the heat link and the selection of the temperature sensor (GRTs have a wide range of responsivities at these temperatures) the radiometer sensitivity can be optimised to meet the requirements of the experiment.

Before we consider the power range of a cryogenic radiometer, there is one important constraint to be considered, namely, the upper and lower temperature limits of the cavity. The upper limit is the critical temperature of the superconductor used for the heater leads and is about 9.3 K. With the development of high-temperature superconductors this limit can of course be increased and the constraint may then arise from the decreasing thermal diffusivity and the radiative losses from the cavity. The lower limit is fixed by the temperature of the heat sink. Without moving into the sophisticated area of $^3$He cryostats or dilution refrigerators, the lowest achievable temperature using a pumped $^4$He bath is about 0.8 K. However, experience has shown that problems can arise if the cavity temperature is in the region of 3.5 K and below because in this temperature region hydrogen gas, the major component of the residual gas in the vacuum chamber, condenses. The exact temperature for condensation to take place is unique to each radiometer depending upon the gas pressure of the system. Particular care is required when designing a cryogenic radiometer to avoid this problem especially if a shutter is to be used in front of the cavity aperture.

Because of these temperature limits, the only way to increase the operating power range is to alter the heat link as the range cannot simply be increased by allowing the cavity to rise to a higher temperature. For the example quoted in this section using a stainless-steel heat link, the power range is about 9 to 1. To increase this power range and yet still maintain a similar radiometer sensitivity, a material whose thermal conductivity is more strongly dependent on temperature is required. Materials such as alumina, magnesia, sapphire and pitch-bonded graphite all have an approximate $T^3$-dependent thermal conductivity over the temperature range 2–10 K. If pitch-bonded graphite is substituted for the stainless steel in the example, then a power range of about 40 to 1 can be achieved while still maintaining the cavity temperature in the range 3.5–10 K. Figure 2.15 shows the identical parameters to Figure 2.14 but using a graphite heat link.

Finally, we will address the problem of the maximum and minimum power that can be measured using a cryogenic radiometer. As the design of radiometers is still in its infancy, there are few experimental results, which can be used to form definite conclusions and hence one can only surmise.

The upper power limit will probably be influenced by three factors:

(1) the boil-off rate of the helium liquid in the reservoir, since all the power absorbed in the cavity will eventually pass into the liquid. However, the advent of mechanical cooling engines capable of lifting several watts at 4 K largely remove this limitation;
(2) the temperature stability of the heat sink—this will be more difficult to achieve with a large heat flow;
(3) a large heat flow might create significant temperature gradients in the cavity.

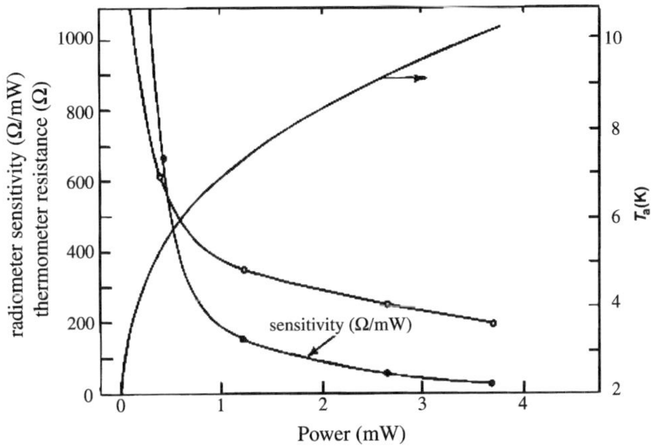


FIG. 2.15. The resistance of the cavity thermometer, o; its temperature, ——; and the radiometer sensitivity in Ω/mW using a graphite heat link.

Probably factor (1) will be the limit for liquid-cooled cryostats. Typically, boil-off rates from cryostats vary over the range 0.03–0.5 l of liquid helium per hour. As the heat of vaporization for helium is 2.6 J/cm$^3$, this would indicate that the upper power limit would be of the order of a few tenths of a watt. For mechanical cooled systems the limitation is likely to be set only by physical size of the cooling engine.

The lower limit will be influenced by a different set of factors, the principal two being:

(1) the sensitivity of the temperature sensing, which may in turn be limited by the thermal anchoring of the heater and thermometer wires to the cavity;
(2) the changes in the radiant losses from the cavity due to fluctuations in the temperature of the surroundings.

At present, commercial systems can measure 1 mW with an uncertainty of a few nW, but systems are under development at NPL and have been demonstrated in prototype form by others (see Section 2.6) that will allow the measurement of 1 μW with uncertainties of 1 pW.

This section completes the general discussion and the following sections will present a range of example applications with references, and discuss in a little more detail some specific examples to highlight the principal applications and their uncertainties, i.e. measurement of blackbody radiation and monochromatic sources for spectral responsivity measurements. The latter will be based on NPL-built instruments as these are most familiar to the author.

## 2.4 Applications of Cryogenic Radiometers

Over the last 20 years there have been many designs of the cryogenic radiometer from different groups to meet an increasing range of applications [38]. The highest accuracy applications of a cryogenic radiometer require the radiometer to be used directly, measuring either total (i.e., broadband) radiation or monochromatic radiation. For total radiation measurements, the source must be within the same vacuum chamber and at temperatures low enough to prevent significant outgassing, as cryogenic systems are very effective cold traps for emitted contaminants. This limits their application in normal laboratory situations.

The following list of examples of the use of cryogenic radiometers is not necessarily comprehensive but gives an overview of the diversity of applications. In reviewing these applications it will rapidly become clear that any current uncertainty limitation is rarely associated with cryogenic radiometry, but almost exclusively the transfer standards or sources being measured.

(1) The current main application for cryogenic radiometers is as a basis for optical radiometric scales in national standards laboratories. Designs include various designs from NPL (in collaboration with Oxford Instruments Ltd. [2, 27], radiometers from Cambridge Research and Instrumentation [39, 40], and Finland [41].

(2) Cryogenic radiometers have been used to measure total radiation from blackbodies as a means of direct calibration [42] and for thermodynamic temperature scale realisations [1]. There is also a design for a space-based instrument capable of measuring the solar constant, or total solar irradiance (TSI), with an uncertainty of more than 10 times lower than the current best achievable [43–45]. A mission proposal (TRUTHS, Traceable Radiometry Underpinning Terrestrial-Helio-Studies) exists which would fly a cryogenic radiometer on a small satellite performing as a primary standard in space for calibrating Earth-viewing remote-sensing instruments. In this case the cryogenic radiometer would be used to make spectral responsivity measurements in space in a similar way to that in a terrestrial laboratory and consequently allow a traceability chain similar to that described in Figure 2.1 to be established in space [45, 46].

(3) The wide dynamic range and spectral non-selectivity of cryogenic radiometers has been used to calibrate space instrumentation such as CERES

[47], for Earth radiation budget experiments, though in this particular case the scale was based on a blackbody source and the cryogenic radiometer served as a transfer standard radiometer.

(4) Cryogenic radiometers have been used to measure spectrally filtered synchrotron radiation from electron storage rings as a basis for calibrating detectors in the deep UV spectral region [48]

The most common use of cryogenic radiometers is to measure monochromatic radiation to establish spectral responsivity scales, the radiation coming from either an intensity-stabilised laser [2] or an incandescent lamp dispersed by a monochromator [40, 49]. In each case, the ultimate goal is to determine the response of transfer standard detectors. These detectors can then be used as the basis not only for spectral responsivity measurements (Chapter 3) but also for source scales (Chapter 5).

### 2.4.1 Example Radiometers

The following examples, although all from NPL, scope out the operating principles and differences between the dominant applications of cryogenic radiometers, i.e. measurement of source radiance/irradiance or spectral power. The reader is directed to the associated references for specific details as this text only serves to provide an overview. As summarised above, there are a number of other specific radiometer designs but these only differ in detail and not in the operating principle. The most significant of these differences relates to the use of monochromator radiation as opposed to lasers for spectral responsivity and this is discussed in Chapter 3.

### 2.4.2 Total Radiation Thermometry

The first successful cryogenic radiometer was that of Quinn and Martin [1]. The Quinn–Martin radiometer or QM radiometer, or total radiation thermometer, as it is sometimes called, was not designed for work on optical radiometric scales, but for the determination of the Stefan–Boltzmann constant (SB) and thermodynamic temperature by total radiation thermometry in the range from $-40°C$ to $+100°C$. The instrument consisted of a blackbody receiver operating at around 2 K, the cryogenic radiometer, and a variable temperature blackbody radiator. The instrument measured the total radiation emitted from the blackbody through a precisely known solid angle defined by two apertures. The SB constant could then be determined from

$$M'(T_{\mathrm{tp}}) = g\sigma T_{\mathrm{tp}}^4 \tag{2.16}$$

where $M'(T_{tp})$ is the total radiant exitance of the blackbody at a temperature $T_{tp}$, $g$ is a geometric factor defining the solid angle of collection, $\sigma$ is the Stefan–Boltzmann constant and $T_{tp}$ is the thermodynamic temperature of the triple point of water, 273.16 K.

Quinn and Martin calculated the uncertainty attributable to their experimental determination and compared their value for $\sigma$ with the theorctical calculation given by fundamental constants in Codata [50],

$$\sigma = 2\pi^5 k^4 / 15 h^3 c^2 \tag{2.17}$$

The experiment was similar to the one carried out earlier by Blevin using an ambient temperature ESR and a blackbody at the freezing point of gold [29], but with a cryogenic radiometer the uncertainties are an order of magnitude smaller.

The experimental value measured for $\sigma$ was

$$5.66959 \pm 0.00076 \times 10^{-8} \ \mathrm{W\,m^{-2}\,K^{-4}}$$

compared with that of Codata:

$$5.670400 \pm 0.000040 \times 10^{-8} \ \mathrm{W\,m^{-2}\,K^{-4}}$$

The agreement of these results within their combined uncertainties confirms that the systematic and experimental uncertainties determined for the QM radiometer are reasonable and that the instrument is truly an absolute radiometer to at least 2 parts in $10^4$.

Thermodynamic temperatures, $T$, were determined by measuring the ratio of the total radiant exitance $M(T)$ to that of the blackbody at the temperature of the triple point of water. From the relation $M(T)/M(T_{tp}) = T^4/(T_{tp})^4$, $T$ can be calculated. The scale could then be disseminated by means of platinum resistance thermometers which were in thermal equilibrium with the blackbody at temperature, $T$ [1].

#### *2.4.2.1 Description of the apparatus*

The total-radiation thermometer is shown schematically in Figure 2.16. It consists of two principal parts, a blackbody radiation source and a cryogenic radiometer, both working over the spectral range 0.8–400 μm. The blackbody radiator A, designed to have a high emissivity, is suspended by stainless-steel wires P from a copper shield S cooled by a liquid-nitrogen reservoir N. The blackbody is surrounded by five copper radiation shields B′, one of which is temperature-controlled B and supported from the copper shield in the same manner as the radiator. The blackbody temperature is measured, with an uncertainty of less than 1 mK, by eight platinum sheathed, capsule type, platinum resistance thermometers (PRTs) L,

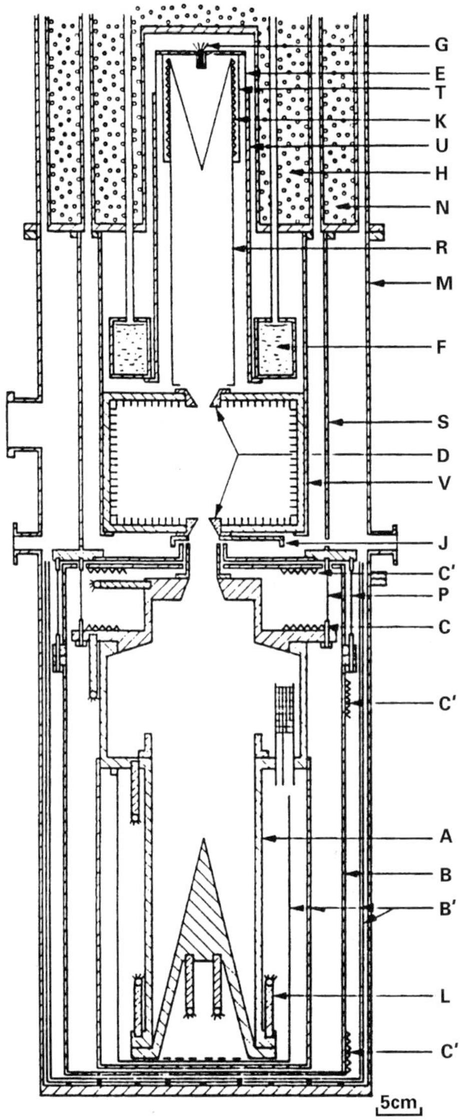


FIG. 2.16. Total-radiation thermometer: A, blackbody radiator; B, temperature-controlled shield; B′, radiation shields; C, blackbody radiator heater; C′, shield heater; D, apertures; E, poorly conducting heat link; F, superfluid-helium reservoir; G, GRT (cavity thermometer); H, helium reservoir; J, shutter; K, cavity heaters; L, PRTs (radiator thermometers); M, vacuum chamber; N, nitrogen reservoir; P, support wires; R, radiometer cavity; S, copper tube; T, cavity radiation shield; U, copper tube; V, radiation trap.

positioned at various points on it. Radiation from the blackbody passes through a pair of apertures D fixed to the radiation trap V which is suspended from the liquid-helium reservoir H; the resulting beam of radiation passes into, and is absorbed by, the cryogenic radiometer cavity R.

The cylindrical blackbody cavity is made from electro-formed copper 0.25 mm thick, diameter 70 mm, length 380 mm, and has a re-entrant cone at the closed end. The interior surface is coated with a 100-μm-thick layer of 3 M Nextel black paint, type C-101. This paint has an average diffuse reflectance of about 4% in the wavelength range 0.5–50 μm; above 50 μm it steadily becomes a specular reflector with a reflectance of about 30% at 500 μm.

The cavity is connected to a copper top plate, 1 mm thick, that contains the thermometer well, by a cylindrical electro-formed copper radiation shield, T. The shield surrounds the heater and so enables condition (3) in Section 2.3.1 to be fulfilled. The top plate is supported from an annular reservoir of superfluid helium F by a composite tube, consisting of a copper tube U, wall thickness 2.5 mm, into which at the upper end is soldered a piece of stainless-steel tube E, 90 mm in diameter, 40 mm in length and 0.25 mm wall thickness. The stainless-steel tube acts as the poorly conducting heat link. The temperature of F, around 2 K, is maintained constant with a stability of a few μK and thus it behaves as a constant-temperature heat sink.

The radiant power absorbed in the cavity creates a temperature gradient across the heat link, and the temperature of the cavity, measured with a GRT G, reaches a steady state when the heat gained by radiation is equal to the heat lost by conduction through the heat link. A shutter J, also anchored to the liquid-helium reservoir, is closed when a steady state is reached and electrical power is substituted using one of six wire-wound heaters K. The electrical power is adjusted to maintain the cavity temperature at the same level and when this is achieved a measurement of the electrical power is then a precise measure of the radiant power $M'(T)$ in Eq. (2.16).

The whole apparatus is enclosed within a vacuum chamber M which is evacuated with a turbomolecular pump to a pressure of approximately $1 \times 10^{-7}$ Pa. The cryogenic radiometer was used for the example in Section 2.3.2 and its sensitivity is given in Table 2.3.

#### *2.4.2.2 Method for Measuring $T$ and $\sigma$*

It is usual to relate $T$ to the International Temperature Scale, at present, the International Temperature Scale of 1990, $T_{90}$. This permits determinations of $T$ from various primary thermometers to be compared directly over the whole temperature range.

To determine $T - T_{90}$ using the total-radiation thermometer, the blackbody temperature is set to a nominal value of $T_{90}$ using the calibrated PRTs and a series of measurements is made to determine $M'(T)$. The blackbody is then set to $T_0$ (273.16 K) and a series of measurements is made to determine $M'(T_0)$. The average of these values $M'(T_0)$ with each individual value of $M'(T)$ is used to calculate a value of $T$ and, by comparison with the corresponding value of $T_{90}$, the difference $T - T_{90}$ can be found. At the time of the measurements made by this instrument the temperature scale was actually the International Practical Temperature Scale of 1968, but the principle is of course the same.

To determine $\sigma$ three series of measurements were made to evaluate $M'(T_0)$, each series comprising about 10 measurements. Before and after these measurements the apparatus was dismantled so that the diameters of the two apertures and their distance could be precisely measured; from these measurements $g$ was determined. To investigate any systematic errors that could arise from the defining geometry, the original aperture set was replaced by a second set with different dimensions and two further series of 10 measurements were made to determine $M'(T_0)$. The value of $\sigma$ was calculated using all the measurements.

#### *2.4.2.3 Results*

Only the results of the determination of $\sigma$ will be presented in some depth as temperature is of peripheral interest in a book on radiometry. It should also be emphasised that it was this measurement of $\sigma$ that demonstrated the potential of cryogenic radiometry and set the scene for its significant contribution to many of the advances described within this book.

A summary of the calculated corrections and uncertainties in the determination of $\sigma$ are given in Table 2.4, where $F(\varepsilon, \alpha)$ is the correction for the thermal-radiation transfer function, $D(T)$ and $L(T)$ are the diffraction and the absorption corrections at the aperture edges, $s$ is the correction for the radiation trap scattering, $M'(T)$ is the uncertainty in the electrical-power measurement, $T_{68}$ is the uncertainty in the practical realisation of the blackbody radiator temperature and $g$ is the uncertainty in the measurement of the geometrical factor. There are also very small corrections for the gas pressure and for the change in the radiant power entering the radiometer cavity from the radiation trap on opening and closing the shutter. A full analysis of the uncertainties is given in the original paper [1].

The thermal-radiation transfer function is the net emissivity-absorptivity of the system and includes the correction for the absorptivity of the cryogenic radiometer cavity. The uncertainty associated with this correction is the only significant uncertainty that directly stems from the cryogenic radiometer.

TABLE 2.4. The Calculated Corrections and Relative Uncertainties in the Determination of $\sigma$ Expressed in Terms of $10^4(\delta\sigma/\sigma)$

| | Large pair of apertures | | Small pair of apertures | |
|---|---|---|---|---|
| | Correction | Uncertainty | Correction | Uncertainty |
| $F(\varepsilon, \alpha)$ | 6.1 | 1.4 | 2.7 | 1.0 |
| $D(T)$ | 1.1 | 0.7 | 2.1 | 1.2 |
| $L(T)$ | 2.4 | 0.2 | 1.8 | 0.15 |
| $s$ | −0.8 | 0.2 | −0.8 | 0.2 |
| $M'(T)$ | — | 0.2 | −0.8 | 0.2 |
| $T_{68}$ | — | 0.1 | — | 0.1 |
| $g$ | — | 0.9 | — | 1.6 |
| Standard deviation of $n$ measured values | — | 0.12 | — | 0.5 |
| Combined | 8.8 | 1.7 | 5.8 | 2.2 |

*Note*: The uncertainties are type B except for the standard deviation of $n$ measured values, which is type A.

TABLE 2.5. Mean Values of $\sigma$ Including the Type A and B Uncertainties, the Values are Expressed in Units $10^{-8}\,W/m^2\,K^4$

| | $\sigma$ | Standard deviation of $n$ measurements (type A) | Standard deviation of type B uncertainties |
|---|---|---|---|
| *Large pair of apertures* | | | |
| 1st series | 5.670 00 | 0.000 05 ($n = 8$) | 0.000 96 |
| 2nd series | 5.669 87 | 0.000 08 ($n = 10$) | 0.000 96 |
| 3rd series | 5.669 85 | 0.000 08 ($n = 9$) | 0.000 96 |
| Mean value | 5.669 91 | 0.000 07 ($n = 27$) | 0.000 96 |
| *Small pair of apertures* | | | |
| 1st series | 5.668 94 | 0.000 27 ($n = 10$) | 0.0012 |
| 2nd series | 5.669 71 | 0.000 35 ($n = 10$) | 0.0012 |
| Mean value | 5.669 33 | 0.000 31 ($n = 20$) | 0.0012 |

The mean values of $\sigma$ for the five series of measurements are given in Table 2.5.

The final value of $\sigma$ was obtained by combining the results for the two sets of apertures with a weighting inversely proportional to the square of their respective standard deviations and is

$$\sigma = (5.66967 \pm 0.00076) \times 10^{-8}\ \mathrm{W\,m^{-2}\,K^{-4}}$$

The electrical-power measurement was made using standard resistances and digital voltmeters whose calibration could be traced back to the NPL maintained standards of the ohm and volt, respectively. Hence to compare this value of $\sigma$ with other values a small adjustment must be made to account for the difference between the SI watt and the NPL watt at the time of the measurement. From the CODATA Bulletin [50]

$$1\ \mathrm{W}_{\mathrm{npl1985}} = 0.999\,985\,92\ \mathrm{W}$$

and hence the value of $\sigma$ has been adjusted to

$$\sigma = (5.669\,59 \pm 0.000\,76) \times 10^{-8}\ \mathrm{W\,m^{-2}\,K^{-4}}$$

A more accurate value of $\sigma$ may be calculated from the theoretical expression given in Eq. (2.17), using the recommended values for the fundamental constants at that time [50]. Then,

$$\sigma = (5.670\,51 \pm 0.000\,19) \times 10^{-8}\ \mathrm{W\,m^{-2}\,K^{-4}}$$

The difference between the two values is 1.1 times the standard deviation of that difference and hence they can be considered as not to be in disagreement. This result confirmed the theoretical calculations that this cryogenic radiometer is an absolute radiometer with a maximum uncertainty of 1.3 parts in $10^4$. Experimental values for $\sigma$ obtained since 1921 (references to these values can be found in Reference [29]) compared to the value derived from fundamental constants are shown in Figure 2.17.

Table 2.5 demonstrates the short-term reproducibility and sensitivity of the cryogenic radiometer. The long-term stability of the radiometer can be deduced from Figure 2.18; this shows two series of measurements, with a time interval of about 4 years between the series, of the radiant power from the blackbody radiator at temperature $T_0$ using the large set of apertures. A considerable proportion of the scatter can be attributed to the degradation of the reflecting surfaces of the apertures leading to a loss in radiant flux from the blackbody as a result of diffraction losses: this is fully explained in Reference [1]. Nevertheless, these results still demonstrate the remarkable stability of a cryogenic radiometer.

It should also be noted from Tables 2.4 and 2.5 that for an improvement in the measured value of $\sigma$, a reduction must be achieved in the type B uncertainties. This is now possible in two areas. Firstly, much experience has been gained in aperture design and measurement [51, 52], which can lead to a reduction in the uncertainties in $D(T)$, $L(T)$ and $g$ and, secondly, the recent introduction of new black coatings, for example, Martin Marietta Infrablack [53] which, if used to coat the radiating blackbody and the radiometer cavity, could reduce the uncertainty in $F(\varepsilon, \alpha)$. Such a radiometer (called the absolute radiation detector, ARD) has been designed and constructed at NPL [54] and following some modifications from its original design is, at the time of writing this book, under evaluation.

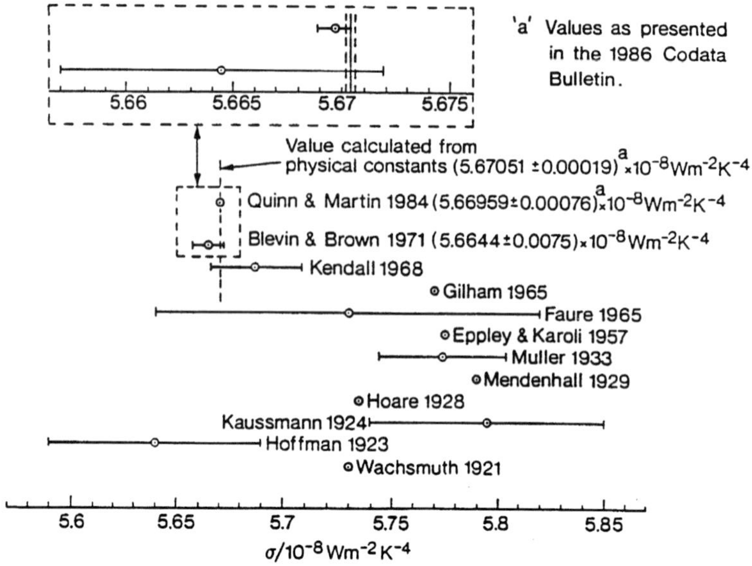


FIG. 2.17. Determinations of $\sigma$ since 1921.

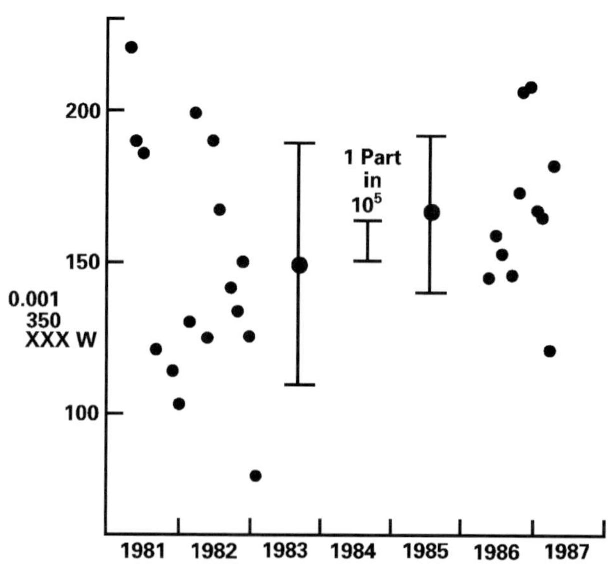


FIG. 2.18. The radiant power from the blackbody radiator at T0. The average value from 1981 to 1983 is 0.001350149 W and from 1986 to 1987 is 0.001350167 W.

To summarise, the above work showed that a cryogenic radiometer could measure radiant power with an accuracy of 1.3 parts in $10^4$ and a sensitivity of 1 to 2 parts in $10^5$.

### 2.4.3 Spectral Responsivity

The application of cryogenic radiometry to optical radiometry was first suggested by Geist in combination with Blevin and Quinn [55], and led to an experiment performed by Zalewski and Martin to compare the self-calibration technique discussed in Section 2.2.1 with the QM radiometer. The experiment required the modification of the QM radiometer to allow laser radiation, rather than blackbody radiation, to be measured. The results of the experiment were not published, but demonstrated the feasibility of using a cryogenic radiometer as the basis for optical radiometric scales [56].

This experiment led NPL to design a new cryogenic radiometer optimised for laser radiation, the so-called primary standard (PS) radiometer [2]. Similar instruments built to the NPL design by Oxford Instruments Ltd, UK, have been used at NIST [25] and PTB [26]. The purpose of this radiometer is to provide the NPL's primary reference standard for optical power measurements. The radiometer is capable of measuring the power of an intensity-stabilised laser beam in the visible region of the spectrum with an uncertainty of about 5 parts in $10^5$, and can be used to calibrate secondary transfer standard detectors, such as silicon photodiodes.

#### *2.4.3.1 Description of the radiometer*

The cavity part of the radiometer has already been illustrated in Figure 2.7; the complete apparatus is shown in Figure 2.19.

The radiometer cavity G is suspended with its associated components from the base plate M of a liquid-helium reservoir. The radiometer operates in vacuo and the vacuum chamber Q is evacuated using a turbo-molecular pump. The laser beam A, the power of which is to be measured, enters the cavity after passing through a window B at the Brewster angle. The cavity, heat sink, heat link and heaters have been fully described in Sections 2.1–2.5; only the radiation trap will be described at this stage.

The main purpose of the radiation trap F is to prevent scattered radiation from passing into the cavity through its 12 mm diameter aperture. The trap is made of 1.5 mm thick, 60 mm diameter copper tube fitted with two internal baffles and coated internally with 3 M Nextel black paint. The trap is bolted to the 4.2 K outer radiation screen and measurements confirm that its temperature is close to 4.2 K.

One of two silicon photodiodes E (quadrant detectors) used to detect the position of the laser beam is mounted in the trap, the other diode E′ is fixed

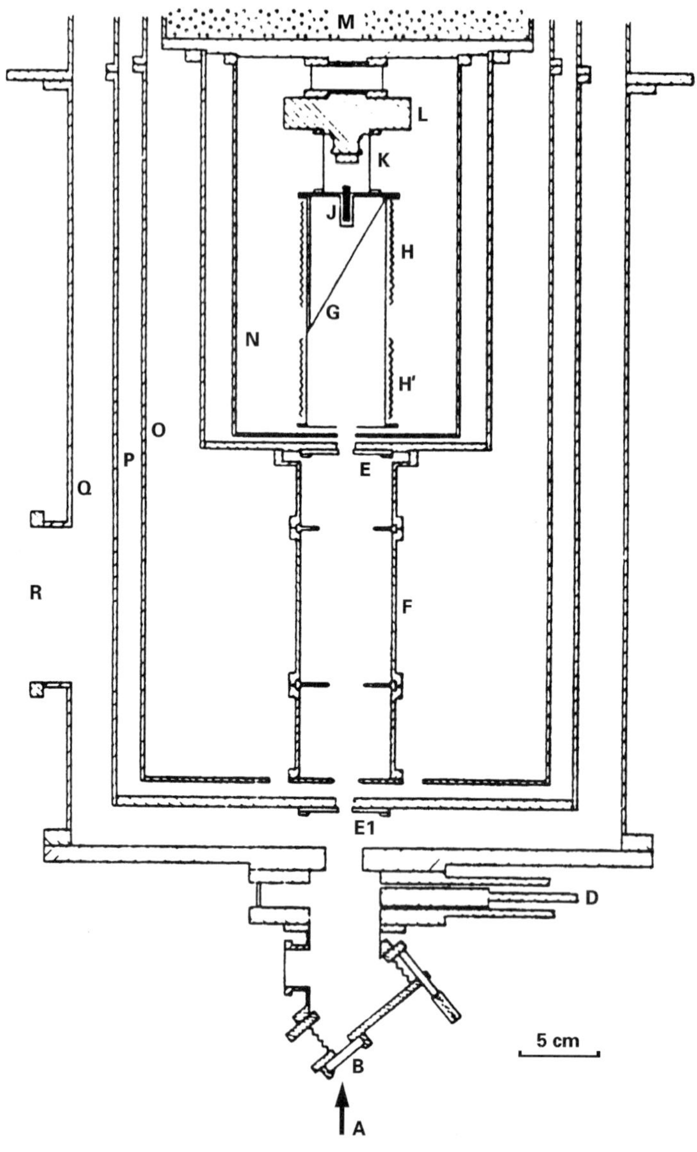


FIG. 2.19. Cryogenic radiometer for optical radiation measurements (PS Radiometer): A, laser beam; B, Brewster angled window; D, Gate valve; E, E′, quadrant detectors; F, radiation trap; G, radiometer cavity; H, H′, electrical heaters; J, thermometer well; K, poorly conducting heat link; L, heat sink; M, helium reservoir; N, radiation shield 4 K; O, radiation shield 60 K; P, radiation shield 77 K; Q, vacuum chamber.

on the outer of the cooled radiation shields. Each diode has a 45 mm diameter active area with a 9 mm diameter central hole and is divided into four distinct quadrants. Cooling the diode E′ to 77 K does not alter its ambient temperature responsivity, but the responsivity of the diode at 4.2 K is reduced by 80%. The response is uniform across the whole surface of the diode to within ±1%.

The centres of the holes of the quadrant detectors are on the vertical axis of the radiometer, and they define the geometric beam of radiation passing through the window that is absorbed in the cavity. The combination of the black coating and baffles in the trap virtually eliminates any scattered or thermal radiation reaching the cavity aperture from outside this geometric beam. The radiant power from the trap at 4.2 K is only 1 nW. However, the power in the geometric beam due to thermal radiation from the window by emission, reflection and transmission (i.e. the effective window emissivity is unity) is not insignificant and must be considered. It can be shown that the fraction $f$ of radiation diffusely emitted by an aperture of radius $R$ and intercepted by a coaxial aperture of equal radius at a distance $D$ (where $R \ll D$) is given to a good approximation by

$$f = R^2/(2R^2 + D^2) \tag{2.18}$$

and so, assuming that the effective emissivity of the window combined with its surroundings is 1, a worst-case scenario, the radiant power $M$ absorbed by the cavity is

$$M = \sigma\pi R^2 f T_{\mathrm{w}}^4 \tag{2.19}$$

where $T_{\mathrm{w}}$ is the temperature of the window and the surroundings. For $T_{\mathrm{w}} = 293\,\mathrm{K}$, $R = 4.5\,\mathrm{mm}$ and $D = 200\,\mathrm{mm}$, the radiant power $M = 1.34 \times 10^{-5}\,\mathrm{W}$ and is part of the background radiation absorbed in the cavity. It is only the change in this power during the period of measurement that has to be considered, that is, that principally due to the change in the window temperature. The objective is to measure the power of the laser beam (about 1 mW) with an uncertainty of about 5 parts in $10^5$. The maximum change in the window temperature that can be tolerated before a correction has to be applied to this power measurement is about 0.3 K, but the maximum change observed in the window temperature over a period of 1 h is only 0.1 K.

Hence, the measurement of the power of the laser beam to the stated uncertainty requires no correction for any other radiative source. However, when measuring sources of lower intensity this background radiation may become significant. The significance is perhaps obvious in terms of noise but can also be critical in terms of dynamic range with a potentially significant background flux saturating the detector's sensitivity. This consequently may lead to the need for active temperature stabilisation of the window and/or its cooling.

### *2.4.3.2 Procedure for the calibration of transfer standards*

Figure 2.20 shows the optical and electrical systems for calibrating transfer standard detectors with such a radiometer. The window is cleaned and its transmittance measured before it is bolted onto the vacuum chamber. The lower vacuum chamber is evacuated until the pressure is low enough for the gate valve to be opened. The laser beam is directed into the cavity with a mirror and through a limiting aperture of about 7 mm in diameter between the mirror and the window. The aperture prevents the major part of any residual scatter around the beam, introduced by the mirror, from entering

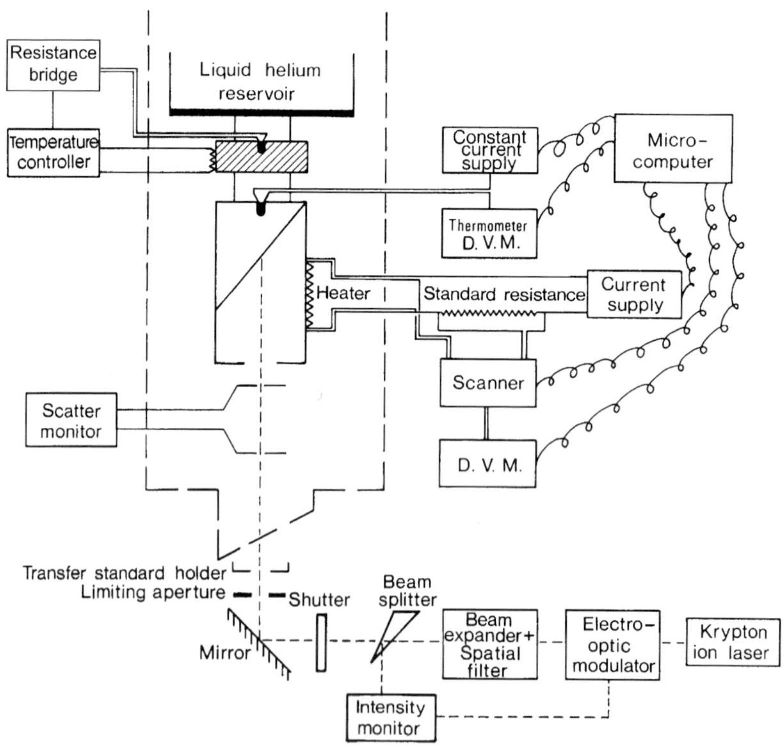


FIG. 2.20. Schematic diagram showing the optical and electrical systems for calibrating secondary transfer standards. Often an additional polariser (not shown) may be required between the shutter and laser stabilisation system to improve the degree of polarisation of the laser.

the radiometer. The beam is aligned to pass along the vertical axis of the cavity by adjusting the mirror until the signal on the quadrant detectors is a minimum. The window angle is then set to the Brewster angle, with respect to that of the polarisation of the incident laser beam. This can be relatively easily achieved by simply viewing the reflected beam. At the appropriate matching angle, the reflectance drops to a minimum, typically $<1 \times 10^{-5}$ of the incident power, providing all components are optimally aligned and cleaned. With the reference temperature block controlled at about 5 K the cavity is allowed to reach thermal equilibrium and its temperature monitored with a GRT. When thermal equilibrium has been achieved, the laser beam is interrupted with a shutter and electrical power substituted in the radiometer and adjusted to obtain the identical temperature rise of the cavity. The electrical power is then equal to the power of the laser beam after correcting for the window transmittance and any scattered laser radiation detected on the quadrant detectors.

Meanwhile, the transfer standards to be calibrated are positioned in the laser beam just below the window and above the limiting aperture and their response determined when irradiated by the laser radiation. The detectors are then removed and the laser power is again measured by the radiometer. The process is repeated at several laser wavelengths so that the spectral responsivity of the transfer standards can be determined. The final part of the calibration procedure is the removal of the window after closing the gate valve to re-determine its transmittance at each laser wavelength. The measurement of window transmittance in this way is relatively simple, as the positioning and angle can be accurately set using the same procedure as described above, i.e. looking for a minimum in the reflectance.

Table 2.6 gives the corrections and uncertainties (at the one-standard-deviation level) that are applied to the measurement of the laser beam power. From the discussions in Section 2.3.1.3, any non-equivalence between electrical and optical heating is negligible and contained within the uncertainties listed, with the dominant contribution being cavity absorptance.

### 2.4.4 Discussion

The two radiometers described were designed to fulfil different objectives. The original radiometer was designed to measure blackbody radiation over a wide spectral range. To meet this requirement the cavity detector was diffusely absorbing with a large entrance aperture and the radiating source was situated within the vacuum chamber. The corrections and uncertainties for this apparatus are given in Table 2.4.

The second radiometer was designed to measure collimated, monochromatic radiation permitting the use of a highly absorbing specular-cavity

TABLE 2.6. The Corrections and Uncertainties in the Measurement of the Power of the Laser Beam, Expressed as Parts in $10^4$ of the Measured Power, for the NPL Helium-cooled PS Radiometer

| | Correction | Uncertainty |
|---|---|---|
| Window transmittance | 2.5 | 0.5 |
| Beam scatter from mirror | 0.5 | 0.15 |
| Absorptance of cavity | 0.2 | 0.1 |
| Electrical-power measurement | — | 0.1 |
| Sensitivity of radiometer | — | 0.3 |
| Changes in thermal and scattered radiation | — | 0.1 |
| Sum in quadrature | 3.2 | 0.6 |

*Note*: It should be noted that uncertainty due to repeatability is not identified specifically here; it is included within the uncertainties of the radiometer sensitivity and changes in thermal background.

detector. The radiating source in this case is situated outside the vacuum chamber and the radiation has to pass through a window before entering the cavity. The corrections and uncertainties for this system are shown in Table 2.6.

By examining the values in the two tables, the uncertainty directly arising from the non-equivalence of electrical to radiative power is seen to be small compared to the uncertainties arising from other sources; for example, in the original radiometer, the uncertainties associated with defining a beam of blackbody radiation using an aperture system, and, in the second radiometer, the uncertainties associated with directing a laser beam into a vacuum system using a mirror and a window. Hence, to emphasize the conclusion already stated in Section 2.4, it is the peripheral areas that require further development before the full potential of cryogenic radiometry can be realised.

As highlighted in Section 2.4, a number of subsequent radiometers following similar design principles have been developed over the years targeted at specific applications. For example it soon became apparent that the PS radiometer described above was relatively large and with the availability of newer black coatings, the cavity could be significantly reduced in size and allow for more horizontal constructions removing the need for the turning mirror.

In addition, it was noted that the availability of liquid helium and its cost was highly varied around the world and that this was often a limitation to potential users. In recognition of this, NPL designed a new smaller radiometer utilising a mechanical cooling engine to cool the radiometer to cryogenic temperatures [27]. The mechanical cooling engine effectively replaces the need for any liquid cryogens and simply requires electricity for full operation. A schematic representation of this radiometer is shown in Figure 2.21.

The operation of this radiometer is similar to that of the PS radiometer and the QM radiometer except that in this case the radiation enters in the horizontal plane rather than vertical as with the others. Since the mechanical cooler only cools to around 15 K it also needs to use high-temperature superconductors for electrical connectors rather than conventional niobium wire. The overall uncertainty of this instrument is presented in Table 2.7, where it can be seen that it has similar performance to those cooled by liquid

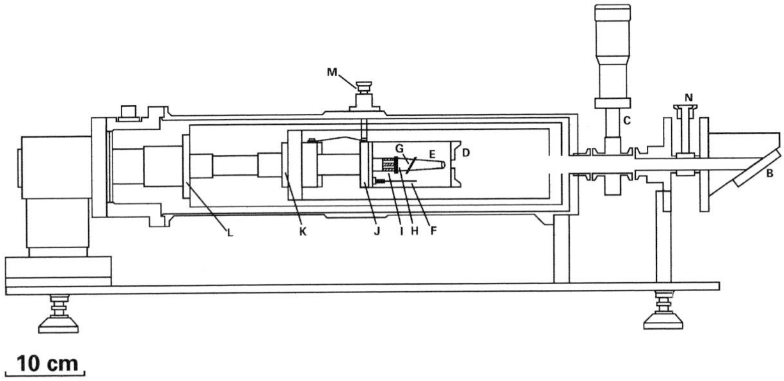


FIG. 2.21. Schematic drawing of the mechanically cooled cryogenic radiometer. A, laser beam; B, Brewster angled window; C, gate valve; D, quadrant detector; E, cavity; F, high temperature super-conducting heater leads; G, heater; H, RhFe thermometer; I, reference heat link; J, reference temperature heat sink; K, second stage cooler; L, first stage cold head; M, thermal shorting mechanism; N, vacuum port for window chamber.

TABLE 2.7. Corrections and Uncertainties of the Power of the Laser Beam Expressed as Parts in $10^4$ of the Measured Power, for the NPL Mechanically Cooled Radiometer

| | Correction | Uncertainty |
|---|---|---|
| Window transmittance | 3.0 | 0.3 |
| Beam scatter | 2.0 | 0.15 |
| Absorptance of cavity | 0.2 | 0.05 |
| Electrical power measurement | | 0.05 |
| Sensitivity of radiometer | | 0.1 |
| Changes in thermal and scattered radiation | | 0.1 |
| Sum in quadrature | | 0.37 |

*Note*: Uncertainty due to repeatability is not identified specifically here; it is included within the uncertainties of the radiometer sensitivity and changes in thermal background.

helium. A new radiometer is currently under construction at NPL utilising a 4 K mechanical cooling engine to cool a helium reservoir which is subsequently pumped to cool the heat sink to around 2.5 K, thus having all the benefits of mechanical cooling but with the very high sensitivity obtainable from lower temperatures. This new radiometer is intended to measure signals of around 1 μW with a resolution of around 1 pW.

## 2.5 Confirmation of Accuracy

Uncertainties of <0.1% are now routinely reported for many radiometric quantities traced to a cryogenic radiometer. However, it should of course be remembered that a cryogenic radiometer on its own is simply a well-characterised instrument and as such, unless linked to a more fundamental concept, has the potential for unknown systematic errors to exist or to develop and that these could then propagate into all other radiometric quantities.

The check mechanism is of course similar for all metrological quantities: intercomparison. For optical radiometric quantities, as is the case for all other SI quantities, such comparisons are organised (at the highest level) by the appropriate Consultative Committee (CC) of the Comitè International des Poids et Mesures (CIPM) (in this case CCPR, Consultative Committee for Photometry and Radiometry).

In terms of cryogenic radiometers, a CCPR supplementary comparison has been completed using trap detectors as a transfer standard. The comparison showed that, provided the radiometers were operated correctly, the results from all cryogenic radiometers, irrespective of the mode of operation or manufacture, agreed within their combined uncertainties [57] (Fig. 2.22). It should be noted however, that in carrying out such a comparison a number of users found that their operational procedures were not adequate, and in some cases requiring fairly large uncertainties to be added for the actual comparison process, demonstrating the importance of such an activity and how easy it can be to introduce significant errors to a measurement even though the basic standard has such a high intrinsic accuracy. In the above example only one participant had a result which deviated from the mean by an amount larger than their declared $2\sigma$ uncertainty and in this case it was later discovered that the vacuum pumping system associated with the cryogenic radiometer system was not in full working order at the time of the comparison and this likely led to uncorrected errors in the instrument operation. Perhaps this also indicates that care needs to be taken when interpreting the results of comparisons which, although now since the implementation of the mutual recognition arrangement (MRA) [58] are

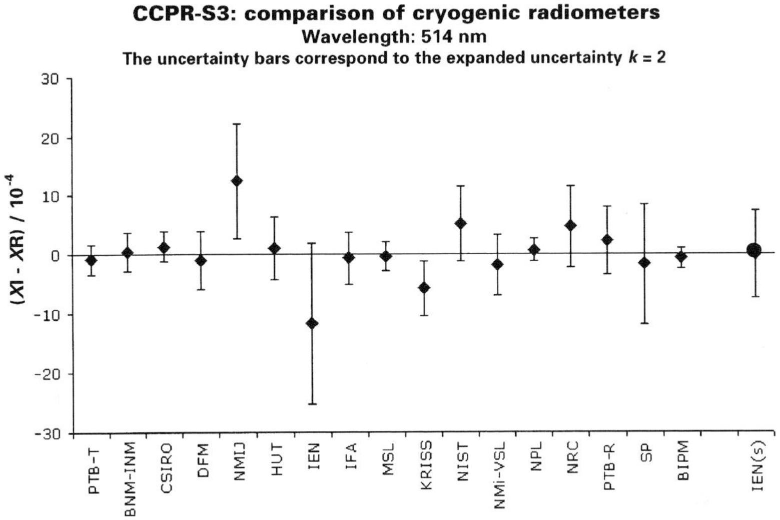


FIG. 2.22. Results of the CCPR supplementary comparison of cryogenic radiometers CCPR-S3 at 514 nm taken from the BIPM MRA database [58]. The results are presented as the relative difference of the participant to a weighted mean. The value for IEN(S) refers to a supplementary bilateral comparison performed after the original comparison but following the same protocol.

performed with significant increased rigour and transparency, are also carried out by NMIs under significant time pressure.

In addition to the above there have also been a series of direct comparisons of cryogenic radiometers performed bilaterally. In these, one radiometer is typically transported to the location of a second and both instruments then used to view the same source under the same conditions, resulting in higher accuracies due to the removal of many of the uncertainties associated with the transfer standard which no longer applied under these circumstances [59–62].

Such comparisons, while obviously important, are time consuming and difficult to organise and only check equivalence of measurement of a particular quantity. In order to be confident of the uncertainty relative to the SI quantity it is essential that a cryogenic radiometer or a quantity it measures can be compared to an independent quantity of similar or lower uncertainty. For the PS radiometer described earlier this was carried out through a comparison to the QM radiometer effectively linking optical radiometric quantities and in particular the candela to the Stefan–Boltzmann constant [63]. However, the design of both these instruments would only allow this experiment to be carried out to an uncertainty of around 2 parts in $10^4$ at best. Therefore, a new instrument, Absolute Radiation Detector (ARD) [54], was built at NPL to do this, with a target uncertainty of around 0.001%. This latter experiment is currently in progress.

## 2.6 New/Complimentary Technologies

Most cryogenic ESRs traditionally used for tasks such as calibrating silicon trap photodiodes are designed to measure radiant power levels of 100 μW to 1 mW with a relative random component of uncertainty, due to the combination of noise and drift during the measurement, of the order of $10^{-5}$. For this a noise floor of 1 nW is adequate, which can easily be achieved for helium-cooled ESRs with relatively conventional thermometry and electronics as described in previous sections. However, there are applications, such as irradiance measurements from low-temperature infrared blackbodies or laser-illuminated diffuse integrating spheres, or radiant power measurements using monochromator-based sources, where the radiant power reaching the ESR is only a few microwatts or even a few nanowatts. For such applications, the noise and drift aspects of the ESR must be more carefully designed, so they do not dominate the uncertainty.

Motivated by such concerns, there have been a number of research activities associated with cryogenic ESRs, not only to improve their accuracy but also to widen their range of applications. These include higher sensitivity temperature sensors, operation at higher temperatures, and advances in the data acquisition and analysis algorithms. In this section we review each of these areas in turn.

### 2.6.1 Use of Superconductive Transition Edge Thermometers

An important aspect to achieving low-noise and low-drift radiant power measurements in an ESR is to obtain very good control of the temperature difference, $\Delta T$, between the receiver cavity and the heat sink. This is usually achieved in cryogenic ESRs through the use of a servo control loop to control the temperature of the heat sink. In some cases where active cavity control is utilised as in some commercial radiometers and in several radiometers developed by NIST, a second servo loop is used to control the temperature of the detecting cavity, in this way minimising time needed for the cavity to respond to changes in input power, in particular the switching between electrical and optical.

The temperature can be actively controlled by a servo loop only as well as it can be sensed. The most commonly used temperature sensor in cryogenic radiometers is the commercially available GRT, discussed in Section 2.3, which has good sensitivity over the entire range from below 2 K up to past 9 K. However, much better sensitivity is conceivable using superconductor-based temperature sensors, operating at or near the superconducting transition temperature, $T_c$.

The simplest way to use a superconductor as a temperature sensor is to monitor the electrical resistance at the superconducting transition. Here the resistance drops suddenly from its normal state value to zero. Thus the resistance slope, $dR/dT$, where $R$ represents the resistance of the sensor and $T$ represents temperature, can be quite high at $T_c$. Low-temperature superconducting-transition-edge bolometers based on this temperature-sensing principle had been discussed for many years, and have been applied to low-temperature heat pulse and pulsed phonon experiments and infrared and submillimeter detection [64]. As a first example of the use of superconducting transition edge thermometers in an ESR, a prototype was developed by NIST in the mid-1990s [65]. The thin-film niobium sensors had a $T_c$ of about 9.19 K with a typical resistive superconducting transition width (10–90%) of 3 mK. The sensitivity figure of merit $(T/R)(dR/dT)$ for the niobium sensors was typically about 6500, whereas the same figure of merit for a GRT (at 4 K) is typically only about 2.2. As a result, the typical NEP improvement of the ESR based on superconductive resistive edge sensors compared to the conventional GRT approach was reported to be over two orders of magnitude. The prototype ESR based on niobium resistive edge sensors had an NEP on the order of a few pW in the measurement of power levels from 0.5 nW to 5 μW, which is quite impressive. However, the particular unit built suffered from drift because of parasitic thermal effects that are not fundamentally related to the use of a resistive edge sensor. With proper redesign of the thermal, electrical, and optical subsystems, it is believed that the drift effects could be eliminated.

In the late 1990s, Cambridge Research & Instrumentation (CRI), makers of the commercial CryoRad and LaseRad ESRs, performed some research aimed at developing a commercial ESR that used resistive transition edge sensors [66]. They developed a prototype ESR that, like NIST, used niobium thin-film transition edge sensors, operating at about 9.135 K. The prototype had a natural time constant ($\tau$) of 3 s and a thermal link conductance ($G$) of $2.9 \times 10^{-4}$ W/K, and achieved receiver temperature stability of 40–80 nK rms over a receiver power range from 25 nW to 1.5 μW while viewing an ambient scene. This is an improvement of nearly two orders of magnitude over their commercial LaseRad ESR. However, being a prototype, not everything was optimised, and the measured system power noise floor was on the order of 0.2 nW rather than a value in the 10–20 pW range that would have been expected if the NEP were dominated by temperature fluctuations.

In the examples above, the operating temperature choice of 9.2 K was made mainly because of the availability and familiarity with niobium thin-film sensors. There are two advantages to operating in the temperature range 2–6 K compared with 9.2 K. One is that the heat capacity of copper decreases significantly with temperature, giving a shorter thermal time constant for a given cavity mass. The other is that it is desirable to operate a few degrees below the superconducting transition of the substitution heater leads, typically niobium–titanium with a $T_c$ near 9.2 K, to ensure that there is no Joule heating in the leads that would otherwise forfeit one of the major advantages of cryogenic ESRs. For resistive-edge sensors, however, this would require a small research program to identify and develop an appropriate resistive edge sensor with a $T_c$ in the 2–6 K temperature range. Materials fabrication and degradation with ageing are some of the major challenges in such a development programme. Possible candidates that show promise from the superconductivity literature are Sn/Au films for $T_c$ in the 2–3 K range [67] and Sn films, given that Sn has a bulk $T_c$ of 3.7 K. Also, meander lines of very thin (25 nm rather than the conventional thickness of 100–200 nm) niobium can have $T_c$s down to below 6 K.

Note that the receiver cavity must operate at a higher temperature than the heat sink, and that this temperature difference determines, in conjunction with the thermal conductance ($G$) of the thermal link between the receiver cavity and the heat sink, the dynamic range of the ESR. This may require a difference between the $T_c$ of the receiver sensor and the $T_c$ of the heat sink sensor. In the NIST example of the niobium resistive edge prototype, the natural variation of the sensor $T_c$s ranged from 9.1904 to 9.1938 K, which, while seemingly quite small, was enough to provide the required dynamic range [65]. For higher dynamic range, a method of producing sensors with suppressed $T_c$ may be required.

Another way of using a superconductor as a temperature sensor is to utilize the temperature-dependence of the inductance in a thin-film strip that occurs just below $T_c$. This variation is greatest in the limit of very thin films, when the kinetic inductance dominates the magnetic inductance. Superconducting thermometers based on kinetic inductance, known as kinetic inductance thermometers, were proposed [68] and developed [69] in the late 1980s at NIST. One goal of the program was to develop a cryogenic ESR based on a kinetic inductance [70]. While nano-Kelvin temperature resolution was demonstrated, which would correspond to NEP of 2 pW for a typical value of $G$, the addition of a radiation absorber proved too complex in the particular design that was developed. The kinetic inductance approach was eventually abandoned in favour of the simpler resistive edge approach discussed above [65].

It should be noted for comparison that the Absolute Cryogenic Radiometer II (ACR II) developed by NIST for the low background infrared (LBIR) facility, based on conventional GRTs measured with commercially available AC bridge electronics, achieves a noise floor in the 7–10 pW range when operating in the relatively quiet environment of a low-background (<20 K) infrared chamber, and is currently limited mainly by the digitization of the electrical power readout electronics [71, 72]. This is roughly a factor of 10 improvement over its commercial predecessor, the ACR I, and is comparable to the values obtained in both examples above using superconducting resistive-edge sensors. The increased performance of the ACR II is attributable to a careful thermal re-design, most notably the replacement of the stainless-steel thermal link between the receiver cavity and the heat sink with one made of Kapton, which greatly reduces the heat capacity in the thermal link. This is a reminder that while the choice of temperature sensor is important, and superconducting transition-edge sensors show great promise in reducing the noise floor, there are many other factors that influence the practical noise floor obtainable in a practical ESR.

### 2.6.2 Operation near Liquid Nitrogen Temperature

Despite the great success of liquid-helium-cooled ESRs operating with a reservoir temperature near 4.2 K or below and its mechanically cooled counterpart, an ESR requiring only liquid-nitrogen cooling (77 K) but with comparable measurement uncertainty could potentially offer much lower operating cost, greater convenience, and wider availability to the radiometric community. Motivated by this prospect, and by the need for an ESR in a liquid-nitrogen-cooled space-simulating radiometric test chamber, NIST began development of an ESR that used high-$T_c$ superconductors as the temperature sensors [73]. It used thin-film sensors of photolithographically patterned striplines of the high-$T_c$ superconductor $YBa_2Cu_3O_{7-\delta}$ (YBCO), for which $T_c = 89$ K. There was a YBCO sensor on both the heat sink and the receiver, and each was operated in an independent servo loop. The receiver natural time constant $\tau$ was about 72 s and thermal link conductance $G$ was about 1 mW/K. The resistive superconducting transition width (10–90%) for both sensors was 0.6 K. While narrow enough to have much greater sensitivity compared with traditional thermometers, this was wide enough that, at this value of $G$, both the heat sink and the receiver could be operated in the superconducting resistive edge mode for power levels up to about 0.1 mW. When dark-tested (with the cavity viewing a 90 K background) in active cavity mode with electrical heater excitation at 10 μW power level, this ESR had a conservatively estimated noise floor of 20 nW and exhibited no perceivable drift over 2 h.

As part of research to develop a commercial version of the liquid-nitrogen-cooled high-$T_c$ ESR, CRI built a prototype ESR that used a YBCO sensor for the receiver cavity temperature control [66]. Using the CRI standard CryoRad electronics, they achieved a temperature control stability of 0.7 μK (rms) for receiver heater powers in the 10–100 μW range. When dark-tested in a 90 K environment, a 1.6 nW noise floor was achieved at a receiver power of 15 μW. This is more than a factor of 10 better than that reported on the NIST prototype [73], although the NIST prototype used less sophisticated control electronics. It is comparable to the noise performance of the CRI commercial bench top liquid-helium-cooled radiometer. However, while showing potential, this was only a prototype, and the high-$T_c$ resistive-edge-based ESR technology has not yet been sufficiently developed to the point of being commercialized.

Another approach that was developed for using a superconductor as a temperature sensor involves monitoring the critical current of the sensor [74]. The critical current is the value of current that drives a superconductor into the normal state at temperatures below $T_c$. It is a strong function of temperature in YBCO a few degrees below $T_c$. It was implemented and tested with the high-$T_c$ ESR described above, but with specially designed room-temperature electronics to provide low-noise measurement of the critical current. This operating mode can be used to increase the maximum measurable radiant flux of the ESR since the greatest variation of critical current with temperature occurs several degrees below $T_c$. Thus, the heat sink can be operated in critical current mode at 85 K, for example, while the receiver is operated using resistive-edge mode within the transition at 89 K.

### 2.6.3 Improvements in Control Algorithms

In an active-cavity type of ESR, a servo loop acts to vary the receiver cavity electrical heater power as necessary to keep the temperature of the cavity constant in time. When the shutter is opened in the conventional mode of operation (Figure 2.23a) the additional optical power absorbed in the receiver immediately begins to raise the cavity temperature. Though the servo loop eventually responds by turning down the electrical power to the appropriate level to return to the original temperature, this can take a significant amount of time. When the shutter is closed the cavity initially cools, and the servo loop again requires a significant fraction of the shutter cycle period to recover from this temperature disturbance.

To solve this problem of large power changes associated with opening and closing the shutter during the shutter cycle, some newly developed ESR's use a feed-forward control algorithm that works by simply not forcing large power changes to occur in the first place. This algorithm takes advantage of

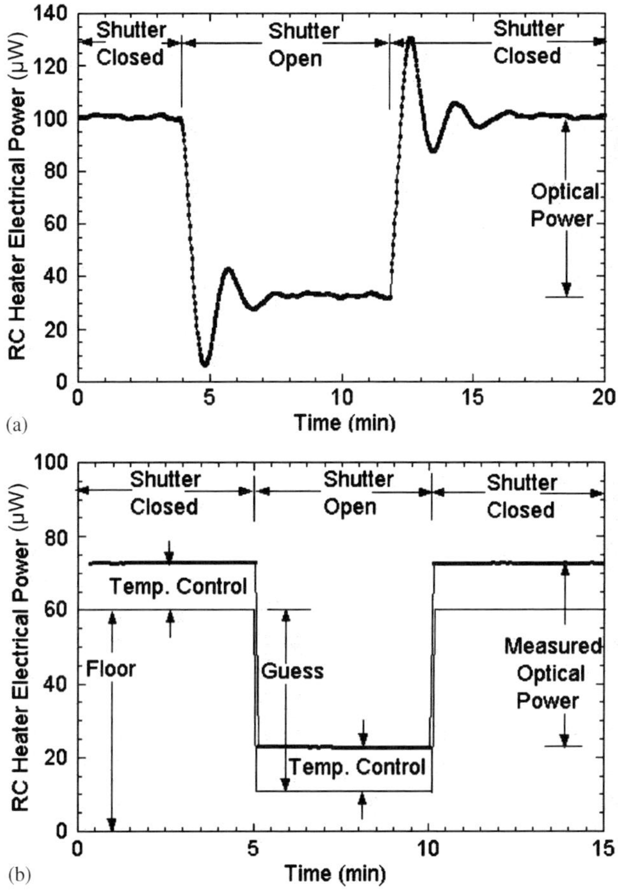


FIG. 2.23. Conventional versus feed-forward active cavity control. (a) Conventional, with control loop parameters optimised for low noise in steady state rather than quick recovery from transitions. (b) Feed forward, in the case of a near-perfect guess.

two pieces of information not otherwise utilized. One is that the approximate level of disturbance can often be guessed fairly accurately. The other is that the time when it occurs, at the shutter transition, is completely predictable and controllable. It is implemented by adding a "floor" to the RC electrical power when the shutter is closed (Figure 2.23). The floor is simply a constant power value that is added to the output of the servo, which can be done digitally in the case of digital servo loop. Then, at the instant that it directs the shutter to open, the digital algorithm directs the floor to be dropped by a "guess" of the amount of radiant power (Figure 2.23b).

The guess is simply another constant power value that represents the expected optical power modulation. For example, it could be simply the value measured in a previous shutter cycle. As the shutter closes, the algorithm puts the floor back up its previous shutter-closed value. There is still a conventional (e.g. PID) temperature control algorithm for the receiver cavity, but it ends up controlling only a relatively small amount of electrical power that rides on top of the floor electrical power. Thus, as in Figure 2.23b, if the guess is perfect the receiver cavity temperature is never destabilized as the shutter is opened or closed, and in effect the temperature control algorithm does not even "see" the shutter opening or closing. Note that the effect of a poor guess is simply that the time waiting for transients to subside is longer. For reasonably good guesses, that time approaches zero. In the limit that the guess approaches zero (a very bad guess), the feed-forward algorithm reduces to the conventional algorithm yielding data more like Figure 2.23a. One benefit of using this feed-forward algorithm is that it keeps the receiver in steady state throughout the shutter cycle. Then the servo tuning parameters can be optimised for low-noise temperature control in the presence of relatively small perturbations encountered in steady state, rather than simultaneously requiring optimal servo tuning parameters for timely return to steady state from large disturbances. Another benefit is that it decreases the time required to make a measurement involving a train of shutter cycles. This translates into more measurements being done in a given amount of time, resulting in lower statistical uncertainty.

Variations of such a feed-forward algorithm have been implemented on ambient temperature ESRs developed for two different space flight missions. One of these is the earth-viewing NIST Advanced Radiometer (NISTAR), a prototype ESR of which was described in Reference [75]. The other is the solar-viewing Total Irradiance Monitor (TIM) ESR on NASA's Solar Radiation and Climate Experiment (SORCE) [76].

### 2.6.4 Improvements in Data Processing Algorithms

It has been common practice to use a simple time-domain analysis to extract the optical power measurement from data such as those in Figure 2.23. The procedure is to compute the difference between the shutter-closed and shutter-open average steady-state values, being careful to avoid the transient regions. The Type A random uncertainty from noise is then quoted conservatively as the quadrature sum of the standard deviations of the shutter-open and shutter-closed values. When significant drift is seen, as in low-power measurements, correction is attempted by drawing a baseline through the data and performing the analysis relative to it, but there can be additional uncertainty as to where to draw the baseline.

An improved data processing algorithm for ESRs was introduced in the TIM ESR [76], and is being adopted by new ESRs being developed at NIST. The main idea is to extract the power measurement from a periodic shutter-modulated data set using digital signal processing in the frequency domain. This enables effective filtering of noise and drift sources at frequencies well away from the shutter frequency, and enables the demodulation of the power to be sensitive to its phase relative to the shutter phase. Phase-sensitive detection is commonly used for demodulation of relatively higher frequency AC signals ($>5$ Hz) using lock-in amplifiers, and has been used in most cryogenic-ESRs utilising AC bridges for monitoring the resistance of receiver cavity and heat sink temperature sensors. However, its use in ESRs to demodulate the relatively low-frequency ($<0.1$ Hz) power signals, while maintaining absolute accuracy, is a new development enabled by the trend towards digital data acquisition and post-processing.

Phase-sensitive detection of the electrical substitution power works as follows. Given a periodic train of shutter-modulated receiver electrical heater power cycles where 50% of the time is spent with the shutter open and 50% of the time is spent with the shutter closed (Figure 2.24a), define $N$ as the number of digitally sampled data points per cycle, and label the sampled power values as $\phi_I$ and the simultaneously sampled shutter position values as $\psi_I$. Note that $I$ is simply an index to keep track of the sampled input values, and the $\psi_I$ values may be, for example, simply 0 for closed and 1 for open. The phase-sensitive detection algorithm consists of multiplying the power values $\phi_I$ by a complex sinusoid $\exp(\mathbf{i}2\pi I/N)$ at the shutter frequency and low-pass filtering with a number of repeated boxcar running averages. The same operation is done on the shutter position values $\psi_I$ to enable the shutter transmittance factor to be divided out. The real part then gives the in-phase demodulated, filtered, electrical substitution heater power, $P_J$. For the case of four repeated boxcar averages, this can be expressed as

$$P_J = \mathrm{Re}\left[\frac{\sum\limits_{M=J-N+1}^{J}\sum\limits_{L=M}^{M+N-1}\sum\limits_{K=L-N+1}^{L}\sum\limits_{I=K}^{K+N-1}\mathbf{e}^{\mathrm{i}2\pi I/N}\phi_I}{\sum\limits_{M=J-N+1}^{J}\sum\limits_{L=M}^{M+N-1}\sum\limits_{K=L-N+1}^{L}\sum\limits_{I=K}^{K+N-1}\mathbf{e}^{\mathrm{i}2\pi I/N}\psi_I}\right] \tag{2.20}$$

where each summation represents another cycle-wide boxcar average. Note that there is a one-to-one correspondence between input index $I$ and output index $J$, and using Eq. (2.20) to compute a demodulated power value referenced to a given time requires averaging over several cycles before and after that time. Figure 2.24b shows an example of applying Eq. (2.20) to demodulate the data of Figure 2.24a. In the case of SORCE, departure from

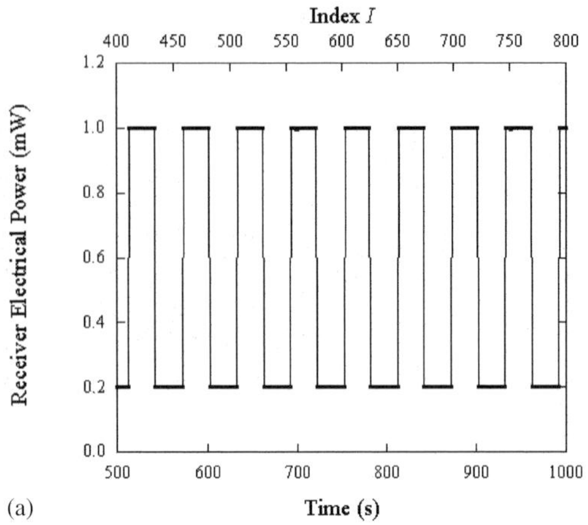


(a)

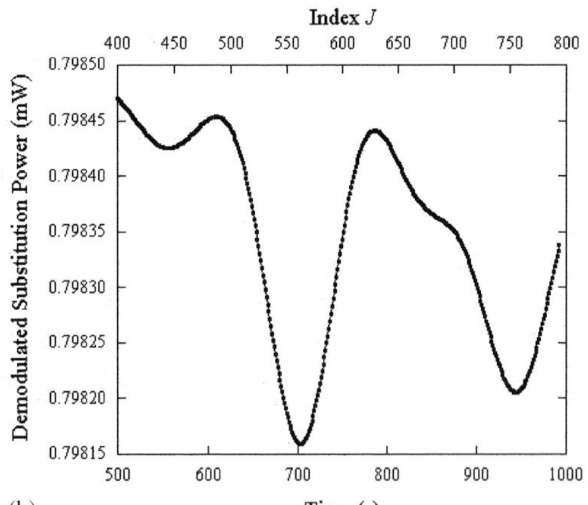


(b)

FIG. 2.24. Phase-sensitive detection example. (a) Input shutter-modulated receiver heater power. Data were taken with $N = 48$ and with a feed-forward algorithm with a floor of 0.9 mW and a power to be measured of about 0.8 mW. (b) Output demodulated power. The appropriate time average of this data set gives the measured electrical substitution power during the time interval of interest, which must of course be at least 4 shutter periods in this case.

steady-state temperature during the shutter cycle, which would cause departure from the square-wave shape of the data of Figure 2.24a, is corrected by including the servo loop gain at the shutter fundamental, resulting in a more complex form of Eq. (2.20) [76].

When using phase-sensitive detection of the electrical substitution power, all of the usual corrections such as cavity absorptance, window transmittance, and non-equivalence must still be applied to the substitution heater power in order to arrive at a correct optical power. However, in the case of the non-equivalence correction, it is argued in Reference [76] that (to a large extent) thermal effects that give rise to non-equivalence are out of phase with the shutter (at least in the case of SORCE) such that most of the non-equivalence effect appears in the out-of-phase (imaginary) component of the demodulated electrical substitution power. Thus, the process of using only the in-phase (real) substitution power via phase-sensitive detection, it is argued, reduces the non-equivalence correction, and accordingly, the uncertainty component resulting from the non-equivalence effect. This is a significant result in that one of the traditional arguments for using a cryogenic ESR has been reduction in the non-equivalence component of uncertainty (Section 2.3.1, Condition 3), whereas TIM on SORCE, an ambient temperature ESR, is evidently able to reduce the non-equivalence uncertainty dramatically simply by using phase-sensitive detection. It remains to be seen how widely accepted and independently validated these ideas become, and to what extent phase-sensitive detection becomes used in cryo-ESRs.

### 2.6.5 Electrically Substituted Bolometers

A complementary technology closely related to the absolute ESRs discussed above is embodied in the NIST cryogenic electrically substituted bolometer (ESB) [77]. This device has found its application as an irradiance responsivity transfer standard radiometer in the laser-illuminated integrating sphere facility at NIST known as the IR-SIRCUS (discussed in Chapter 4), since it uniquely satisfies five competing requirements: (a) spectrally flat response from visible (400 nm) through thermal infrared (20 μm) wavelengths, (b) linearity from the noise floor to 1 mW, (c) noise floor below 100 pW, (d) response time of tens of ms so as to be compatible with chopping above 10 Hz, and (e) active area of 5 mm diameter or larger. The noise requirement comes from the fact that diffuse radiant power levels of only 100 nW or so need to be measured to 0.1% or better. Semiconductor photodetectors satisfy (b)–(e) but not (a). Absolute cryogenic ESRs generally satisfy (a), (b), and (e), but usually not (c) and never (d) since the cavity thermal mass limits the time constant to several seconds at best. Room temperature thermal detectors such as thermopiles, pyroelectric detectors, and, in particular, the electrically calibrated pyroelectric radiometer (ECPR) [31, 32] have noise floors of 10 nW or higher when the other requirements are satisfied, thus cannot satisfy (c). Conversely, liquid-helium-cooled bolometers have low enough noise to satisfy (c) in the short term but they are highly non-linear, violating (b), and tend to drift in the longer term. The NIST ESB is a low noise liquid helium-cooled silicon bolometer that has a spectrally flat, low thermal mass to enable chopping at 15 Hz and incorporates the type of chopper-synchronized electrical substitution used in the ECPR to make it both linear and stable. Thus, it satisfies all of the requirements above for use as spectrally flat transfer standard. Like the ECPR, it implements electrical substitution with chopper synchronized electrical heater pulses applied to its gold-black absorber. It is the temperature stabilizing aspect of electrical substitution, rather than its ability to provide electrical calibration, that is important for the ESB. This is because the ESB is used as a transfer standard and calibrated against a trap detector in irradiance mode to set its absolute scale.

An ambient-temperature electrically substituted bolometer was developed recently as a stable, electrically calibrated detector for the solar irradiance monitor (SIM) instrument on SORCE [78]. This is a flat plate detector at the exit slit of a solar-viewing spectrometer, in an aluminized sphere for enhanced absorptance, and is coupled with a paired reference flat plate detector that is kept in the dark. This ESR is reported to have a 1 nW noise floor, very good for an ambient thermal detector, and uses a phase-sensitive detection algorithm similar to that described above.

## References


1. T. J. Quinn and J. E. Martin, *Phil. Trans. Roy. Soc. London* 316, 85–181 (1985).
2. J. E. Martin, N. P. Fox, and P. J. Key, *Metrologia* 21, 147–155 (1985).
3. E. F. Zalewski and J. Geist, Silicon photodiode absolute spectral response self-calibration, *Appl. Opt.* 19, 1214–1216 (1980).
4. J. Geist, E. F. Zalewski, and A. R. Shaeffer, Spectral response self-calibration and interpolation of silicon photodiodes, *Appl. Opt.* 19, 3795–3799 (1980).
5. J. Geist, A. J. D. Farmer, P. J. Martin, F. J. Wilkinson, and S. J. Collocott, Elimination of interface recombination in oxide passivated silicon $p^+n$ photodiodes by storage of negative charge on the oxide surface, *Appl. Opt.* 21, 1130–1135 (1982).
6. E. F. Zalewski, Recent developments in the techniques for the self-calibration of silicon photodiodes, *Proc. 10th IMEKO Int. Symposium on Photon Detectors*, 127–136 (1982).
7. P. J. Key, N. P. Fox, and M. L. Rastello, Oxide-bias measurements in the silicon photodiode self-calibration technique, *Metrologia* 21, 81–87 (1985).
8. L. P. Boivin and T. C. Smith, *Appl. Opt.* 17, 3067–3075 (1978).
9. J. Geist and H. Baltes, *Appl. Opt.* 28, 3929–3939 (1989).

*10.* J. Geist, D. Chandler-Horowitz, A. M. Robinson, C. R. James, R. Köhler, and R. Goebel, *J. Res. Natl. Inst. Stand. Technol.* 96, 463–492 (1991).
*11.* E. F. Zalewski and C. R. Duda, *Appl. Opt.* 22, 2867–2873 (1983).
*12.* N. P. Fox, *Metrologia* 28, 197–202 (1991).
*13.* T. R. Gentile, J. M. Houston, and C. L. Cromer, *Appl. Opt.* 35, 4392–4403 (1996).
*14.* A. Bittar, *Metrologia* 32, 497–500 (1995/1996).
*15.* T. Kubarsepp, P. Karha, and E. Ikonen, *Appl. Opt.* 39, 9–15 (2000).
*16.* J. Gran and A. Sudbo, *Metrologia* 41, 204–212 (2004).
*17.* J. Geist, G. Brida, and M. L. Rastello, *Metrologia* 40, S132–S135 (2003).
*18.* K. Ångström, *Nova Acta Soc. Sci. Upsal. Ser. 3* 16, 1 (1893).
*19.* F. Kurlbaum, *Ann. Phys.* 287, 591 (1894).
*20.* F. Hengstberger, "Absolute Radiometry." Academic Press, New York, 1989.
*21.* D. C. Ginnings and M. L. Reilly, *Temp. Meas. Control Sci. Ind.* 4, 339–348 (1972).
*22.* R. E. Bedford and C. K. Ma, *J. Opt. Soc. Am.* 66, 724 (1976).
*23.* J. C. De Vos, *Physica* 20, 669 (1954).
*24.* T. J. Quinn, "Temperature," p. 307. Academic Press, London, 1983.
*25.* J. M. Houston, C. L. Cromer, J. E. Hardis, and T. C. Larason, *Metrologia* 30, 285–290 (1993).
*26.* Fu Lei and J. Fischer, *Metrologia* 30, 291–296 (1993).
*27.* N. P. Fox, P. R. Haycocks, J. E. Martin, and I. Ul-haq, *Metrologia* 32, 581–584 (1995/1996).
*28.* R.A. Smith, F.E. Jones, R.P. Chasemar, "The Detection and Measurement of Infra-red Radiation," p. 47. Clarendon, London, 1957.
*29.* W. R. Blevin and W. J. Brown, *Metrologia* 7, 15–29 (1971).
*30.* R. C. Brown, P. J. Brewer, and M. J. T. Milton, *J. Mater. Chem.* 12, 2749–2754 (2002).
*31.* R. J. Phelan and A. R. Cook, Electrically calibrated pyroelectric optical-radiation detector, *Appl. Opt.* 12, 2494–2500 (1973).
*32.* J. Geist and W. R. Blevin, Chopper-stabilized null radiometer based upon an electrically calibrated pyroelectric detector, *Appl. Opt.* 12, 2532–2535 (1973).
*33.* L. P. Boivin and F. T. McNeely, *Appl. Opt.* 25, 554 (1986).
*34.* J. S. Blakemore, *Rev. Sci. Instrum.* 33, 55 (1989).
*35.* R. L. Rusby, *Temp. Meas. Control Sci. Ind.* 5, 829 (1982).
*36.* B. W. A. Ricketson, *Platinum Met. Rev.* 33, 55 (1989).
*37.* Y. S. Touloukian, R. W. Powell, C. Y. Ho, and P. G. Klemens, "*Thermophysical Properties of Matter*," Vol. 1. Plenum Press, New York, 1970.
*38.* N. P. Fox, *Metrologia* 32, 535–543 (1995/1996).

39. C. C. Hoyt and P. V. Foukal, *Metrologia* 28, 163–167 (1991).
40. C. A. Schrama, R. Bosma, K. Gibb, H. Reijn, and P. Bloembergen, *Metrologia* 35, 431–435 (1998).
41. T. Varpula, H. Seppa, and J.-M. Saari, *IEEE Trans. Instrum. Meas.* 38, 558–564 (1989).
42. R. U. Datla, K. Stock, A. C. Parr, C. C. Hoyt, P. J. Miller, and P. V. Foukal, *Appl. Opt.* 31, 7219–7225 (1992).
43. J. E. Martin and N. P. Fox, *Metrologia* 30, 305–308 (1993).
44. J. E. Martin and N. P. Fox, *Solar Phys.* 152, 1–8 (1994).
45. N. P. Fox, J. Aiken, J. J. Barnett, X. Briottet, R. Carvell, C. Frohlich, S. B. Groom, O. Hagolle, J. D. Haigh, H. H. Kieffer, J. Lean, D. B. Pollock, T. Quinn, M. C. W. Sandford, M. Schaepman, K. P. Shine, W. K. Schmutz, P. M. Teillet, K. J. Thome, M. M. Verstraete, and E. Zalewski, Traceable radiometry underpinning terrestrial- and helio-studies (TRUTHS), *Proc. SPIE* 4881, 395–406 (2003).
46. N. P. Fox, J. Aiken, J. J. Barnett, X. Briottet, R. Carvell, C. Frohlich, S. B. Groom, O. Hagolle, J. D. Haigh, H. H. Kieffer, J. Lean, D. B. Pollock, T. Quinn, M. C. W. Sandford, M. Schaepman, K. P. Shine, W. K. Schmutz, P. M. Teillet, K. J. Thome, M. M. Verstraete, and E. Zalewski, Traceable radiometry underpinning terrestrial- and helio-studies (TRUTHS), *Adv. Space Res.* 32, 2253–2261 (2003).
47. P. V. Foukal and J. Jauniskis, *Metrologia* 30, 279–283 (1993).
48. A. Lau-Främbs, H. Rabus, U. Kroth, E. Tegeler, G. Ulm, and B. Wende, *Metrologia* 32, 571–574 (1995/1996).
49. L. P. Boivin and K. Gibb, *Metrologia* 32, 565–570 (1995/1996).
50. P. J. Mohr and B. N. Taylor, *Rev. Mod. Phys.* 77, 1–107 (2005).
51. T. M. Goodman, J. E. Martin, B. D. Shipp, and N. P. Turner, *Inst. Phys. Conf. Ser.* 92, 121 (1989).
52. J. E. Martin, N. P. Fox, N. Harrison, B. Shipp, and M. Anklin, *Metrologia* 35, 461–464 (1998).
53. S. M. Pompea, D. W. Bergener, D. F. Shepard, S. Russak, and W. L. Wolfe, *Opt. Eng.* 23, 149 (1984).
54. J. E. Martin and P. R. Haycocks, *Metrologia* 35, 229–233 (1998).
55. W. R. Blevin, J. Geist, and T. J. Quinn, Private communication.
56. E. F. Zalewski, P. J. Key, J. E. Martin, and J. B. Fowler, Private communication.
57. http://kcdb.bipm.org.
58. http://www.bipm.org/en/convention/mra/.
59. R. Köhler, R. Goebel, R. Pello, O. Touayar, and J. Bastie, *Metrologia* 32, 551–555 (1995/96).
60. R. Goebel, R. Pello, P. Haycocks, and N. P. Fox, *Metrologia* 33, 177–179 (1996).

61. R. Goebel, R. Pello, K. D. Stock, and H. Hofer, *Metrologia* 34, 257–259 (1997).
62. K. D. Stock, H. Hofer, M. White, and N. P. Fox, *Metrologia* 37, 437–439 (2000).
63. N. P. Fox and J. E. Martin, *Appl. Opt.* 29, 4686–4693 (1990).
64. P. L. Richards, Bolometers for infrared and millimeter waves, *J. Appl. Phys.* 76, 1–24 (1994).
65. C. D. Reintsema, J. A. Koch, and E. N. Grossman, High precision electrical substitution radiometer based on superconducting-resistive-transition edge thermometry, *Rev. Sci. Instrum.* 69, 152–163 (1998).
66. S. Libonate and P. Foukal, Advanced absolute radiometers using superconducting transition thermometers, *Metrologia* 37, 369–371 (2000).
67. W. A. Simmons and D. H. Douglass,, Jr., Superconducting transition temperature of superimposed films of tin and silver, *Phys. Rev. Lett.* 9, 153–155 (1962).
68. D. G. McDonald, Novel superconducting thermometer for bolometric applications, *Appl. Phys. Lett.* 50, 775–777 (1987).
69. J. E. Sauvageau and D. G. McDonald, Superconducting kinetic inductance bolometer, *IEEE Trans. Magn.* 25, 1331–1334 (1989).
70. J. E. Sauvageau, D. G. McDonald, and E. N. Grossmann, Superconducting kinetic inductance radiometer, *IEEE Trans. Magn.* 27, 2757–2760 (1991).
71. A. C. Carter, T. M. Jung, A. Smith, S. R. Lorentz, and R. Datla, Improved broadband blackbody calibrations at NIST for low-background infrared applications, *Metrologia* 40, S1–S4 (2003).
72. A. C. Carter, S. R. Lorentz, T. M. Jung, and R. U. Datla, ACR II: Improved absolute cryogenic radiometer for low background infrared calibrations, *Appl. Opt.* 44, 871 (2005).
73. J. P. Rice, S. R. Lorentz, R. U. Datla, L. R. Vale, D. A. Rudman, M. Lam Chok Sing, and D. Robbes, *Metrologia* 35, 289–293 (1998).
74. M. Lam Chok Sing, J. P. Rice, C. Dolabdjian, and D. Robbes, Low-noise temperature control using a high $T_c$ superconducting sensor compared to a conventional PRT method, *Proc. European Conf. on Applied Superconductivity*.
75. J. P. Rice, S. R. Lorentz, and T. M. Jung, The next generation of active cavity radiometers for space-based remote sensing, Preprint volume of the 10th Conference on Atmospheric Radiation (American Meteorological Society, 28 June – 2 July 1999), 85–88 (1999).
76. G. M. Lawrence, G. Rottman, J. Harder, and T. Woods, Solar total irradiance monitor (TIM), *Metrologia* 37, 407–410 (2000) More details are found in the SORCE Algorithm Theoretical Basis Document available at: http://eospso.gsfc.nasa.gov/eos_homepage/for_scientists/atbd/.

77. J. P. Rice, An electrically substituted bolometer as a transfer-standard detector, *Metrologia* 37, 433–436 (2000).
78. J. Harder, G. M. Lawrence, G. Rottman, and T. Woods, Solar spectral irradiance monitor (SIM), *Metrologia* 37, 415–418 (2000).

# 3. REALIZATION OF SPECTRAL RESPONSIVITY SCALES

L. P. Boivin
National Research Council of Canada, Ottawa, Canada

## 3.1 Introduction

Many different types of detectors are used for a wide variety of radiometric and photometric measurements. These detectors have various properties which must be known in order to perform these measurements accurately. The single most important property of a detector is its spectral responsivity. It is the only detector property which is the object of regular international comparisons. Its accurate measurement is the basis of detector-based radiometry and photometry, where primary radiometric and photometric standards are realized using detectors whose spectral responsivities have been accurately measured using absolute radiometers. The spectral responsivity of a detector is the ratio of the output response of the detector to the input monochromatic radiant power. It is called spectral responsivity because in most cases the responsivity of a detector is wavelength-dependent. For example, the responsivity of photodiodes is usually expressed in terms of amperes (of output photocurrent) per watt (of incident radiation). When using the term responsivity, or spectral responsivity, it is usually implied that the values are absolute. In some cases, the responsivity values have been normalized at some wavelength (say to unity at 550 nm), or using a more complicated procedure. We then refer to these normalized values as the relative spectral responsivities of the detector. To put these concepts in mathematical terms: if the monochromatic incident radiant power is denoted by $P(\lambda)$ (in W) and the corresponding output photocurrent (say) of the detector is $I(\lambda)$ (in A), then the absolute spectral responsivity $S(\lambda)$ is given by $S(\lambda) = I(\lambda)/P(\lambda)$ and has units of amperes per watt. The quantity $S_{\text{rel}}(\lambda)$ given by $S_{\text{rel}}(\lambda) = S(\lambda)/S(\lambda_0)$, for example, is a relative spectral responsivity quantity which is dimensionless and has a value of unity at $\lambda_0$.

The above concepts apply to the radiant power responsivity, where the incident beam underfills the detector aperture, and the detector measures the total power in the beam. In some cases, the irradiance responsivity is required, where the beam overfills the detector aperture, and the detector measures the irradiance in the beam. In such cases, $P(\lambda)$ is replaced by $E(\lambda)$, the irradiance (in W/cm$^2$) and the irradiance responsivity is measured in A/W/cm$^2$.



EXPERIMENTAL METHODS IN THE PHYSICAL SCIENCES, vol. 41
ISSN 1079-4042 DOI: 10.1016/S1079-4042(05)41003-6

This chapter will be concerned with the measurement of absolute radiant power responsivity; the measurement of irradiance responsivity will be mentioned only briefly; it is one form of dissemination of spectral responsivity scales that is beyond the scope of this chapter. Also, we shall be concerned mostly with the measurement of *absolute* spectral responsivity; for the purposes of this chapter, relative responsivity measurements will be discussed in the sections dealing with the interpolation or extrapolation of absolute measurements to other spectral regions.

In this chapter, we shall discuss the realization of spectral responsivity scales using laser, monochromator and synchrotron radiation-based methods. In each case, we shall review the apparatus, procedures and transfer standards used. Interpolation and extrapolation techniques will be discussed, and sources of uncertainty analysed in some detail. In all cases, we consider only steady, i.e., continuous radiation; pulse radiometry will not be covered.

## 3.2 Realization of Spectral Responsivity Scales

A spectral responsivity scale is *realized* by the first-level spectral responsivity calibration of *transfer standard radiometers* using absolute radiometers. These so-called *transfer standards* are used for important radiometric or photometric applications:

- the dissemination of spectral responsivity scales,
- the realization of luminous intensity scales (the candela),
- the realization of spectral irradiance scales,
- international intercomparisons.

Detectors that presently are suitable for use in transfer standards include silicon, germanium and InGaAs photodiodes, and certain types of thermal detectors (e.g. thermopiles). For the realization of spectral responsivity scales, transfer standards do not usually incorporate any optical filters and so can be used over a fairly wide spectral range. We can denote these as broadband transfer standards. When an optical filter is used (for example, an interference filter or a $V(\lambda)$ filter), we usually denote these as *filter radiometers*, and they are used, for example, to realize a luminous intensity scale or a spectral irradiance scale. In this chapter, we shall be concerned with broadband-type transfer standards, used in the realization and dissemination of spectral responsivity scales. The design, properties and calibration of filter radiometers will be discussed in Chapter 4.

Let us clarify what we mean by absolute radiometer, in the context of this chapter. *Absolute radiometry* refers to the measurement of optical radiation from first principles, i.e., without using optically calibrated detectors or radiometers. Absolute radiometry can be source- or detector-based. Source-based absolute radiometry using either electron storage rings or blackbody radiators is used today mostly for the realization of spectral radiance and irradiance standards. Detector-based absolute radiometry uses so-called *absolute radiometers*, which are self-calibrating and thus require no external calibration. The two principal types of absolute radiometers are electrical substitution radiometers (ESRs) and self-calibrated silicon photodiodes [1, 2]. Trap detectors [3–5] are a special case of self-calibrated diodes. The purpose of this chapter is not to discuss the properties of absolute radiometers, but rather to discuss how they are used to calibrate transfer standards, i.e., to realize spectral responsivity scales. There are novel methods for measuring the absolute quantum efficiency (hence, the responsivity) of certain types of detectors, but these are still at the experimental stage and are not used for practical spectral responsivity scale realizations. An example is the use of correlated photons in parametric fluorescence; this topic is discussed in Chapter 5. The most accurate absolute radiometer in use today is the cryogenic radiometer, and most National Measurements Institute (NMIs) now use a cryogenic radiometer as the basis of their radiometric and/or photometric scales. Because of this, we shall assume the use of a cryogenic radiometer in the following sections. Cryogenic radiometers are ESRs operating at liquid-helium temperature. The properties of cryogenic radiometers are discussed in detail in Chapter 2. The properties and applications of room-temperature-type ESRs are discussed in detail in an earlier book [6].

In order to calibrate transfer standards spectrally using an absolute radiometer, a source of monochromatic radiation is required. There are two general approaches used to provide this monochromatic radiation, one using lasers and the other using monochromators in conjunction with broadband sources. A special case of the monochromator approach is where the source is synchrotron radiation. Because of the very different methodology used in that case, this approach is treated as a separate case below. The applications of synchrotron radiation in modern radiometry are also treated in Chapter 5.

### 3.2.1 Laser-Based Methods

Laser-based methods are the ones most commonly used for calibrations performed at the highest level of accuracy [7–30]. The main advantages of laser-based calibrations are:

- very accurately known wavelengths,
- very small bandwidth, i.e., a high spectral purity,

- a high signal-to-noise level; this is particularly important when using absolute radiometers,
- the laser radiation can be stabilized to a high degree using electro-optic or acousto-optic modulators,
- very low stray light level,
- good beam geometry—low beam divergence and easy alignment.

#### *3.2.1.1 Apparatus*

Figure 3.1 shows a typical configuration for laser-based spectral responsivity measurements. Some of the types of lasers used for this application are: helium–neon, argon, krypton, helium–cadmium, Nd:YAG, Ti:sapphire. In all the cases discussed here, CW radiation is assumed. Table 3.1 lists the more common types of lasers used and the corresponding emission lines.

When a tunable laser is used (e.g. Ti:sapphire), a beam splitter is inserted in the beam to direct part of the beam to a wavemeter to measure accurately the wavelength of the laser radiation. As shown in the figure, wedged beam splitters are used to isolate the reflections from the two surfaces and avoid interference effects. The laser beam passes through an electro-optic stabilizer; the polarizer shown is usually part of the stabilizer and must be adjustable (in azimuth) in order to minimize the reflection from the Brewster window. The spatial filter is used to clean up and condition the beam. The output is a fairly large diffraction pattern (Airy pattern) and an aperture is used to select only the central bright spot, which is focussed on the cryogenic radiometer or the transfer standard. Part of the converging beam is reflected

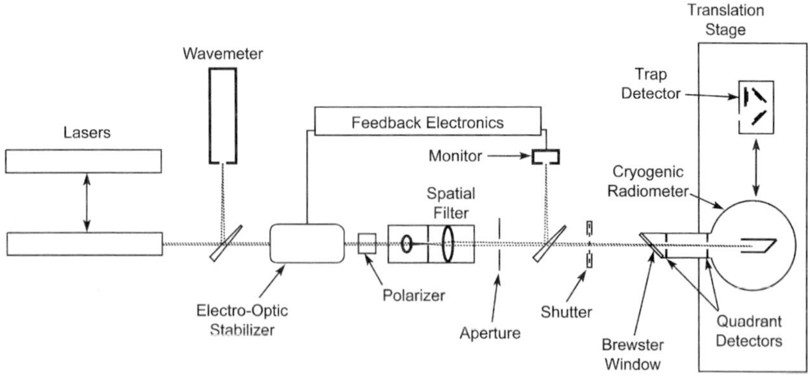


Fig. 3.1. Typical configuration for cryogenic radiometer-based spectral responsivity measurements using lasers.

TABLE 3.1. Principal Laser Lines Used for Cryogenic Radiometer Calibrations

| Spectral domain | Wavelength (nm) | Type of laser |
|---|---|---|
| UV | 238.3 | Argon ion (doubled) |
| | 257.3 | Argon ion (doubled) |
| | 325 | Helium–cadmium |
| | 350.7 | Krypton ion |
| | 356.4 | Krypton ion |
| VIS | 406.74 | Krypton ion |
| | 441.57 | Helium–cadmium |
| | 476.2 | Krypton ion |
| | 487.99 | Argon ion |
| | 514.53 | Argon ion |
| | 532.08 | Nd:YAG (doubled) |
| | 568.2 | Krypton ion |
| | 632.82 | Helium-neon |
| | 647.1 | Krypton ion |
| | 676.4 | Krypton ion |
| Near-IR | 799.3 | Krypton ion |
| | 700–1050 (tuning range) | Ti:sapphire |
| | 1064 | Nd:YAG |
| | 1249–1335 (tuning range) | Laser diode |
| | 1460–1600 (tuning range) | Laser diode |
| FIR | 9.2–10.8 μm (tuning range) | $CO_2$ |
| | 10.6 μm | $CO_2$ |

by means of a wedged beam splitter to a monitor detector; the signal from this monitor is used both as feedback to the stabilizer and to apply a correction for residual drift, if necessary. The main beam passes through a shutter and enters the cryogenic radiometer or the transfer standard. Figure 3.1 shows a typical set-up, where both the cryogenic radiometer and the transfer radiometer(s) are mounted on a translation stage which allows radiometer interchange. The figure shows a trap detector as transfer radiometer, but various types of radiometers are also used. In other set-ups, a mirror is used to redirect the beam to a vertical axis cryogenic radiometer that does not move; the transfer radiometers are mounted on a translation or rotation stage, which allows the radiometers to be brought into or out of the beam in front of the cryogenic radiometer.

#### *3.2.1.2 Measurements*

Most of the laboratories using cryogenic radiometers for this type of measurements use one of the two types of commercially available cryogenic radiometers, the Radiox, made by Oxford Instruments (UK) or the LaseRad, made by Cambridge Research and Instrumentation (USA); both of these companies no longer manufacture off-the-shelf cryogenic radiometers. The measurement sequence is the same with both in principle, although the measurement procedure is quite different, mostly because of the different electronics used with these radiometers. The Radiox has a longer time constant (1–3 min) compared to ~30 s with the LaseRad. The Radiox is operated by manually switching between radiant and electrical modes of heating; in the electric heating mode, special algorithms are used [14, 27, 31] to determine the equivalent electrical power. The LaseRad is operated in a servo mode, where the electrical power applied to the cavity is automatically adjusted to maintain the cavity at a constant temperature. The incident radiant power is obtained as the difference in servo power as the shutter is opened or closed. Once the laser has been stabilized at a suitable power level (usually between 0.5 and 1.0 mW), the radiant power is measured by the cryogenic radiometer; the translation stage is then displaced sequentially to each transfer radiometer, and their net responses are measured, from which the absolute spectral responsivities are obtained.

Corrections have to be applied to the measured optical power for the following factors: cryogenic radiometer effects (non-equivalence of electrical and radiant heating and imperfect cavity absorptance), residual drift in laser power, scattered radiation and transmittance of the Brewster window. The main cryogenic radiometer effects are non-equivalence effects and imperfect absorptance of the cavity; see Chapter 2 for a discussion of these. The ratio of monitor detector signals, measured during the cryogenic and transfer radiometer measurements respectively, can be used to correct for laser power drift. Scattered radiation is an important source of uncertainty; corrections for scattered radiation can be derived. This involves measurements using quadrant detectors mounted inside the cryogenic radiometer [14] and measurements with the transfer radiometer at different distances along the axis [21, 27] and measurements using different aperture diameters in front of the radiometers [21]. In practice, it is best to obviate the need for scattered radiation corrections by making the scattered radiation component very small and the same for both the cryogenic radiometer and transfer radiometers. Scattered radiation effects can be minimized by using suitable beam-shaping techniques [21, 32], operating in a dust-free environment [18]. The effect of scattered radiation will be very small if limiting apertures of the same diameter are used in front of the cryogenic and transfer radiometers, and if these apertures are in the same plane, orthogonal to the optical axis [18].

Minimizing the Brewster window reflectance and measuring its transmittance are important parts of these measurements. The Brewster window is usually mounted on a separate assembly, which can be detached from the cryogenic radiometer so that its transmittance can be measured separately. The assembly usually incorporates a vacuum bellows and gimbal adjustments to allow the orientation of the window to be adjusted with respect to the laser beam, while the polarizer at the output of the electro-optic stabilizer, or a separate polarizer down stream, is used to ensure that the radiation incident on the window is perfectly linearly polarized and parallel to the plane of incidence. If a steering mirror or beam splitter is used after the polarizer to direct the beam into the cryogenic radiometer, it is important that it be aligned properly with respect to the azimuth of the polarized radiation to avoid introducing elliptical polarization in the beam [21]. It is also important to have exactly the same optical configuration in the transmittance measurement as in the actual measurements in order to avoid errors.

Figure 3.2 shows the set-up used at PTB (Physikalisch-Technische Bundesanstalt, Germany) [18] to allow adjustment and measurement of the Brewster window transmittance in situ, i.e. without having to remove the Brewster assembly in order to measure it.

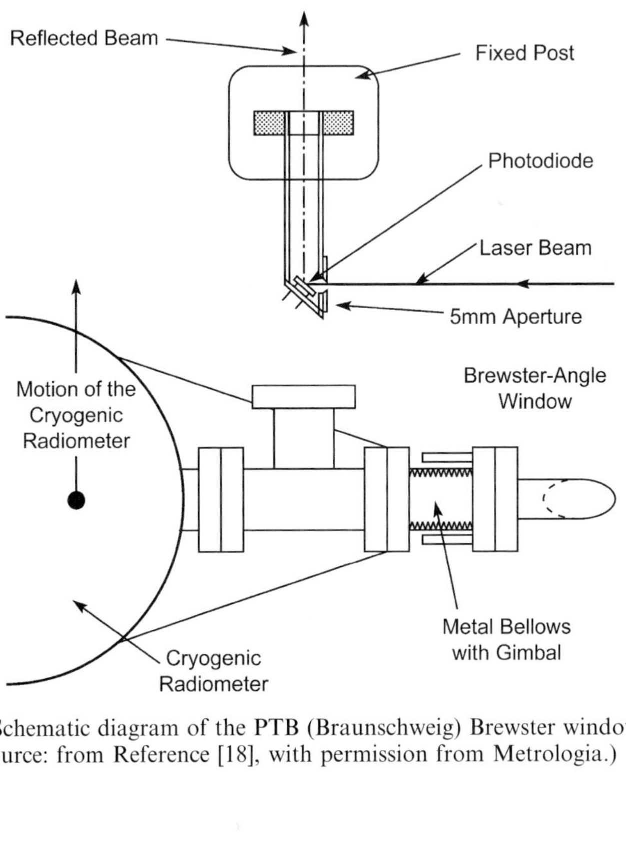


FIG. 3.2. Schematic diagram of the PTB (Braunschweig) Brewster window configuration. (Source: from Reference [18], with permission from Metrologia.)

The apparatus shown schematically in Figure 3.1 is typical for measurements in the UV visible and near-IR. For measurements further in the IR, using $CO_2$ lasers (wavelength of $\sim 10\,\mu m$), modified apparatus is used [23, 26]. Some modifications include: the use of ZnSe instead of fused silica for the windows and beam splitters, the use of choppers and lock-in amplifiers for the transfer standards, acousto-optic modulators for stabilization, auxiliary He–Ne laser beams inserted coaxially with the IR laser beam for alignment purposes, etc. Another special case is that of the calibration of transfer standards used for fibre-optic communication systems; here, diode lasers are used as sources and the wavelengths used are nominally 1300 and 1550 nm [13, 24, 25]. In most cases, the laser radiation is coupled to a short length of optical fibre using conventional fibre couplers, and the beam emerging from the fibre recollimated using a microscope objective. For fibre-coupled sources with random polarization, the Brewster window on the cryogenic radiometer can be replaced by a near-normal incidence window; this would make the window transmittance measurement less sensitive to small geometrical variations and surface contamination. Some set-ups [13] do not use optical fibres but only a focusing lens and beam-shaping apertures. As in other systems, stabilizers and polarizers are used to condition the air beam, but spatial filters are not necessary. In some cases [24, 25], 632.8 nm radiation is also injected into the fibre using fibre-optic couplers, for alignment purposes.

### *3.2.1.3 Transfer radiometers*

Laser-based cryogenic radiometer facilities are capable of measuring monochromatic laser radiation with an overall accuracy of the order of $\pm 0.01\%$. In principle, it should be possible then to measure the responsivity of a transfer radiometer with an accuracy approaching this value. It is obvious that only radiometers incorporating detectors having excellent radiometric properties can be used. The necessary properties are

- high degree of spatial uniformity across the detector surface;
- high degree of linearity over a wide dynamic range: this is particularly important with laser-based systems;
- high degree of stability vs. ageing and fatigue effects;
- small temperature coefficients of responsivity;
- available in relatively large sizes (5–10 mm diameter).

In the spectral range 250–1000 nm, certain types of silicon photodiodes are available which are suitable for this application. The type of silicon photodiode now in use in most labs is the Hamamatsu type S1337-1010.

This is a 10 mm × 10 mm PIN diode, and it is usually purchased without a window and mounted in a simple radiometer head incorporating a temperature sensor, such as a thermistor or RTD (Resistance Temperature Detector) element. In some cases, a window is included in the radiometer head, which can then be hermetically sealed. This is desirable in order to minimize ageing effects due to contamination or reversible changes caused by variations in relative humidity [33]. If a window is used, it will usually be UV-grade fused silica, and must have a small wedge angle and be slightly tilted with respect to the detector surface in order to avoid interference effects [7] due to the coherence of the radiation. For radiometer windows, wedge and tilt angles less than 1° are sufficient to avoid these effects, while introducing negligible polarization effects. The uniformity of selected S1337 diodes is excellent, typically better than ±0.1% in the central 6 mm region. Below 400 nm, the uniformity is poorer and usually degrades with time, especially without windows. Angular variation of responsivity and polarization sensitivity of single detectors is not a problem in the present application, where measurements are made at quasi-normal incidence using quasi-collimated radiation. These issues will be discussed below, in the context of trap detectors and monochromator-based measurements.

The long-term stability of the S1337 is also very good. For example, the working standards used at the NRC (National Research Council, Canada) were found to change by <0.1% per year in the range 400–950 nm, and by 0.1–0.2% per year in the range 250–400 nm. In the UV, stability of the detectors will depend very much on the irradiation history. The UV radiation dose and wavelength will affect the responsivity of these detectors, mostly in the UV, but also to a lesser extent above 400 nm [34, 35]. They should not be exposed to radiation below 250 nm. For measurements in the range 250–400 nm, Hamamatsu S1227 and S5227 silicon diodes are also commonly used. They are UV-optimized versions of the S1337.

Work at NRC [36] and elsewhere indicated that the S1337 exhibited a sharp drop in responsivity (saturation) at a photocurrent of approximately 1 mA, irrespective of wavelength and beam diameter, for beam diameters 4 mm or greater. More recent work [37–39] with trap detectors (see below) fabricated using S1337 detectors indicated a more complex behaviour when very small beam sizes were used (0.5–2 mm). With small beam sizes, the onset of saturation occurs at lower photocurrents. From this work, one can conclude that with beam diameters of 2 mm or greater, and photocurrents of 0.5 mA or less, non-linearity effects should not exceed 0.01% . For greater photocurrents or smaller beam sizes, the S1227 or S5227 are preferable since they remain linear to a much higher photocurrent [37, 38]. Alternatively, one should consider applying a small amount of reverse bias to the S1337. It has been shown [39] that applying as little as 6 V reverse bias to S1337 traps resulted in linear operation to within 0.05% for a beam diameter of 2 mm and a photocurrent up to 14 mA.

The most important type of transfer radiometer used in the UV to near-IR spectral range for laser-based measurements is the so-called trap detector. This type of radiometer consists of a number of detectors arranged in series optically, in a light trapping configuration, in order to virtually totally absorb the incident radiation. Trap detectors were first developed as absolute radiometers [3–5]. For a silicon photodiode, we can write

$$S(\lambda) = \lambda\eta_i(\lambda)(1 - \rho(\lambda))/K = \lambda\eta_e(\lambda)/K \tag{3.1}$$

where $S(\lambda)$ is the absolute spectral responsivity of the diode in $AW^{-1}$, $\lambda$ the wavelength of the radiation in nanometres, $\rho(\lambda)$ the reflectance of the diode, $\eta_i(\lambda)$ the internal quantum efficiency, $K$ a constant and $\eta_e(\lambda) = (1 - \rho(\lambda))\eta_i(\lambda)$ the external quantum efficiency. The value of the constant $K$ is given by $K = hc/ne = 1239.5\,\mathrm{W\,nm/A}$ for wavelengths in air, where $h$ is Planck's constant, $c$ the velocity of light, $n$ the refractive index of air and e the charge of the electron. For certain types of diodes such as the S1337, $\eta_i(\lambda)$ is very close to unity in the VIS-near-IR spectral range. Thus, by arranging a number of diodes in a light-trapping configuration, the incident radiation can be virtually totally absorbed by the group of diodes; this type of detector assembly is referred to as a *trap detector*. For such a trap detector then $\eta_e \cong 1$. The total photocurrent for the group of detectors is summed, and the responsivity of this trap detector is then given by

$$S_{\mathrm{trap}}(\lambda) \cong \lambda/K \tag{3.2}$$

Thus, this trap detector becomes an absolute radiometer (to within a small residual error) since its responsivity is known a priori. Some of the first traps to be used widely were available commercially as QED-200 [3] and made use of inversion layer diodes. Later, traps incorporating S1337 became widely used, and are probably today the most commonly used. Traps based on S1337 diodes do not have an external efficiency as high as that of the QED-200, and so, inherently, are not as accurate as absolute detectors; however, their so-called "quantum deficiency" can be measured and corrections applied, so that S1337 traps can be quite accurate as absolute radiometers. With respect to the QED-200, they have the advantage of not requiring reverse bias and being usable over a wider spectral range. Although trap detectors were first developed as absolute detectors, it became apparent that they were also excellent transfer radiometers for laser-based calibrations using cryogenic radiometers. Some of the advantages of trap detectors over single detectors as transfer radiometers are: better uniformity; improved long-term stability; and no inter-reflection problems. Another possible advantage is simpler modelling of the spectral responsivity for interpolation purposes. This will be discussed further below. For monochromator-based measurements, trap detectors have other advantages (and disadvantages) that will be discussed below.

Many different configurations of trap detectors have been studied [5, 40–44]: reflection traps, transmission traps, wedge traps and tunnel traps. The two most common configurations are the three-detector reflection trap and the six-detector transmission trap, as shown in Figure 3.3.

In the reflection trap, the incident radiation is reflected successively from the three detectors at 45°–45°–0°–45°–45° until the residual radiation emerges from the trap back along the direction of incidence. With the transmission trap, the radiation is transmitted through the trap co-axially with the incident beam, after undergoing six successive reflections at 45°. Figure 3.4 compares the total reflectance (or transmittance) of the two traps shown in Figure 3.3, compared to that of a single detector at normal incidence (S1337 detectors assumed). Above 400 nm, the residual reflectance is $<1\%$. One important property of these two designs (and of most trap designs) is that the trap is polarization-independent, i.e. the spectral responsivity is independent of the state of polarization of the incident radiation. This is true in the ideal case of perfectly collimated radiation incident exactly along the optical axis of a perfect trap in which the individual diodes have identical radiometric properties and are perfectly aligned. In practice, such traps have a small residual polarization sensitivity due to a departure from these ideal conditions. Polarization independence is an important requirement for traps-as-transfer radiometers since laser radiation is usually highly polarized. Even though the design is polarization-insensitive, the trap must be fabricated very carefully. It has been shown [45] that small alignment errors can cause non-negligible polarization sensitivity. Transmission traps are less sensitive than reflection traps in this respect. Trap detectors do

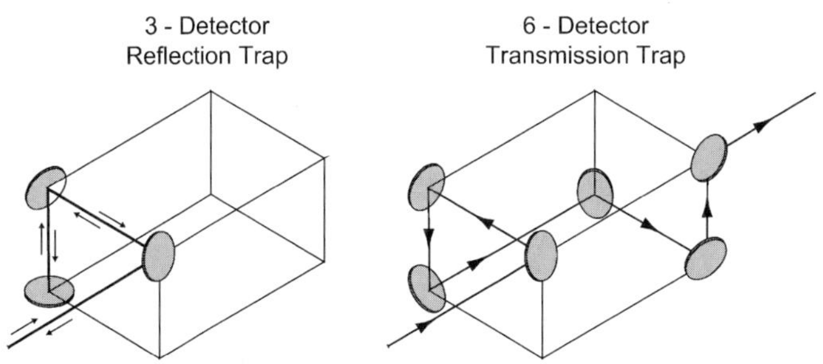


FIG. 3.3. Configuration of a 3-detector reflection trap and a 6-detector transmission trap.

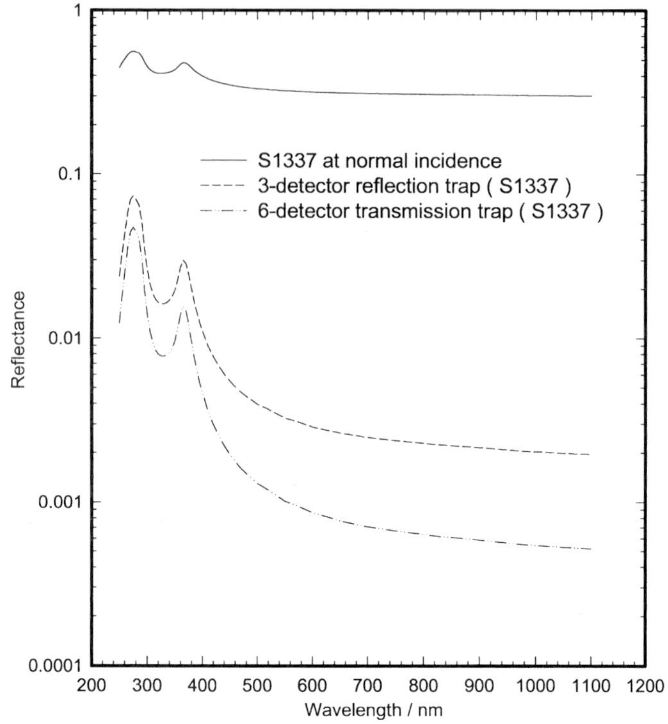


FIG. 3.4. Total reflectance of a three-detector reflection trap and a 6-detector transmission trap (29 nm oxide thickness assumed).

not include windows. In some cases, they are calibrated inside vacuum but then used in air. Calculations and measurements have shown [46] that the effect on responsivity of air vs. vacuum operation is a few parts in $10^5$ at most.

The UV is a particularly difficult spectral range for detector-based radiometry. Stability of detectors to UV irradiation is a particular problem, especially in laser-based cryogenic radiometry. The Hamamatsu S1337, S1227 and specially the S5227 have shown relatively good stability in the UV, but only above 250 nm. For wavelengths below 250 nm, PtSi detectors [31, 35] and special silicon photodiodes [31, 35, 47] are preferable. In some cases, GaAsP detectors are used because they are quite stable to UV irradiation. They are not very uniform, unfortunately, even in trap configurations, so that they are not suitable for measurements at the highest accuracy and do not appear to be used directly for laser-based calibrations with cryogenic radiometers.

In the infrared, there appear to be relatively few direct realizations of spectral responsivity scales using lasers and cryogenic radiometers. Some notable examples of such realizations, and the types of detectors used, are:

- Calibration of InGaAs detector based radiometers using Ti:sapphire, Nd:YAG and diode lasers and a cryogenic radiometer at PTB [29]. The radiometers were either 5 mm diameter single InGaAs detectors or trap detectors using 10 mm diameter detectors. All detectors were windowless. The wavelengths used spanned the near-IR range: 960, 1064, 1311, and 1550 nm.
- Calibration of 14BT-type thermopiles using a cryogenic radiometer at 1.318 μm (Nd:YAG laser) at NPL (National Physical Laboratory, Teddington, Middlesex, UK ) [48].
- Calibration of lithium tantalate pyroelectric detectors at 10.6 μm using a $CO_2$ laser and cryogenic radiometer at NIST (National Institute of Standards and Technology, Gaithersburg, MD, USA) [23]. The 1 cm × 1 cm pyroelectric detectors are coated with black paint. The uniformity of the detectors is not very good, and this is probably the major limitation to the overall uncertainty of this calibration (0.48% standard uncertainty).
- Calibration of "sphere radiometers" at 10.6 μm using a $CO_2$ laser and cryogenic radiometer at NPL [26]. The sphere radiometers developed specially for this application incorporate a liquid-nitrogen-cooled HgCdTe detector and a 5 cm diameter integrating sphere. These radiometers have good uniformity, large area, are inherently fast and are effectively windowless (no coherence effects).
- Calibration of sphere radiometers and an ECPR (electrically calibrated pyroelectric radiometer) using laser diodes and a cryogenic radiometer at IFA (Instituto de Fisica Aplicada, Spain) [25, 49]. The sphere radiometer is an assembly consisting of a 3 mm InGaAs detector in conjunction with a 50 mm diameter Spectralon sphere, and the ECPR is a Laser Precision RS5900. The laser wavelengths used are 1300, 1480, and 1550 nm.
- Calibration of cooled germanium detectors and an ECPR at 1550 nm using a laser diode and a cryogenic radiometer, at the BNM (Bureau National de Métrologie, France) [13]. The germanium detectors have 5 mm diameter and are cooled to −5°C or −20°C. The overall standard uncertainty is quoted as ±0.5%. This relatively large uncertainty is attributed to poor beam quality coupled to the fact that the transfer radiometer and cryogenic radiometer apertures are not in the same plane and do not have the same diameter.

Figure 3.5 compares the spectral responsivity curves of the principal types of detectors used as transfer standards for the realization of spectral

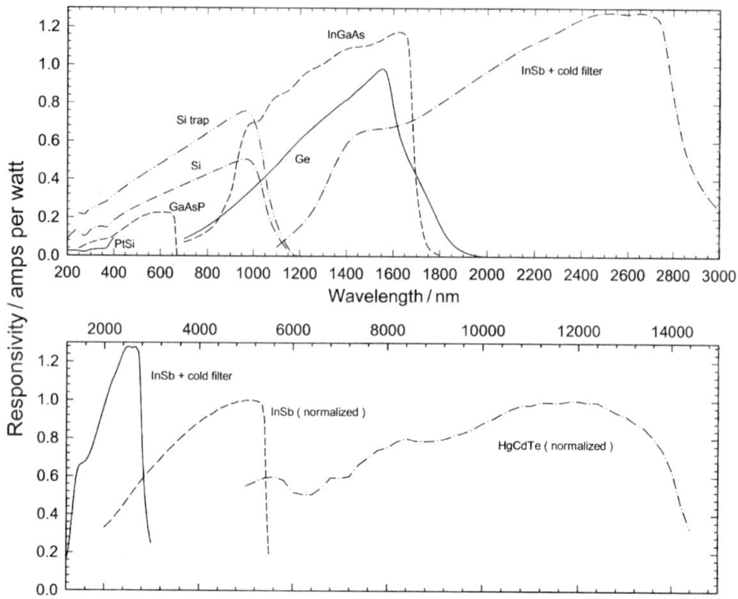


FIG. 3.5. Spectral responsivity curves of the principal types of detectors used as transfer standards for cryogenic radiometer-based calibrations.

responsivity scales from the UV to the far-IR. Not shown on this graph are thermopiles, bolometers and pyroelectric detectors, which would span the whole spectral range from 200 nm to at least 15,000 nm. Below 200 nm, PtSi detectors and special types of silicon photodiodes can be used.

### *3.2.1.4 Interpolation and extrapolation to other wavelengths*

In the above sections, we have discussed the direct calibration of transfer radiometers using a cryogenic radiometer and laser-based apparatus. In the realization of a spectral responsivity scale in any spectral region, it is necessary to calibrate the radiometers over the full spectral range, without gaps. This is impossible to do with laser-based calibrations, which are carried out at discrete wavelengths. To obtain full spectral coverage, it is necessary to interpolate between the laser wavelengths or to extrapolate the scale beyond the minimum or maximum wavelengths of the lasers. In this section, we discuss techniques for doing this. Interpolation or extrapolation can be achieved by means of mathematical techniques or by means of auxiliary measurements using spectrally neutral thermal detectors.

*3.2.1.4.1 Auxiliary measurements.* The most common approach to extend spectral responsivity scales to wavelengths other than those used in the laser-based calibrations is to generate a scale of relative spectral responsivity using monochromator-based apparatus and spectrally neutral detectors as reference detectors. Figure. 3.6 shows a typical configuration. The monochromator can be double or single (as shown); double monochromators will give higher spectral purity, but with lower throughput. Both grating and prism instruments are used. The filter wheel shown is used with grating instruments to allow the insertion of order-sorting filters to eliminate stray light due to higher-order radiation. The source used will depend on the spectral range; tungsten–halogen sources are best for the range 400–2500 nm. In the UV, the most common sources are xenon arcs and argon mini-arcs; xenon lamps are preferable to mercury vapour lamps because their spectral distribution is more continuous, with few discharge lines. Deuterium lamps, on the other hand, have a relatively weak output, which can be a problem when using thermal detectors. In the IR, beyond ~2–3 μm, sources used include argon maxi-arcs [50] and Oppermann sources [51]. The exit slit of the monochromator is usually replaced by a circular aperture, which is imaged onto the detectors by means of an optical system. The output optics usually consists of mirror systems in order to allow operation over large spectral ranges without de-focusing effects and to minimize inter-reflection effects which lens systems introduce. The mirror systems usually have to be custom-designed to reduce the aberrations that would result from operating at fairly large off-axis angles due to the

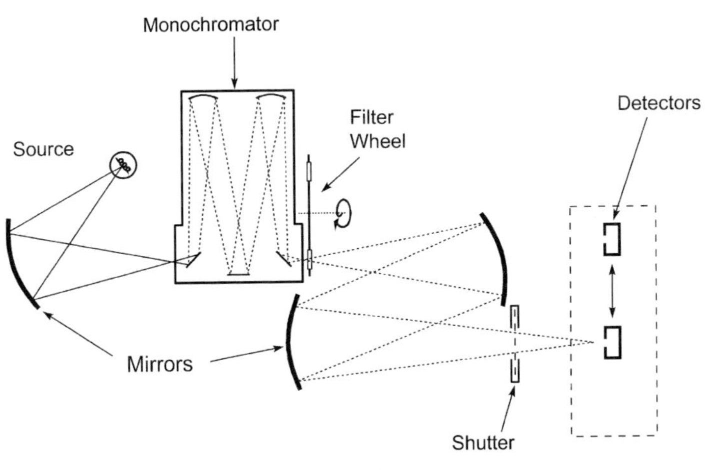


Fig. 3.6. Schematic diagram of a typical monochromator set-up for detector calibrations.

high-throughput monochromators (e.g. $f/4$) normally used for this application. Some of the types of mirror systems used include spherical–torroidal combination [52], spherical–spherical combination [50], Cassegrain systems and paraboloid–paraboloid combination [51]. The transfer standard(s) and reference detector are intercompared directly, wavelength by wavelength, by means of a linear or rotation stage.

The reference detector is a thermal detector whose spectral responsivity is nearly constant over a large spectral range. Suitable thermal detectors that have been used include thermopiles [50], pyroelectric detectors [51] and bolometers. In some cases, the reference thermal detector has been calibrated directly at a laser wavelength using a cryogenic radiometer or a trap detector, so that the calibration of the transfer standard by the reference detector gives directly the absolute spectral responsivity values. In other cases, the calibration with the reference detector gives only a relative spectral responsivity scale. It is necessary to perform an auxiliary absolute calibration of the transfer standard at one or a few wavelengths using another transfer standard, or directly using a cryogenic radiometer, to obtain the absolute full spectral scale. In either case, it is very important to ensure the spectral flatness of the reference detector used. In practice, this means measuring the spectral variation of the absorber reflectance and window transmittance of the reference detector used and applying a correction. An example is shown in Figure 3.7. Thin-film thermopiles were used at NRC [53] to extend the calibration of InSb working standards from 1500 to 3000 nm. At 1500 nm, the relative calibration with the thermopiles was spliced to an absolute calibration of the InSb using Ge working standards. A correction was applied to the thermopile spectral responsivity. The absorptance of the black coating and the sapphire window transmittance were measured spectrally and the product of these, normalized to unity at the splice wavelength of 1500 nm, was applied as a correction. Figure 3.7 shows that this correction is quite small.

Pyroelectric detectors have some advantages over thermopiles. They have higher NEP and are not subject to drift and background thermal effects. In order to reduce spectral effects, pyroelectric detectors are sometimes used as cavity detectors [51, 54]. Figure 3.8 shows the configuration of one such pyroelectric radiometer used at the NPL, incorporating a hemispherical reflector to reduce spectral non-blackness to about $\pm 0.25\%$ for the spectral range from 1 to 20 μm. Another potential problem in using thermal detectors is nonlinearity effects. The spectral throughput of monochromator systems varies considerably over a large spectral range. For double monochromators, the dynamic range can be even greater. Thus, it is important to verify the linearity of the thermal detectors used as reference; thermal detectors do not have as high a dynamic range as good-quality photovoltaic

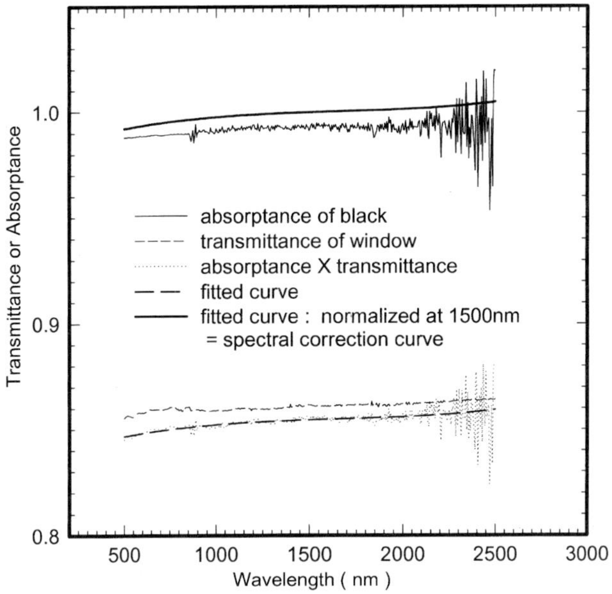


FIG. 3.7. Derivation of a correction for spectral selectivity of a thermopile from measurements of reflectance of absorber and transmittance of window. (Source: from Reference [53].)

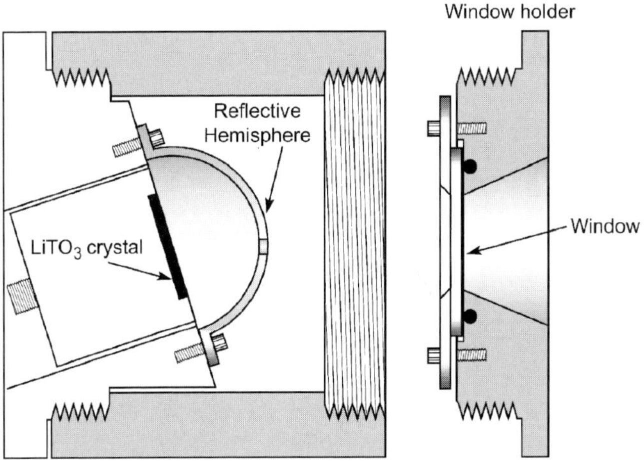


FIG 3.8. Configuration of a cavity pyroelectric detector used at NPL. (Source: Theo Theocharous, National Physical Laboratory, UK.)

detectors. Reference [55] gives a review of various techniques for verifying the linearity of detectors. Monochromator-based systems have other sources of uncertainty that must be considered: wavelength errors, bandwidth effects, stray light, and geometrical effects (non-collimated radiation). These effects will be discussed below.

*3.2.1.4.2 Mathematical interpolation.* For the spectral range from approximately 400 to 1000 nm, and for transfer radiometers incorporating silicon photodiodes, various interpolation formulae have been developed to obtain spectral responsivity values at wavelengths other than the laser wavelengths. These techniques are based on detailed knowledge of the physics of silicon photodiodes. The interpolation formulae are fitted to the discrete measurement points. The brief treatment below follows the treatment used by Werner et al. [28] at PTB, which itself is based on a model developed by Gentile et al. [17] at NIST, in turn based on work carried out by Geist and several coworkers and described in several papers. The paper by Geist and Baltes [56] gives a detailed treatment on this subject and a useful bibliography.

From Eq. (3.1) given above, developing a mathematical model for $S(\lambda)$ involves modelling the external quantum efficiency $\eta_e(\lambda)$, which requires separate modelling of the diode reflectance $\rho(\lambda)$ and the internal quantum efficiency $\eta_i(\lambda)$. The reflectance is relatively easy to model and fit for S1337-based detectors, since it has a monotonic variation in the range 400–1000 nm. Werner et al. have used a function of the type

$$\rho(\lambda) = a \exp\left(\frac{b}{\lambda}\right) + c\lambda + d \tag{3.3}$$

where $a$, $b$, $c$, $d$ are fitting parameters. The model for $\eta_i(\lambda)$ used by Werner et al. is similar to that used by Gentile et al., except that it has an additional term to account for backreflection from the rear surface of the diode at long wavelengths:

$$\begin{aligned}\eta_i(\lambda) = P_f + \frac{(1-P_f)}{\alpha(\lambda)T}\{1-\exp[-\alpha(\lambda)T]\} - \frac{(1-P_b)}{\alpha(\lambda)(D-T)}\{\exp[-\alpha(\lambda)T]\\ -\exp[-\alpha(\lambda)D]\} - P_b\exp[-\alpha(\lambda)h] + R_{back}\exp[-\alpha(\lambda)h]P_b\end{aligned} \tag{3.4}$$

The fitting parameters and their physical interpretation are: $P_f$, the collection efficiency at the $SiO_2$/Si interface; $T$, depth of the pn junction; $P_b$, the bulk value of collection efficiency; $D$, depth at which $P_b$ is reached; $R_{back}$, the reflection coefficient of the rear surface. Non-fitted parameters are the total thickness $h$ of the diode, and $\alpha(\lambda)$, the Si absorption coefficient. The values of $\alpha(\lambda)$ can be obtained from the literature [57]. Werner et al. have used an interpolation function for $\alpha(\lambda)$ obtained by fitting to literature values the following expression for $\alpha(\lambda)$:

$$\alpha(\lambda) = A_1 \exp[A_2/(\lambda - \lambda_0)] + A_3\lambda + A_4/\lambda + A_5 \quad (3.5)$$

where the $A_x$ are fitting parameters.

The interpolation uncertainty when using the above technique appears to be approximately 0.01% ($k = 1$). Some work has been carried out to extend these interpolation techniques to the ultraviolet region [58–60]. The situation is much more difficult in that spectral range. Some workers [60] have used an interpolation formula similar to Eq. (3.4) above, in which $\eta_i(\lambda)$ is replaced by $\eta_i'(\lambda) = q(\lambda)\eta_i(\lambda)$, where $q(\lambda)$ is the quantum yield of silicon, and have developed interpolation formulae for $q(\lambda)$. Interpolating the reflectance $\rho(\lambda)$ is also much more difficult in the UV because of the complexity and large values of the reflectance in that spectral region (Fig. 3.4). As a result, interpolation of spectral responsivity values with an uncertainty better than 0.5% ($k = 1$) is difficult to achieve in the UV, and interpolation by means of auxiliary measurements as described in the previous section is probably preferable.

### *3.2.1.5 Sources of uncertainty*

A full discussion on analysis of uncertainties is beyond the scope of this section; this topic is covered in detail in Chapter 6. Here, we will present a brief introduction to uncertainty analysis and discuss the major uncertainty components in the realization of spectral responsivity scales using laser-based methods. Additional uncertainties incurred during scale extensions by means of auxiliary measurements will be discussed only briefly since they are essentially the same as for monochromator-based scale realizations which will be discussed in the next section. Uncertainties are classified as being either type A or type B, depending on how they are evaluated. Simply put, for type A uncertainties, the method of evaluation is by statistical analysis of data. For type B uncertainty, the method of evaluation is by means other than statistical analysis. In many cases, random effects (such as electrical noise) will be type A uncertainties, whereas systematic effects (such as a monochromator wavelength error) will be type B uncertainties. The total uncertainty in a measurement will be calculated from several component uncertainties, some of type A, some of type B. Some of these components will be wavelength dependent, some not. If the various uncertainty components are uncorrelated, then the total uncertainty is calculated as the quadrature sum of the individual components. This is usually the assumption made in this kind of measurements. If there are $N$ uncertainty components, designated by $\delta_i$ with $i = 1$ to $N$, then the total uncertainty $\bar{\delta}$ is given by

$$\bar{\delta} = \sqrt{\sum_{1}^{N} \delta_i^2}$$

The magnitude of these uncertainty components can be expressed in absolute or relative terms. If there are correlations between some of the uncertainty components, then the calculation of total uncertainties is more complicated [12]. The total uncertainty can be expressed as a standard uncertainty or an expanded uncertainty. This relates to the confidence level in the statement of uncertainty. With the standard uncertainty, there is a probability of approximately 67% that the "true" value will be within the bounds of the specified uncertainty. With the expanded uncertainty, the confidence level is increased, depending on the coverage factor. With a coverage factor $k = 2$, the probability is increased to approximately 95%. Expanded uncertainties are obtained from the standard uncertainties by multiplying by 2. Another consideration involved in uncertainty analysis is the type of probability distribution assumed. Most workers assume a normal (Gaussian) distribution. In the discussions in this chapter, we shall not address specifically the type of distribution, but only the type of uncertainty (A or B). Most workers have reported standard uncertainties ($k = 1$) and we shall assume this in the discussion below. The major sources of uncertainty in laser-based measurements are the following:

- Cryogenic radiometer effects. These include:
  - The measurement of electrical power (B). Determined by the calibration accuracy of DVMs and standard resistors used to measure the electrical substitution power; most workers are reporting uncertainty values in the range 0.001–0.003%.
  - Cavity absorptance (B). For the UV-VIS-near-IR, uncertainty values quoted are in the range 0.001–0.003%, and in the far-IR (10 μm), values quoted vary from 0.01–0.02%.
  - Non-equivalence effects (B). This uncertainty refers to the slightly different response of the cryogenic radiometer to equal optical or electrical power inputs; most cryogenic radiometer cavities incorporate either a second temperature sensor or a second heating element at a different location in the cavity; comparison of measurements made with the two sensors or heaters allows an estimate of non-equivalence effects to be made; the reported values vary between 0.001 and 0.005%.
  - Repeatability(A). The repeatability of a measurement of optical power by a cryogenic radiometer is dependent on several factors: the sensitivity (NEP) and drift of the cryogenic radiometer, the stability and power

level of the laser. With laser-based measurements at power levels of 0.5–1 mW, NEP effects are ~0.001%. Quoted values for repeatability are in the range 0.003–0.03% and are thus limited by wavelength-dependent laser power and stability.

- Measurement of Brewster window transmittance (B). This is one of the most important source of uncertainty in laser-based measurements. Brewster windows must be adjusted carefully to minimize the reflection loss. Their transmittance is measured in a separate operation, which involves venting and separating the window assembly from the cryogenic radiometer, while maintaining the same conditions that exist during actual radiometric measurements. Figure 3.2 shows a Brewster assembly used at PTB [18] that allows in situ measurement of the window transmittance. Measurements often take several days, during which surface contamination of the window must be avoided. Special care must be taken when using auxiliary beam-steering mirrors during radiometric measurements but not during transmittance measurements. This situation occurs usually when using vertical-axis cryogenic radiometers. In such cases, the mirror must be carefully aligned so as to not alter the state of polarization of the radiation, on which the window transmittance depends [21]. The uncertainty in the transmittance measurement can be estimated by comparing measurements carried before and after the cryogenic radiometer measurements, and by comparing measurements made using slightly different beam or measurement geometries [21, 27], or at different positions on the window. Uncertainties are obviously wavelength-dependent, and quoted values [8–10, 17, 18, 21, 23–30] range from about 0.003% in the VIS, to 0.03% in the UV and IR and 0.1 to 0.2% in the farIR [23, 26].
- Scattered radiation (B). This is also an important source of uncertainty in laser-based measurements. The small diameter laser beam (1–2 mm waist diameter) entering the radiometers is surrounded by a halo of scattered radiation. This scattering is caused by imperfections in the optical elements, edges of apertures used for beam shaping, inter-reflections in optical components, and dust on the optical elements or in the air. Although it is not scattered radiation, we also include in this category Gaussian beam truncation effects. The distribution of scattered radiation in the halo will vary along the optical axis. If the cryogenic radiometer aperture and the transfer radiometer aperture do not have the same diameter and are not in the same plane, they will not measure the same amount of scattered radiation, resulting in error. Scattered radiation from the Brewster window can also cause error. Some cryogenic radiometers include quadrant detectors [14, 21, 29] close to the cavity, which allow the measurement of scattered radiation in order to apply a correction or estimate the uncertainty associated with scattered radiation. Corrections for scattered light

(or uncertainty estimates) can be determined by measuring the variation along the optical axis of the signal of a detector with a fixed aperture (referred to by some as the "distance effect") and by measuring at various positions along the axis the signal variation when the aperture diameter varies. Scattering effects are very wavelength-dependent. The best approach to minimizing scattered light effects is to use a measurement set-up where the cryogenic radiometer and transfer radiometers have apertures of the same diameter located in the same plane orthogonal to the optical axis (see Reference [18], for example). Estimated residual (i.e. after corrections) uncertainties for scattered radiation vary from about 0.002 to 0.005% in the range 400–900 nm to ~0.03% in the UV and near-IR to 0.1% in the far-IR (10 μm) [8–10, 17, 18, 21, 23–30].

- Transfer standard effects:
  - Linearity (B). This can be of some concern in laser-based measurements, where relatively large power levels (0.5–1 mW) and small beam sizes (0.5–2 mm) are used. As discussed above, S1337-type detectors and traps are susceptible to linearity problems [36–39]. For some types of detectors, non-linearity effects can be wavelength-dependent [36], geometry-dependent (i.e. overfill vs. underfill [36, 61, 62]), or even temperature-dependent [36, 63]. For example, some types of InGaAs detectors show nonlinear behaviour at all wavelengths in the overfill mode, and even in the underfill mode below 1000 nm [61]. Large area Ge detectors will exhibit progressive decrease in responsivity above 1 mW due to series resistance effects [63]. Although some workers apply corrections for non-linear behaviour, it is preferable to avoid this necessity by using detectors under conditions that ensure linear behaviour. Many methods can be used for determining linearity corrections or estimating uncertainties due to residual nonlinearity effects [55]. Typical uncertainties quoted for nonlinearity effects are 0.001–0.003% in the range 400–900 nm, increasing to 0.01% in the UV, for Si-based radiometers; in the near-IR, 0.01% for Ge-based radiometers [63] and 0.001% [29] to 0.03% [25] for InGaAs-based radiometers; and <0.01% in the far-IR for HgCdTe [26].
  - Non-uniformity (A or B). The effect of detector non-uniformity on measurement uncertainty can be estimated by determining the change in responsivity when the beam is displaced by a small amount (±0.5 mm typically) in several directions on the detector surface [11, 26, 48]; or by determining the change in responsivity when beams of different sizes are used; or a combination of these two. Sometimes non-uniformity effects are included with measurement repeatability. The lowest uncertainties are seen with silicon trap detectors in the range 500–900 nm, with values of ±0.002–0.003%, increasing to 0.03% in the UV at 250 nm. For single

Si diodes, in the best cases, uncertainties range from about 0.006 to 0.06%. In the near-IR, available Ge and InGaAs detectors are not nearly as uniform as Si detectors. The use of trap configurations can improve the uniformity [63, 64]. Alternatively, by using radiometer configurations incorporating detectors with integrating spheres, it is possible to significantly reduce uncertainties due to non-uniformity [25, 26, 48, 61].

- Temperature effects (A). Since the spectral responsivity of any detector is temperature dependent to some extent, it is important to record the temperature of a radiometer during calibration.

Figure 3.9 shows the temperature coefficients of responsivity for typical photovoltaic detectors, as well as the corresponding spectral responsivity curves. Temperature sensitivity is relatively small over most of the spectral ranges of the detectors but increases rapidly in the bandgap region. Although it is possible to directly control the radiometer temperature using circulating water jackets or thermoelectric-type elements [65], this does not appear to be the most common situation. Usually, a temperature sensor element such as an RTD is incorporated in the radiometer and the range of temperatures of the radiometer

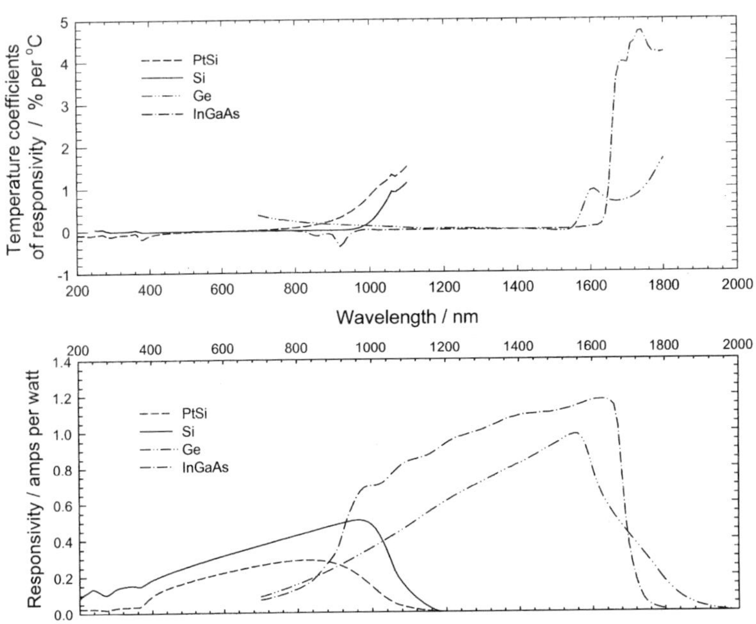


FIG. 3.9. Temperature coefficients of responsivity of some typical photovoltaic detectors; the corresponding spectral responsivity curves are also shown.

during calibration is recorded. Using these data (including the uncertainty of the temperature sensor) and the known temperature coefficients of the type of detector used (as shown in Fig. 3.9) an uncertainty due to temperature sensitivity can be derived, and in some cases a temperature correction applied. For S1337 Si diodes in the range 300–900 nm, or good-quality InGaAs detectors in the range 1000–1600 nm, temperature coefficients are <0.02%/°C, so that temperature corrections and residual temperature uncertainties are small. For Si traps, temperature effects are even smaller; measurements done at PTB [66] and BIPM (Bureau International des Poids et Mesures, Paris) and shown in Figure 3.10 show that Si traps have temperature coefficients <0.005%/°C throughout the range 300–900 nm. Typical uncertainties quoted for temperature effects are: 0.02–0.001% for S1337 detectors in the range 250–800 nm [11, 21, 66]; 0.001% for S1337 traps in the range 476–676 nm (Kr laser) [21]; 0.009% for HgCdTe-based radiometers at 10 μm [26].

○ Polarization sensitivity (B). The responsivity of photovoltaic detectors varies with the angle of incidence and state of polarization of the radiation. For single detectors used at normal incidence with quasi-collimated laser radiation, polarization effects are negligible. In many cases, detectors are tilted slightly in order to avoid inter-reflection problems. In such cases, there is a small but not negligible angular and polarization dependence. Figure 3.11 shows the ratio of responsivity at

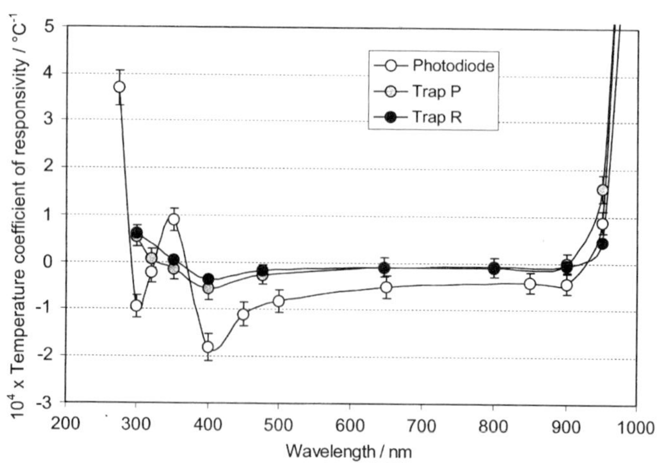


Fig. 3.10. Temperature coefficient of a silicon trap detector, compared to that of a single detector. (Source: R. Goebel and M. Stock, BIPM, *Report on the key comparison CCPR-K2.b of spectral responsivity measurements in the wavelength range 300 nm to 1000 nm*, with permission from Metrologia.)

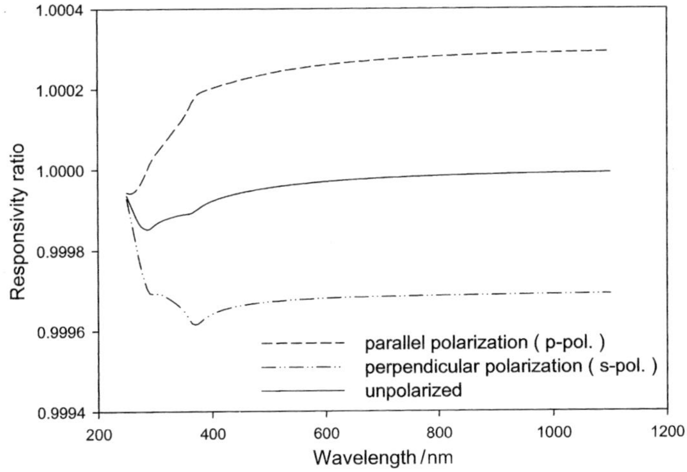


FIG. 3.11. Ratio of responsivity at 2° incidence to that at normal incidence for an Hamamatsu S1337 silicon photodiode.

2° incidence to that at normal incidence for s-polarized p-polarized and unpolarized radiation, for a windowless S1337 (29 nm oxide assumed).

The magnitude of the effect varies between −0.04% and +0.02% for s- and p-polarizations respectively. For S1337 diodes, and many types of Si diodes, corrections can be calculated for operation at small tilt angles [67], with residual uncertainties of at most 0.003%. For other types of detectors, corrections would have to be determined (or uncertainties estimated) by measuring the variation in detector signal as a function of angle of incidence and azimuth angle. As mentioned above, traps can exhibit some polarization sensitivity due to small fabrication defects or detectors having small differences in properties. The associated uncertainty can be estimated by measuring the response variation as the trap is rotated about an axis defined by the incident radiation [68].

- Repeatability (A). This addresses the day-to-day repeatability of calibrations, including removal, remounting and realignment of the radiometers. Many factors contribute to this uncertainty component, some of which are also considered separately: stability of the laser source; detector non-uniformity in conjunction with small alignment differences; detector noise and drift; temperature differences. Obviously, repeatability will be wavelength dependent. Some values quoted are: 0.005% in the VIS using traps [21]; 0.02–0.03% in the near-IR using InGaAs-based radiometers [25, 29]; 0.02% using HgCdTe-based radiometers [26] to 0.3% using pyroelectric detectors [23] in the far-IR (10 μm).

- Stability (A). This addresses long-term changes in the responsivity due to natural ageing effects, fatigue effects due to irradiation (mostly in UV), surface contamination effects and gradual accumulation of water vapour in dewars used in conjunction with liquid-nitrogen-cooled detectors. Changes in spectral responsivity of detectors due to UV irradiation has been investigated in some detail [34, 35]. In certain cases, the long-term stability of detectors to UV irradiation can be improved by pre-ageing the detectors using UV irradiation [34]. Surface contamination effects are usually reversible to some degree, by cleaning the detectors [33]. Apparent ageing effects in narrow spectral ranges due to water vapour absorption inside dewars have been reported [69] for liquid-nitrogen-cooled InSb and HgCdTe detectors; these effects are usually reversible by re-evacuation of the dewars. Most workers do not include in the calibration uncertainty budget a component associated with detector stability since it is not relevant to the calibration per se, and to do so requires detailed calibration histories of the detectors. However, in some cases it is appropriate to do so, where detectors are being calibrated as part of an intercomparison involving many calibrations over a fairly long period of time. Some values quoted are: 0.005% for Si traps in the range 476–676 nm (Kr laser), presumably over a time frame of 1–2 years [21]; 1 month stability values of 0.02% in the range 476–800 nm, and 0.23% at 257 nm, for S1337 Si diodes [11].
- Electrical (B). Included here are uncertainties associated with the calibration of DVM's and amplifiers used to make measurements of the response of transfer radiometers. Type A uncertainties are usually negligible compared to type B effects. For DC measurements using operational amplifiers, uncertainties are quite small. Values between 0.001 and 0.015% are quoted, from the UV to the near-IR. In the IR and far-IR, where pyroelectric, InSb or HgCdTe are used, AC techniques are often used, using choppers and lock-in detection. In such cases, uncertainties can be somewhat higher; values quoted range between ~0.02% [48] and 0.3% [23].

Overall uncertainties are usually obtained by the quadrature sum of the uncertainty components discussed above. Typical overall uncertainties quoted for laser-based realization of spectral responsivity scales using cryogenic radiometers are as follows:

- Using S1337 traps in the range 400–950 nm, values quoted range from a low of 0.007% [27] to a high of 0.03% [17], typical values being around 0.01% [21]. In the UV range 257–360 nm, values quoted are in the range 0.02% [48] to 0.1% [28];

- Using single Si detectors (usually S1337), some values quoted are 0.25% at 257 nm [11], 0.02% [11] or 0.1% [70] in the range 400–800 nm;
- In the near-IR (1000–1550 nm), using InGaAs single detectors, values quoted are in the range 0.05–0.15% [25, 29, 48];
- In the IR at 10 μm, some values quoted are 0.13% using HgCdTe-based radiometers [26] and 0.48% using pyroelectric detectors [23].

#### *3.2.1.6 Interpolation/extrapolation uncertainties*

The above uncertainties apply to the calibration of transfer radiometers at discrete laser wavelengths. The interpolation and extrapolation to other wavelengths using mathematical techniques or auxiliary measurements in order to obtain continuous spectral responsivity scales will introduce additional uncertainty components:

- Mathematical interpolation (B). This technique has been discussed briefly above, and has been used mostly in the spectral range from about 400 to 1015 nm. Interpolation uncertainty values quoted are 0.03% from 406 to 920 nm [17] for traps; better than 0.01% from 400 to 950 nm increasing to ~0.05% at 1015 nm, for traps [28]; 0.05% for traps and single detectors in the range 450–900 nm [NPL, unpublished].
- Extrapolation using auxiliary measurements. This involves the measurement of the relative spectral responsivity of the transfer radiometers using thermal detectors in conjunction with monochromator-based apparatus. The relative responsivity data is then spliced to the absolute calibration at one of the laser-measured points, or the thermal reference detector is itself calibrated in absolute terms at one or more of the laser wavelengths, either directly using the cryogenic radiometer or using a transfer radiometer (usually a trap) itself calibrated using the cryogenic radiometer. The types of reference detectors used include thermopiles, bolometers and pyroelectric detectors. The principal additional uncertainty components associated with such scale extensions are as follows:
  - Monochromator wavelength uncertainty (B). The responsivity of photovoltaic detectors varies strongly with wavelength (Fig. 3.5).

    The responsivity uncertainty corresponding to a monochromator wavelength uncertainty can be obtained directly from the slopes of the spectral responsivity curves. Figure 3.12 shows the percentage responsivity variation per nanometre for various types of detectors. In the bandgap region, wavelength sensitivity can be quite large. Also, most detectors used in the UV-spectral range exhibit strong wavelength dependence, which can equal or exceed 1–2%/nm in at least part of

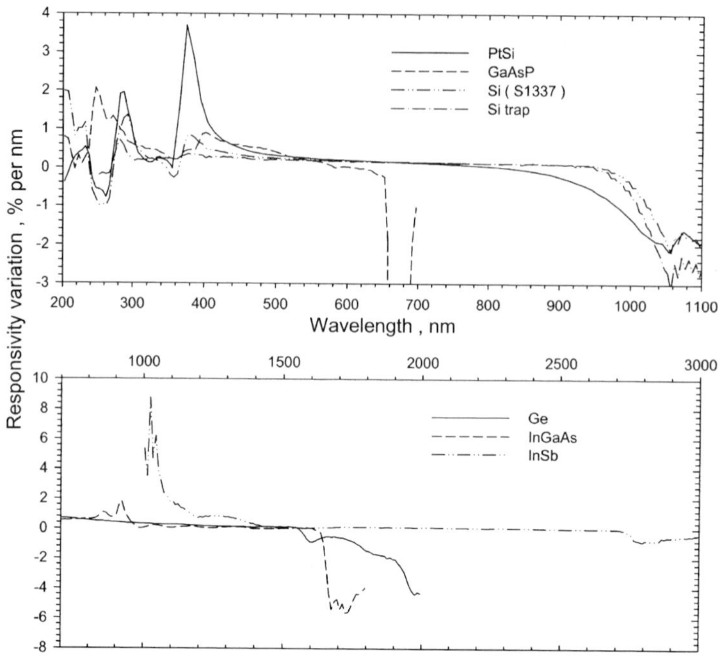


FIG. 3.12. Wavelength effects: % responsivity variation per nanometer for typical detectors.

the spectral range. Wavelength effects can dominate the uncertainty budgets for the realization of a relative scale in the UV. NPL [48] reports uncertainties varying from 0.5 to 0.1% due to wavelength effects for the realization of a scale of relative spectral responsivity in the range 200–400 nm.

- Monochromator bandwidth (B). The uncertainty caused by the finite spectral bandpass of the monochromator is not as straightforward to estimate as that due to wavelength effects. Several papers have described different approaches for doing these calculations [71, 72]. Figure 3.13 shows calculated bandwidth errors corresponding to several types of detectors, for a 5 nm bandwidth FWHM (full-width at half-maximum).

  It can be shown [72] that bandwidth effects are negligible where the spectral responsivity curve is quasi-linear. Also, in general, bandwidth effects vary approximately as the square of the bandwidth [72]. The relative spectral distribution of the monochromator output influences bandwidth effects, and can cause non-negligible bandwidth errors even in a spectral region where the inherent bandwidth effects for a detector would be small (due to quasi-linear responsivity variation, for example).

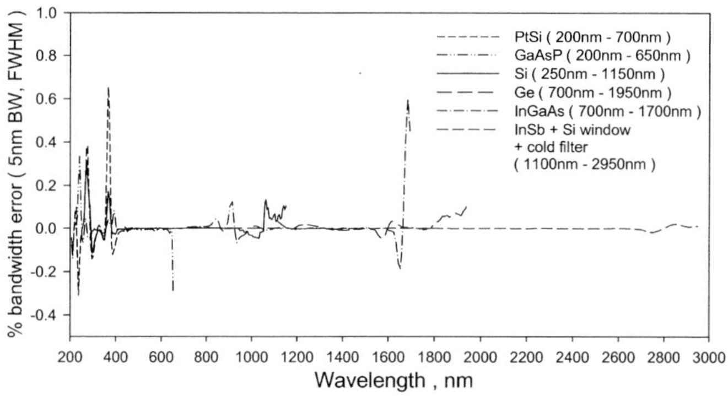


FIG. 3.13. Bandwidth errors corresponding to a bandwidth of 5 nm FWHM for various types of detectors.

Calculations such as those shown in Figure 3.13 can be used to derive corrections or estimate uncertainties for bandwidth effects. Some workers include bandwidth effects with wavelength effects in their uncertainty budgets.

- Monochromator stray light (B). This refers to out-of-band radiation originating inside the monochromator due to grating defects, scattering, multiple diffraction. Higher-order diffracted radiation associated with grating instruments is usually not included since in most cases it can be completely eliminated using so-called order sorting filters. Same-wavelength scattered and inter-reflected radiation is also not included. It is usually possible to make the uncertainty due to scattered light negligible by careful design of the measurement set-up. Stray light is usually negligible when using double monochromators. It is not always possible to use double monochromators because of signal-to-noise ratio (SNR) considerations; monochromators have low throughput and thermal detectors have relatively large NEP. The situation is particularly difficult in the UV, where stray light tends to be higher than in other spectral regions and where some of the detectors used (Si, PtSi) strongly bias long wave stray light. Figure 3.14 shows an example of monochromator stray light: the single monochromator has a UV grating and a Xenon source; the detector is a Si diode (S1337). The stray light varies from about 15% at 200 nm to <0.2% at 350 nm. This figure illustrates the use of sharp cut filters to measure stray light in order to apply corrections and to estimate uncertainties due to stray light. Stray light uncertainties are usually small (0.02% [48] to 0.1% [73] in the near-IR), but can be appreciable in the UV (e.g. 0.6% at 200 nm [48]).

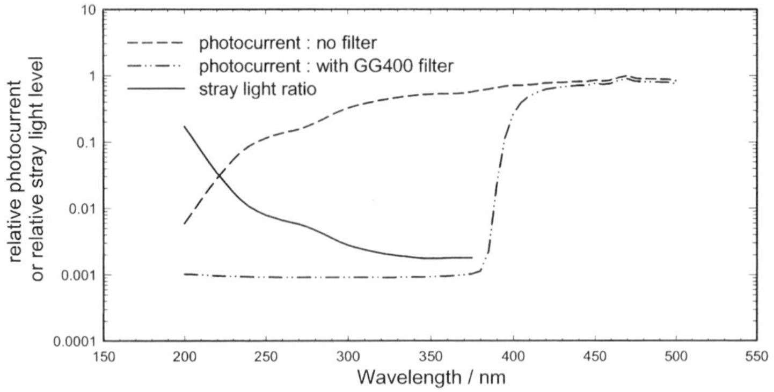


FIG. 3.14. Relative stray light for a single monochromator, for a xenon source and a silicon photodiode.

- Monochromator beam geometry (B). The radiation emerging from a monochromator is not collimated, and the beam focused on the test detector has a finite angular spread, usually defined in terms of its *f*-number. Also, the radiation usually has a significant degree of polarization. We denote by *effective responsivity* the ratio of the responsivity of the detector to this converging (or diverging) beam to the responsivity for the same optical power in a collimated beam at normal incidence. In the usual case, spectral responsivity scales are defined with respect to normal incidence. The departure of the effective responsivity from unity must then be corrected for, or more commonly, treated as an uncertainty. Figure 3.15 shows the effective responsivity for a Si S1337 diodes at several wavelengths, for unpolarized radiation, as a function of the semi-angular spread of a uniform beam whose axis is normal to the detector. For an *f*/8 beam, the beam geometry effect is no more than 0.01% throughout the range 250–1000 nm, but for an *f*/4 beam, the effect can be as large as 0.06%. If the distribution of radiation in the beam is rotationally symmetric about the axis, it is clear that the effective responsivity will be the same whatever the state of polarization of the radiation.
- Reference detector spectral selectivity (B). Even if a correction is applied for the spectral selectivity of absorbers and windows of the thermal detector used as reference, there is a residual uncertainty, which can be estimated from the measurements of window transmittance and absorber reflectance. For example, for the realization of relative responsivity scales using cavity pyroelectric detectors as reference (Fig. 3.8), NPL [48] quotes uncertainties of 0.06% for absorber non-blackness (200–2500 nm) and 0.2–0.04% in the range 200–1640 nm for window transmittance. At NRC, the uncertainties in the correction

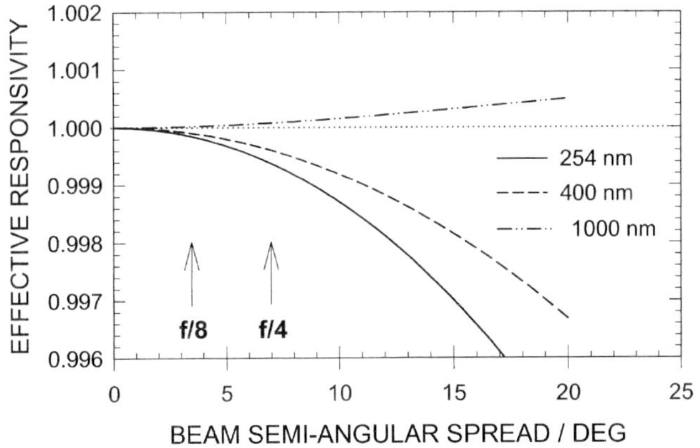


FIG. 3.15. Effective responsivity of an Hamamatsu S1337 diode, for unpolarized incident radiation, as a function of *f*-number, for several wavelengths.

factors for absorber reflectance and window transmittance (Fig. 3.7) of the thermopiles used for InSb scale extension varies from 0.08% at 1500 nm to 1% at 3000 nm.

- Detector linearity (reference and test) (B). At the power levels used in monochromator-based measurements (a few microwatts), non-linearity is not a problem with most photovoltaics. However, some InGaAs detectors exhibit non-linearities below 1000 nm in the underfill or overfill mode [61], and above 1000 nm in the overfill mode [61, 62], even at relatively low photocurrents. Large area Ge detectors can also show small non-linearity effects, viz. 0.05% below 0.1 mA [50]. The use of AC lock-in techniques in conjunction with pyroelectric reference detectors can result in linearity-related uncertainty; NPL quotes a value of 0.06% [48].
- Detector uniformity (reference and test) (A or B). This uncertainty will be determined mostly by non-uniformity in the test radiometer, if the non-uniformity of the reference detector does not vary very much with wavelength. This will be the case where the absorber is carbon black or a black paint, but may not be the case with metal blacks such as gold-black, unless used in a cavity configuration [51, 54]. Uncertainties due to non-uniformities are often not given explicitly, but are included in repeatability and alignment uncertainties.
- Detector temperature variation (reference and test) (A). Uncertainties for the transfer radiometers are as above for the laser-based measurements. The temperature coefficients of the reference thermal detectors are generally larger than those of photovoltaics (except in bandgap

regions), but the resultant uncertainties are very small: for relative measurements, the only temperature variation of the reference detector that matters is the variation during any spectral scan, not day-to-day variation. Temperature effects are not usually listed explicitly but are often included in the repeatability component.

- Repeatability (A). This catch-all uncertainty component includes many sources that contribute to day-to-day variations in the measurement results: source drift and noise; detector noise; electronics noise and drift; detector uniformity; alignment; temperature variations. Some of the factors affecting repeatability are sometimes treated separately. Repeatability tends to be the worst in the UV. Some values quoted: 1.5% at 200 nm, and between 0.03% and 0.06% from the near-UV to the IR (>200–2500 nm) [48].
- Electrical effects (A or B). This includes uncertainties associated with voltmeter and amplifier calibrations; typical values quoted vary between 0.01% and 0.05%. The use of AC lock-in techniques can also add other uncertainty components, such as that due to chopper stability (e.g. 0.03–0.06% [48]), but some of these can be included with repeatability.

The overall uncertainties of the full spectral scales obtained from the laser-based scales by interpolation (mathematical mostly) and extrapolation (auxiliary measurements mostly) varies very much from laboratory to laboratory, depending on the techniques used. In the range 400 to ~1000 nm, where Si detectors and traps are used, the use of mathematical modelling adds uncertainties of at most 0.05% to the laser-based scales, which have typical overall uncertainties of 0.01%. In the UV, near-IR and FIR, it appears that the combination of auxiliary relative measurements and their concomitant uncertainties with the laser-based measurements results in scales with overall uncertainties of ~0.2% in the best cases.

### 3.2.2 Monochromator-Based Methods

In the previous section, we have discussed the realization of spectral responsivity scales using cryogenic radiometers and laser-based apparatus. This is the approach giving the highest level of accuracy, at least in the 400–1000 nm spectral range using silicon diode-based transfer radiometers. However, laser sources have certain important limitations, the principal ones being:

- the spectral coverage is fairly limited; several lasers must be used and interpolation and extrapolation techniques must be used to obtain full spectral coverage;

- the measurements cannot easily be automated, as far as spectral scanning is concerned;
- the coherence of the radiation used can cause problems due to interference effects in detector windows and other optical elements;
- high acquisition cost, high maintenance cost and high manpower cost.

Because of these limitations of laser-based systems, a few laboratories [50, 52, 73, 74] have developed monochromator-based facilities for the first-level calibration of transfer radiometers using cryogenic radiometers. In this section, we shall consider only systems using conventional sources; monochromator-based systems using synchrotron sources will be discussed in the next section.

#### *3.2.2.1 Apparatus and procedures*

Figure 3.16 gives a schematic diagram of the apparatus used at VSL (Van Swinden Laboratory, the Netherlands) [50], which is very similar to the set-up used at NRC [52]. A high-throughput (*f*/4.8) double monochromator is used in conjunction with tungsten halogen, argon maxi-arc or xenon-arc sources to provide monochromatic radiation from 200 nm to 20 μm. The monochromator is used in the subtractive dispersion mode to provide the dispersion of a single monochromator but the spectral purity of a double monochromator. It can also be used as a single monochromator for spectral regions where the output of the double monochromator is too low.

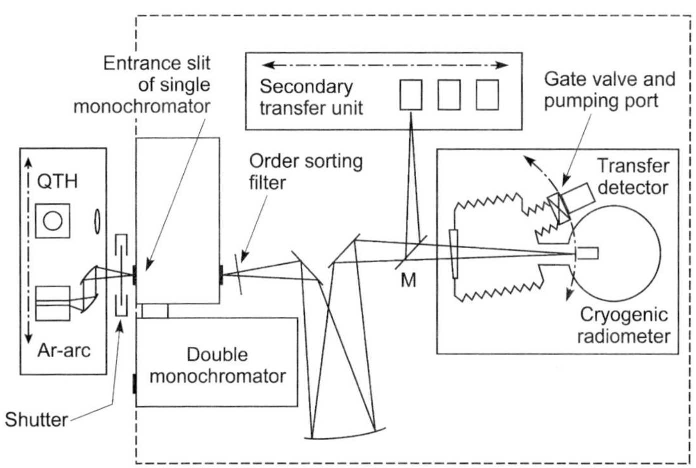


FIG. 3.16. Schematic diagram of the monochromator-based cryogenic radiometer facility at the Van Swinden laboratory. (Source: from Reference [50], with permission from Metrologia.)

The beam emerging from the monochromator passes through an order sorting filter and is imaged onto the radiometers by means of a convex–concave spherical mirror combination (+ folding mirror) designed to minimize spherical aberration. A second folding mirror (M) can be inserted into the beam to direct it towards a secondary transfer unit, which allows the calibration of other radiometers (working standards, intercomparison artefacts, etc.) to be calibrated using the transfer radiometers, once these have been calibrated using the cryogenic radiometer.

A pivoting vacuum chamber is linked by means of a large bellows to a fixed window assembly through which the beam from the monochromator enters; the cryogenic radiometer and the transfer standard are mounted to a flange which rotates about an axis through the window in order to bring each radiometer alternatively into the beam. The window is tilted slightly to avoid inter-reflections, and also incorporates a small wedge angle to allow the use of lasers for comparison purposes. The window transmittance does not have to be measured because both radiometers are inside vacuum. VSL uses $CaF_2$ and KRS-5 windows for spectral ranges 200–8000 nm and 2000–20,000 nm, respectively. NRC uses a Schott WG225 window. This window material limits the spectral range to 250–2500 nm approximately. It is necessary to use this material to eliminate background effects caused by the shutter, which is located in front of the window in the NRC set-up. The transfer radiometer is attached to a gate valve, which allows radiometer exchange without breaking the main vacuum. The transfer radiometers and cryogenic radiometer have equal diameter apertures, which are located at exactly the same distance from the window to virtually eliminate scattered radiation effects. NIST [73] uses a different configuration for the radiometer chamber. Both transfer radiometer and cryogenic radiometer are located inside a vacuum chamber, but the transfer radiometer is mounted in front of the cryogenic radiometer, on a translation stage perpendicular to the optical axis, which allows the transfer radiometer to be moved in and out of the beam. The vacuum chamber itself can translate along the optical axis. This allows the cryogenic and transfer radiometer apertures to be located in the same plane during measurements, and so avoid scattered radiation effects. NIST uses a fused silica window (spectral range used is 900–1800 nm) and a single monochromator.

The cryogenic radiometers used for monochromator-based measurements have a different configuration from those used for laser-based measurements because of the angular spread of the beam entering the radiometer. Figure 3.17 shows the configuration of the NRC cryogenic radiometer (the one at VSL is almost identical). In order to prevent vignetting of the beam, the cavity must be located much further out than for the laser-based instruments. This necessitates adding a liquid-nitrogen shield protruding

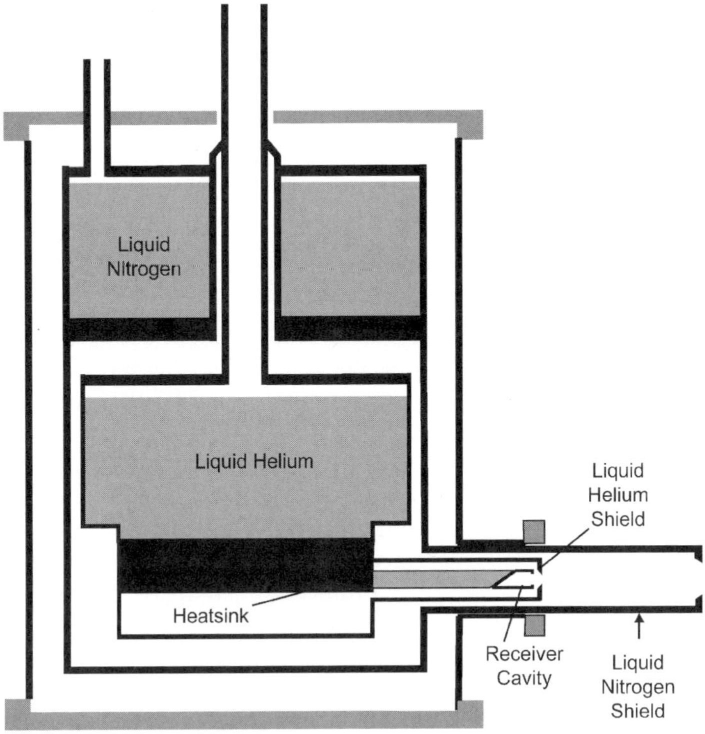


FIG. 3.17. Configuration of the NRC cryogenic radiometer.

outside the cryostat in order to reduce drift. The cavity mass is reduced as much as practical in order to enhance sensitivity (i.e., reduce NEP) and reduce the time constant; these are highly desirable properties of cryogenic radiometers used in this application.

The measurements are fully automated and, in principle, the transfer radiometer can be calibrated at arbitrary wavelength intervals over a certain spectral range (as determined by the source/grating combination). In practice, cryogenic radiometer time constant considerations impose a relatively large wavelength interval in most spectral ranges. For example, at NRC, a 50 nm interval is used to calibrate Si-based transfer radiometers in the range 450–1000 nm. This calibration is repeated four times during the day (48 individual measurements) for a total measurement time of 4 h. Cryogenic radiometer NEP effects are much more important with monochromator-based measurements than with laser-based measurements because of the much lower radiant power available. At NIST, the typical radiant power is ~1 μW, peaking at about 2 μW, with a bandwidth of ~4 nm; at NRC, the maximum radiant power is ~10 μW for the tungsten source, or 30 μW with the Xe source, with minimum usable powers of about 1–2 μW, and a BW of 5 nm. At VSL, the maximum power is ~ 60 μW using tungsten, or 80 μW using Xe, with bandwidths of 5 nm in the UV and VIS and 10 nm in the near-IR.

The wavelength accuracy of the monochromators is verified either by scanning spectral lines from low-pressure discharge lamps [50, 52], or by measuring the transmittance of special materials having well-known sharp absorption bands, e.g. Nd:Yb:Sm glass [73] or polystyrene films [50]. In the latter cases, it is important that the reference data on the absorption lines correspond to the same bandpass as that of the monochromator since the position of the lines is bandpass-dependent. Similarly, when scanning emission lines from spectral lamps, the slit width must be the same as in the actual measurements. It is easier to check wavelength accuracy with the slits closed down, but there is no guarantee that a wavelength shift will not occur on opening the slits to full working width, due to imperfections in the slit assembly. On the other hand, determining the wavelength error to ±0.05 nm with a bandpass of 5 nm is not trivial, especially with the rounded-top slit function [68] one gets when using entrance slit/exit aperture configurations. NRC [68] developed an algorithm to determine the effective readout wavelength $\lambda_0$ of the monochromator with a resolution of ~0.02 nm with rounded-top slit functions and a bandpass of 5 nm. Another procedure which is simpler and gives good results (better than 0.1 nm typically) is to estimate $\lambda_0$ from the weighted mean of the measured slit function data $I(\lambda)$:

$$\lambda_0 = \sum \lambda I(\lambda) \Big/ \sum I(\lambda) \tag{3.6}$$

Once the wavelength errors have been measured over the whole spectral range, wavelength correction formulae can be derived to apply wavelength corrections in real-time during calibrations.

#### *3.2.2.2 Transfer standards*

Transfer radiometers developed for use with monochromator apparatus generally use the same types of detectors as for laser-based measurements, but the housings can be somewhat different. In this section, we will discuss briefly some of these. The transfer radiometers used at NRC [52, 61, 68] and VSL [50, 74] are all vacuum compatible, so that they can be mounted on the same vacuum chamber as the cryogenic radiometer. The single and trap detectors used by both labs are very similar. Figure 3.18 shows the configuration of the single detector radiometer used at NRC [52]; it incorporates a tilted windowless S1337 diode, a separate wedged fused silica window and an RTD temperature sensor. The NRC trap design [68] is shown in

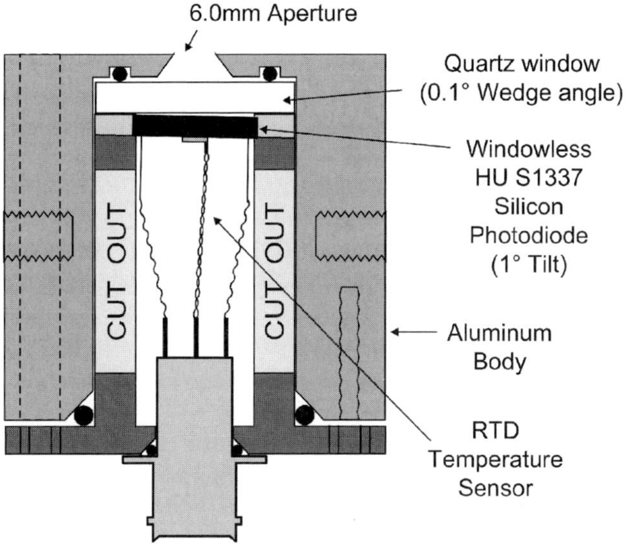


FIG. 3.18. Configuration of the NRC single-detector transfer radiometer for cryogenic radiometer-based calibrations.

Figure 3.19. The trap is windowless and is designed to accommodate without vignetting a 5 mm diameter *f*/8 beam focused at the 6 mm diameter entrance aperture of the trap. In order to allow this wide field of view, large area (18 mm × 18 mm) windowless silicon photodiodes (Hamamatsu S6337) are mounted in a compact three-detector reflection trap configuration. Traps are particularly advantageous in the UV because of their reduced sensitivity to wavelength errors and bandwidth effects, which are significant with silicon-based radiometers. Figure 3.20 compares bandwidth and wavelength effects for a single silicon diode and a Si trap.

For the UV, NRC uses radiometers such as shown in Figure 3.18, but using special UV-enhanced silicon photodiodes or PtSi diodes. For the near-IR (900–1500 nm), VSL uses the same radiometer configuration, but the detectors are large-area (10 mm × 10 mm) germanium detectors, and the window material is Schott RG715 glass, which is used to prevent fatigue effects caused by short-wavelength radiation [75]. The spatial uniformity within the central 6 mm diameter region is about $\pm 0.5\%$ in the best cases. For the far-IR (1000–20000 nm), VSL uses a special thin-film thermopile (IPHT-Jena type TS-76) mounted in a housing similar to that shown in Figure 3.18. The housing window is KRS-5 and Kr gas is used to thermally isolate the sensor. The centre to edge variation in responsivity of the 7 mm diameter sensors is about 4%. NIST [73] uses three types of detectors for the

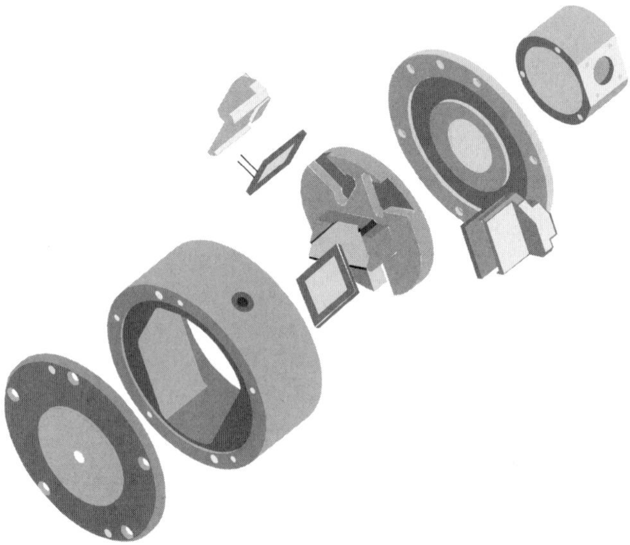

FIG. 3.19. Configuration of the NRC reflection trap radiometer for cryogenic radiometer-based calibrations.

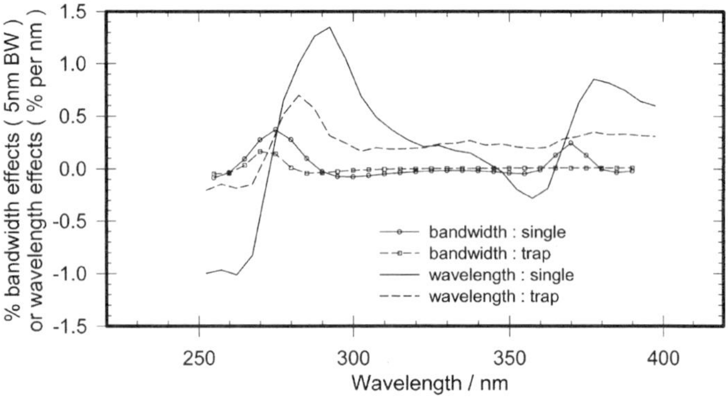


FIG. 3.20. Comparison of bandwidth and wavelength effects in the UV for single and trap detectors (Hamamatsu S1337).

near-IR: 5 mm diameter InGaAs detectors (Ge Power Devices), mounted in a standard NIST base with temperature control (26.0°C); 5 mm diameter Ge diodes (EG&G Judson), with on-chip temperature control (set to −30°C during calibrations); 5 mm diameter pyroelectric detectors (Oriel). In order to be able to make use of the high accuracy of the cryogenic radiometer, it is

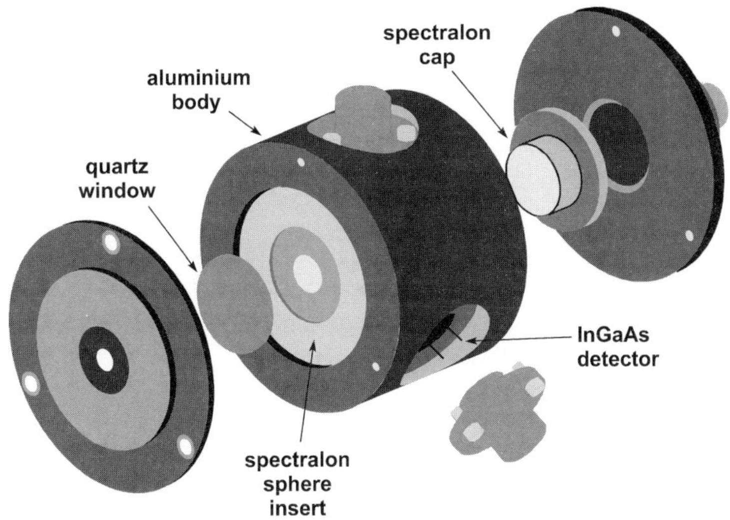


FIG. 3.21. NRC InGaAs-detector-based sphere radiometer for cryogenic radiometer calibrations in the near-IR.

important to have transfer radiometers having the best possible radiometric properties; in particular, they should have good spatial uniformity. This is not a problem with Si diodes and traps, but is a limitation with IR detectors. NRC developed sphere radiometers specifically for monochromator-based cryogenic radiometer calibrations. Figure 3.21 shows the configuration of these radiometers [61]. The radiometer incorporates a small diameter Spectralon integrating sphere and three 5 mm diameter InGaAs detectors mounted on the side of the housing. The radiometer has a responsivity of about 25% that of a single InGaAs detector; the spatial uniformity over the 6 mm diameter aperture is about $\pm 0.1\%$ from 900 to1600 nm.

### *3.2.2.3 Interpolation and extrapolation to other wavelengths*

In principle, it should be possible with monochromator-based systems to calibrate the transfer radiometers over their full spectral range, at the desired wavelength interval, without requiring auxiliary interpolation or extrapolation techniques. In many cases, it is not practical to do this, because of the relatively long measurement times with cryogenic radiometers. At NRC, measurements in the range 400–1000 nm using Si-based radiometers, and 1000–1600 nm using InGaAs-based radiometers are carried out at 50 nm intervals; VSL also uses 50 nm intervals in those spectral ranges. Although 50 nm intervals is satisfactory for certain applications (e.g. international comparisons), it is not satisfactory for dissemination of spectral responsivity scales, where intervals of 10 nm or even 5 nm are required.

Also, wider spectral ranges are sometimes required. For example, at NRC, in the near-IR, the basic scale realization using sphere radiometers and the cryogenic radiometer is carried out in the range 1000–1600 nm only, due to throughput limitations of the monochromator; however, the disseminated scale covers the spectral range 700–1800 nm, so that scale extrapolation is required. We will discuss here the approach used at NRC for interpolation and extrapolation. The approach used by other labs, where needed, is very similar; in particular, the approach used at NRC for scale extrapolation using spectrally flat thermal detectors is very similar to that used by several labs in conjunction with laser-based measurements, as described above.

The silicon diode-based radiometers have a quasi-linear spectral responsivity variation in the spectral range from ~400 to 950 nm. It is thus possible to use conventional mathematical interpolation techniques to obtain data at 10 nm intervals, say, from the measured data at 50 nm intervals; it is not necessary to use the more complicated modelling techniques used with laser-based methods. NRC uses cubic spline or third-degree Lagrange polynomial fits applied to contiguous four-point segments covering the spectral range of interest. Applying high-order polynomial fits to the whole spectral range is more complicated and the results are not as good, especially close to the ends of the interval. The procedure adds an interpolation uncertainty to the interpolated points but not to the reference ( = measured) points, as would be the case if other fitting techniques were used (such as least-squares fitting). The interpolation uncertainty will be discussed briefly in the next section. Below 400 nm, these interpolation techniques do not work as well, even for more closely spaced measurement points. Interpolation errors for interpolated values at 5 nm intervals from measured data at 10 nm can be as large as 0.5%. This is due to the complex shape of the spectral responsivity curve of Si detectors below 400 nm (see Fig. 3.9). An alternative method for interpolation or extrapolation in the UV is auxiliary measurements using spectrally flat detectors. NRC uses thin film, fast thermopiles to extend or interpolate the scale below 400 nm, and extend the scale from 950 to 1100 nm. The sphere radiometers used in the near-IR at NRC for scale realization are not very suitable for mathematical interpolation because they are InGaAs detector based and the spectral responsivity curve is not suitably quasi-linear. For scale interpolation and dissemination, germanium working standards are calibrated at 50 nm intervals using the sphere radiometers; the germanium-based scale is then interpolated using the above cubic spline technique in the range 900–1500 nm, and extrapolated above and below that spectral range by means of auxiliary measurements using thermopiles.

#### 3.2.2.4 Sources of uncertainty

The main sources of uncertainty in the realization of a spectral responsivity scale using monochromator apparatus and cryogenic radiometers will be reviewed only briefly here, since most uncertainty components were discussed above in detail.

- Cryogenic radiometer effects. These are essentially the same as for laser-based measurements: electrical substitution power measurements (B); cavity absorptance (B); non-equivalence effects (B); and repeatability or NEP effects (A). NEP effects can be more important with monochromator-based measurements because of the low optical power available in some spectral ranges. NIST [73] gives an overall cryogenic effects uncertainty of 0.22%, which is dominated by noise effects. NRC and VSL do not include repeatability with cryogenic radiometer effects. VSL and NRC give values of 0.01% and 0.017% as uncertainty associated with cryogenic effects.
- Monochromator effects
  - Wavelength effects (B). As discussed above, these are monochromator and detector dependent. NIST [73] quotes an uncertainty of 0.15% including both wavelength and bandwidth effects, for InGaAs detectors in the range 900–1600 nm. VSL [74] quotes 0.01% for Si traps (450–650 nm) and 0.05% for Ge detectors (900–1500 nm) [50]. NRC quotes 0.015% for Si detectors and traps (450–650 nm) [68] and 0.01% (1000–1600 nm) or 0.1% (900–1000 nm) for InGaAs-based radiometers [61].
  - Bandwidth effects (B). These are also monochromator and detector dependent. NIST [73] uses a bandpass of ~4 nm; uncertainty associated with bandwidth and wavelength effects is estimated at 0.15% for the calibration of InGaAs detectors (900–1600 nm). VSL [50, 74] uses a bandwidth of 5 nm in the UV and VIS, 10 nm in the IR, and estimates bandwidth effects to be negligible for Si-based detectors (450–650 nm) or Ge detectors (900–1500 nm). NRC uses a bandwidth of 5 nm in all cases. Bandwidth effects are estimated to be: 0.01% for Si detectors and traps (450–650 nm) [68]; up to 0.04% or 0.08% for Si traps and detectors, respectively, in the range 300–400 nm; 0.01% (1000–1600 nm) or 0.05% (900–1000 nm) for InGaAs-based radiometers [61].
  - Stray light (B). Both NRC and VSL use double monochromators and estimate stray light effects to be negligible. NIST [73] uses a single monochromator and estimates stray light effects at 0.1%.
  - Beam geometry effects (B). These are the effects caused by the finite angular spread and degree of polarization of the monochromator radiation. This source of uncertainty has been discussed in the previous section. NRC, VSL and NIST all estimate these effects at 0.01% or less.

- Transfer standard effects. These would be essentially the same as for laser-based measurements, but evaluated in the context of monochromator-based measurements:
  - Linearity (B). This is generally less of a concern here, because power levels are much smaller and beam sizes on the detectors larger. For Si-based radiometers, this source of uncertainty is generally assumed to be negligible. In the near-IR, VSL [50] quotes an uncertainty of 0.05% for large area Ge detectors, up to a photocurrent of 0.1 mA. NRC quotes uncertainties of 0.05% for InGaAs-based sphere radiometers in the range 900–1000 nm, and 0.01% in the range 1000–1600 nm [61]. At NIST, uncertainty due to non-linearity of InGaAs, Ge and pyroelectric detectors is negligible compared to other components. VSL quotes a non-linearity component of 0.2% for the thin-film thermopiles used in the range 1–20 μm [50].
  - Non-uniformity (A or B). This component is sometimes quoted explicitly, and sometimes included implicitly with repeatability or alignment effects. VSL quotes a value of 0.01% for measurements using Si-based radiometers [74], 0.2% for Ge radiometers (900–1500 nm) and 1.0% for thermopiles [50]. NRC includes uniformity effects with repeatability, except in the near-IR, where a value of 0.02% is quoted for sphere radiometers (900–1600 nm) [61]. NIST quotes a value of 0.22% for diode positioning and alignment effects for InGaAs detectors [73].
  - Temperature effects (A). Temperature effects are often not quoted explicitly, but included in the repeatability component; this is true in particular for Si-based measurements where temperature effects are quite small. In the near-IR, where InGaAs, Ge or pyroelectric detectors are used, temperature effects are more important. NRC does not use temperature control and quotes uncertainties of 0.2 or 0.02% for the ranges 900–1000 nm or 1000–1600 nm, respectively, using InGaAs-based sphere radiometers [61]. NIST uses temperature control of its radiometers and does not quote a temperature-related uncertainty explicitly. VSL does not use temperature control and quotes an uncertainty of 0.1% for its Ge detectors (900–1500 nm) [50].
  - Polarization sensitivity (B). Polarization effects are included in the beam geometry effects in the previous section.
- Repeatability (A). This addresses the day-to-day repeatability of calibrations, including removal, remounting and realignment of the radiometers. Many factors contribute to this uncertainty component, some of which are also considered separately: stability of the sources; detector non-uniformity in conjunction with small alignment differences; detector noise and drift; temperature differences. NRC estimates values between 0.02 and 0.04% in the range 300–1000 nm, using Si-based radiometers and Xe or tungsten

sources; 0.04% (900–1000 nm) or 0.02% (1000–1600 nm) using InGaAs-based sphere radiometers and tungsten sources [61]. VSL quotes values of 0.007% (450–650 nm) using Si artefacts [74]; 0.02% (900–1500 nm) using Ge detectors [50]; and 0.3–6.0% (1–20 μm) using thermopiles [50]. At NIST, because of the very low optical power used (1 μW typically), the repeatability is dominated by the noise of the ACR (0.2%) [73].

- Electrical (B). Included here are uncertainties associated with the calibration of DVM's and amplifiers used to make measurements of the response of transfer radiometers. Values quoted are usually small; 0.01% at NRC [61, 68], and VSL [50, 74] for the UV, VIS and near-IR. NIST quotes uncertainties of 0.04 and 0.1% for photocurrent measurements of test and monitor detectors, respectively, for InGaAs detectors in the range 950–1600 nm [73]. VSL gives values between 0.1% and 1% for thermopile voltage response measurements in the range 1–20 μm [50].

Overall uncertainties are obtained as usual by the quadrature sum of the individual components. As with laser-based measurements, lowest uncertainties are quoted for the range 450–700 nm, using Si traps; 0.02% at VSL [74] and 0.04% at NRC [68]. In the near-IR, the lowest uncertainties are those quoted by NRC: 0.04% for sphere radiometers (1000–1600 nm) [61].

#### *3.2.2.5 Interpolation/extrapolation uncertainties*

It is often necessary to obtain a scale over a wider spectral range and at a finer wavelength interval than that obtained using direct monochromator-based calibrations using the cryogenic radiometer. In the range 400–1000 nm using Si-based radiometers, modelling techniques can be used for interpolation purposes. The uncertainties associated with this approach are about 0.05% in the worst case. We have discussed in the previous section a simple cubic spline method, which is suitable for monochromator-based measurements and which requires no a priori knowledge of detector physics. This method has been used at NRC for interpolations in the range 400–950 nm for Si-based radiometers, in the range 700–1500 nm for germanium radiometers and 1200–2700 nm for InSb radiometers.

The interpolation uncertainty is estimated as follows. Interpolated values corresponding to parabolic, Lagrange cubic and Lagrange 4th degree are calculated; using the Lagrange cubic values as true values, the maximum differences from the true values corresponding to the other interpolations are taken as uncertainties. Figure 3.22 shows the interpolation uncertainties thus obtained for Hamamatsu S1337 Si diodes interpolated at 10 nm intervals from source data at 50 nm intervals, in the range 400–950 nm.

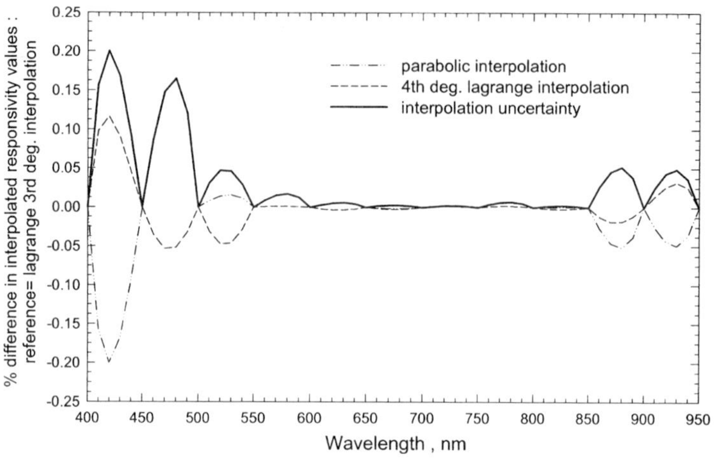


FIG. 3.22. Interpolation uncertainties for Si detectors (S1337) interpolated at 10 nm intervals from source data at 50 nm intervals.

Interpolation uncertainty is smaller than 0.05% from 500 to 950 nm, but can be as large as 0.2% in the range 400–450 nm. For germanium and InSb detectors, interpolation uncertainties are typically <0.05% (700–1500 nm) or 0.1% (1500–2500 nm) respectively, for interpolations at 10 nm from data at 50 nm intervals. Where scale extensions are carried out by means of auxiliary measurements using thermal detectors, the situation is exactly the same as for laser-based measurements, and uncertainties estimated in the same way.

### 3.2.3 Methods Using Synchrotron Radiation

A few laboratories realize spectral responsivity scales in the UV, VUV and soft X-ray ranges using monochromator-based apparatus, cryogenic radiometers and synchrotron radiation as a source. Although this is really a subset of the monochromator-based measurements described above, it is discussed here in a separate section because of the highly specialized apparatus and techniques used. Synchrotron radiation is the wide spectrum radiation emitted by electrons in so-called electron storage rings. A discussion of the properties of synchrotron radiation is beyond the scope of this chapter, but is covered in detail in Chapter 5. We will consider here the facilities of only two laboratories, the NIST SURF III facility and the PTB BESSY II facility. Japan also has a facility [76] using undulator radiation and a ESR for UV-detector calibration. Interpolation or extrapolation techniques will not be discussed since they do not appear to have been used or needed.

### *3.2.3.1 NIST SURF III storage ring*

Figure 3.23 shows a schematic diagram of the apparatus set-up at beamline 4 of the SURF III facility at NIST [47, 77] for cryogenic radiometer-based measurements. SURF III is operated with an electron energy of 331 MeV and a typical beam current of 180 mA. The radiation from the storage ring is focussed onto the entrance slit of a 2 m normal incidence single monochromator by means of a grazing incidence plane-toroidal mirror combination. The beam emerging from the monochromator goes through a mirror box, which contains refocussing mirrors to image the beam 1:1 onto the radiometers, as well as a $CaF_2$ beam splitter and monitor detector. The whole system, including radiometers, is under vacuum. The part of the beam up-stream from the exit slit is held under ultrahigh vacuum, and is separated from the downstream portion by means of a $CaF_2$ window; the downstream portion is under high vacuum and has its own pumping station. This arrangement greatly reduces the time required to exchange detectors and re-evacuate the detector box. The monitor detector is used to correct for drift in optical power. The calibration procedure is the same as was discussed above for the realization of a scale in the near-IR. The test radiometers are mounted on an X–Y stage in front of the cryogenic radiometer, which allows the radiometers to be scanned in a plane perpendicular to the optical axis, to measure spatial uniformity and also to move the radiometers out of the beam for optical power measurement by the cryogenic radiometer. The whole detector box can also translate along the optical axis, so that both the test and cryogenic radiometer apertures are located at the same position along the optical axis during measurement.

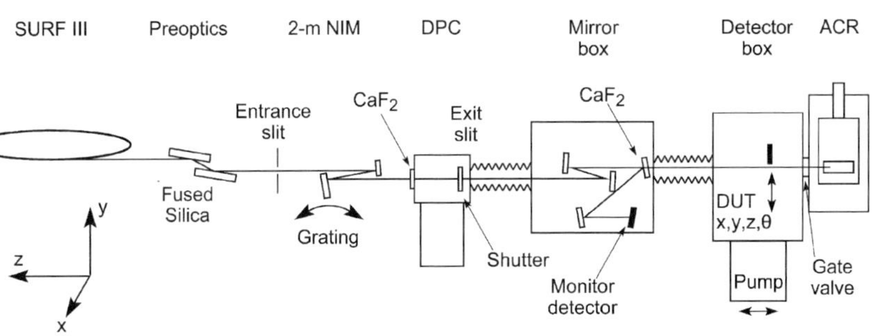


FIG. 3.23. Schematic diagram of the apparatus set-up at the NIST SURF III storage ring for cryogenic radiometer-based calibrations. (Source: from Reference [77], with permission from the Optical Society of America.)

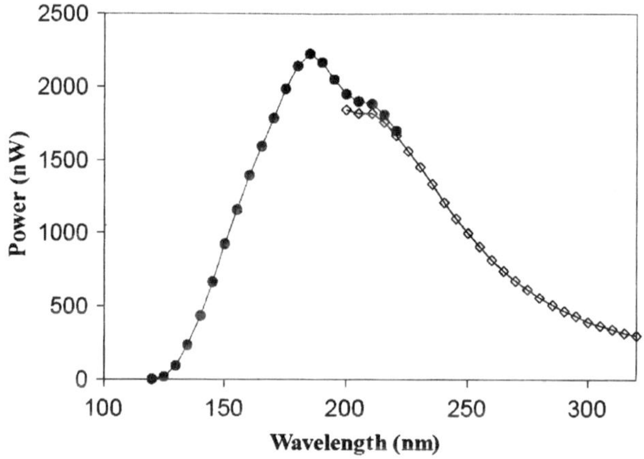


FIG. 3.24. Output power incident on the radiometers at the NIST SURF III beamline 4 (Fig. 3.23). (Source: from Reference [47], with permission from the American Institute of Physics.)

The cryogenic radiometer is essentially the same type as that used by NRC and VSL, is operated in the active (servo) mode and has a time constant of a few seconds [77]. The facility is used in the range 125–320 nm, with peak power of ~2.5 μW at a wavelength of ~180 nm. Figure 3.24 shows the output power distribution; the low wavelength cut-off is set by the $CaF_2$ window.

A fused quartz window, which can be inserted into the beam just behind the exit slit, is used as order sorting filter above 200 nm. The wavelength calibration of the monochromator is carried out by scanning a holmium oxide wavelength standard (NIST SRM#2034); the residual wavelength uncertainty is 0.2 nm.

### *3.2.3.2 PTB BESSY II storage ring*

Much of the work for cryogenic radiometer-based detector calibrations using synchrotron radiation was first carried out by PTB at the BESSY I facility in Berlin [31, 78–82]. A new storage ring BESSY II was developed and replaced BESSY I in 2000. The apparatus used for detector calibrations were moved to BESSY II and improved. See Chapter 5, Section 5.3, for more information on synchrotron radiation in general, and BESSY II in particular.

BESSY II is normally operated with an electron beam energy of 1700 MeV. At PTB's laboratory at BESSY II several beam lines are used. One is dedicated to source calibrations and three to cryogenic radiometer-based detector calibrations in several spectral domains: in the UV–VUV

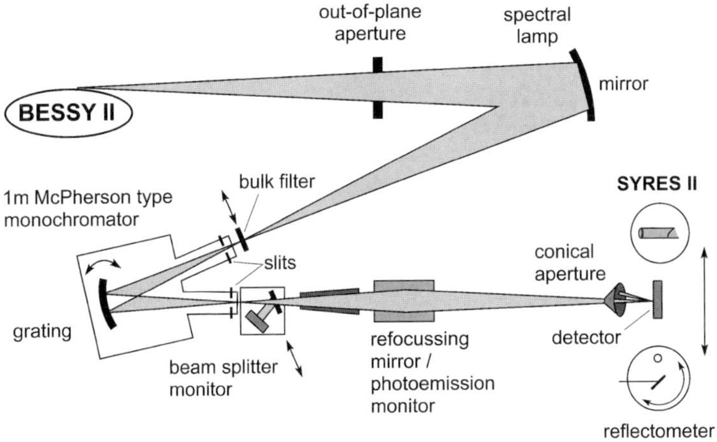


FIG. 3.25. Schematic diagram of the apparatus set-up at PTBs UV-VUV beamline for cryogenic radiometer-based calibrations at BESSY II. (Source: from Reference [83], with permission from Elsevier.)

(40–400 nm) [83, 84], in the soft X-ray range (0.7–35 nm) [85, 86] and in the hard X-ray range [87]. We will discuss mostly the UV–VUV apparatus here since it is more relevant to the spectral domain considered in this chapter. Figure 3.25 shows a schematic diagram of the UV–VUV apparatus at BESSY II.

The entire system is under ultrahigh vacuum. A 1 m normal incidence monochromator is used. In order to cover the spectral range and suppress higher orders, several spherical gratings and pre-mirrors with optimized coatings are used; higher order suppression is also achieved using various filters and a rare-gas filter. The radiation emerging from the monochromator is imaged using an elliptical mirror onto either the cryogenic radiometer (SYRES II), or the test detector, or a reflectometer. This apparatus can also be used to measure the reflectance of various samples. Above 120 nm, a beam splitter and photodiode are used as monitor to correct for time variation of the radiant power incident on the detectors. Below 170 nm, monitoring is achieved by measuring the photoemission from the elliptical refocusing mirror. For the X-ray beamlines, the electron beam current is used for monitoring purposes. The radiant power incident in first spectral order on the detectors is shown in Figure 3.26 for an electron current of 100 mA.

The cryogenic radiometer used at the UV–VUV beamline at BESSY II (SYRES II) is an improved version of the one (SYRES) developed originally for the BESSY I facility, whereas SYRES itself now is used in the soft and

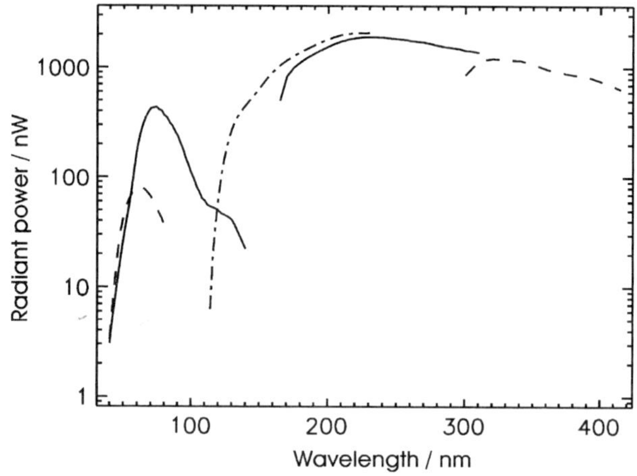


FIG. 3.26. Radiant power incident in first spectral order on detectors for PTBs UV-VUV beamline at BESSY II for an electron current of 100 mA. (Source : M. Richter, Physikalisch-Technische Bundesanstalt, Berlin, private communication.)

hard X-ray regime. Both can measure optical powers in the range from 0.1 to 10 μW with a relative standard uncertainty below 0.2%. The time constant is fairly long, between 20 and 40 s. The SYRES II radiometer is composed of two parts, the cryostat itself, containing the absorber cavity, and the beamline interface, containing cooled baffles, which is specific to the UV–VUV beamline. The cryogenic radiometer is operated in the dynamic substitution mode (i.e., servo or feedback mode) similar to the mode of operation of the CRI cryogenic radiometers. The mode of operation of SYRES, including corrections for drift, etc. is described in detail in Reference [31]. For SYRES II, the operational characteristics are similar.

#### 3.2.3.3 *Transfer radiometers*

Several types of detectors have been investigated for use in the UV, VUV and soft X-rays with synchrotron radiation, but only a few of these types are suitable as transfer standards. The main problems are uniformity and stability to UV irradiation. NIST has investigated [47, 77, 88] silicon p-on-n (Hamamatsu S1337 and S5227), nitrided silicon n-on-p (IRD AXUV-100G), silicon n-on-p with protective window (UDT UV100), Schottky-type PtSi-n-Si, GaN, GaP and GaAsP. NIST found that the nitrided silicon n-on-p and PtSi were much more stable to irradiation at 135 nm

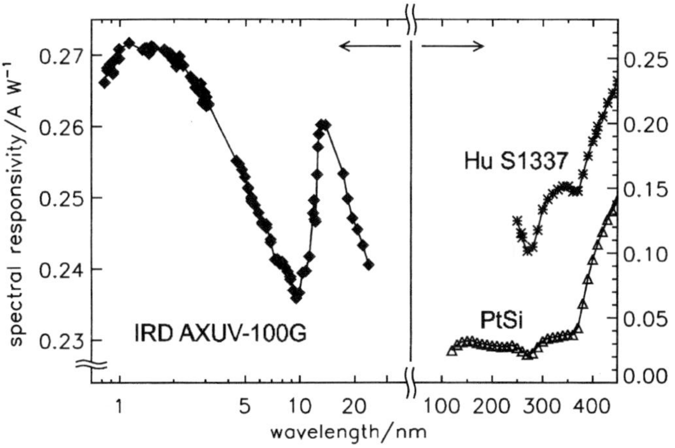


FIG. 3.27. Spectral responsivity curves of some types of detectors investigated at the PTB facility and found suitable for use in the UV–VUV and soft X-rays regions. (Source: from Reference [31], with permission from the Optical Society of America.)

than the Hamamatsu S1337 and S5227, which are good choices for transfer standards above 250 nm.

PTB has investigated mostly silicon p-on-n (Hamamatsu S1337 and S5227), nitrided silicon n-on-p (IRD AXUV100, IRD UVG100, IRD SXUV100) and Schottky-type PtSi-n-Si, Au-GaAsP, GaP and GaAlN. They have carried out extensive tests on UV and VUV stability of various types of detectors [31, 35, 89–91]. They found that PtSi detectors are good candidates for transfer standards below 250 nm because of their good uniformity and stability to UV irradiation. Above 250 nm, Hamamatsu S5227 are the best candidates as UV standards. At very short wavelengths (i.e., in the soft X-ray range below 30 nm), nitrided silicon n-on-p (e.g. IRD AXUV-100G) appeared to be good choices, but recent measurements [86] indicated that significant ageing and degradation can occur for this type of detector, even if rarely used and carefully stored. n-on-p diodes with TiSi passivation (e.g. IRD SXUV100) have been shown to be stable under intense soft X-ray irradiation [92]. Figure 3.27 compares the spectral responsivity curves of three types of detectors.

### 3.2.3.4 *Sources of uncertainty*

- Measurement of radiant power with the cryogenic radiometer. NIST quotes an overall uncertainty of 0.3% for cryogenic radiometer effects [47]. A detailed budget is not given, but most of the uncertainty is of type A and results from the noise floor of the radiometer and the low optical power used in the measurements. PTB gives a detailed analysis of uncertainties for the measurement of optical power using the cryogenic radiometer [85]. The budget is given for 13 nm radiation at a power level of 0.2 μW. The overall uncertainty is 0.11%, and the major contributions are uncertainties in measured heating power difference (type A, 0.1%) and in radiant energy conversion efficiency of the absorber (type B, 0.03%). The overall uncertainty drops to 0.03% for radiant powers above 1 μW.

- Spectral responsivity measurements. This includes several sources of uncertainty, listed below. PTB values given below are based on measurements made at 13 nm. Detailed budgets for calibrations in the UV–VUV are similar to the ones given for BESSY I [81].
  - Monochromator effects:
    - Wavelength errors and bandwidth effects (B). The magnitude of these effects can be estimated as discussed earlier. NIST treats these together, and quotes a value of 0.3%. It is not clear for which detectors this applies to; it is probably a worst case estimate. At PTB, all detector-dependent uncertainties are based on the types of detectors shown in Figure 3.27 above. Uncertainties associated with wavelength and bandwidth effects are 0.01% or less.
    - Stray and scattered radiation (B). Both NIST and PTB use single monochromators, so that stray radiation effects must be considered. Higher order radiation caused by the gratings cannot always be completely eliminated. Residual stray light is estimated at 0.1% at the NIST facility. At PTB, there is an uncertainty of 0.2% due to scattered radiation; this is the largest uncertainty component for detector calibrations.
    - Beam geometry (angular spread and polarization) (B). These effects are deemed negligible at NIST because of the small angular spread and normal incidence of the radiation. Uncertainty due to angle of incidence effects is estimated to be 0.005% at PTB.
  - Detector effects. The principal effects here are non-uniformity, nonlinearity, temperature variation and stability. Most of these are not listed separately by NIST or PTB; these effects are either implicitly included with other effects or deemed to be negligible. PTB quotes an uncertainty of 0.01% due to temperature variation of the detectors ($\pm 2$ K).
  - Electrical effects (monitor and test detector photocurrent measurement, DVM calibration). Both type A and type B effects are included here. NIST quotes values of 0.1% for both monitor and test detector. PTB quotes an uncertainty of 0.1% for the measurement of diode photocurrent and 0.06% for DVM calibration.

- Repeatability (A). This includes several effects not already covered by the repeatability of the cryogenic radiometer measurements: test detector or cryogenic radiometer positioning, in conjunction with detector non-uniformity; source noise and imperfectly compensated drift, etc. NIST gives a value of 0.2% for these effects. PTB does not give this uncertainty explicitly; it is included implicitly in the uncertainties in the measurement of diode photocurrent.

The overall uncertainty for transfer standard calibrations at NIST using the cryogenic radiometer-based facility at the SURF III beamline is 0.5% in the range 125–320 nm. At the BESSY II facility at PTB, the overall uncertainty is 0.26% for the soft X-ray beamline. For the UV–VUV beamline, overall uncertainties are similar (0.2–0.6%).

## 3.3 Dissemination of Spectral Responsivity Scales

A spectral responsivity scale is realized and maintained by means of transfer radiometers, which have been calibrated spectrally using cryogenic radiometers and laser or monochromator-based apparatus, in conjunction with spectral interpolation or extrapolation. These transfer radiometers can then be used to calibrate other radiometers or working standards, which are used for various other radiometric applications; this is referred to as the dissemination of the spectral responsivity scale. The radiometers calibrated using the transfer radiometers include so-called filter radiometers, and working standards used with routine detector calibration facilities. The dissemination of spectral responsivity scales will not be discussed here. It involves very similar apparatus and procedures to those discussed above for monochromator-based measurements. Filter radiometers are radiometers incorporating detectors, optical filters, precision apertures and, in some cases, temperature control elements. Filter radiometers are used for the realization of spectral irradiance scales and luminous intensity or luminance responsivity scales. The design and properties of some types of filter radiometers are discussed in Chapter 4. A special case of dissemination of spectral responsivity scales is the realization of a scale of irradiance or radiance responsivity, where the artefacts are filter radiometers that are calibrated in the overfill mode (as opposed to the underfill mode, which has been the case in this chapter). There are two main methods for doing this. One method involves the use of large area uniform sources [93–95]. This approach will be discussed in Chapter 4. The other method involves the raster scanning of a small beam across the aperture of the radiometer. This method is described in several papers [96–98].

## References


1. J. Geist, E. F. Zalewski, and R. Schaefer, Spectral response self-calibration and interpolation of silicon photodiodes, *Appl. Opt.* 19, 3795–3799 (1980).
2. R. L. Booker and J. Geist, Induced junction (inversion layer) photodiode self-calibration, *Appl. Opt.* 23, 1940–1945 (1984).
3. E. F. Zalewski and C. R. Duda, Silicon photodiode device with 100% external quantum efficiency, *Appl. Opt.* 22, 2867–2873 (1983).
4. J. M. Houston and E. F. Zalewski, Photodetector spectral response based on 100% quantum efficient detectors, *Proc. SPIE* 1109, 268–277 (1989).
5. N. P. Fox, Trap detectors and their properties, *Metrologia* 28, 197–202 (1991).
6. F. Hengstberger, "Absolute Radiometry." Academic Press, San Diego, 1989.
7. L. P. Boivin, Some aspects of radiometric measurements involving gaussian laser beams, *Metrologia* 17, 19–25 (1981).
8. N. P. Fox and J. E. Martin, Comparison of two cryogenic radiometers by determining the absolute spectral responsivity of silicon photodiodes with an uncertainty of 0.02%, *Appl. Opt.* 29, 4686–4693 (1990).
9. R. U. Datla, K. Stock, A. C. Parr, C. C. Hoyt, P. J. Miller, and P. V. Foukal, Characterization of an absolute cryogenic radiometer as a standard detector for radiant power measurements, *Appl. Opt.* 31, 7219–7225 (1992).
10. K. D. Stock and H. Hofer, Present state of the PTB primary standard for radiant power based on cryogenic radiometry, *Metrologia* 30, 291–296 (1993).
11. Fu Lei and J. Fischer, Characterization of photodiodes in the UV and visible spectral region based on cryogenic radiometry, *Metrologia* 30, 297–303 (1993).
12. International Organization for Standardization, "Guide to the Expression of Uncertainty in Measurement." Geneva, Switzerland, 1995.
13. O. Touayar, J. M. Coutin, and J. Bastie, The use of INM cryogenic radiometer to calibrate transfer standard detectors at 1550 nm, *Proc. SPIE* 2550, 65–73 (1995).
14. T. R. Gentile, J. M. Houston, J. E. Hardis, C. L. Cromer, and A. C. Parr, NIST high-accuracy cryogenic radiometer, *Appl. Opt.* 35, 1056–1068 (1996).
15. T. C. Larason, S. S. Bruce, and C. L. Cromer, The NIST high accuracy scale for absolute spectral response from 406 nm to 920 nm, *J. Res. Natl. Inst. Stand. Technol.* 101, 133–140 (1996).

16. S. P. Morozova, V. A. Konovodchenko, V. I. Sapritsky, B. E. Lisiansky, P. A. Morozov, U. A. Melenevsky, and A. G. Petic, An absolute cryogenic radiometer for laser calibration and characterization of photodetectors, *Metrologia* 32, 557–560 (1996).
17. T. R. Gentile, J. M. Houston, and C. L. Cromer, Realization of a scale of absolute spectral response using the NIST high accuracy cryogenic radiometer, *Appl. Opt.* 35, 4392–4403 (1996).
18. K. D. Stock and H. Hofer, PTB primary standard for optical radiant power: transfer-optimized facility in the clean-room centre, *Metrologia* 32, 545–549 (1996).
19. R. Köhler, R. Goebel, R. Pello, O. Touayar, and J. Bastie, First results of measurements with the BIPM cryogenic radiometer and comparison with the INM cryogenic radiometer, *Metrologia* 32, 551–555 (1996).
20. R. Goebel, R. Pello, R. Köhler, P. Haycocks, and N. Fox, Comparison of the BIPM cryogenic radiometer with a mechanically cooled cryogenic radiometer from the NPL, *Metrologia* 33, 177–179 (1996).
21. R. Köhler, R. Goebel, and R. Pello, Experimental procedures for the comparison of cryogenic radiometers at the highest accuracy, *Metrologia* 33, 549–554 (1996).
22. R. Köhler, R. Goebel, M. Stock, and R. Pello, An international comparison of cryogenic radiometers, *Proc. SPIE* 2815, 22–30 (1996).
23. T. R. Gentile, J. M. Houston, G. Eppeldauer, A. L. Migdall, and C. L. Cromer, Calibration of a pyroelectric detector at 10.6 μm with the NIST high-accuracy cryogenic radiometer, *Appl. Opt.* 36, 3614–3621 (1997).
24. E. G. Atkinson and D. J. Butler, Calibration of an InGaAs photodiode at 1300 nm with a cryogenic radiometer and a diode laser, *Metrologia* 35, 241–245 (1998).
25. P. Corredera, J. Campos, M. L. Hernanz, J. L. Fontecha, A. Pons, and A. Corróns, Calibration of near-infrared transfer standards at optical-fibre communication wavelengths by direct comparison with a cryogenic radiometer, *Metrologia* 35, 273–277 (1998).
26. E. Theocharous, T. R. Prior, P. R. Haycocks, and N. P. Fox, High-accuracy, infrared, spectral responsivity scale, *Metrologia* 35, 543–548 (1998).
27. K. M. Nield, J. F. Clare, J. D. Hamlin, and A. Bittar, Calibration of a trap detector against a cryogenic radiometer, *Metrologia* 35, 581–586 (1998).
28. L. Werner, J. Fischer, U. Johannsen, and J. Hartmann, Accurate determination of the spectral responsivity of silicon trap detectors between 238 nm and 1015 nm using a laser-based cryogenic radiometer, *Metrologia* 37, 279–284 (2000).

*29.* L. Werner, R. Friedrich, U. Johannsen, and A. Steiger, Precise scale of spectral responsivity for InGaAs detectors based on a cryogenic radiometer and several laser sources, *Metrologia* 37, 523–526 (2000).

*30.* A. Corróns, J. L. Fontecha, P. Corredera, J. Campos, A. Pons, and M. L. Hernanz, Ultraviolet calibration of detectors with respect to a cryogenic radiometer, *Metrologia* 37, 555–558 (2000).

*31.* H. Rabus, V. Persch, and G. Ulm, Synchrotron radiation operated cryogenic electrical-substitution radiometer as the high accuracy primary detector standard in the ultraviolet, vacuum-ultraviolet, and soft X-ray spectral ranges, *Appl. Opt.* 36, 5421–5440 (1997).

*32.* D. J. Butler, R. Köhler, and G. W. Forbes, Diffraction effects in the radiometry of coherent beams, *Appl. Opt.* 35, 2162–2166 (1996).

*33.* R. Köhler, R. Goebel, R. Pello, and J. Bonhoure, Effects of humidity and cleaning on the sensitivity of Si photodiodes, *Metrologia* 28, 211–215 (1991).

*34.* R. Goebel, R. Köhler, and R. Pello, Some effects of low-power ultraviolet radiation on silicon photodiodes, *Metrologia* 32, 515–518 (1996).

*35.* L. Werner, Ultraviolet stability of silicon photodiodes, *Metrologia* 35, 407–411 (1998).

*36.* L. P. Boivin, Automated absolute and relative spectral linearity measurements on photovoltaic detectors, *Metrologia* 30, 355–360 (1993).

*37.* R. Goebel and M. Stock, Nonlinearity and polarization effects in silicon trap detectors, *Metrologia* 35, 413–418 (1998).

*38.* K. D. Stock, S. Morozova, L. Liedquist, and H. Hofer, Nonlinearity of the quantum efficiency of Si reflection trap detectors at 633 nm, *Metrologia* 35, 451–454 (1998).

*39.* P. Balling and P. Kren, Linearity limits of biased S1337 trap detectors, *Metrologia* 38, 249–251 (2001).

*40.* J. L. Gardner, Transmission trap detectors, *Appl. Opt.* 33, 5914–5918 (1994).

*41.* J. L. Gardner, A four-element transmission trap detector, *Metrologia* 32, 469–472 (1996).

*42.* J. Lehman, J. Sauvageau, I. Vayshenker, C. Cromer, and K. Foley, Silicon wedge-trap detector for optical fibre power measurements, *Meas. Sci. Technol.* 9, 1694–1698 (1998).

*43.* J. H. Lehman and C. L. Cromer, Optical tunnel-trap detector for radiometric measurements, *Metrologia* 37, 477–480 (2000).

*44.* G. P. Eppeldauer, Opto-mechanical and electronic design of a tunnel-trap Si radiometer, *J. Res. Natl. Inst. Stand. Technol.* 105, 813–828 (2000).

*45.* R. Goebel, S. Yilmaz, and R. Pello, Polarization dependence of trap detectors, *Metrologia* 33, 207–213 (1996).

46. R. Goebel, S. Yilmaz, and R. Köhler, Stability under vacuum of silicon trap detectors and their use as transfer instruments in cryogenic radiometry, *Appl. Opt.* 35, 4404–4407 (1996).
47. P.-S. Shaw, T. C. Larason, R. Gupta, S. W. Brown, R. E. Vest, and K. R. Lykke, The new ultraviolet spectral responsivity scale based on cryogenic radiometry at SURF III, *Rev. Sci. Instrum.* 72, 2242–2247 (2001).
48. N. P. Fox, E. Theocharous, and T. H. Ward, Establishing a new ultraviolet and near-infrared spectral responsivity scale, *Metrologia* 35, 535–541 (1998).
49. P. Corredera, M. L. Hernanz, J. Campos, A. Corrons, A. Pons, and J. L. Fontecha, Absolute power measurements at wavelengths of 1300 nm and 1550 nm with a cryogenic radiometer and a tuneable laser diode, *Metrologia* 37, 519–522 (2000).
50. C. A. Schrama, P. Bloembergen, and E. W. M. van der Ham, Monochromator-based cryogenic radiometry between 1 μm and 20 μm, *Metrologia* 37, 567–570 (2000).
51. D. H. Nettleton, T. R. Prior, and T. H. Ward, Improved spectral responsivity scales at the NPL, 400 nm to 20 μm, *Metrologia* 30, 425–432 (1993).
52. L. P. Boivin and K. Gibb, Monochromator-based cryogenic radiometry at the NRC, *Metrologia* 32, 565–570 (1996).
53. L. P. Boivin, Properties of indium antimonide detectors for use as transfer standards for detector calibrations, *Appl. Opt.* 37, 1924–1929 (1998).
54. J. H. Lehman, Pyroelectric trap detector for spectral responsivity measurements, *Appl. Opt.* 36, 9117–9118 (1997).
55. W. Budde, "Optical Radiation Measurements—Physical Detectors of Optical Radiation," 72–80. Academic Press, New York, 1983.
56. J. Geist and H. Baltes, High accuracy modelling of photodiode quantum efficiency, *Appl. Opt.* 28, 3929–3939 (1989).
57. E. D. Palik, "Handbook of Optical Constants." Academic Press, New York, 1985.
58. N. M. Durant and N. P. Fox, A physical basis for the extrapolation of silicon photodiode quantum efficiency into the ultraviolet, *Metrologia* 30, 345–350 (1993).
59. A. Bittar, Extension of a silicon-based detector spectral-responsivity scale into the ultraviolet, *Metrologia* 32, 497–500 (1996).
60. T. Kübarsepp, P. Kärhä, and E. Ikonen, Interpolation of the spectral responsivity of silicon photodetectors in the near ultraviolet, *Appl. Opt.* 39, 9–15 (2000).
61. L. P. Boivin, Properties of sphere radiometers suitable for high-accuracy cryogenic radiometer-based calibrations in the near-infrared, *Metrologia* 37, 481–484 (2000).

62. P. Corredera, M. L. Hernanz, M. González-Herráez, and J. Campos, Anomalous non-linear behaviour of InGaAs photodiodes with overfilled illumination, *Metrologia* 40, S150–S153 (2003).
63. K. D. Stock, R. Heine, and H. Hofer, Spectral characterization of Ge trap detectors and photodiodes used as transfer standards, *Metrologia* 40, S163–S166 (2003).
64. N. P. Fox, Improved near-infrared detectors, *Metrologia* 30, 321–325 (1993).
65. L. P. Boivin, Current work at the NRCC on absolute radiometer based calibrations in the infrared, *Inst. Phys. Conf. Ser.* 92, 81–88 (1989).
66. L. Werner, J. Fischer, U. Johannsen, and J. Hartmann, Accurate determination of the spectral responsivity of silicon trap detectors between 238 nm and 1015 nm using a laser-based cryogenic radiometer, *Metrologia* 37, 279–284 (2000).
67. L. P. Boivin, Spectral responsivity of various types of silicon photodiodes at oblique incidence: comparison of measured and calculated values, *Appl. Opt.* 40, 485–491 (2001).
68. L. P. Boivin, Measurements using two types of transfer radiometer developed for a monochromator-based cryogenic radiometer facility, *Metrologia* 35, 363–368 (1998).
69. E. Theocharous and N. P. Fox, Reversible and apparent ageing effects in infrared detectors, *Metrologia* 40, S136–S140 (2003).
70. J. Campos, A. Pons, and P. Corredera, Spectral responsivity scale in the visible range based on single silicon photodiodes, *Metrologia* 40, S181–S184 (2003).
71. J. Campos, A. Corrons, A. Pons, P. Corredera, J. L. Fontecha, and J. R. Jimenez, Spectral responsivity uncertainty of silicon photodiodes due to calibration spectral bandwidth, *Meas. Sci. Technol.* 12, 1926–1931 (2001).
72. L. P. Boivin, Study of bandwidth effects in monochromator-based spectral responsivity measurements, *Appl. Opt.* 41, 1929–1935 (2002).
73. P.-S. Shaw, T. C. Larason, R. Gupta, S. W. Brown, and K. R. Lykke, Improved near-infrared spectral responsivity scale, *J. Res. Natl. Inst. Stand. Technol.* 105, 689–700 (2000).
74. C. A. Schrama, R. Bosma, K. Gibb, H. Reijn, and P. Bloembergen, Comparison of monochromator-based and laser-based cryogenic, radiometry, *Metrologia* 35, 431–435 (1998).
75. P. Lecollinet and J. Bastie, Fatigue effects in germanium photodetectors, *Metrologia* 30, 351–354 (1993).
76. T. Saito, I. Saito, T. Yamada, T. Zama, and H. Onuki, UV detector calibration based on ESR using undulator radiation, *J. Electr. Spectr. Relat. Phenom.* 80, 397–400 (1996).

77. P.-S. Shaw, K. R. Lykke, R. Gupta, T. R. O'Brian, U. Arp, H. H. White, T. B. Lucatorto, J. L. Dehmer, and A. C. Parr, Ultraviolet radiometry with synchrotron radiation and cryogenic radiometry, *Appl. Opt.* 38, 18–28 (1999).
78. G. Ulm and B. Wende, Radiometry laboratory of PTB at BESSY, *Rev. Sci. Instrum.* 66, 2244–2247 (1995).
79. M. Richter, U. Johannsen, P. Kuschnerus, U. Kroth, H. Rabus, G. Ulm, and L. Werner, The PTB high-accuracy spectral responsivity scale in the ultraviolet, *Metrologia* 37, 515–518 (2000).
80. G. Ulm, Radiometry with synchrotron radiation, *Metrologia* 40, S101–S106 (2003).
81. T. Lederer, H. Rabus, F. Scholze, R. Thornagel, and G. Ulm, Detector calibration at the radiometry laboratory of PTB in the VUV and soft X-ray spectral ranges using synchrotron radiation, *Proc. SPIE* 2519, 92–107 (1995).
82. A. Lau-Främbs, U. Kroth, H. Rabus, E. Tegeler, G. Ulm, and B. Wende, First results with the new PTB cryogenic radiometer for the vacuum-ultraviolet spectral range, *Metrologia* 32, 571–574 (1996).
83. M. Richter, J. Hollandt, U. Kroth, W. Paustian, H. Rabus, R. Thornagel, and G. Ulm, The two normal-incidence monochromator beam lines of PTB at BESSY II, *Nucl. Instr. Meth. A* 467–468, 605–608 (2001).
84. M. Richter, J. Hollandt, U. Kroth, W. Paustian, H. Rabus, R. Thornagel, and G. Ulm, Source and detector calibration in the UV and VUV at BESSY II, *Metrologia* 40, S107–S110 (2003).
85. F. Scholze, J. Tümmler, and G. Ulm, High-accuracy radiometry in the EUV range at the PTB soft X-ray beamline, *Metrologia* 40, S224–S228 (2003).
86. F. Scholze, G. Brandt, P. Müller, B. Meyer, F. Scholz, J. Tümmler, K. Vogel, and G. Ulm, High-accuracy detector calibration for EUV metrology at PTB, *Proc. SPIE* 4688, 680–689 (2002).
87. M. Krumrey and G. Ulm, High-accuracy detector calibration at the PTB four-crystal monochromator beamline, *Nucl. Instr. Meth. A* 467–468, 1175–1178 (2001).
88. L. R. Canfield, R. E. Vest, R. Korde, H. Schmidtke, and R. Desor, Absolute silicon photodiodes for 160 nm to 254 nm photons, *Metrologia* 35, 329–334 (1998).
89. K. Solt, H. Melchior, U. Kroth, P. Kuschnerus, V. Persch, H. Rabus, M. Richter, and G. Ulm, PtSi-n-Si Schottky-barrier photodetectors with stable spectral responsivity in the 120 nm to 250 nm spectral range, *Appl. Phys. Lett.* 69, 3662–3665 (1996).
90. M. Richter, U. Kroth, A. Gottwald, Ch. Gerth, K. Tiedtke, T. Saito, I. Tassy, and K. Vogler, Metrology of pulsed radiation for 157-nm lithography, *Appl. Opt.* 41, 7167–7172 (2002).

91. P. Kuschnerus, H. Rabus, M. Richter, F. Scholze, L. Werner, and G. Ulm, Characterization of photodiodes as transfer detector standards in the 120 nm to 600 nm spectral range, *Metrologia* 35, 355–362 (1998).
92. F. Scholze, R. Klein, and T. Bock, Irradiation stability of silicon photodiodes for extreme-ultraviolet radiation, *Appl. Opt.* 42, 5621–5626 (2003).
93. K. R. Lykke, P.-S. Shaw, L. M. Hanssen, and G. P. Eppeldauer, Development of a monochromatic uniform source facility for calibration of radiance and irradiance detectors from 0.2 μm to 12 μm, *Metrologia* 36, 141–146 (1999).
94. G. Eppeldauer and M. Racz, Spectral power and irradiance responsivity calibration of InSb working standard radiometers, *Appl. Opt.* 39, 5739–5744 (2000).
95. S. W. Brown, G. P. Eppeldauer, and K. R. Lykke, NIST facility for spectral irradiance and radiance responsivity calibrations with uniform sources, *Metrologia* 37, 579–582 (2000).
96. P. Toivanen, F. Manoochehri, P. Kärhä, E. Ikonen, and A. Lassila, Method for characterization of filter radiometers, *Appl. Opt.* 38, 1709–1713 (1999).
97. C. A. Schrama and H. Reijn, Novel calibration method for filter radiometers, *Metrologia* 36, 179–182 (1999).
98. C. A. Schrama and E. W. M. van der Ham, Sampling period criterion in a scanning-beam technique, *Appl. Opt.* 39, 1500–1504 (2000).

# 4. TRANSFER STANDARD FILTER RADIOMETERS: APPLICATIONS TO FUNDAMENTAL SCALES*

George P. Eppeldauer, Steven W. Brown, Keith R. Lykke
National Institute of Standards and Technology, Gaithersburg, Maryland, USA

## 4.1 Introduction

Traditionally, spectral radiant power responsivity measurements propagate the reference scale from radiant-power-measuring primary standard radiometers to instruments that make other types of radiometric and photometric measurements. Often, the most frequently measured radiometric quantities are irradiance and radiance, not optical power. Irradiance is used to describe the amount of radiant flux falling on a surface or reference plane. Radiance is the most commonly used radiometric quantity to characterize and measure extended sources such as lamps, screens, and blackbody radiators (see Chapter 1). Similarly, the most common photometric measurements are measurements of illuminance, the amount of optical radiation falling on a surface weighted by the response of the human eye, and luminance, the amount of radiation from a source weighted by the response of the human eye.

Transfer standard filter radiometers are used to maintain and propagate reference irradiance and radiance responsivity scales to other radiometric and photometric calibration facilities and to validate irradiance and radiance responsivity scales realized in different ways. In this chapter, we discuss the design, characterization, and calibration of transfer standard filter radiometers used to measure sources over the spectral range from the ultraviolet to the mid-infrared. In particular, we will discuss the use of filter radiometers in the visible spectral region for the validation and propagation of fundamental irradiance and radiance responsivity scales. In Section 4.2, we discuss basic principles involved in the design of filter radiometers. In Section 4.3, we describe design considerations of filter-based irradiance meters. In Section 4.4, radiance meters are discussed.

The construction of a reference standard radiometer depends not only on the type of radiometric application (e.g. radiance or irradiance measurements), but also on the calibration method developed for the instrument.

*Contribution of the National Institute of Standards and Technology.



EXPERIMENTAL METHODS IN THE PHYSICAL SCIENCES, vol. 41
ISSN 1079-4042 DOI: 10.1016/S1079-4042(05)41004-8

The basic idea of modern radiometric measurements is to calibrate the responsivity of the transfer standard radiometer with the simplest possible geometry against the highest-level radiometer standard and then to use the transfer standard radiometer for real applications. In Section 4.5, calibration and characterization issues are presented.

Use of filter radiometers reduces the number of steps in the calibration chain making it possible to significantly reduce the uncertainty in a variety of source measurements. Because of the shorter calibration chain, the calibration cost is reduced along with the measurement uncertainty in many source measurement applications. In Section 4.5, we describe applications of filter radiometers, including the derivation and maintenance of Systeme International (SI) units, the kelvin and the candela, from the reference spectral responsivity scales.

Detectors used in irradiance and radiance meters commonly have a broad spectral response; for example, silicon detectors respond to radiant flux from below 200 to beyond 1100 nm. Consequently, irradiance and radiance meters designed to measure sources with a broad spectral distribution are often equipped with optical filters for spectral selectivity. In these instruments, spectral characteristics such as the center wavelength and bandpass of the filters are important design issues that impact the overall performance of the instrument.

Filter radiometers are used in a wide variety of applications, for example to achieve the spectral selectivity necessary to provide information about a chemical process or to determine the temperature of a blackbody [1]. Filter radiometers are often designed to achieve a desired spectral response; luminance meters, illuminance meters, and colorimeters are filter radiometers with spectral responsivities designed to match target spectral distributions. Luminance and illuminance meters are designed to mimic the Commission Internationale de l'Eclairage (CIE) photopic spectral luminous efficiency function, $V(\lambda)$, while colorimeters are multi-channel filter radiometers with individual channels designed to approximate the CIE color matching functions [2].

In remote-sensing applications, filter radiometers are commonly used to monitor environmental parameters and physical processes. In these applications, filters are chosen based on their intended application. To monitor atmospheric processes, filters are often chosen to coincide with atmospheric absorption features [3] or to reject absorption features [4]. To monitor processes on land or in the oceans, filters are typically chosen to coincide with atmospheric transmittance windows to minimize effects of the intervening atmosphere on the measurement.

Another important field where irradiance meters play a large role is in the measurement of ultraviolet (UV) radiation [5]. A typical problem in this field is the measurement of the changes in the level of terrestrial solar UVB radiation to assess the effects of ozone depletion and its impact on biological systems. In this field, the measurement uncertainties of filter radiometers with UV filters are high because of the instability and lower performance of the optical components (as compared to the visible range), such as sources, diffusers, filters, and detectors. Also, the intensity of natural UV radiation is low compared with the visible background. Because a meaningful global measurement program requires lower than 1% measurement uncertainties, careful design and calibration of these instruments is critical if they are to provide physically meaningful data.

Consideration of optical detector and filter elements is implicit in filter–detector-based optical metrology. Application-specific design issues—dependent on the radiometric quantity to be measured—primarily revolve around the geometrical constraints of the measurement. In radiant power responsivity measurements, the total power (radiant flux) leaving a source is measured by the detector. The unit of radiant power is the watt (W). The radiation to be measured has to under-fill the active area of the radiation-measuring detector. The geometrical design of power-measuring detectors and radiometers is simple. Usually, they do not need any input optics or defining apertures. An example is shown in Figure 4.1(a), where the total radiant power $P$ in a laser beam is measured by a detector. Laser power measurement is a practical example for detector radiant power responsivity measurement.

In irradiance measurements, the total radiant flux in a defined area in a reference plane is to be determined. The irradiance responsivity is the output current or voltage of the detector divided by the irradiance at the reference plane of the irradiance meter. A simplified scheme of irradiance measurements is shown in Figure 4.1(b). In the example shown, the meter's reference plane is the plane including the aperture with area $A_1$. The radiant flux leaving the overfilled aperture (with a total radiant power $P$) must be entirely measured by the detector. In this case, the irradiance in the reference plane is equal to the measured power divided by the aperture area.

In radiance measurements, the radiant flux from a certain area within an extended source is measured. Figure 4.1(c) shows an example for the radiance ($L$) measurement of a sphere source. For comparison, the same detector and aperture with area $A_1$ can be used in the radiance meter as in Figure 4.1(b) for irradiance measurement. The difference between the two measurements is that the radiance meter is equipped with input optics (e.g. a lens) to image a spot from the source surface onto the detector. The target spot, subtended by the meter, must be located inside of the source aperture that has an area $A_2$. The radiance measurement angle is determined by the input optics that includes the field stop aperture with area $A_1$.

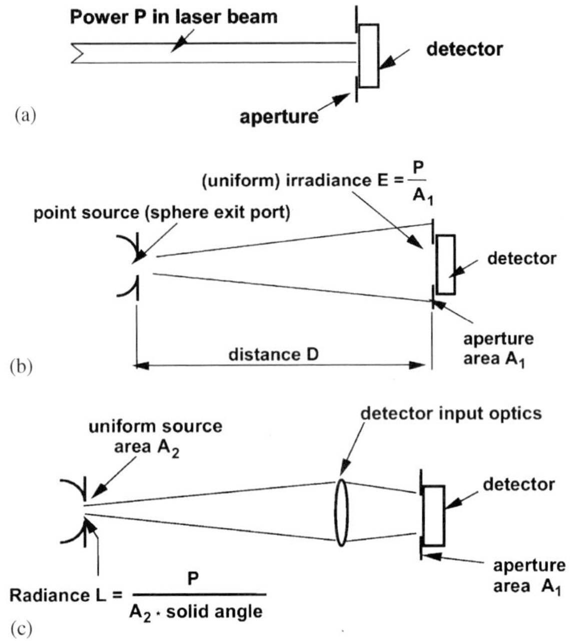


FIG. 4.1. (a) Simplified scheme of radiant power measurement. (b) Simplified scheme of irradiance measurements. (c) Simplified scheme of radiance measurements.

Calibrating a filter radiometer, designed for the measurement of the required radiometric quantity with the simplest possible geometry against the highest level radiometer standard, is one of the fundamental underlying principles of modern radiometric measurements. For radiant power, the detector responsivity is equal to its output current or voltage divided by the measured radiant power. The incident radiant power is determined from calibrations against a reference radiometer such as an electrical substitution cryogenic radiometer or silicon trap-detector. The irradiance responsivity scale is derived from the radiant power responsivity scale. In an ideal case, the irradiance responsivity of an irradiance meter is equal to the radiant power responsivity of its detector times the area of the detector aperture.

Radiance-measuring radiometers always need input optics that can image the radiant power from a target area of the source onto the detector. Note that the detector itself still measures optical power. In scale derivations, spatially uniform, Lambertian and extended sources (e.g. integrating spheres, see below) that overfill the radiometer entrance pupil are used as transfer calibration sources. The radiance responsivity of the meter is equal to the detector output current or voltage divided by the source radiance within the target area. Knowing the measurement geometry, the radiance of the transfer calibration source can be determined using an irradiance meter. Using this method, the radiance responsivity scale is derived from the irradiance responsivity scale.

## 4.2 Common Design Considerations for Filter Radiometers

Selection of the appropriate detectors and filters for a given radiometric application is a critical step in the design of a filter radiometer to achieve low uncertainties in the propagation of reference spectral responsivity scales. A preamplifier, when it is used with a selected detector, should be designed to obtain low measurement uncertainty even at high signal gains. Also, the signal-to-noise ratio should be optimized for the output of the filter radiometer [6]. In many cases, detectors and filters have temperature dependences to their detector responsivity or filter transmittance. If the temperature sensitivity of the filter radiometer is significant in the designed wavelength range, temperature monitoring or control is needed. Typically, temperature control is simpler in practice than temperature monitoring which requires subsequent correction or temperature normalization to the data [7].

### 4.2.1 Detector Considerations

Spectral responsivity, spectral response linearity, long-term stability and temperature dependence are all important considerations in choosing the appropriate detector for the envisioned application. When radiant power responsivity scales are derived from a primary standard cryogenic radiometer, high spatial-response uniformity is needed as well. Of course, for scale realizations and propagations, the highest quality photodetectors are required to minimize the responsivity uncertainty of the device.

In the visible region, silicon detectors are most often used. For standard applications, Hamamatsu S1337 (or the equivalent S1336 or S6337) photodiodes are commonly used. A physical model of the internal quantum efficiency (IQE) has been developed for this type of photodiode and can be used for extrapolation and interpolation of its spectral responsivity [8]. The IQE is the ratio of the number of collected electrons to the number of photons absorbed by the detector after the front surface reflection loss. A device's external quantum efficiency (EQE) is defined as the IQE times (1−reflectance). The reflection loss can be minimized in a device by arranging a number of photodiodes such that the reflection from one photodiode is incident on the next one in the "chain." In this configuration, the total device reflectance is equal to the reflectance from a single photodiode element to the *n*th power, where *n* is the number of photodiode elements in the device. A detector arranged in this type of configuration is known as a trap detector. Because of their high IQE, connecting the S1337 detectors in parallel and measuring the sum of the photocurrents gives close-to-unity EQE. A trap detector built from S1337 photodiodes is frequently called a "quantum-flat detector" because of its extremely high EQE. The EQE of a tunnel-trap detector built from six S1337 photodiodes is a constant, 0.998, within 0.1%, between 500 and 900 nm. These photodetectors have exhibited excellent long-term radiometric stability and spatial response uniformity. A maximum-to-minimum change in the spectral power responsivity of less than 0.03% was measured on the NIST Spectral Comparator Facility (SCF) on four S1337 photodiodes from 1992 to 2002 between 405 and 920 nm. On average, the maximum-to-minimum variation in the spatial response of these $1 \times 1$ cm detectors was 0.3%.

The photodiodes of a trap detector should be specially selected for equal shunt resistance to keep the resultant shunt resistance reasonably high. High photodiode shunt resistance is needed in those applications where high sensitivity (small amplification for the amplifier input noise), low current-to-voltage conversion uncertainty, and excellent linearity are needed. In addition to excellent radiometric characteristics, the Hamamatsu S1227 (or S1226) silicon photodiodes have shunt resistances in the 1–10 GΩ range. These are more than an order of magnitude larger shunt resistances than that of the S1337. Compared with the S1337 photodiode, they have a suppressed responsivity in the near-infrared range and are frequently used for photometric and other visible applications. The noise-equivalent photocurrent that can be measured with a S1227 photodiode is $10^{-16}$ A [9].

Many applications require measurements in the UV and infrared (IR) regions, outside the spectral range appropriate for silicon detectors. Passivated detectors that are resistant to UV damage should be used for filter radiometers operating in the UV range. Such detectors are thick oxide nitrided Si (UVG-100), the thin oxide nitrided Si (AXUV-100G) photodiodes (commercially available from IRD) and platinum-silicide photodiodes [10]. Silicon photodiodes with platinum–silicide (Pt–Si) front windows have about a decade lower responsivity than the nitrided Si photodiodes.

In the IR range, germanium (Ge), indium gallium arsenide (InGaAs), and extended-InGaAs photodiodes are conveniently used in filter radiometers between 0.8 and 2.5 μm. Indium antimonide (InSb) detectors are used for the 1 to 5 μm spectral range and mercury cadmium telluride (MCT) detectors are used for the 2 to 26 μm spectral range.

IR detectors are often operated at reduced temperatures to reduce the background noise and improve their operational performance, for example, to increase their shunt resistance (discussed below). Cooling is especially important for Ge and InGaAs photodiodes because they have low shunt resistances at room temperature. As a result of cooling, the shunt resistance can be increased significantly. As an example, the shunt resistance of a 5 mm diameter Ge photodiode increased from 17 kΩ at 25°C to 20 MΩ at −30°C. The shunt resistance change of a similar size InGaAs photodiode for the same temperature change was about 40 times smaller. However, the 25°C value was 2 MΩ [11]. Usually, a one-to four-stage thermoelectric (TE) cooler and a temperature-sensing thermistor are built into a photodiode can. The can is hermetically sealed with a window. The thermistor and the TE cooler thermal characteristics must be known to properly set the required (cold) constant temperature.

InSb detectors are typically placed in an evacuated dewar and operated at liquid-nitrogen temperatures (77 K). InSb detectors have an instability that occurs when short wavelength radiation is allowed to impinge on the detector. In this case, charges build up and are trapped in the device. Charge continues to build up until electric discharge occurs, resulting in a non-stable signal component, called InSb "flashing" [12], superimposed on the useful signal. Consequently, care should be taken when purchasing and using InSb detectors. The detector should have internal masking to minimize flashing and filtering to prevent short wave radiation from reaching the detector. Frequently, cold cut-on (high-pass) filters are used inside the dewar to block short wavelength radiation and minimize InSb flashing. The cut-on filter will set the low-end wavelength limit for the InSb filter radiometer. The upper cut-off wavelength of an InSb radiometer should be controlled by cold filters. An example for a 2.7 μm cut-off filter was shown in Chapter 3, Figure 3.5. An example for bandpass filter application will be shown later in this chapter (Figs. 4.21 and 4.22).

The maximum-to-minimum spatial response non-uniformity is within 1% even for large area (7 mm diameter) InSb detectors [13]. Less than 1% responsivity changes were obtained when the irradiance from blackbody radiators was measured over a time period of 1 year.

Detectors with known relative spectral responsivity are often used to extend the reference radiant power responsivity scale to the UV and IR ranges. Detectors with constant, flat responsivity versus wavelength are the most suitable devices for responsivity scale extensions. With the selection of high-quality pyroelectric detectors, pyroelectric radiometers can be developed for accurate responsivity scale transfer [14]. Using these pyroelectric radiometers, the sub-0.1% ($k = 1$) scale uncertainty in the silicon wavelength range will increase to 0.25% between 1 and 2 μm and the uncertainty will reach 0.34% at 2.5 μm [15]. In the UV spectral range between 125 and 320 nm, the relative combined standard uncertainty in the responsivity scale is less than 0.5% when UV damage-resistant detectors are calibrated against a cryogenic radiometer [16].

### 4.2.2 Filter Considerations

Filters are critical components of filter radiometers. Two types of filters are commonly used: colored glass filters and multilayer interference filters. Colored glass filters are used to block short- or long-wavelength radiation and to design instruments with a wide spectral bandpass, e.g. photometers and colorimeters. Interference filters are used in general when glass filters are not available or when a narrow spectral bandpass is required.

The color of filter glasses is determined by dissolving or suspending coloring agents in the glass. The optical density, or specific spectral transmittance of a filter, is controlled by the concentration of coloring agents in the glass or its thickness. Glass filters of different thicknesses that are cemented together are widely used in instruments such as in photometers and colorimeters.

A large variety of filters with different spectral transmittance curves can be realized with custom-made interference filters. The stability of interference filters has been a long-standing issue. To improve their radiometric performance, the high-reflectance multilayer coatings used in Fabry–Perot type filters can be sealed to protect the coatings from moisture. Also, the filters can be constructed from hard, durable, non-hygroscopic materials. In addition, the stability of interference filters has improved in the last several years with the development of ion-beam-assisted deposition (IBAD) of the dielectric layers that comprise the filter. IBAD filters are less sensitive to humidity, UV exposure, and temperature changes than conventional multilayer dielectric interference filters.

The transmission of interference-type filters depends on incident beam geometry. As the angle of beam incidence increases from the normal of the filter surface, the center wavelength shifts to shorter wavelengths. The lower the effective index of refraction, the greater the shift. Also, the filter transmittance is temperature dependent. It is important to keep the temperature coefficient of spectral transmittance low. The peak transmission wavelength increases and decreases with temperature in a linear fashion; nearly all optical filters exhibit a positive linear temperature coefficient.

Different types of filters have been used in the power and irradiance measuring radiometers discussed in this chapter. The out-of-band transmission of bandpass filters can range from $10^{-3}$ to $10^{-7}$. Consideration of the out-of-band transmission is very important when filters are used with broadband sources and detectors. Flatness, parallelism, surface quality, and pinholes are important issues mostly for imaging applications. In order to keep measurement uncertainty low, high-performance interference filters are to be purchased and parallel beams and temperature control for the filter(s) should be applied.

Inter-reflections between the aperture and filters or between the filter and the detector can result in a responsivity that is dependent on measurement geometry. Inter-reflections between filter components is also a common problem. In order to avoid inter-filter reflections when more filters are used, the individual filters (filter layers) should be glued together using optical cement. In general, the lowest measurement uncertainty is obtained when the same beam geometry is used at both calibration and applications. The simplest solution is to use a "point-source" geometry.

### 4.2.3 Temperature Control

In many applications, both the photodiodes and the filters need temperature control to keep the responsivity change of the filter radiometer negligibly small when operated under different ambient temperatures. The temperature-dependent, radiant power responsivity changes of two silicon photodiodes frequently used in radiometers and photometers, Hamamatsu S1226 and S1337, are shown in Figure 4.2. At 700 nm (close to the long end of the visible range), the S1226 photodiode has a temperature coefficient of responsivity less than 0.05%/°C. The temperature dependence increases nearly quadratically to longer wavelengths. At 1000 nm, the photodiode has a temperature dependence greater than 0.5%/°C. A 4°C temperature difference between an instrument's calibration and application will produce

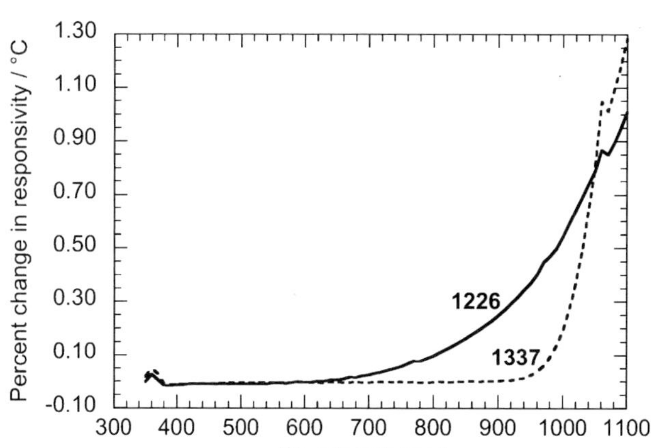


FIG. 4.2. The temperature coefficients of radiant power responsivity versus wavelength of Hamamatsu S1226 and S1337 silicon photodiodes.

0.2% responsivity change at 700 nm. In contrast, the Hamamatsu S1337 photodiode, widely used to propagate the spectral radiant power responsivity scale, has a negligible responsivity temperature coefficient below 940 nm. Instruments made from these photodiodes do not need temperature control up to this wavelength limit.

Filters can have a significant temperature dependence as well. For example, an illuminance meter measured the constant illuminance from a Standard Illuminant A source between 22°C and 29°C. A temperature coefficient of 0.083%/°C was obtained for the illuminance responsivity of the photometer. Figure 4.3 shows the linear curve fit to the measured data. The 0.083%/°C temperature coefficient was dominated by the temperature dependence of the cut-on edges of the individual filters used in the $V(\lambda)$ filter combination in front of the Hamamatsu S1226 silicon photodiode. The photodiode has a contribution to the temperature-dependent illuminance responsivity at wavelengths longer than 630 nm (see below).

Filters in radiometers are typically not cooled, but are maintained at a slightly elevated temperature using a separate temperature control loop. Temperature stabilization at a slightly elevated temperature rather than a lower temperature is typically used to avoid condensation on the filters. The filter radiometer illustrated in Figure 4.4 is a common design, where the temperature of the photodiode and the filter are controlled independently from each other.

Instruments using colored glass filters can be extremely stable if proper care is taken in the filter selection. For example, Figure 4.5 shows the measured relative stability of a group of standard illuminance meters over a 9-year period. The optical component packages in each illuminance meter contain a Hamamatsu S1226 silicon photodiode, a spectral filter, and an

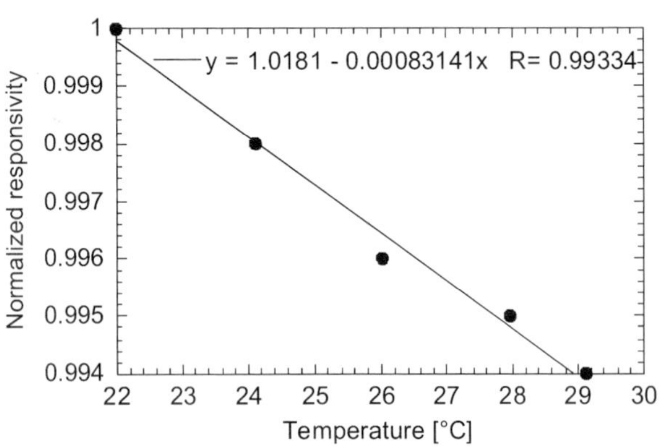


FIG. 4.3. Temperature-dependent responsivity of an illuminance meter.

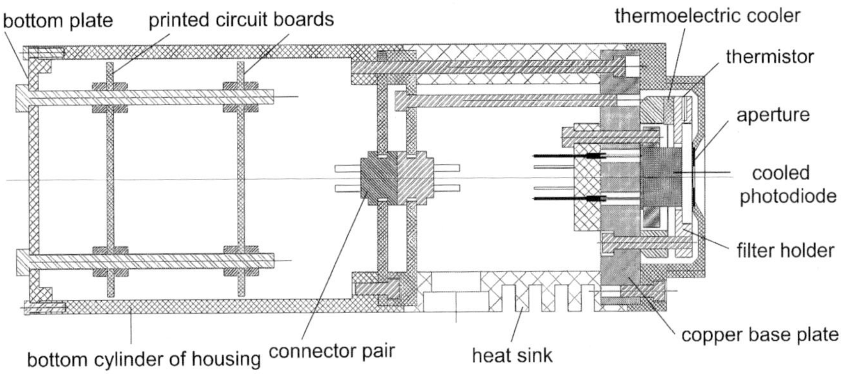


FIG. 4.4. Design of irradiance-meter head using single-element photodiode.

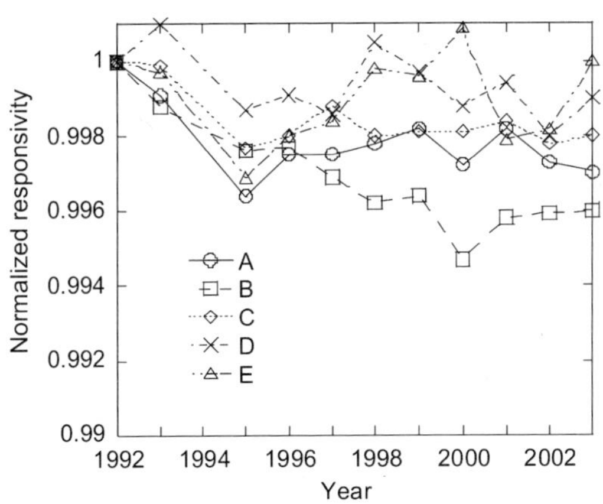


FIG. 4.5. Nine-year long stability of illuminance responsivity of a group of illuminance meters.

aperture in front of the filter. The filter between the aperture and the photodiode matches the spectral responsivity of the radiometer to the standardized CIE $V(\lambda)$ sensitivity of the human visual system [17]. The temperature of all filter-photodiode packages was monitored and corrections were applied for the illuminance responsivities as necessary.

The relative uncertainty of the yearly responsivity scale realizations was also 0.1%. Five illuminance meters were stable to within the repeatability of the measurements. Three instruments showed dramatic temporal changes in their relative responsivity. Each layer in the filter combinations used in the five stable radiometers shown in Figure 4.5 was an individually polished color glass filter [18]. Filter combinations for the three less stable illuminance meters were purchased from other commercial sources.

### 4.2.4 Preamplifier Design

The preamplifier is needed to measure the short-circuit current from the photodiode and to amplify the signal to obtain a high enough signal-to-noise ratio at the output. The preamplifier is the dominant amplification stage where the fundamental signal-to-noise ratio is determined. The preamplifier always measures the DC output of the photodetector. In general, the preamplifier characteristics must be matched to the impedance (shunt resistance and junction capacitance) of the detector if high measurement performance is to be achieved [6].

The simplest version of a widely used photocurrent meter is shown in Figure 4.6. The photocurrent of photodiode P is measured by a short-circuit current-meter. The current-to-voltage conversion in the current meter is realized by an operational amplifier (OA) and a feedback resistor ($R$) that has a parallel (stray) capacitance ($C$). This circuit diagram looks very simple. However, when low-uncertainty photocurrent measurements are needed over a wide dynamic signal range with a large signal-to-noise ratio in the output voltage $V$, the fundamental gains of the photocurrent meter need to be known. The fundamental gains are the signal gain, the loop gain, and the closed-loop voltage gain; each should be optimized for high accuracy measurements. The frequency dependent knowledge of the fundamental gains is especially important when the optical radiation is modulated. Signal chopping can be applied to tune the signal frequency outside of the $1/f$ noise range of low-frequency (DC) measurements. Also, in a measurement where the background signal component is large, such as in the infrared range, a

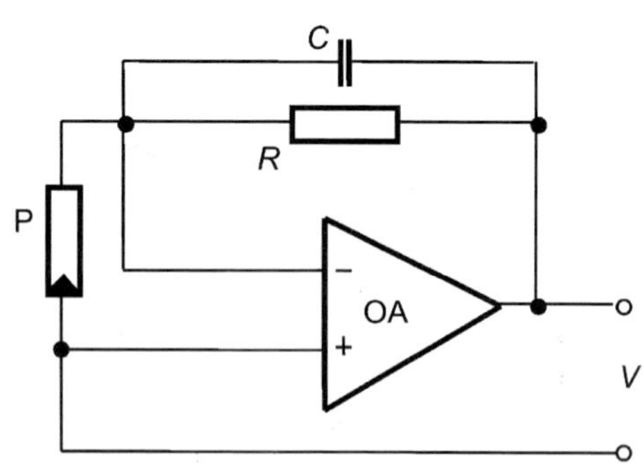


FIG. 4.6. Circuit diagram of a simple photocurrent meter.

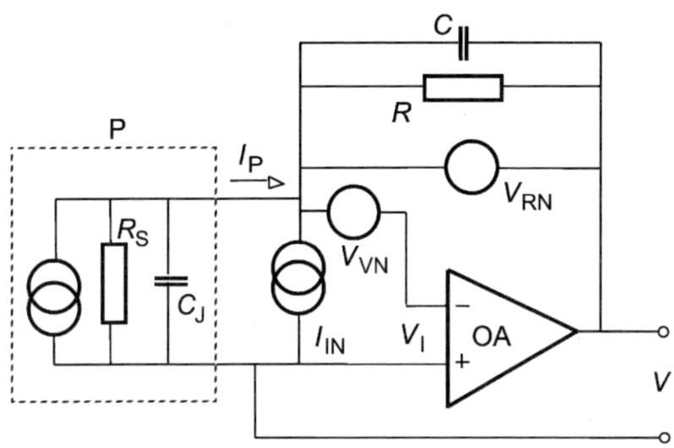


FIG. 4.7. Equivalent circuit of a photodiode short-circuit current meter.

modulated signal can be easily separated from the DC signal produced by the background radiation.

The equivalent circuit of the photocurrent meter of Figure 4.6 is shown in Figure 4.7. P has a shunt resistance $R_S$ and a junction capacitance $C_J$ which together produce the photodiode impedance $Z_d$. The photocurrent $I_P$ is converted into an output voltage $V$ through the feedback impedance of the OA. The feedback impedance $Z$ is the parallel connection of $R$ and $C$. The OA input voltage $V_I$ is the voltage between the inverting negative and non-inverting inputs. $V_I$ is small because of the large OA open-loop gain, $A$. The open-loop gain is the ratio of the output voltage $V$ to the input voltage $V_I$.

As the maximum of $V$ is 10 V, and $A$ is about $10^6$, $V_I$ is equal to or smaller than 10 μV. This very small voltage drop on P produces a small load resistance (input resistance of the current meter) $R_I$ for the photodiode [19]:

$$R_I = \frac{R}{A} \tag{4.1}$$

In case of low signal frequencies (DC photocurrent measurements), the requirement for linear short-circuit current measurements is $R_I \ll R_S$. Only low-frequency (DC) design considerations that can be applied for filter radiometers using silicon, UV, or near-to-mid-IR photodiodes are discussed in this chapter. The feedback impedance $Z$ and the impedance of the photodiode $Z_d$ create a feedback network from the OA output to the OA input. The voltage attenuation of the feedback network is $\beta$. Consequently, the characteristics of the three primary components in the analog control loop, the photodiode, the feedback components, and the OA must be considered jointly to optimize the preamplifier design.

The photocurrent-to-voltage conversion is given by the ratio of the output voltage $V$ to the input photocurrent $I_P$:

$$A_I = \frac{V}{I_P} = R \frac{1}{1 + G^{-1}} \quad (4.2)$$

where $G$ is the loop gain. The photocurrent-to-voltage conversion is given by $R$, which is the signal gain, when $G \gg 1$ at the signal frequency. This is a very important design requirement for the analog control loop when low photocurrent measurement uncertainty is needed from a filter radiometer. For example, for a 0.02% photocurrent-to-voltage conversion uncertainty, $G$ should be 5000 or higher at the signal frequency (that corresponds to the electrical bandwidth of a DC mode measurement).

The DC loop gain is the product of $A_0$ and $\beta_0$ at low frequencies [19]:

$$G = A_0 \beta_0 \quad (4.3)$$

where $A_0$ is the DC open-loop voltage gain of the OA and

$$\beta_0 = \frac{R_S}{R_S + R} \quad (4.4)$$

is the DC feedback attenuation. From Eqs. (4.2)–(4.4), the uncertainty of the photocurrent-to-voltage conversion depends on the photodiode shunt resistance. If a photodiode with high $R_S$ is selected, the feedback attenuation $\beta_0$ will be close to unity, resulting in high $G$. Usually, the feedback resistor $R$ is changed in decade increments to cover a wide dynamic signal range. The feedback resistors ranging from 1 kΩ to 100 GΩ are often used in preamplifiers to measure photocurrent (and hence radiant power) over a wide dynamic range.

A photocurrent meter (which is an analog control loop) can operate in the linear regime when these feedback resistors are used. At $R = 10^{12}\,\Omega$, the current-to-voltage conversion becomes non-linear. Using $R = 10^{11}\,\Omega$, the settling time of the output voltage $V$ can be 2 min [19]. With the right design, $R_S$ and $R$ should be similar to each other and $R = 10\,\text{G}\Omega$ is usually enough to achieve very low noise in the output voltage $V$ of the current meter [9]. With a maximum $R = 10\,\text{G}\Omega$ selection, the settling time will be less than 17 ms [19]. The loop-gain must be high enough at the signal frequency for all $R$ selections to perform current-to-voltage conversions with the required low uncertainty.

The DC closed-loop voltage gain, $A_{V_0}$ which is the reciprocal of $\beta_0$ (the inverse DC feedback attenuation),

$$A_{V_0} = \frac{R_S + R}{R_S} \quad (4.5)$$

determines the amplification of the input voltage noise $V_{VN}$ of the OA. It is important to choose an operational amplifier for the photocurrent meter with low input voltage noise. Equations (4.2), (4.3), and (4.5) comprise the basic gain equations of the preamplifier.

In general, an OA has to be selected which satisfies the noise, drift, input current, and speed (bandwidth) requirements. The $1/f$ (peak-to-peak) noise components of operational amplifiers between 0.1 and 10 Hz are available from data catalogs. The $1/f$ voltage noise is about 1 μV for the OPA627BM and 2.5 μV for the OPA111BM. The input current is about 5 pA for the OPA627BM and about 1 pA for the OPA111BM. The OPA128LM operational amplifier, which has a $1/f$ voltage noise of 4 μV and a $1/f$ current noise about an order of magnitude lower than that of the other two models, can be used where smaller than 75 fA input current is needed. This current will produce a small voltage drop on the high source resistance (the parallel connected $R$ and $R_S$). When $R = 100\,\mathrm{G}\Omega$ is selected, $R_S$ can be close to 100 GΩ to obtain a large loop gain and low closed-loop voltage gain. However, care should be taken because the resistor noise can dominate the total noise at the output of the current meter and also the measurement will be very slow because of the long settling time. The high resistor noise can be decreased by making the electrical bandwidth very small. For slow DC measurements (the electrical bandwidth is smaller than 0.1 Hz) the amplifier noise is not available from data sheets; therefore, the total noise has to be measured [20]. In an optimum design, the resistor noise and the amplified $1/f$ noise are equalized for the output of the photocurrent meter at a given electrical bandwidth.

In Figure 4.7, $V_{RN}$ is the Johnson (resistor) noise. This "white" noise shows up directly at the OA output (without any amplification). The current noise $I_{IN}$ is caused by the fluctuation of the OA input current. Both $I_{IN}$ and $V_{VN}$ have "white" and $1/f$ noise components superimposed on each other.

The photocurrent $I_P$ produces a shot noise $I_{PN}$, which is not shown in Figure 4.7. The current $I_{PN}$ is converted to the OA output similar to $I_P$. This current noise component is $(2eI_P\Delta f)^{1/2}$, where $e$ is the elementary electron charge, $1.60 \times 10^{-19}$ C, $I_P$ is the photocurrent, and $\Delta f$ is the electrical bandwidth. A $10^{-14}$ A photocurrent produces an rms current noise of $I_{PN} = 7.1 \times 10^{-18}$ A at $\Delta f = 16$ mHz, which corresponds to an integration time constant of 10 s.

At the OA output, the signal-produced voltage $V$ has to be always much larger than the superimposed total noise voltage originating from the above four noise components. The signal-to-noise ratio at the current meter output has to satisfy the uncertainty requirement of a measurement.

The DC signal gain ($V_0/I_P$) of a preamplifier must be known or calibrated against high accuracy standards, such as resistor standards or current-source standards. In order to avoid tedious and limited accuracy signal-gain calibrations, the feedback resistors ($R$) should be purchased with a 0.01% uncertainty (tolerance) of their decade nominal values. Such resistors are commercially available from 1 kΩ to100 MΩ with a resistance temperature coefficient of 10 ppm/°C. The $R = 1\,\text{G}\Omega$ signal gain can be calibrated against the $R = 100\,\text{M}\Omega$ signal gain by measuring the same constant input current (or optical radiation when the detector is connected to the preamplifier input). In order to utilize the 0.01% uncertainty of resistors $R$, a beam shutter should be used at the input of the filter radiometer. The output offset voltage of the preamplifier can be canceled out from the measurement by subtracting the dark reading (shutter closed) from the signal + dark reading (shutter open) at the output of the preamplifier (or the following digital voltmeter).

Usually, integrating-type digital voltmeters measure the preamplifier output voltage for the duration of 100 power line cycles (NPLC = 100) of 1.7 s to avoid long measurement times. An rms photocurrent noise floor of 1.5 fA was measured with NPLC = 100 on an optimized silicon photodiode current meter when $R$ was selected to a maximum of $10^{10}$ V/A to avoid long settling times and high resistor noise [9]. Well-designed silicon photodiode current meters can be used as building blocks of high-performance filter radiometers, including photometers, pyrometers, and tristimulus colorimeters. These radiometers have a signal (radiant power or photocurrent) dynamic range of greater than 14 decades.

Photocurrent meters designed for AC measurements can measure weaker optical signals with shorter measurement times than DC photocurrent meters [9, 19]. A practical sensitivity limit in AC (chopped radiation) measurement mode with a 10 s integration time constant of 0.1 fA can be achieved [9]. The fundamental preamplifier gain equations (4.2), (4.3) and (4.5), are frequency-dependent. For AC measurements, the parallel $C$ capacitors are to be calculated for all $R$ selections to optimize the overall preamplifier performance.

## 4.3 Design Considerations of Irradiance Meters

The type of radiometric applications of the meter, as well as the calibration measurement conditions, need to be considered in the design and construction of an irradiance meter. In an ideal irradiance meter, the spectral irradiance responsivity $s_E(\lambda)$ is proportional to the projected area of the aperture in the measurement direction

$$s_E(\lambda) = A_d \cos(\alpha)\, s_{avg}(\lambda) \tag{4.6}$$

In Eq. (4.6), $s_{avg}(\lambda)$ is the average spectral power responsivity over the aperture area (A/W), $A_d$ ($m^2$) the aperture area , and cos($\alpha$) the projection of the aperture area in the measurement direction, where $\alpha$ is the angle between the incident radiant flux and the normal of the aperture plane. This is known as the cosine law of detector irradiance measurements. Note that the projected area is a maximum at perpendicular incidence ($\alpha = 0^\circ$) and zero at $\alpha = 90^\circ$. If the incident beam is not parallel, $\alpha$ is no longer a constant, but can take a range of values depending on the measurement geometry. In this case, the integral of cos($\alpha$) will be measured for all incident angles of the beam.

When irradiance is measured from an extended source, the angular responsivity of the irradiance meter has to follow the cosine function within the angular range determined by the size of the extended source and the detector aperture. In general, the FOV of the meter has to be larger than the largest convergence angle of the beam to be measured and overfills the aperture of the meter. This means that all of the radiation incident on the overfilled aperture of the irradiance meter must be measured by the underfilled active area of the detector.

Different kinds of apertures can be used in irradiance meters. The aperture edge should be always very thin to minimize radiation scatter. Also, the reflectance from the aperture surface should be well designed if the aperture reflectance is high. Otherwise, low-reflectance apertures should be used. Bi-metal apertures, such as black nickel-coated copper can be thin (e.g. 0.13 mm) and still convenient to use. A short bevel can be made between the aperture and the front surface of the radiometer to keep the stray radiation low.

One of the common approaches to the calibration of irradiance meters is to calibrate the spectral power responsivity of the irradiance meter and measure the aperture area. In this approach, the irradiance responsivity is equal to the spectral power responsivity times the measured aperture area. Irradiance measuring transfer standard radiometers that convert power responsivity into irradiance responsivity must have apertures of known area in front of the detector and good spatial uniformity.

The effective aperture area depends on the radiometer design and the geometry of the experimental setup and may not be equal to the geometrically measured aperture area. In some cases, the effective area depends on the beam geometry or source to instrument distance. To minimize the uncertainty associated with the effective area, in scale derivations and transfer to test artifacts, the measurement geometry should be specified. The geometrically measured aperture area can be equal to the effective aperture area if the beam incident on the instrument aperture is approximately parallel. In order to obtain this well-collimated beam shape, a "point-source"-type irradiance calibration geometry should be used.

Test irradiance meters that are calibrated by substitution against a reference irradiance meter do not need to have their aperture areas accurately measured. However, knowledge of the location of their reference planes is important if the measurement geometry changes. In general, the front surface of the aperture is the reference plane of an irradiance meter.

### 4.3.1 Mechanical and Optical Design

In order to understand design issues of irradiance measurements, consider ideal and real beam shapes and simplified detector input geometries shown in Figure 4.8. For an ideal irradiance meter, a thin aperture should be positioned on the active area of the detector. In this ideal case, the shape of the beam that overfills the aperture can change from a collimated, parallel beam to a large angle of $\beta$ without any change in the irradiance responsivity. This measurement geometry is shown in Figure 4.8(a). The irradiance from different shapes of beams can be measured accurately with this ideal detector-input arrangement.

Typically, the detector has a protective window on the front and a spectrally selective filter is placed between the defining aperture and the reference detector. In most real instruments, therefore, a separation, or gap, exists between the aperture and the detector. The advantage of a point-source-like irradiation is that it produces a uniform irradiance with a well-collimated beam shape within the aperture of the irradiance meter. When this irradiance measuring geometry is used, shown in Figure 4.8(b), a separation or gap between the aperture and the detector does not affect the performance of the instrument.

If the incident beam is not collimated, multiple beam reflections between the detector surface and optical element between the detector and the

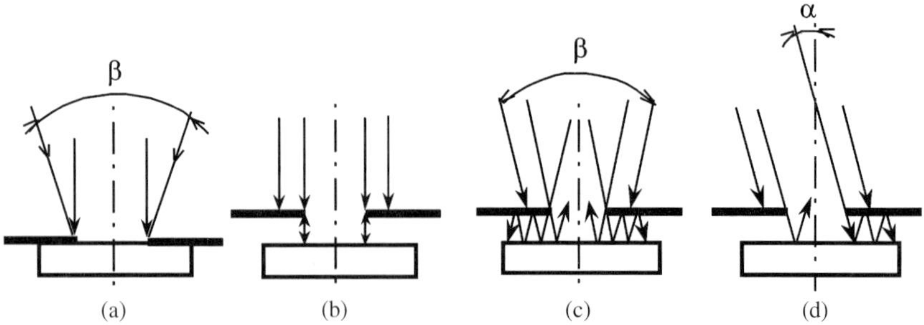


FIG. 4.8. Traditional input optics of irradiance meters with different incident beams. (a) Bare detector with a thin aperture on the top. (b) The parallel incident beam is perpendicular to the plane of the aperture. (c) The incident beam is converging. (d) The parallel incident beam is not perpendicular to the plane of the aperture.

aperture, or between the detector and the aperture itself can occur. This situation is shown in Figure 4.8(c) where the incident beam has a range of angles $\beta$. The reflectance pattern and the irradiance response can change with changing $\beta$. In this case, the instrument responsivity can be dependent on the geometry of the measurement. As a consequence, the instrument will no longer strictly follow the cosine law of detector irradiance measurements. A third situation is shown in Figure 4.8(d), where the parallel incident beam is no longer perpendicular to the aperture plane but is incident at an angle $\alpha$ relative to the normal of the aperture plane. In this situation, unwanted inter-reflections can also occur. Again, both the internal reflectance pattern and the responsivity deviation from the cosine function can be dependent on beam geometry.

For example, consider the angular responsivity of a real filter-type irradiance meter. The irradiance meter has a filter with a peak at 860 nm and a full-width, half-maximum bandpass of 60 nm. The separation between the aperture and the detector surface is approximately 4 cm. Figure 4.9 shows the responsivity of the radiometer as it was rotated around the center of its aperture in both the vertical and horizontal planes. The source was a Wi-41G lamp located 3 m away from the 4 mm diameter detector aperture. The responsivity deviation from the (ideal) cosine function is less than 0.05% within a beam convergence angle of $\pm 1^\circ$ (relative to the normal of the aperture plane). The deviation increases to 0.2% at about $\pm 2.5^\circ$. The two lobes located $\pm 2.5^\circ$ from normal incidence ($0^\circ$ radiometer rotation) where the responsivity is increased result from an increase in detector output signal caused by multiple internal reflections. It follows from the measurements that the geometry will affect the response of this instrument for beam geometries with convergence angles greater than $1^\circ$. To improve irradiance

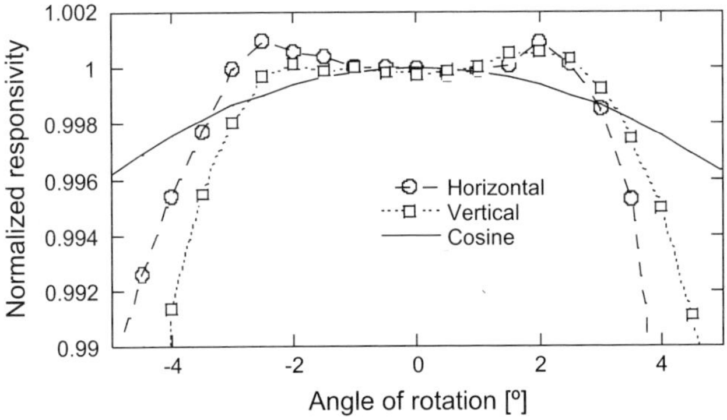


FIG. 4.9. Angular responsivity of an irradiance measuring filter radiometer where the aperture–filter–detector separations are large.

measurement uncertainty, better design is needed to increase the acceptance angle of irradiance measuring filter radiometers.

A properly designed irradiance meter follows the cosine law of detector irradiance measurements within a specified angular range or acceptance angle. This is known as the instrument's field-of-view (FOV). An irradiance meter designed for high-accuracy measurements can only be used to measure sources within this FOV. The magnitude of the FOV depends on the application. The FOV should be small for "point source" and collimator measurements to reject radiation outside of the radiometer FOV. A larger FOV is needed for extended source measurements, especially if the source is close to the radiometer. Examples of different types of irradiance meters constructed with or without a diffuser, where the cosine responsivity is maintained within larger acceptance angles (or FOV), are discussed below.

### 4.3.2 Bandpass Type Irradiance Meters

The bandpass of an irradiance meter can be limited for different reasons. When broadband radiation is measured, the spectral responsivity of the irradiance meter must be standardized to obtain uniform measurement results. A typical example is the photopic filter for illuminance measurements. In other broadband source measurements, such as blackbody radiators, it makes the measurement simpler and faster if the spectral bandwidth of the detector is limited. For instance, it is possible to determine the temperature of a blackbody if its broadband radiation is measured only in a limited spectral range. In this case, the calibration of the bandpass radiometer is easier and faster, especially when tunable lasers are used for the radiometer calibration. The use of bandpass radiometers is especially important in the infrared, where the spectral range is much broader than in the visible.

#### *4.3.2.1 Irradiance meter design for single detectors*

In a well-designed irradiance meter, the input optical components (especially the detector and the aperture) are located close to each other to minimize the effects of internal reflections for different input beam geometries. Such an irradiance meter that can be built with different kinds of single element photodiodes is shown in Figure 4.4. Both filters and transmitting diffusers can be positioned in front of the photodiode. A thin aperture touches the front surface of the filter (or diffuser). The diffuser–input irradiance meter design issues will be discussed in Section 4.3.3. The aperture, filter, and photodiode are positioned as close to each other as possible.

The FOV of the irradiance meter, without using a diffuser, depends on the size of the detector, the gap between the detector and detector window, the thickness of the filter or filter combination, the gap between the filter and

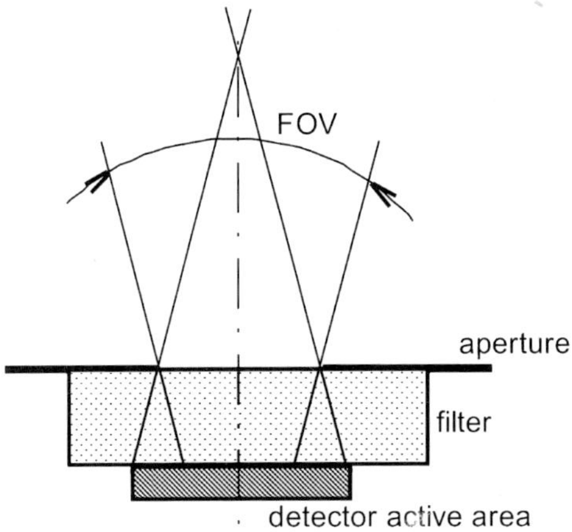


FIG. 4.10. FOV determination of a filter radiometer.

detector window, the gap between the filter and the aperture, and the size of the aperture. The gap between the filter and aperture should be equal to zero. In the best design, the aperture is thin and it touches the front surface of the filter. As the thickness of the filter or filter combination is given for a certain application, only the detector and aperture sizes can be selected to approach the desired FOV. Figure 4.10 shows the unvignetted FOV.

In the radiometer design shown in Figure 4.4, the radiometer has two independent temperature control loops. The first loop controls the temperature of the photodiode. Photodiodes with a thermoelectric cooler and a thermistor inside the sealed-can can be cooled to low temperatures for high-sensitivity applications. The bottom of the photodiode-can is attached to a copper base plate that delivers the dissipated heat to the heat sink of the housing. The second temperature control stabilizes the temperature of the filter/diffuser holder. The holder is pulled with three nylon screws against three thermoelectric coolers that are attached to the copper base plate through a copper ring. The leads of the TE cooler and the thermistor are attached through connectors and cable to an outside temperature controller. Selection of high stability and accurate temperature controllers is important to keep the measurement uncertainty low.

The photodiode leads are connected to the pins of a connector pair. The opposite side of the connector is mounted to the front panel of the bottom cylinder of the housing where the current-to-voltage converter (preamplifier) is located. Two printed circuit boards are used for the circuit components of the preamplifier including the operational amplifier, the feedback components, and the range switching reed relays. The rotary switch for gain control and two more connectors are located on the bottom plate of the housing. One connector connects the power supply to the radiometer head. The second connector can be used to remote control the signal gains of the converter.

#### *4.3.2.2 Filter-trap type irradiance meter*

Internal reflections between the input aperture and the detector can be avoided if tunnel-trap detectors are used in place of the single element detectors. A significant advantage of a tunnel-trap detector is that no light is retro-reflected from the detector to the input aperture [21]. The scheme of a tunnel-trap detector is shown in Figure 4.11 [22]. An aperture on the front of the instrument defines the reference plane that is set to be in the same plane as the front panel. A five-channel, temperature-stabilized filter wheel is located between the aperture and the front of the tunnel-trap detector. The length of the filter packages is limited to 10 mm by the thickness of the filter wheel. A rotating knob located on the backside of the cover box is mounted on the shaft of the wheel for filter selection.

The space for the wheel, between the aperture and the trap detector, is about 12 mm. This separation does not decrease the FOV of this irradiance meter. An expanded view of the tunnel trap detector configuration is shown

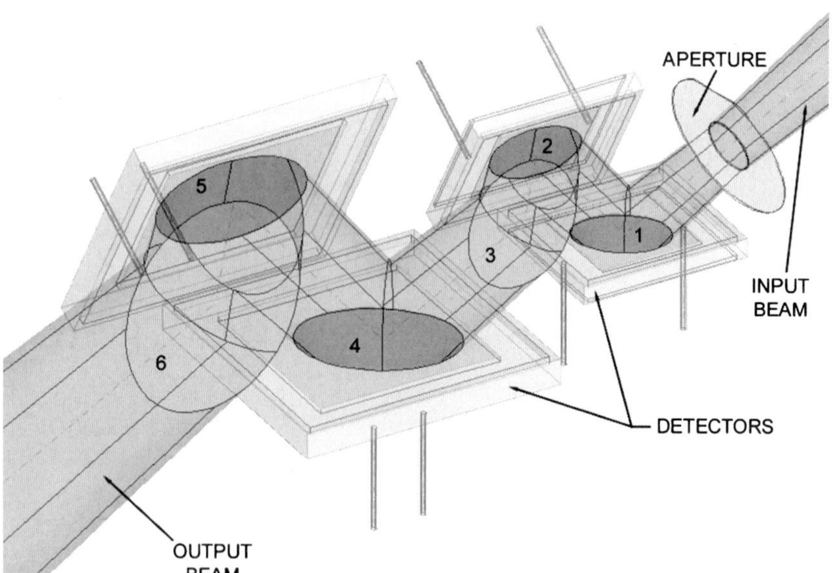


FIG. 4.11. Beam propagation inside the tunnel-trap detector.

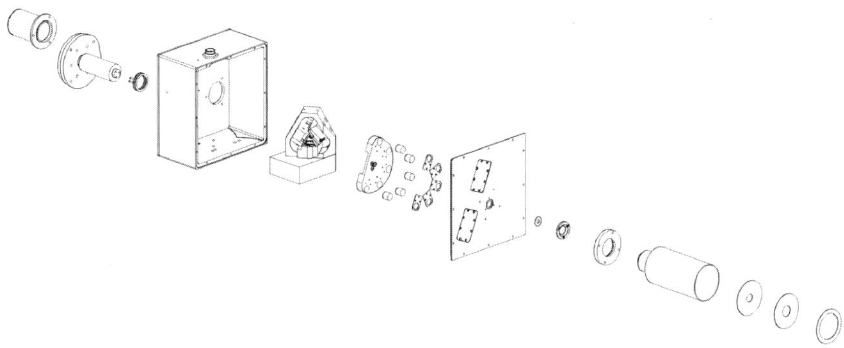

FIG. 4.12. Expanded view of the filter–trap radiometer layout.

in Figure 4.12. Two 10 × 10 mm and four 18 × 18 mm silicon photodiodes, equivalent to the Hamamatsu S1337 model, were packed in a light trapping arrangement. Temperature control for the photodiodes is not needed because the temperature coefficient of responsivity of these photodiodes is close to zero in the visible range and up to 950 nm. However, in order to minimize voltage gain for the input noise and drift of the operational amplifier in the photocurrent measuring circuit of the trap detector, the S1337 photodiodes were selected for high and equal shunt resistance. This selection was necessary because the six photodiodes were connected in parallel.

Figure 4.13 shows the normalized angular responsivity of the tunnel trap detector when it is equipped with a circular aperture of 5 mm diameter. The responsivity deviation from the expected cosine function is <0.02% within a 6° FOV. At 8° FOV, the deviation is still <0.05%.

### 4.3.3 Diffuser-Type Irradiance Meters

A transmitting diffuser between the aperture and the detector is often used in the case where a relatively large FOV is needed. The diffuser can eliminate the changing reflectance patterns for different input beam shapes or source sizes. Glass diffusers or thin Teflon diffusers are often used for the visible and near-IR wavelength ranges. Diffusers can have very different angular transmittance characteristics. In principle, the goniometric Bidirectional Transmittance Distribution Functions (BTDF) of diffusers should be measured. For example, the BTDF of two different diffusers, a 3.2-mm-thick Spectralon diffuser and a 2.5-mm-thick flashed opal glass type diffuser, are illustrated in Figure 4.14 [23]. The BTDF was measured at 600 nm with normal beam incidence. The Spectralon diffuser showed a large forward scatter between ±10°, while the normalized transmittance of the

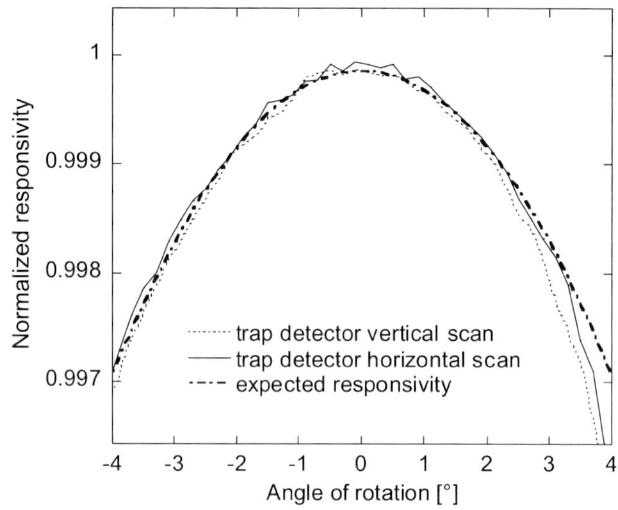


FIG. 4.13. Normalized angular responsivity of the tunnel-trap irradiance meter.

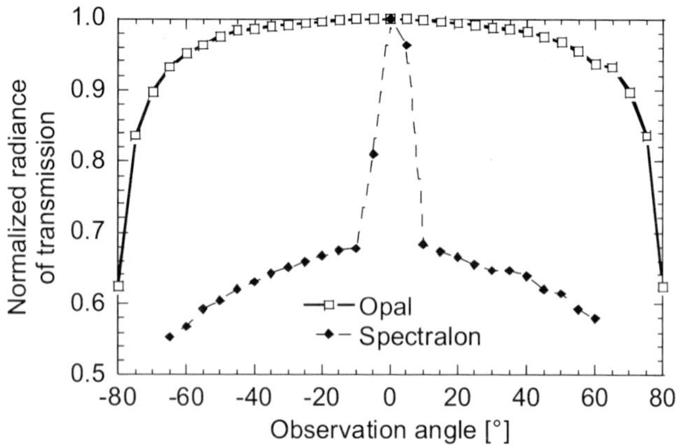


FIG. 4.14. BTDF of flashed opal glass and Spectralon diffusers at normal beam incidence.

flashed opal glass was constant in a reasonably wide angular range. Figure 4.15 shows that the ideal Lambertian BTDF curve (solid circle) and the measured flashed opal glass transmittance (dashed circle) curve are close to each other. Consequently, the flashed opal glass diffuser is the better choice for accurate irradiance meter developments when a relatively large FOV is required.

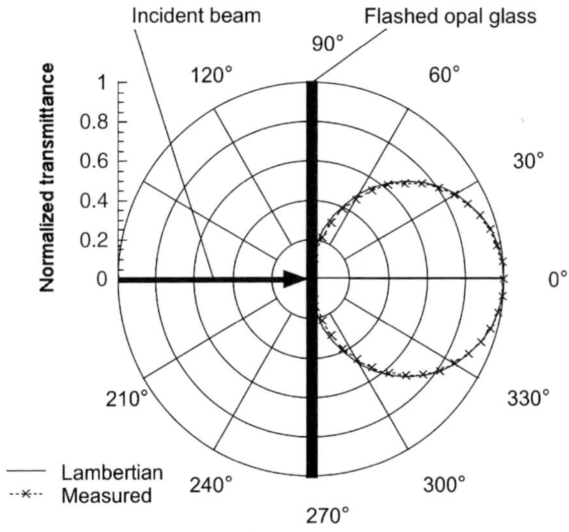


FIG. 4.15. Comparison of the angular transmittance distribution of a flashed opal glass diffuser (dashed) to an ideal Lambertian distribution (continuous).

The spectral dependence of the BTDF should also be considered; the BTDF of the above-measured flashed opal glass diffuser is significantly different at 1550 nm. Also, ground glass diffusers are not as good as flashed opal glass diffusers. They can be used only in the visible and within a smaller FOV. Unfortunately, there is not enough measurement data for different kinds of diffusers available. Therefore, goniometric characterizations should be made to select the best diffuser for a given application where low irradiance measurement uncertainty is needed.

Usually, optimization of the input geometry of diffuser-type irradiance meters is performed experimentally [24]. The angular responsivity depends upon the diameter of the aperture and the quality of the diffuser. When filters are used, they should be positioned between the diffuser and the detector and all three components (aperture, filter, and diffuser) should be included in the optimization. As the diffuse spectral reflectance and transmittance of the flashed opal glass diffusers is wavelength-dependent, the optimum cosine responsivity depends on the wavelength [24]. Consequently, the input geometry has to be optimized for the spectral distribution of the source to be measured.

To illustrate the effect of a transmitting diffuser application, the angular responsivity of a silicon irradiance meter using the design of Figure 4.4 built with selected aperture and diffuser to maintain the closest responsivity to the cosine function was measured with and without the diffuser. The dashed

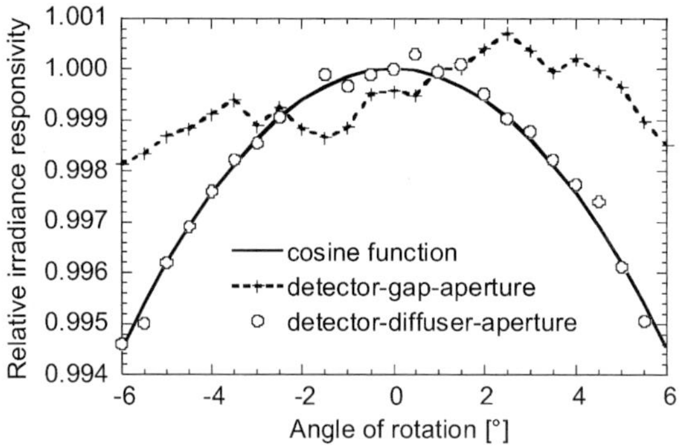


FIG. 4.16. Measured angular responsivity of a silicon irradiance meter with and without diffuser.

curve on Figure 4.16 shows how different the angular responsivity was from the ideal cosine function when a gap of about 7 mm was applied between the detector and the aperture. The open circles were obtained with the optimized aperture–diffuser combination. The small structure in the data points was caused by measurement noise including lamp instability and dust particles in the beam. The aperture was irradiated by a Wi-41G lamp from a distance of 3 m. The deviation from the cosine function is approximately equal to the measurement uncertainty of 0.03%. The deviation from the cosine function is still <0.15% at $\pm 12^\circ$ incidence angles.

A side effect of the cosine response optimization in diffuser-type irradiance meters is a change in the spatial responsivity distribution. The responsivity at the aperture edges will be smaller than in the center of the aperture. As an example, the response uniformity of the active area of a cosine optimized InGaAs irradiance meter is shown in Figure 4.17 when a 6.4 mm diameter aperture and a diffuser that transmits 40% at 1400 nm were selected. The spatial responsivity scans shown in Figure 4.17 were made at 1500 nm with a beam diameter of 1.1 mm. Each distance step of the scan was 0.5 mm along the horizontal $x$- and $y$-axis of the aperture relative to the aperture center. The $y$-axis shows the output signal of the InGaAs irradiance meter as normalized to the center position of the aperture. As shown in Figure 4.17(a), the responsivity decrease at the edge of the aperture can be as high as 20%. In Figure 4.17(b), the radiation scattered from the area surrounding the aperture can be seen for the same radiometer. Diffuser-input irradiance meters are not recommended for accurate radiant power measurements because of their high spatial non-uniformity of responsivity.

The effective area of an irradiance meter can be calculated as the ratio of the integrated irradiance responsivity to the radiant power responsivity in

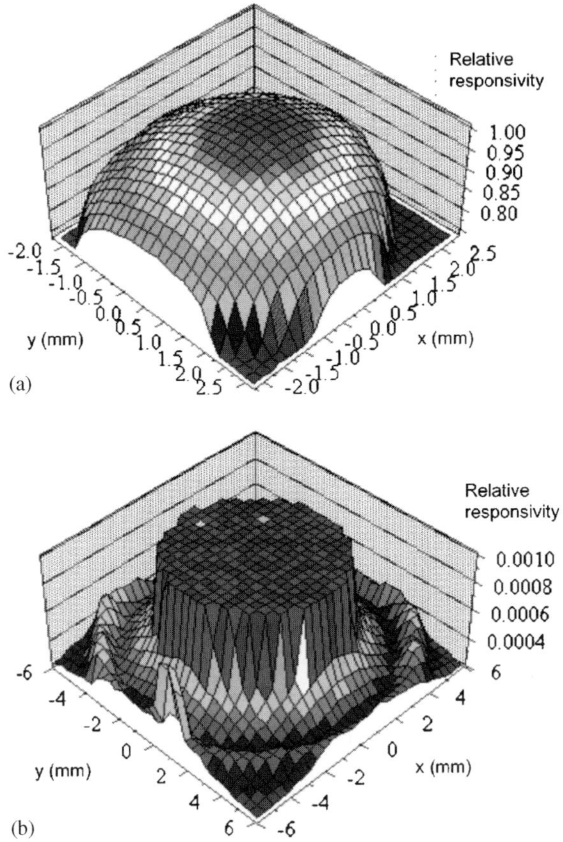


FIG. 4.17. Spatial responsivity distribution of a diffuser-input InGaAs irradiance meter. (a) the uniformity of the sensitive area; (b) the stray radiation around the radiometer aperture.

the center of the aperture. The effective area is utilized in flux transfers when the measured irradiance in the meter's aperture plane is transferred to another device, e.g., a sphere source to calibrate the radiance of its exit port [25]. Flux transfer is used when the measurement geometry is different from "point source" geometry. The result of the above spatial responsivity change will be a decreased effective aperture area for irradiance measurements. Consequently, for irradiance meters with a diffuser input, measurement of the real, geometrical aperture area is meaningless.

The spectral characteristics of silicon and InGaAs irradiance meters are shown as well. The spectral characteristics were determined by the photodiode and the flashed opal glass diffuser used in the meters. In Figure 4.18,

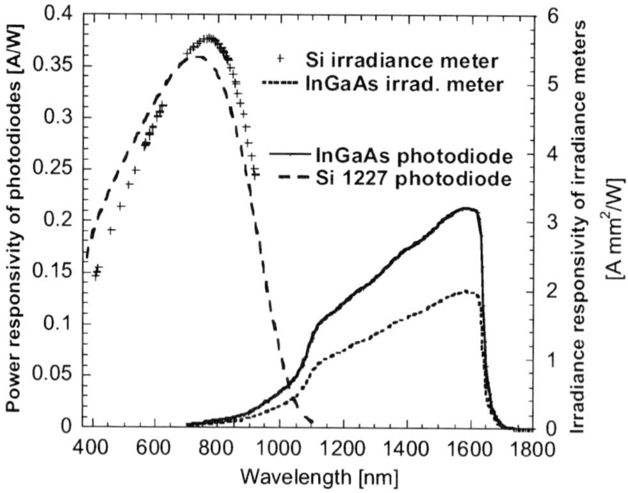


FIG. 4.18. Spectral responsivities of the Si and InGaAs photodiodes and the corresponding irradiance meters.

the spectral responsivity of the silicon and InGaAs photodiodes and the irradiance meters are shown. The magnitudes of the irradiance responsivities are similar for the two meters because the larger 8.0 mm diameter aperture of the Si meter produces higher irradiance responsivity than the smaller 6.4 mm diameter aperture of the InGaAs meter. Sources with either monochromatic or known (e.g., Planckian) spectral power distributions can be measured with these broadband irradiance meters. Similarly to InSb filter radiometers (as discussed in Section 4.3.5), use of bandpass filters between the diffuser and the detector makes it possible to measure the band-weighted irradiance of broadband extended sources.

### 4.3.4 Sphere-Type Irradiance Meters

When irradiance measurements are needed for even larger FOVs than those of diffuser-input irradiance meters, sphere-input irradiance meters can be used. The spatial uniformity of responsivity will not be degraded in well-designed sphere-input radiometers. Accordingly, these radiometers can be used for both irradiance and radiant power measurements with low measurement uncertainties. Another advantage of sphere-input radiometers is that the entrance port with the input aperture can be made large to measure the incident radiant power of large-diameter beams.

As an example, the construction of a sphere-input irradiance meter is shown in Figure 4.19. In this radiometer, an integrating sphere is placed

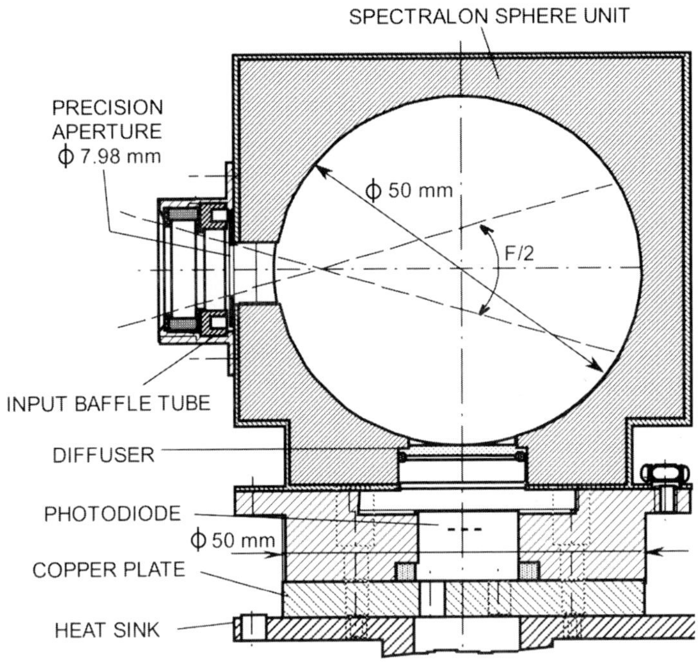


FIG. 4.19. Sphere-input irradiance meter.

between the transmitting diffuser and the precision aperture. The sphere produces a spatially uniform input surface within the aperture area. In the design shown, the shape of the incident beam can change between $f/2$ and parallel. The FOV is limited by the precision aperture and the front aperture of the input baffle tube. The sphere wall is made of machined Spectralon (LabSphere, Inc., North Sutton, NH), a diffusing material with a reflectance of 99% over much of the visible. The irradiance on any part of the sphere wall is almost the same and proportional to the input flux (radiant power) penetrating the aperture. The plane-parallel transmitting diffuser, located in the exit port of the sphere, is part of the sphere wall. It is irradiated by the multiple reflections within the sphere. Any filters in the sphere-input irradiance meter should be located between the transmitting diffuser and the detector.

The efficiency of the flux transfer between the transmitting diffuser and the photodiode is determined from the geometry between them. The flux coupling efficiency, $\kappa$, for a 5-mm-diameter photodiode, an exit port diameter of 10.5 mm, and a separation between the diffuser and the photodiode of 6.5 mm, is equal to 8.5%. The efficiency, $\tau$, of the 50 mm diameter sphere with its two ports and a transmitting diffuser with a transmittance of 36% located in the exit port, is equal to 21% [26, 27]. The overall radiant flux efficiency is $\kappa\tau = 0.018$, resulting in a signal attenuation of 56. The high-signal attenuation results in low radiometer responsivity. The decreased responsivity can be a disadvantage of sphere-input radiometers relative to the diffuser–input irradiance meters discussed above.

High throughput is sacrificed in sphere-input radiometers in order to achieve a spatially uniform receiving area and a large angular irradiance responsivity that follows the cosine function. Optimization of the spatial response uniformity and wide FOV, or large angular responsivity, in sphere-input radiometers are related tasks. Usually, the spatial non-uniformity is 0.1% or less, a factor of 3–5 better than that of the best quality silicon photodiodes. This is an important transfer standard feature for converting radiant power responsivity into irradiance responsivity.

The spatial response uniformity of the sphere-type irradiance meter of Figure 4.19 is shown in Figure 4.20. The scanning was made at 1500 nm by a 1.1-mm-diameter spot with 0.5 mm step sizes. The 0.1% maximum-to-minimum spatial responsivity change has a slope caused by the 90° asymmetrical arrangement of the entrance and exit ports. The spatial uniformity of the responsivity could be further improved with a symmetrical multiple-detector arrangement. Spectralon has a significant retro-reflectance; at normal beam incidence to the aperture plane, more radiation leaves the sphere than at larger incidence angle. Therefore, for non-normal incident radiation, more reflected radiation is captured by the sphere and higher detector signal can be measured. Consequently, the angular responsivity will not follow the cosine response law. Instead, it will have two lobes similar to those of

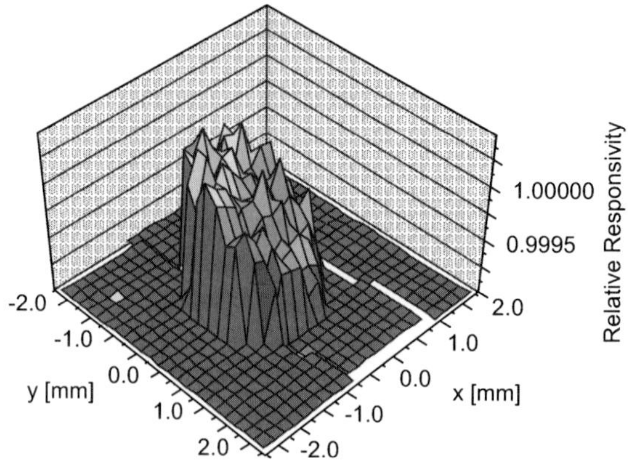


FIG. 4.20. Spatial uniformity of responsivity of a sphere-input Ge irradiance meter.

Figure 4.9. Tilting the plane of the input aperture can improve the situation. Even with a tilted aperture, the angular responsivity can be 0.2 to 0.3% different from the three-dimensional cosine function because of the asymmetric entrance and exit port arrangement. In order to obtain the best angular responsivity (the closest to the three-dimensional cosine function), either symmetrical entrance and exit port arrangements or additional baffles might be used.

With the right design, the angular range of sphere-input irradiance meters where the angular responsivity can follow the cosine function can be larger than that of diffuser-type irradiance meters. The FOV can be increased until the projected (from the incident radiation) and viewed (by the detector) areas will not overlap on the sphere wall. In the example of Figure 4.19, the *f*/2 corresponds to a FOV of about 30°. Time consuming selection of diffusers and apertures (like for the diffuser-type irradiance meters) and optimization for source wavelength is not needed here. Also, the highly non-uniform spatial responsivity of the cosine-optimized diffuser-type irradiance meters does not exist with the sphere–type irradiance meters.

### 4.3.5 Dewar-Type Infrared Filter Radiometers

The most popular IR detectors, InSb and HgCdTe (MCT), require operational temperatures close to 77 K for high radiometric performance. InSb detectors are usually photovoltaic (PV) devices (photodiodes) that can measure optical radiation from approximately 1 to 5.5 μm. MCT detectors are either PV or photoconductive (PC) and have a significant response to radiation over the range from 2 to 26 μm. The PC MCT detectors need bias currents. The cut-off wavelength can be tuned by changing the ratio of the composite material components of the detector. The spatial response non-uniformity of MCT detectors can be large. For low uncertainty measurements, they should be calibrated and used only in irradiance measurement mode to average out the spatial response non-uniformity [28]. In addition, the incident beam should be spatially uniform over the active area of the detector.

Filters used with the IR detectors discussed above should be maintained at low temperature as well to improve the noise floor by rejecting most of the ambient background radiation from the detector. IR detectors are mounted in cryogenic dewars, where vacuum gives the temperature insulation between the cold detector and the warm external dewar wall. The detectors are mounted on a cold platform (finger). The position of the detector can be different after each cooling cycle. A typical change in the detector position, after a cooling cycle, is half a millimeter. Determination of the reference plane of a dewar-type irradiance meter with a low uncertainty is not easy because of this significant distance change relative to the front surface of the dewar. Also, a plane-parallel dewar window that is typically used to measure non-coherent radiation can shift the focus of a converging incident beam. These additional problems are to be taken into consideration in the design of the input optics when low measurement uncertainty is needed.

Using a cold low-pass filter, the upper cut-off wavelength of an InSb radiometer can be limited to, e.g. 2.5 μm. In this case, the background-radiation-produced signal-component can be significantly decreased. InSb radiometers with low background signal components can be operated in DC measurement mode. When an InSb radiometer has a wider spectral bandwidth in the longer wavelength region, the DC background signal component can be large. In this case, the signal to be measured must be chopped to distinguish the signal from the background component. A 0.2 $pW/Hz^{1/2}$ noise equivalent power (NEP) can be achieved in AC measurement mode when the background component of the detector signal is well attenuated using small FOV and narrowband cold filters [13].

In field calibrations, most IR detectors are used in irradiance (overfilled) mode [13] even if the responsivity scale was extended to the IR in radiant power measurement mode (e.g., using pyroelectric transfer standard detectors). Accordingly, a large IR detector with spatially uniform responsivity should be used for IR irradiance measurements. A 7-mm-diameter PV InSb detector with a shunt resistance of about 100 kΩ and a spatial response non-uniformity of less than 1% can be an optimum choice to obtain high irradiance responsivity and low preamplifier output noise. An aperture of known area (e.g., 6.4 mm diameter) can be positioned close (within 1 mm) to the detector to convert power responsivity into irradiance responsivity. As an example, the input design of an InSb filter radiometer is shown in Figure 4.21. This radiometer has a precision aperture of 6.4 mm diameter. It can measure both radiant power and irradiance. Only one cold filter is shown, located inside the cylindrical holder of the cold FOV limiter. Usually, at least two cold filters are needed to achieve 5–6 orders of magnitude blocking outside of the bandpass interval. The efficient blocking is needed when broadband sources (such as blackbodies) are measured. The cold filters and the small FOV produce a high background rejection. Using a second cold aperture in the FOV limiter, positioned 50 mm away from the precision aperture, a 17° unvignetted and a 31° total FOV was realized. With the input geometry shown, a maximum feedback resistor of 1 MΩ could be used in the current-to-voltage converter attached to the backside of the dewar.

The InSb radiometer shown was calibrated in irradiance measurement mode. "Point source" like geometry was used to avoid changing internal

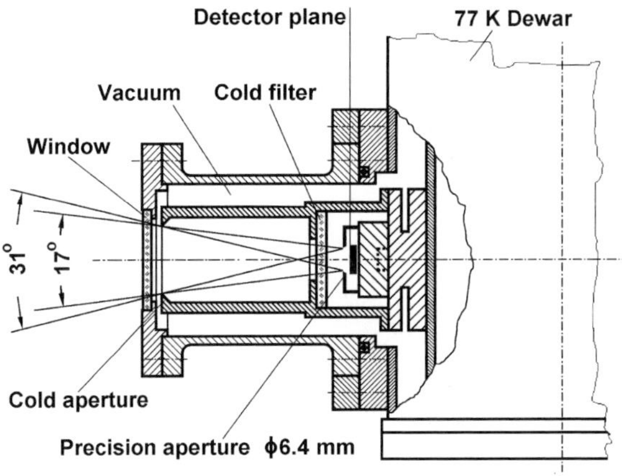


FIG. 4.21. Input-design of a Dewar-type InSb filter radiometer.

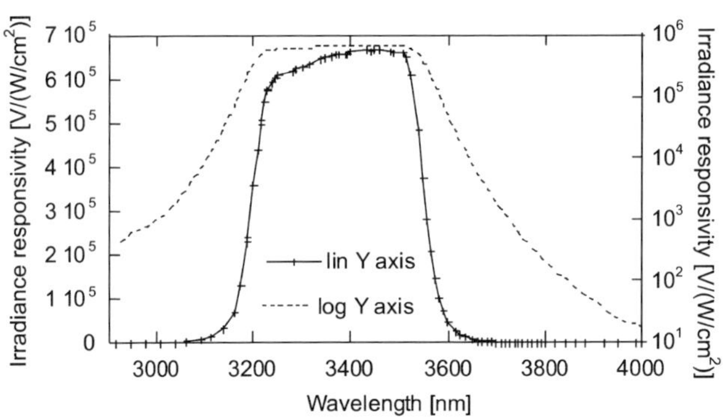


FIG. 4.22. Irradiance responsivity of an InSb filter radiometer.

reflectance patterns. The separation between the radiometer reference plane (which is different from the detector plane shown) and the source aperture was determined from a curve fit during the "point source" type irradiance measurements. The inverse square law was utilized in the curve fit. The measured irradiance responsivity of the InSb filter radiometer equipped with two bandpass filters of the same model is shown in Figure 4.22. The linear Y scale shows that the FWHM bandwidth is about 350 nm. The logarithmic Y scale illustrates the high out-of-band signal blocking. High-sensitivity InSb filter radiometers can be used for accurate irradiance calibration of infrared collimators even at very low irradiance levels.

## 4.4 Design Consideration of Radiance Meters

As discussed in Section 4.1, radiance meters use additional elements to define a measurement solid angle. These instruments measure a target area within a source. Wavelength-selective radiance meters are used to measure broadband sources such as blackbody radiators. As with irradiance meters, typically, the bandpass of these radiance meters is realized with filters. Ideally, the responsivity calibration method of a radiance meter has been considered in its design. For example, when the spectral radiance responsivity is derived from the spectral power responsivity of the detector used in the radiance meter, the front geometry of the radiance meter must be well defined. In general, the signal component from the surrounding source-area outside the instrument's FOV should be highly attenuated by the radiance meter. Otherwise, measurement errors will occur that depend on the out-of-field area of the source.

### 4.4.2 Input Tube Attachment

A radiance meter can be an irradiance meter extended with radiance measuring input optics. The simplest approach to produce a well-defined measurement angle is to place a second aperture at a fixed, known distance in front of the defining aperture in the irradiance meter. An input tube with two apertures, known as a Gershun tube [2], can be attached to the front of a filter radiometer for radiance mode measurements. Additional baffles are often placed between the two apertures to decrease unwanted signal components from stray and out-of-FOV radiation. The two apertures will determine the radiance measurement angle. Figure 4.23 shows the FOV scheme of the Gershun tube. The diameter of the detector aperture is $d$. This aperture in itself could be used for irradiance measurement. The diameter of the front aperture is $D$. The radiant power $P$ entering the detector in radiance measurement mode is

$$P = L\omega_{\mathrm{m}}A \tag{4.7}$$

where $L$ is the radiance of the target surface measured by the meter, $\omega_{\mathrm{m}}$ the viewing solid angle (measurement angle) of the meter, and $A$ the aperture area in front of the detector.

Using the symbols of Figure 4.23,

$$A = \frac{d^2\pi}{4} \tag{4.8}$$

and

$$\omega_{\mathrm{m}} = 2\pi\left(1 - \cos\frac{\gamma}{2}\right) \tag{4.9}$$

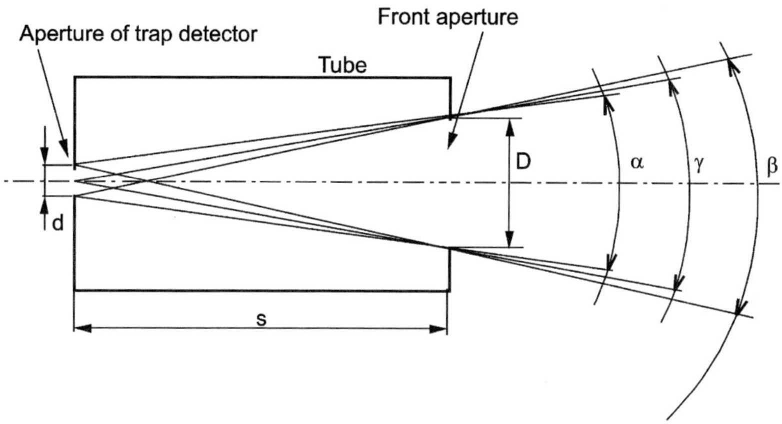


FIG. 4.23. Radiance measuring input tube attachment.

When $D = 11.29\,\text{mm}$, and the separation $s$ between the two apertures is 165 mm, the nominal (or effective) viewing angle can be calculated:

$$\gamma = 2\tan^{-1}\frac{D}{2s} = 3.919^\circ \tag{4.10}$$

From the values given above and for $d = 5\,\text{mm}$,

$$A = 0.1964\ \text{cm}^2, \quad \omega_\text{m} = 0.003674\ \text{sr}$$

When the detector calibrated for power responsivity measures radiant power, the radiance of the target surface can be calculated from Eq. (4.7). Determination of $\omega_\text{m}$ from the nominal viewing angle $\gamma$ is accurate only for small $A$. If we integrate the radiant power reaching the detector (from a uniform source of constant radiance $L$) for an ideal $\gamma$ angle ($d = 0$) and the real case ($d > 0$), using the same detector aperture ($A$), the integrated difference (ideal radiance minus real radiance) will be $-0.012\%$ for $d = 5\,\text{mm}$ and $-0.006\%$ for $d = 3.5\,\text{mm}$. These radiance (luminance) measurement errors are very small. For a detector aperture with $d > 5\,\text{mm}$, the integrated difference should be applied as a correction factor for the measured radiance.

The unvignetted FOV of the Gershun tube is

$$\alpha = 2\tan^{-1}\frac{D-d}{2s} \tag{4.11}$$

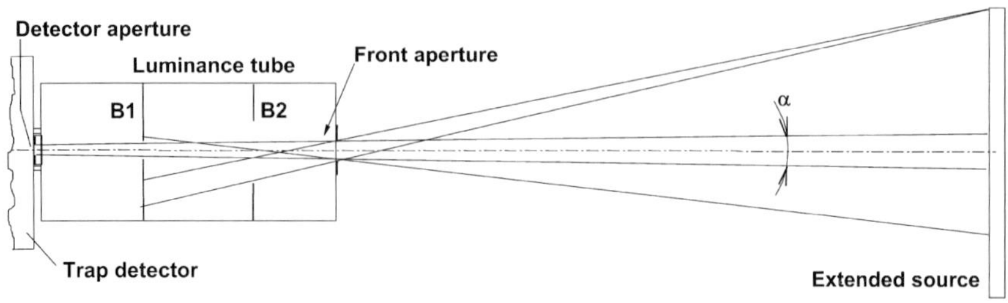


FIG. 4.24. Baffle arrangement (B1, B2) of the radiance/luminance tube in Figure 23.

For the above geometry (with $d = 5\,\text{mm}$), $\alpha$ is 2.18°. The full radiance-measurement angle is

$$\beta = 2\tan^{-1}\frac{D+d}{2s} \tag{4.12}$$

For the above tube design, $d = 5.65°$. This full radiance-measurement angle must be smaller than the FOV of the detector. With this design, the un-vignetted target-spot diameter (the plateau of the output signal) will be 40% of the full spot-diameter seen by the radiance/luminance meter.

The design of the baffle arrangement inside of the tube is shown separately in Figure 4.24 to illustrate the separation of rays from an extended source. Baffles B1 and B2 minimize the effect of stray light and reject radiation from light sources outside of the radiometer FOV. For simplicity, three rays are shown outside of the two FOV rays (producing $\alpha$) to illustrate this situation for a large extended source.

For the tube design discussed above [24, 29], B1 had a diameter of 8.0 mm and it is located at a distance not too far from the detector aperture to avoid any clipping in the FOV. This baffle was fabricated with the same procedure as the other two FOV-limiting apertures. They were all made of copper, coated with black nickel. The aperture thickness is 0.13 mm. The position of B2 is not critical. It has a larger diameter and stops radiation entering the front aperture at large incidence angles.

### 4.4.3 Lens Input Optics

Lens input optics are used when small target areas are needed. As an example, Figure 4.25 shows the simplified optical/mechanical design of a filtered silicon radiance meter. The radiance input optics can be attached to the illuminance measuring base-unit using a threaded or bayonet-type mount. The beveled aperture mirror serves as a field stop, and it is positioned at the focus of a camera lens. Beveling minimizes stray radiation.

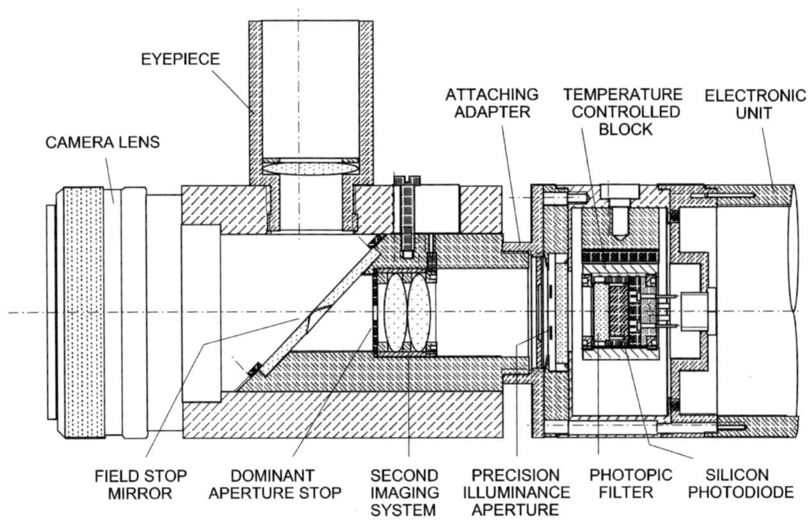


FIG. 4.25. Optical/mechanical scheme of luminance optics attached to an illuminance meter.

The camera lens has a broadband antireflecting (AR) coating, resulting in high transmittance in the visible range. The surroundings of the source target area can be viewed through the eye-piece. The target radiation is imaged to the same center position of the photodiode through a second imaging lens system. The second imaging system produces a well-defined FOV and a very efficient out-of-FOV blocking. The original precision aperture, used for irradiance measurements, stays in front of the photodiode. It is underfilled (not used as a field stop) in radiance measurement mode. There is a dominant aperture stop in front of the second imaging lens system in order to keep the flux response of the optics constant for different target distances of the adjustable-focus camera lens. The filter package and the silicon photodiode are usually temperature controlled to 26°C. The filter–detector package can also be cooled when the cavity holding the filter and detector is sealed from the environment. This is the case for the instrument shown schematically in Figure 4.25. There is a window glued on the front of the temperature-controlled unit. The back of the filter–detector space is o-ring sealed to a case mount.

The relative responsivity of this radiance meter as a function of rotation angle of the meter is shown in Figure 4.26. A 3-mm-diameter source-spot, about 2 m away from the radiance meter, was measured. The pivot point of the luminance meter was the center of the field stop. The figure shows that

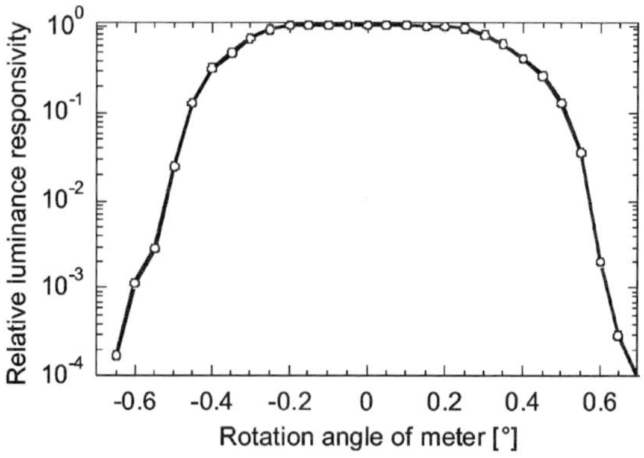


FIG. 4.26. Measurement profile of a radiance/luminance meter.

the luminance measurement angle of the meter was about 1° and the out-of-FOV blocking was larger than four decades.

The responsivity change of the meter was 0.1% when the target distance was changed between 1 and 2 m. This change was equal to the relative measurement uncertainty associated to the results of the responsivity change measurements.

## 4.5 Calibration

A calibration relates the measured quantity from a radiometer, usually current or voltage, to the radiometric quantity being measured, e.g., radiance or irradiance, through the instrument's responsivity. Often, establishment of traceability to the international system of units (SI) or to primary national or international radiometric standards is included as part of the calibration. Traceability is defined as the "property of the result of a measurement or the value of a standard whereby it can be related to stated references, usually national or international standards, through an unbroken chain of comparisons all having stated uncertainties." [30] NIST has a detailed policy outlining its role with respect to traceability [31]; many national metrology institutions have similar policies [32–34]. Determination of an instrument's responsivity and an evaluation of the associated uncertainties are required in a calibration for an instrument to be traceable to NIST, other national metrology laboratories, and the SI.

The Mutual Recognition Arrangement (MRA) for national measurement standards and for calibration and measurement certificates issued by National Metrology Institutes (NMIs) was signed in 1999 by the directors of the NMIs of 38 member states of the Metre Convention and representatives of two international organizations. The objectives of the MRA were to establish the degree of equivalence of national measurement standards maintained by NMIs; to provide for the mutual recognition of calibration and measurement certificates issued by NMIs; and to thereby provide governments and other parties with a secure technical foundation for wider agreements related to international trade, commerce, and regulatory affairs [35].

Under the MRA, the metrological equivalence of national measurement standards are to be determined by a set of Key Comparisons chosen and organized by the Consultative Committees of the International Committee for Weights and Measures (Comite' International des Poids et Mesures, CIPM) working closely with Regional Metrology Organizations (RMOs) [36]. An understanding of the measurement equation and an evaluation of the full uncertainty budget of the measurement by each participating laboratory are critical components of a Key Comparison.

A measurement equation is a mathematical expression describing the relationship between the measured quantity, the source radiometric quantity, and the instrument responsivity. In its simplest form, the measurement equation can be written as

$$i = \int L(\lambda)s(\lambda)\,\mathrm{d}\lambda \tag{4.13}$$

where $i$ is the measured output current, $L(\lambda)$ the source spectral radiance, $s(\lambda)$ the absolute radiance responsivity, and $\lambda$ the wavelength. The detector output current is typically converted into voltage using a current-to-voltage converter, as described in Section 4.2.4. Of course, Eq. (4.13) could have been written in terms of source irradiance and absolute irradiance responsivity as well.

This simplified measurement equation ignores many of the detector characteristics that need to be quantified to establish the uncertainty in the calibration (and subsequent measurements of a source). Typically, the instrument's response linearity, temperature dependence, polarization dependence, out-of-FOV blocking, and other parameters need to be measured. In addition, the current-to-voltage converter and the multimeter used to measure the signal need to be characterized and their contribution to the overall uncertainty established. The uncertainty of the calibration source needs to be included in the uncertainty budget as well. Finally, both short-term stability (repeatability) and long-term stability (to monitor degradation in time) should be considered in establishing an instrument's combined standard uncertainty. Depending on the long-term stability, an instrument may need to be re-calibrated every 6 months, once a year, or once every several years. In general, until an instrument's behavior is well understood, a yearly calibration is recommended.

There are two types of uncertainty components, designated type A and type B. Type A uncertainties are evaluated using statistical methods while type B uncertainties are evaluated using models or other external information. The term "standard uncertainty" refers to an estimated standard deviation. Type B components of standard uncertainty are evaluated to be equivalent to one standard deviation. Assuming each uncertainty component is independent from the others (the components are uncorrelated), the combined standard uncertainty is the root sum square of the individual uncertainty components. Often, the different variables are not completely independent from one another, and correlations between these variables need to be taken into account [37, 38]. Correlations are discussed in detail in Chapter 6.

The expanded uncertainty is the product of the combined standard uncertainty and a coverage factor $k$, where the value of $k$ is chosen based on the desired level of confidence. Typically the expanded uncertainty is reported with $k = 2$, corresponding to a confidence level of 95%. A confidence level of 95% means that there is a 1 in 20 chance that a measurement will fall outside the interval. A coverage factor $k = 3$ corresponds to a confidence level of 99%, meaning there is a 1% chance that a measurement will fall outside the stated interval. In reporting the uncertainty for a measurement, the components of standard uncertainty should be listed and their designation stated (A or B). It should be clear if the stated uncertainty is the combined standard or expanded uncertainty. If an expanded uncertainty is reported, the value of the coverage factor needs to be given as well.

Characterization of an instrument and an understanding and evaluation of all meaningful sources of uncertainty are critical for a proper calibration. A calibration without an associated uncertainty table is limited. The evaluation and expression of uncertainty is difficult and time-consuming; it is not unusual to have incomplete or inaccurate information in an uncertainty table. Determining how best to express a particular uncertainty component can be confusing. There are a variety of useful documents that provide definitions and recommendations for describing and establishing the uncertainties encountered when calibrating a radiometer [39, 40].

There are two widespread calibration approaches for irradiance and radiance meters. In the first approach, the relative spectral responsivity (RSR) is measured and the system is calibrated using a standard source. This is the most common calibration approach. The relative spectral responsivity values are typically determined using broadband illumination sources combined with a tunable, spectrally selective instrument; lamp-monochromator systems are commonly used [41]. The second approach is to use a quasi-monochromatic source of known irradiance or radiance and directly determine the radiometer's absolute spectral responsivity (ASR). The absolute responsivity is determined in one step. Broadly tunable lasers are commonly used as a source in the second approach, with the lasers replacing the lamp–monochromator system described above. Using tunable lasers and integrating spheres with different size apertures, it is straightforward to achieve the correct beam geometry, flux level, and bandwidth [42–46].

Laser-based radiometric calibration facilities have been developed at several national metrology institutes, including NPL [44, 47], PTB [48], HUT [45] and NIST [42–44, 46, 49, 50]. A brief description is given of the NIST laser-based calibration facility, the facility for Spectral Irradiance and Radiance Responsivity Calibrations using Uniform Sources (SIRCUS) [42, 43]. In the SIRCUS facility, shown in Figure 4.27, high-power, tunable lasers are introduced into an integrating sphere producing uniform, quasi-Lambertian, high radiant flux sources. A wavemeter measures the wavelength of the radiation with an uncertainty of 0.01 nm or less; Fabry–Perot interferometers are used to monitor the mode stability of the laser radiation. Reference standard irradiance detectors, calibrated directly against national primary standards for spectral power responsivity, are used to determine the

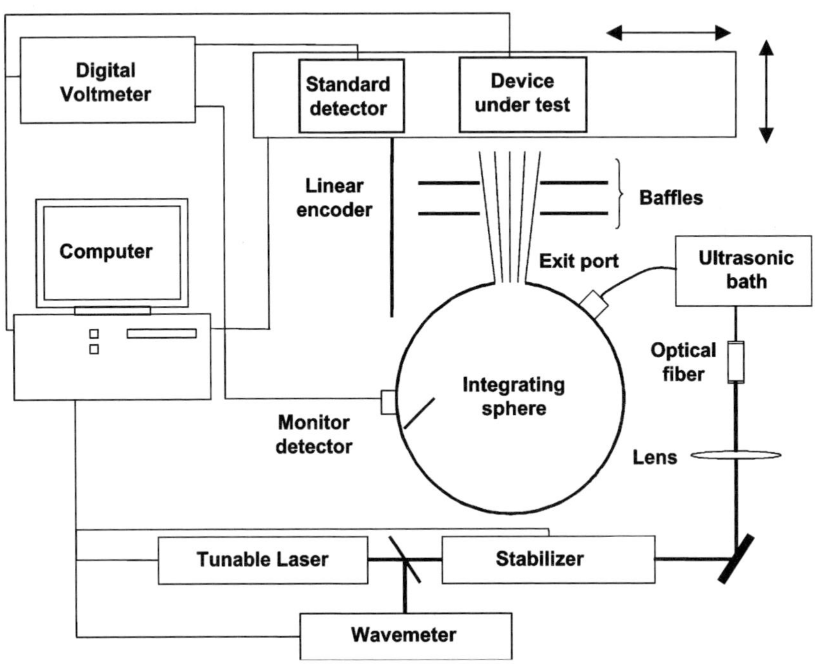


FIG. 4.27. Schematic diagram of the NIST SIRCUS laser-based calibration facility.

irradiance at a reference plane. The source radiance can be readily determined from the measurement geometry as well. Instruments are calibrated directly in irradiance or radiance mode with uncertainties approaching those available for spectral power responsivity calibrations. Ultimately, lasers determine the spectral coverage available in laser-based calibration facilities while the quality and characteristics of the reference standard detectors determine the achievable uncertainty.

In SIRCUS, the laser radiation is introduced into an integrating sphere, often using an optical fiber. Occasionally, a collimator coupled to the sphere is used as a calibration source. Speckle in the image from the source, originating from the coherent nature of the laser radiation, is effectively removed by either rastering the beam inside the sphere with a galvanometer-driven mirror or by placing a short length of optical fiber in an ultrasonic bath. Liquid light guides with a simple transducer attached to the side of the light guide also work well [48]. A monitor photodiode is located on the sphere to correct for any radiant flux changes in the sphere output between measurements with the reference instrument and the device under test. A number of different lasers are used to cover the spectral range from about 210 to 5000 nm, including continuous-wave (cw) dye lasers, solid-state Ti:sapphire lasers, as well as quasi-cw primary, doubled, tripled, quadrupled systems and optical parametric oscillator systems.

Different integrating spheres are used, depending on the radiometric calibration and the wavelength of calibration. Small-diameter integrating spheres—typically diameters of 25 to 50 mm—equipped with precision apertures with diameters ranging from 3 to 8 mm are typically used for irradiance responsivity calibrations. Larger diameter spheres—30 cm diameter—with 5–10 cm diameter exit ports are used for radiance measurements. The spheres are made of either sintered polytetrafluoroethylene-based coating [51] that has high diffuse reflectance from about 250 nm to 2.5 μm or diffuse gold [52] for calibrations from 1 to 20 μm. Typical irradiance levels at 1 m using a 25-mm-diameter integrating sphere with a 5-mm-diameter aperture range from approximately 1 to 10 $\mu W/cm^2$. Radiance levels between 1 and 5 $mW/cm^2/sr$ are standard for a 300-mm-diameter sphere with a 75 mm output port. These radiance and irradiance levels can be continuously adjusted down to zero output, allowing for linearity measurements over many orders of magnitude. Note that the exit apertures are normally calibrated at the NIST facility for aperture area measurement [53].

The irradiance measurement equation, for "point-source" geometry, is

$$I(\lambda) = E_1(\lambda)d_1^2 = E_2(\lambda)d_2^2 \tag{4.14}$$

where $I$ is the radiant intensity of the source at a given laser wavelength $\lambda$, $E_1$ the irradiance measured at the irradiance standard transfer detector located a distance $d_1$ from the source aperture, and $E_2$ the irradiance at the test detector's reference plane located a distance $d_2$ from the source aperture. $E_2$ is determined from the known irradiance $E_1$ multiplied by the ratio of the two distances squared. The irradiance responsivity of the test detector is

$$s_E(\lambda) = \frac{i}{E_2} \tag{4.15}$$

where $i$ is the current measured from the test device.

For radiance responsivity calibrations, the sphere radiance is the radiant intensity divided by the source aperture area and the radiance responsivity of the test radiometer is the measured current from the test instrument divided by the source radiance:

$$s_L(\lambda) = \frac{i}{L} \tag{4.16}$$

For both irradiance and radiance meters, a component approach to determine the relative spectral responsivity is commonly used. In this approach, the spectral power responsivity of the detector is measured, along with the transmittance or reflectance of all intervening optical elements, including filters. The relative spectral responsivity is then determined by multiplying the transmittance (or reflectance) of the individual components along with the detector responsivity. Differences between the component approach and system-level measurements can occur because of different measurement geometries and also because the former does not take into account inter-reflections between optical elements that are present in the complete system. For example, Figure 4.28 shows the calculated (by component) and measured relative spectral responsivity of three channels of a telescope, the Robotic Lunar Observatory (ROLO). Spectral shifts, changes in the spectral shape, and width are readily observable. The out-of-band response also differs significantly between the two approaches.

There is an additional approach for irradiance meters, called the scanned aperture method [45, 54–56]. In this approach, a beam much smaller than the defining aperture of the instrument is raster-scanned over the instrument's defining aperture. The effective aperture area can be determined by the ratio of the total signal summed over the scanned area (total irradiance responsivity) to the product of the total beam power $\phi$(W) and the average spectral responsivity $\bar{s}(\lambda)$, within the active area of the defining aperture [54]:

$$A_{\text{eff}} = \frac{\sum \text{i}(x, y)\Delta x \Delta y}{\phi \cdot \bar{s}(\lambda)} \tag{4.17}$$

where $\Delta x$ and $\Delta y$ are the small steps taken when scanning over the aperture. The spectral irradiance responsivity is the product of the spectral power

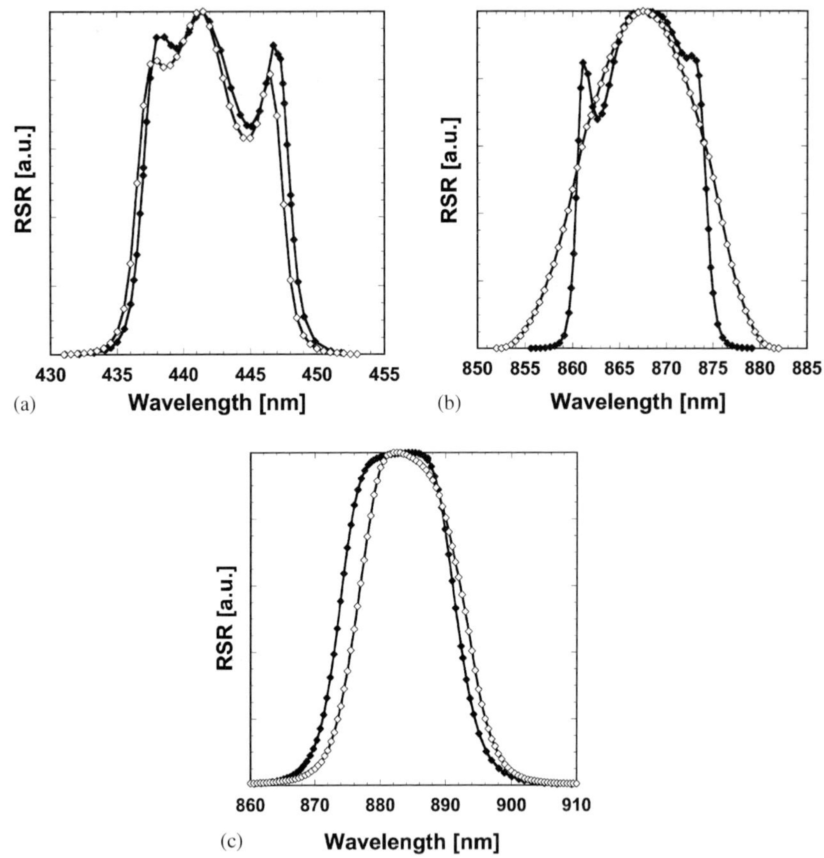


FIG. 4.28. Comparison between calculated (open diamonds) and measured (closed diamonds) relative spectral responsivity of three ROLO filter channels.

responsivity in the center of the aperture and the effective area. The scanned aperture method using a lamp monochromator system has been compared with a direct calibration using tunable laser sources over the visible spectral region using an unfiltered silicon irradiance meter [42] and in the UV using 365 nm irradiance meters [54]. Calibrations using the two different approaches agreed to within 1%.

### 4.5.1 Comparison between Lamp-Monochromator and Laser-Based Calibration Approaches

The general characteristics of laser- [43] and lamp-based [41] calibration facilities at NIST are listed in Table 4.1. The high optical power available

TABLE 4.1. Comparison of the Operating Characteristics of a Laser-Based and a Lamp-Monochromator-Based Calibration Facility

| Parameter | Laser-based facility | Lamp/monochromator |
|---|---|---|
| Optical power | 300 mW | 1 μW |
| Bandwidth | $<$ 0.001 nm | 1 to 5 nm |
| Wavelength uncertainty | $<$ 0.01 nm | 0.1 nm |
| Power responsivity calibration | Yes | Yes |
| Uncertainty | 0.1% | 0.1% |
| Irradiance responsivity calibration | Yes | Yes |
| Uncertainty | 0.1% | 0.5% |
| Radiance responsivity calibration | Yes | No |
| Uncertainty | 0.1% | |
| Digital imaging systems | Yes | No |

with laser-based calibration systems enables radiometers to be calibrated directly for either radiance or irradiance responsivity. The low-wavelength uncertainty of the laser is instrumental in reducing the calibration uncertainty for filtered instruments. Some of the advantages of the laser-based calibration approach are illustrated by the calibration of a Photo-Electric Pyrometer (PEP) used to radiometrically determine the temperature of a blackbody [57]. For accurate radiance temperature determinations, the instrument's spectral out-of-band responsivity needs to be measured as well as its in-band responsivity. The instrument is equipped with a narrow bandpass filter (~1 nm FWHM) for spectral selectivity. Figure 4.29 shows the relative spectral responsivity of the PEP determined on SIRCUS compared with the relative spectral responsivity determined using a conventional lamp–monochromator system. As shown in Figure 4.29(a), the spectral responsivity measured with the lamp–monochromator system is dominated by the spectral bandwidth of the source, and deconvolution of the spectrum using the source slit scatter function is required. In contrast, the fine detail in the spectral responsivity is easily measured with the tunable laser facility because of the monochromatic nature of the source. Note that there are several overlying data points at each wavelength along both the rising and falling edges, demonstrating the extreme wavelength stability and repeatability of the laser-based calibration. Because of the low flux, the out-of-band responsivity is limited to approximately $10^{-6}$ with the lamp–monochromator system (Fig. 4.29(b)). In contrast, the out-of-band responsivity can be measured to the $10^{-9}$ level in the laser-based facility.

Because of the coherent, narrowband nature of the laser radiation, oscillations or interference fringes in the responsivity arising from multiple reflections from surfaces of different optical components can occur. For example,

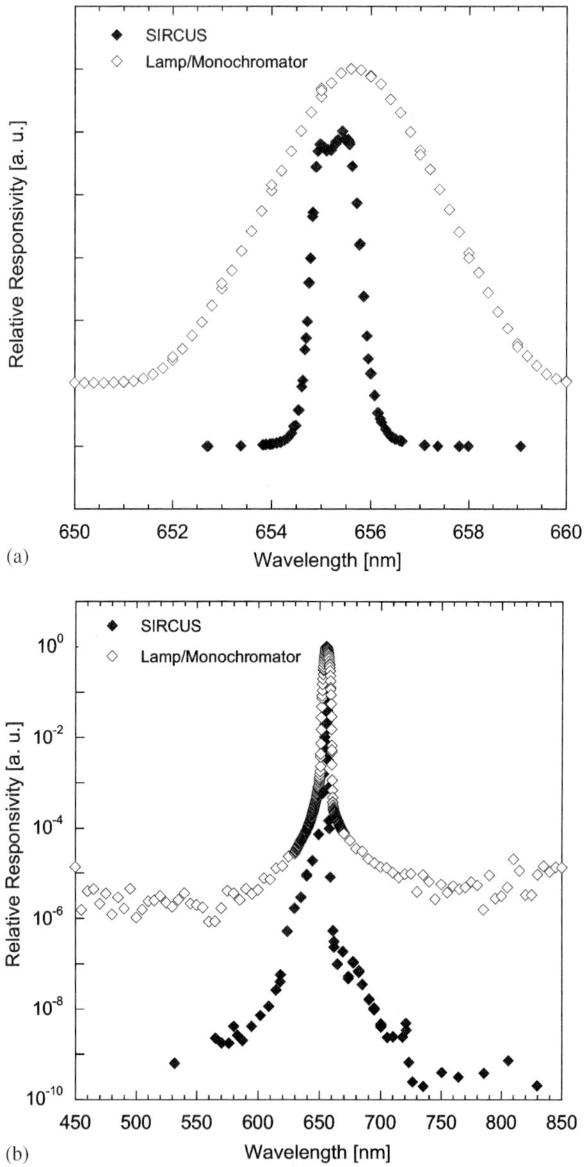


FIG. 4.29. Relative spectral responsivity of the PEP measured on the laser-based and the lamp-monochromator-based facilities at NIST: (top) linear scale; (bottom) log scale.

Figure 4.30 shows the responsivity of a filter radiometer determined using a laser-based source. What appears to be "noise" in the in-band responsivity (Fig. 4.30(a)) is shown to be interference fringes by examining a narrow spectral region in detail and tuning the excitation wavelength in very small increments (Fig. 4.30(b)). Interference fringes are often observed when calibrating radiometers with windowed detectors. The interference fringes can be greatly reduced or eliminated by proper design of the radiometer keeping the calibration in mind, e.g. through the use of wedged windows.

### 4.5.3 Traceability to SI units

Historically, radiometric and photometric measurements and calibrations have been based on standard sources. Spectral radiance and irradiance quantities as well as photometric quantities were based on the properties of blackbody radiators. Beginning in the early 1970s, detectors began to be incorporated into photometric and radiometric standards [58]. These developments were facilitated by improvements in detector material properties (especially silicon detectors) and devices [59] along with the development of high-accuracy electrical substitution radiometers that operated at cryogenic temperatures [60, 61]. Derivations of radiometric and photometric quantities using detectors can often be maintained with lower uncertainties than with source-based primary standards. Consequently, most national metrology institutions have moved or are moving toward detector-based measurements of radiometric, photometric, and colorimetric quantities [58].

The SI unit that serves as the base for all photometric quantities is the candela. The candela was defined in 1948, based on the radiation emitted from a platinum blackbody operating at the freezing-point temperature of molten platinum [62]. In 1979, the candela was redefined in terms of optical power as [63]

> The candela is the luminous intensity, in a given direction, of a source that emits monochromatic radiation of frequency $540 \times 10^{12}$ hertz and that has a radiant intensity in that direction of (1/683) watt per steradian.

The luminous intensity is the luminous flux emitted per unit solid angle in a given direction; the luminous flux is the radiant flux weighted by the photopic efficacy function, $V(\lambda)$. This definition of the candela in terms of optical power and a recognized weighting function provided the opportunity to base the measurement of photometric quantities on the properties of well-designed detectors. NIST and other national laboratories moved rapidly to establish detector-based photometric scales [62–64]. The implementation of the detector-based photometric scale reduced the uncertainty in the

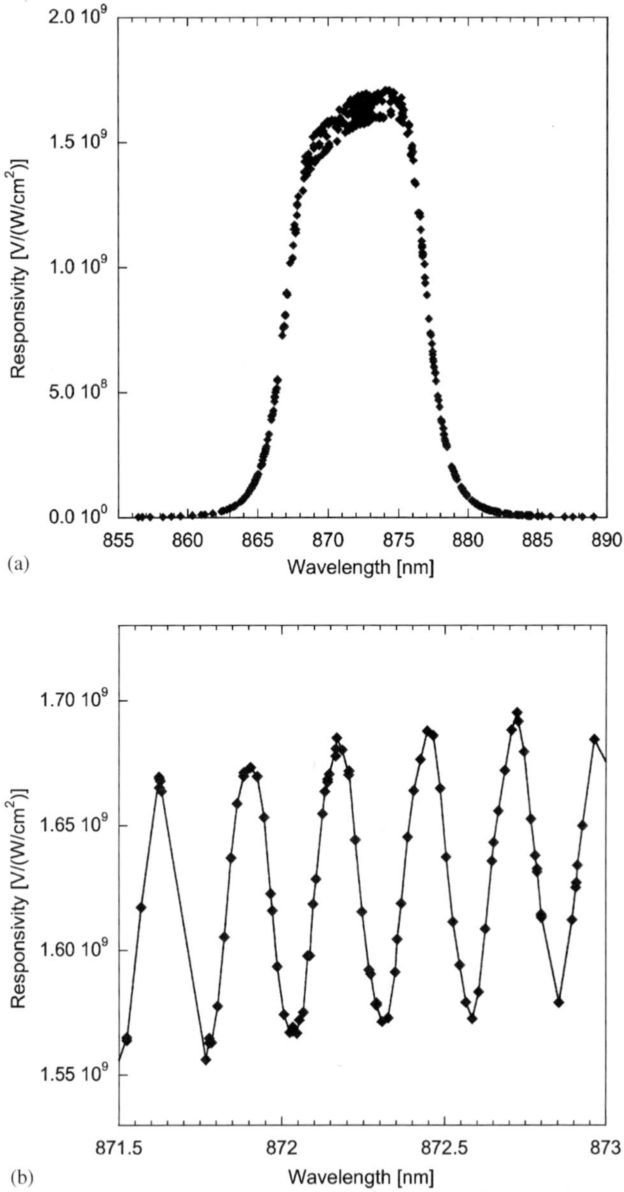


FIG. 4.30. (a) Absolute spectral responsivity of a filter radiometer measured in a laser-based calibration facility. (b) Expanded view of a small spectral region, demonstrating the presence of interference fringes.

derivation of photometric quantities by factor of two at NIST [65]. International Key Comparisons of luminous intensity, luminous flux and luminous responsivity were held in the 1990s; results are available from the Bureau International des Poids et Mesures (BIPM) [35]. Refer to Chapter 7 for additional information on photometry.

Lamp-illuminated integrating spheres or lamp standards of spectral irradiance are often used to calibrate radiometers. The spectral radiance values assigned to lamp-illuminated integrating sphere sources (ISSs) have been determined by reference to the gold-point and other fixed-point blackbodies and the international temperature scale of 1990 (ITS-90) [66, 67]. Given a source spectral radiance $L(\lambda, T)$ and a filter radiometer's absolute spectral responsivity, a comparison can be made between predicted and measured signal from the filter radiometer when measuring the source, with the predicted signal $S_P$ given by

$$S_P = \int s(\lambda)\varepsilon L(\lambda, T)\, d\lambda \tag{4.18}$$

where $\varepsilon$ is the emissivity of the blackbody radiator. For a source such as a blackbody, there exists a well-known relationship between the temperature and the spectral radiance, given by Planck's equation [1, 50].

The spectral responsivities of filter radiometers calibrated on the laser-based calibration facility are determined by direct substitution against a cryogenic radiometer. The irradiance responsivity is determined by measuring a precision aperture that is installed on the front surface of the radiometer. To determine the source radiance, the source exit port area and the distance between the source and detector apertures needs to be measured. Consequently, measurement of a blackbody source radiance using filter radiometers calibrated against, or traceable to, a cryogenic radiometer is based on electrical substitution radiometry and dimensional metrology, independent of the ITS-90 definition of the Kelvin. By comparing the predicted vs. measured signals when the filter radiometer measures the radiometric output of a fixed-point blackbody (e.g. a gold-point blackbody) whose temperature is defined by ITS-90, the ITS-90 temperature scale can be compared to the SI units of length (m) and power (W).

Shortly after the development of tunable dye lasers in the late 1970s, an intercomparison was made between electrical-substitution-based and blackbody-based radiometry using silicon filter radiometers to measure a tungsten halogen lamp [68]. Spectroradiometric determinations of the thermodynamic freezing and melting temperatures of gold, silver, and aluminum blackbodies followed [1, 50]. In 1991, an approach was outlined to derive the radiation temperature scale from 1000 to 3000 K using filter radiometers calibrated for absolute spectral responsivity against standard detectors

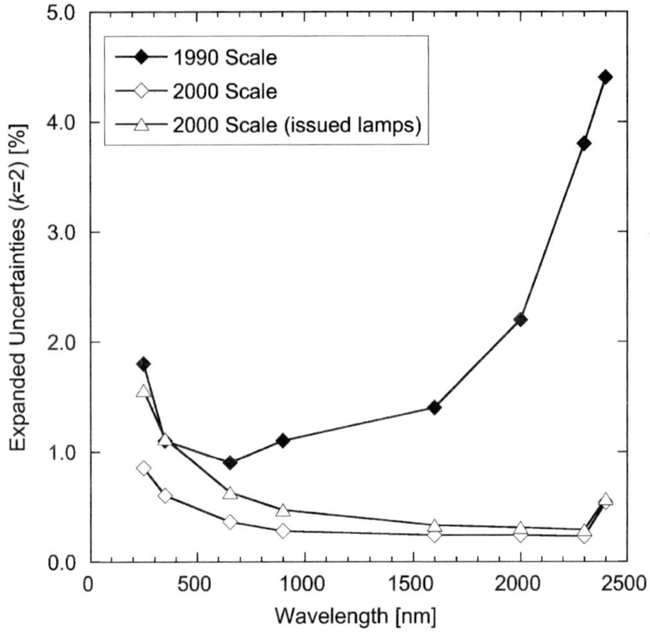


FIG. 4.31. Comparison of the expanded ($k = 2$) uncertainties of the NIST 1990 and 2000 spectral irradiance scale realizations.

traceable to a cryogenic radiometer [69]. The approach was realized at NIST in 2000 with the development of the detector-based spectral irradiance scale [70]. In Figure 4.31, the combined expanded ($k = 2$) uncertainties of the 1990 (source-based) and the 2000 (detector-based) spectral irradiance scale realizations are shown. The detector-based spectral irradiance scale has an uncertainty factor of 2 or more less than the ITS-90 (source-based) irradiance scale. The reduction in uncertainty increases up to a factor of 10 in the infrared spectral region.

A pyrometer—denoted the Advanced Pyrometer-1 (AP1)—was built at NIST to radiometrically measure blackbody temperatures [71]. Spectral selection is achieved using two narrowband interference filters with nominal center wavelengths at 650 nm and an additional infrared blocking filter to suppress radiation beyond 800 nm. The filters are temperature-stabilized near room temperature and the hermetically sealed silicon photodiode has a two-stage thermo-electric cooler for operation at −15°C. It has a NEP of 10 fW. The AP1 was calibrated for absolute spectral radiance responsivity on SIRCUS. As shown in Figure 4.32, it has a peak responsivity between 647 and 652 nm, a full-width half-maximum bandwidth of about 10 nm and

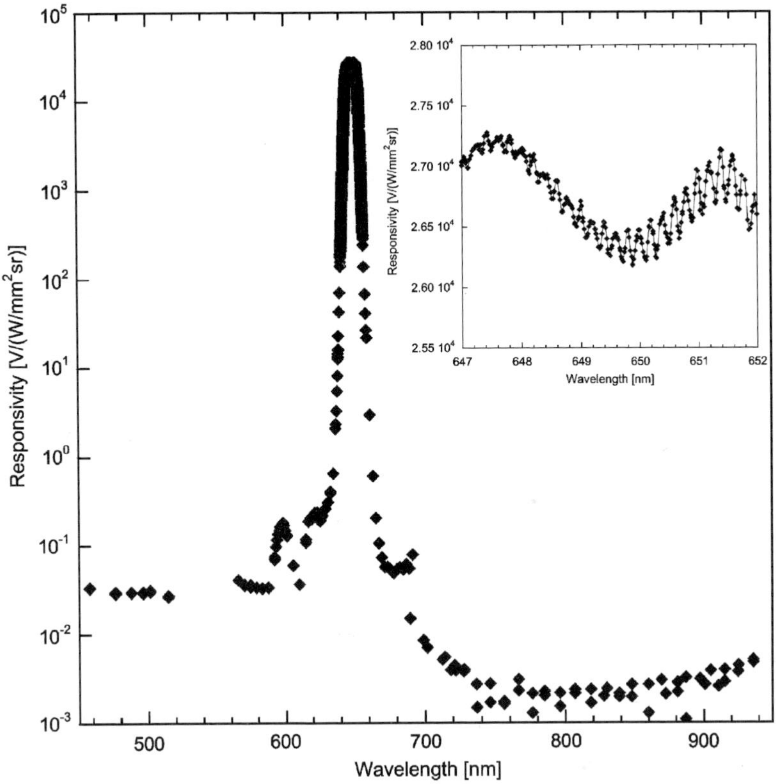


FIG. 4.32. AP-1 responsivity measured on SIRCUS. Inset shows interference fringes in the measured absolute spectral responsivity.

out-of-band blocking better that $10^{-7}$. Interference fringes with a magnitude of 0.5% were observed in the responsivity (see Figure 4.32, inset). The absolute spectral responsivity was therefore measured with 0.03 nm resolution. The AP1 subsequently measured the melt and freeze cycles of silver and gold fixed-point blackbodies. The signal from the AP1 is converted to temperature using Planck's equation [72].

The noise-equivalent temperature at the gold (and silver) freezing temperature is $\approx 2\,\text{mK}$, and the noise will not be the dominant component of the total temperature uncertainties. The expanded uncertainty ($k = 2$) in the radiometric measurement of the gold (or silver) freezing-point blackbody is approximately 0.15%. The radiometric uncertainties can be related to the uncertainties of the temperature determinations from the derivative of the Wien approximation, which shows the relationship between the uncertainty

TABLE 4.2. Summary ITS-90 and Spectroradiometric Determinations of Ag- and Au-Blackbody Freezing Point Temperatures

| Material | $T_{ITS\text{-}90}$ (K) | $u$ ($k = 2$) (K) | $T_{API}$ (K) | $u$ ($k = 2$) (K) | $T_{NPL}$ (K) | $u$ ($k = 2$) (K) |
|---|---|---|---|---|---|---|
| Ag | 1234.93 | 0.080 | 1234.956 | 0.106 | 1235.009 | 0.088 |
| Au | 1337.33 | 0.100 | 1337.344 | 0.121 | 1337.300 | 0.098 |

in radiance, $L$, to the uncertainty in blackbody temperature, $T$:

$$\frac{\Delta L}{L} = \frac{c_2}{\lambda} \frac{\Delta T}{T^2} \tag{4.19}$$

In Eq. (4.19), $c_2$ is the second Planck's constant, and $\lambda$ is the wavelength. Using Eq. (4.19), an uncertainty of 0.1% in radiance responsivity at 650 nm will lead to a temperature uncertainty of 80 mK in the measurement of the temperature of the gold-point blackbody.

In the ITS-90, the assigned temperatures for the Al, Ag, and Au freezing points result from thermometry using ratio pyrometry from the mean of two different and conflicting constant-volume gas thermometry measurements at lower temperatures. There are thermodynamic temperature uncertainties of the freezing points of the primary metal blackbodies that arise primarily from the uncertainties in the lower temperature gas thermometry. Although the thermodynamic temperature uncertainties of the Au- and Ag-freezing temperatures are not stated in the ITS-90, they have been listed in Table 4.2 for comparison. Radiometric determinations of the Au- and Ag-freezing points reported by Fox et al. from the NPL in 1991 [1] are included as well.

Note that the uncertainties in the freezing temperatures of primary gold and silver-point blackbodies determined using thermodynamic measurements (the ITS-90) and radiometric measurements, shown in Table 4.2, are similar. Laser-based calibration facilities can be used to calibrate optical pyrometers directly, eliminating the dependence on a blackbody and the ITS-90. As discussed in Yoon et al. [71], this can result in greatly reduced uncertainties in the measurement of thermal sources at higher temperatures.

Irradiance and radiance scales based on ITS-90 and electrical substitution radiometry have been intercompared, both within a calibration facility [49, 68, 73] and between laboratories [74]. The two approaches agreed to within 0.5% over a wide temperature range.

## 4.6 Conclusions

Modern, low-uncertainty detector-based reference scales for radiometric calibrations are those of spectral irradiance responsivity. Reference spectral irradiance responsivity scales can be realized by radiant power/irradiance measuring trap detectors in the silicon wavelength range. These silicon trap detectors are the highest-level transfer standards that can be used as reference detectors for all low-uncertainty spectral responsivity calibrations in the ultraviolet, visible and near-infrared range. For specific measurement tasks and responsivity scale propagations, filter radiometers can be used as transfer/working standards. These filter radiometers can be developed for many different measurement configurations. They are robust, simple to use, and can measure different radiometric (or photometric) quantities, such as power, irradiance, or radiance (luminous flux, illuminance, or luminance). The filter radiometers can be calibrated against the highest level (cryogenic radiometer or trap detector) standards (with the simplest possible geometry) and used as reference devices for other calibration facilities or field measurements. The result is a short calibration chain with low measurement uncertainty.

## References


1. N. P. Fox, J. E. Martin, and D. H. Nettleton, Absolute spectral radiometric determination of the thermodynamic temperatures of the melting/freezing points of gold, silver and aluminium, *Metrologia* 28, 357–374 (1991).
2. R. McCluney, "Introduction to Radiometry and Photometry." Artech House, Norwood MA, 1994.
3. A. Dahlback, Measurements of biologically effective UV doses, total ozone abundances, and cloud effects with multichannel, moderate bandwidth filter instruments, *Appl. Opt.* 35, 6514–6521 (1996).
4. C. Wehrli, Calibrations of filter radiometers for determination of atmospheric optical depth, *Metrologia* 37, 419–422 (2000).
5. T. C. Larason and C. L. Cromer, Sources of error in UV radiation measurements, *J. Res. Natl. Inst. Stand. Technol.* 106, 649–656 (2001).
6. G. Eppeldauer, "Optical Radiation Measurement with Selected Detectors and Matched Electronic Circuits Between 200 nm and 20 μm." NIST Technical Note 1438, NIST, 2001.
7. G. Eppeldauer, Temperature monitored/controlled silicon photodiodes for standardization, *Proc. SPIE* 1479, 71–77 (1991).
8. T. R. Gentile, J. M. Houston, and C. L. Cromer, Realization of a scale of absolute spectral response using the NIST high-accuracy cryogenic radiometer, *Appl. Opt.* 35, 4392–4403 (1996).
9. G. Eppeldauer, Noise-optimized silicon radiometers, *J. Res. Natl. Inst. Stand. Technol.* 105, 209–219 (2000).

*10.* K. Solt, et al., PtSi-n-Si Schottky barrier photodetectors with stable spectral responsivity in the 120–250 nm spectral range, *Appl. Phys. Letts.* 69, 3662–3664 (1996).

*11.* G. Eppeldauer, Electronic characteristics of Ge and InGaAs radiometers, *Proc. SPIE* 3061, 833–838 (1997).

*12.* G. P. Eppeldauer, A. L. Migdall, and L. M. Hanssen, InSb working standard radiometers, *Metrologia* 35, 485–490 (1998).

*13.* G. P. Eppeldauer and M. Racz, Spectral power and irradiance responsivity calibration of InSb working standard radiometers, *Appl. Opt.* 39, 5739–5744 (2000).

*14.* J. Lehman, et al., Domain-engineered pyroelectric radiometer, *Appl. Opt.* 38, 7047–7055 (1999).

*15.* G. P. Eppeldauer, M. Racz, and L. M. Hanssen, Spectral responsivity determination of a transfer-standard pyroelectric radiometer, *Proc. SPIE* 4818, 118–126 (2002).

*16.* P.-S. Shaw, et al., The new ultraviolet spectral responsivity scale based on cryogenic radiometry at Synchrotron Ultraviolet Radiation Facility III, *Rev. Sci. Instrum.* 72, 2242–2247 (2001).

*17.* "CIE Publication 69: Methods of Characterizing Illuminance Meters and Luminance Meters." Commission Internationale de L'Eclairage (CIE), Vienna, 1987.

*18.* G. P. Eppeldauer and M. Racz, Design and characterization of a photometer-colorimeter standard, *Appl. Opt.* 43, 2621–2631 (2004).

*19.* G. Eppeldauer, Chopped radiation measurement with large area Si photodiodes, *J. Res. Nat. Stand. Technol.* 103, 153–162 (1998).

*20.* G. Eppeldauer and J. E. Hardis, Fourteen-decade photocurrent measurements with large-area silicon photodiodes at room temperature, *Appl. Opt.* 30, 3091–3099 (1991).

*21.* J. L. Gardner, Transmission trap detectors, *Appl. Opt.* 33, 5914–5918 (1994).

*22.* G. P. Eppeldauer and D. C. Lynch, Opto-mechanical and electronic design of a tunnel-trap Si- radiometer, *J. Res. Nat. Stand. Technol.* 105, 813–828 (2000).

*23.* G. Eppeldauer, Near infrared radiometer standards, *Proc. SPIE* 2815, 42–54 (1996).

*24.* G. Eppeldauer, M. Racz, and T. Larason, Optical characterization of diffuser-input standard irradiance meters, *Proc. SPIE* 3573, 220–224 (1998).

*25.* J. H. Walker, R. D. Saunders, J. K. Jackson, and D. A. McSparron, Spectral irradiance calibrations, in "NBS Measurement Services." U.S. Nat. Bur. Stand., 1987.

26. D. G. Goebel, Generalized integrating sphere theory, *Appl. Opt.* 6, 125 (1967).
27. "A Guide to Integrating Sphere Theory." www.labsphere.com, Labsphere Inc., No. Sutton NH, 2004.
28. H. Gong, L. M. Hanssen, and G. P. Eppeldauer, Spatial and angular responsivity measurements of photoconductive HgCdTe LWIR radiometers, *Metrologia* 41, 161–166 (2004).
29. D. W. Allen, G. P. Eppeldauer, S. W. Brown, E. A. Early, B. C. Johnson, and K. R. Lykke, Calibration and characterization of trap detector filter radiometers, *Proc. SPIE* 5151, 471–479 (2003).
30. International Organisation for Standardization (ISO), "International Vocabulary of Basic and General Terms in Metrology." Geneva, Switzerland, 1993.
31. http://ts.nist.gov/traceability.
32. National Physical Laboratory, United Kingdom, www.npl.co.uk.
33. Physikalish Technische Bundesanstalt, Germany, www.ptb.de.
34. National Research Council, Canada, www.nrc.ca.
35. Bureau International des Poids et Mesures (BIPM), www.bipm.org.
36. BIPM Key Comparison Database, kcdb.bipm.org.
37. J. L. Gardner, Correlated color temperature-uncertainty and estimation, *Metrologia* 37, 381–384 (2000).
38. J. L. Gardner, Correlations in primary spectral standards, *Metrologia* 40, S167–S176 (2003).
39. International Organisation for Standardization (ISO), "Guide to the Uncertainty in Measurement." Geneva, Switzerland, 1991.
40. C. L. Wyatt, V. Privalsky, and R. Datla, Recommended practice; symbols, terms, units, and uncertainty analysis for radiometric sensor calibration, NIST Handbook 152 (1998)
41. T. C. Larason, S. S. Bruce, A. C. Parr, "Spectroradiometric Detector Measurements." NIST Special Publication 250-41, U.S. Government Printing Office, Washington, DC, 1998.
42. S. W. Brown, G. P. Eppeldauer, and K. R. Lykke, NIST Facility for spectral irradiance and radiance responsivity calibrations with uniform sources, *Metrologia* 37, 579–582 (2000).
43. G. P. Eppeldauer, et al., Realization of a spectral radiance responsivity scale with a laser-based source and Si radiance meters, *Metrologia* 37, 531–534 (2000).
44. V. E. Anderson, N. P. Fox, and D. H. Nettleton, Highly stable, monochromatic and tunable optical radiation source and its application to high accuracy spectrophotometry, *Appl. Opt.* 31, 536–545 (1992).

45. M. Noorma, et al., Characterization of filter radiometers with a wavelength-tunable laser source, *Metrologia* 40, S220–S223 (2003).
46. A. R. Schaefer and K. L. Eckerle, Spectrophotometric tests using a dye-laser-based radiometric characterization facility, *Appl. Opt.* 23, 250–256 (1984).
47. N. P. Fox, et al., High-accuracy characterization and applications of filter radiometers, *Proc. SPIE* 2815, 32–41 (1996).
48. A. Sperling, *personal communication*.
49. A. R. Schaefer, R. D. Saunders, and L. R. Hughey, Intercomparison between independent irradiance scales based on silicon photodiodes physics, gold-point blackbody radiation, and synchrotron radiation, *Opt. Eng.* 25, 892–896 (1986).
50. K. D. Mielenz, R. D. Saunders, and J. B. Shumaker, Spectroradiometric determination of the freezing temperature of gold, *J. Res. Natl. Inst. Stand. Technol.* 95, 49–67 (1990).
51. Spectralon, Labsphere, Inc., North Sutton, NH.
52. Infragold, Labsphere, Inc., North Sutton, NH.
53. J. Fowler and M. Litorja, Geometric area measurements of circular apertures for radiometry at NIST, *Metrologia* 40, S9–S12 (2003).
54. T. Larason, et al., Responsivity calibration methods for 365-nm irradiance meters, *IEEE Trans. Instrum. Meas.* 50, 474–477 (2001).
55. P.-S. Shaw, R. Gupta, and K. R. Lykke, Characterization of an ultraviolet and vacuum ultraviolet irradiance meter with synchrotron radiation, *Appl. Opt.* 41, 7173–7178 (2002).
56. C. A. Schrama and H. Rejin, Novel calibration method for filter radiometers, *Metrologia* 36, 179–182 (1999).
57. C. E. Gibson, B. K. Tsai, A. C. Parr, "Radiance Temperature Calibrations." NIST Special Publication 250-43, U.S. Government Printing Office, Washington, DC, 1998.
58. A. C. Parr, "A National Measurement System for Radiometry, Photometry, and Pyrometry Based upon Absolute Detectors." NIST Technical Note 1421, 1996.
59. E. F. Zalewski and C. R. Duda, Silicon photodiode device with 100% external quantum efficiency, *Appl. Opt.* 22, 2867–2873 (1983).
60. T. J. Quinn and J. E. Martin, A radiometric determination of the Stefan-Boltzmann constant and thermodynamic temperatures between −408°C and +1008°C, *Phil. Trans. Roy. Soc. London* A 316, 85 (1985).
61. J. E. Martin, N. P. Fox, and P. J. Key, A cryogenic radiometer of absolute radiometric measurements, *Metrologia* 21, 147 (1985).
62. C. L. Cromer, G. Eppeldauer, J. E. Hardis, T. C. Larason, and A. C. Parr, National Institute of Standards and Technology detector-based photometric scale, *Appl. Opt.* 32, 2936–2948 (1993).

*63.* Y. Ohno, "NIST Measurement Services: Photometric Calibrations." NIST Special Publication 250-37, 1997

*64.* C. L. Cromer, G. Eppeldauer, J. E. Hardis, T. C. Larason, Y. Ohno, and A. C. Parr, The NIST detector-based luminous intensity scale, *J. Res. Natl. Inst. Stand. Technol.* 101–102, 109–131 (1996).

*65.* Y. Ohno, Improved photometric standards and calibration procedures at NIST, *J. Res. Natl. Inst. Stand. Technol.* 102, 323–331 (1997).

*66.* H. Preston-Thomas, The International Temperature Scale of 1990 (ITS-90), *Metrologia* 27, 3–10 (1990).

*67.* J. H. Walker, R. D. Saunders, and A. G. Hattenburg, "Spectral Radiance Calibrations." Natl. Bur. Stand. Spec. Publ. 250-1, Washington: U.S. Government Printing Office, 1987.

*68.* A. R. Schaefer and R. D. Saunders, Intercomparison between silicon and blackbody radiometry using a silicon photodiode/filter radiometer, *Appl. Opt.* 23, 2224–2226 (1984).

*69.* R. D. Saunders, et al., Realization of new NIST radiation temperature scales for the 1000 K to 3000 K region using absolute radiometric techniques, *Temperature, Measurement and Control in Science and Industry* 323, 221–225 (1993).

*70.* H. W. Yoon, C. E. Gibson, and P. Y. Barnes, Realization of the National Institute of Standards and Technology detector-based spectral irradiance scale, *Appl. Opt.* 41, 5879–5890 (2002).

*71.* H. W. Yoon, et al., Temperature scales using radiation thermometers calibrated from absolute irradiance and radiance responsivity, in *NCSL International Workshop and Symposium*, Orlando, FL, 2003.

*72.* H. W. Yoon, et al., The realization and the dissemination of the detector-based kelvin, submitted to *Proc. Tempmeko 04*, Dubrovnik, Croatia, 2004.

*73.* H. W. Yoon and C. E. Gibson, A comparison of the absolute detector-based spectral radiance assignment with the current NIST assigned spectral radiance of tungsten-strip lamps, submitted to *Proc. Tempmeko 04*, Dubrovnik, Croatia, 2004.

*74.* G. Machin, et al., A comparison of ITS-90 and detector-based scales between NPL and NIST using metal-carbon eutectics, submitted to *Proc. Tempmeko 04,* Dubrovnik, Croatia, 2004.

# 5. PRIMARY SOURCES FOR USE IN RADIOMETRY*

Jörg Hollandt, Joachim Seidel

High-Temperature and Vacuum Physics Department, Physikalisch-Technische Bundesanstalt, Berlin, Germany

Roman Klein, Gerhard Ulm

Photon Radiometry Department, Physikalisch-Technische Bundesanstalt, Berlin, Germany

Alan Migdall

Optical Technology Division, National Institute of Standards and Technology, Gaithersburg, Maryland, USA

Michael Ware

Department of Physics and Astronomy, Brigham Young University, Provo, Utah

## 5.1 Introduction

While absolute detectors are routinely used to maintain fundamental units and scales in radiometry and photometry, well-characterized sources have an irreplaceable niche in their use to characterize optical-sensing instrumentation. To service this need, most national metrology institutes and other issuers of optical standards furnish a variety of calibrated sources for use by their customers. These sources, usually some type of lamp including deuterium lamps or an arc source, are characterized for their spectral radiometric output or output in appropriate photometric units and thereby provide a mechanism for the transferal of these scales to other users [1].

Blackbody sources and electron storage rings are unique sources for use in radiometry because their spectral output characteristics can be calculated from fundamental principles based upon the knowledge of physical parameters that describe them. While the details of the theory of these sources can be found in the existing literature, their use in radiometry for the provision and distribution of radiometric scales is of considerable importance to users of radiometric techniques [2–5] and hence they are covered here. Blackbody sources offer a convenient mechanism of supplying spectral radiometric scales for users based upon a calibration of the temperature of the source. Electron storage rings as synchrotron radiation (SR) sources for

*Parts of this chapter are a contribution of the National Institute of Standards and Technology.



EXPERIMENTAL METHODS IN THE PHYSICAL SCIENCES, vol. 41
ISSN 1079-4042 DOI: 10.1016/S1079-4042(05)41005-X

radiometry, due to their complexity and cost, are limited to large laboratories that use their radiation for other purposes and hence are not in widespread use by the radiometric community. They are of considerable importance however for metrology institutes because these sources give calculable radiation output over a wide spectral region from the hard X-ray region to the infrared spectral region. The application issues associated with blackbodies and SR are discussed in Sections 5.2 and 5.3, respectively.

Parametric down-conversion of a single photon into a correlated pair of daughter photons is a new and novel source with applications in radiometry. While its absolute spectral distribution is not characterized simply as in the case of blackbodies and SR, it offers a unique way of performing detector characterization and radiance measurement that is of use to radiometry. This technique is discussed in Section 5.4.

## 5.2 Thermal Sources

At any temperature above 0 K (absolute zero), any non-transparent body emits thermal radiation. For an arbitrary body, the emitted radiant power cannot be predicted as a function of temperature and wavelength without further knowledge of material properties. However, for a hypothetical "blackbody" in thermodynamical equilibrium, which completely absorbs all incoming radiation (absorptivity $\alpha = 1$), Kirchhoff found in 1860 that the spectrum of emitted radiation becomes a function of wavelength and temperature alone, independent of the shape and material of the body. Based on the second law of thermodynamics, Kirchhoff concluded that the emissivity $\varepsilon$ has to be equal to the absorptivity $\alpha$ for every body under the same conditions of temperature, wavelength, and direction. The emissivity, which indicates the ability of a body to emit thermal radiation, therefore is unity for a blackbody, and cannot be greater than that for any other body at the same temperature. Based on Kirchhoff's findings, the search for an analytical description of the blackbody radiance spectrum became a major challenge to theoretical physics in the latter part of the 19th century. Although general properties of the blackbody spectrum such as the Stefan–Boltzmann law (1879/1884) and Wien's displacement law (1893) were discovered during the next decades, the correct mathematical expression describing the spectrum could not be derived with the help of thermodynamics or classical statistical mechanics. Thus, the "mystery of thermal radiation" was the most prominent unresolved problem of physics at the end of the 19th century.

Soon after its foundation as the first national metrology institute in 1887, the German Physikalisch-Technische Reichsanstalt (PTR) started to investigate the problem of thermal radiation. This was not only fundamental research, but also the search for a "universal" lighting standard, because the rapidly growing German gas and electrical lighting industries were very interested in such an ideal radiator as a reference source for optical radiation measurement [6–8]. The PTR scientists mainly involved in this research were Ferdinand Kurlbaum, Otto Lummer, Werner Pringsheim and Wilhelm Wien. For their purpose, they had to develop absolute radiation detectors and nearly ideal blackbody radiators. In 1892, Kurlbaum discovered the electrical substitution principle for radiant power measurement, which was independently also found by Ångstrom almost simultaneously [9]. In 1895, Wien and Lummer described in detail the possibility of generating blackbody radiation with cavity radiators. Remarkably enough, an isothermal cavity had already been pointed out as a blackbody equivalent by Kirchhoff in 1860, but this had obliviously been forgotten. The first cavity radiators were operated at temperatures ranging from −182 °C, the boiling point of liquid nitrogen, to 700 °C, the boiling point of liquid sulphur. Investigations of these cavity radiators led Wien to his radiation law in 1896. The electrically heated cavity radiator, which extended the temperature range up to 1500 °C, was developed by Lummer and Kurlbaum during the next 3 years (Fig. 5.1).

Employing this high-temperature cavity radiator, Lummer and Pringsheim succeeded to complete "absolute" measurements of the blackbody radiation spectrum in 1899. With the measurements, they did not attain their original objective, the experimental confirmation of Wien's radiation law. Instead, they set up a milestone on the way to the correct blackbody radiation law, which was found by Planck in the following year, and was first presented in a session of the German Physical Society in Berlin on 19 October 1900 [10, 12]. Eight weeks later, on 14 December 1900, in another session of the Physical Society [11, 12], Planck presented the derivation of his radiation law based on the "quantum hypothesis". This was the birth of quantum physics. In addition, Planck's law and the work of the PTR group also provided the basis of radiometry as a new field of metrology.

For nearly a century until the 1980s, when the electron storage ring BESSY was shown to be another primary radiation source standard emitting calculable SR (Section 5.2), high-temperature blackbodies, which were not fundamentally different from the Lummer–Kurlbaum blackbody, were the only available primary radiator standards for the realization of radiometric units in the infrared, the visible and the ultraviolet spectral regions.

### 5.2.1 Fundamentals of Blackbody Radiation

#### *5.2.1.1 The laws of ideal blackbody radiation*

The spectral radiant energy density $u_\nu$ may be considered as the fundamental quantity describing blackbody cavity radiation, and in fact this was

***Der electrisch geglühte „absolut schwarze“ Körper und seine Temperaturmessung; von O. Lummer und F. Kurlbaum.***

**(Vorgetragen in der Sitzung vom 6. Mai 1898.)**

FIG. 5.1. Lummer and Kurlbaum presented the electrically heated "absolutely black" body they had developed at the PTR and an account of the measurement of its temperature at the meeting of the German Physical Society on 6 May 1898. The design of modern blackbodies is not fundamentally different from this blackbody.

the quantity evaluated by Planck. With respect to the propagation of radiation, however, the spectral radiance $L_\nu$ (or $L_\lambda$) is the more appropriate quantity. Since the two quantities are closely related, $L_\nu = cu_\nu/(4\pi)$, Planck's radiation law is usually stated for spectral radiance today:

$$L_{\nu,\mathrm{BB}}(\nu, T) = (2h\nu^3/c^2)/\{\exp[h\nu/(kT)] - 1\} \tag{5.1}$$

In Eq. (5.1), the spectral density is expressed with respect to the radiation frequency $\nu$, so that $L_{\nu,\mathrm{BB}}(\nu, T)\,\mathrm{d}\nu$ becomes the radiance emitted in the narrow frequency band between $\nu$ and $\nu + \mathrm{d}\nu$. Using $L_\nu|\mathrm{d}\nu| = L_\lambda|\mathrm{d}\lambda|$ and $\nu\lambda = c$, the spectral radiance $L_{\lambda,\mathrm{BB}}$ with respect to wavelength $\lambda$ is found to be

$$L_{\lambda,\mathrm{BB}}(\lambda, T) = (2hc^2/\lambda^5)/\{\exp[hc/(\lambda kT)] - 1\} \tag{5.2}$$

The symbols used in Eqs. (5.1) and (5.2) have their usual meaning: $h = 6.626\,0693 \times 10^{-34}$ J s is Planck's quantum of action, $k = 1.380\,6505 \times 10^{-23}$ J/K the Boltzmann constant, and $c$ the speed of light, i.e., $c = c_0 = 299\,792\,458$ m/s in vacuum and $c = c(\lambda) = c_0/n(\lambda)$ in a medium with refractive index $n$ (which in general depends on the wavelength).

The shape of the blackbody spectrum is essentially the same for all temperatures because $L_{\lambda,\mathrm{BB}}(\lambda, T)$ can be obtained from $L_{\lambda,\mathrm{BB}}(\lambda', T')$ by means of a simple scaling law:

$$L_{\lambda,\mathrm{BB}}(\lambda, T) = (T/T')^5 L_{\lambda,\mathrm{BB}}(T\lambda/T', T') \tag{5.3}$$

This is the general form of the displacement law derived by Wien in 1893, and implies that $L_{\lambda,\mathrm{BB}}(\lambda, T)$ can also be expressed as a function of $\lambda T$ and $T$ instead of $\lambda$ and $T$.

It is customary to define two "radiation constants", the first radiation constant for spectral radiance, $c_{1L} = 2hc_0^2 = 1.191\,042\,82 \times 10^{-16}$ W m$^2$/sr, and the second radiation constant $c_2 = hc_0/k = 1.438\,7752 \times 10^{-2}$ m K, and to use these in Eq. (5.2) above.

Instead of the spectral radiance, the spectral exitance $M_{\lambda,\mathrm{BB}}$ (or $M_{\nu,\mathrm{BB}}$) is also considered, which gives the rate at which cavity radiant power is emitted through a hole in the cavity wall, propagating in all directions. Since such a hole ideally has a Lambertian (isotropically diffuse) emission characteristic, its spectral emittance is simply

$$M_{\lambda,\mathrm{BB}}(\lambda, T) = \pi L_{\lambda,\mathrm{BB}}(\lambda, T) = [c_1/(n^2\lambda^5)]/\{\exp[c_2/(n\lambda T)] - 1\} \tag{5.4}$$

Here, $c_1 = \pi c_{1L} = 2\pi hc_0^2 = 3.741\,771\,38 \times 10^{-16}$ W m$^2$ is the first radiation constant (for spectral emittance—the qualification usually being omitted in this case).

In Eqs. (5.1)–(5.4), refraction in the medium the blackbody radiation propagates in is implicitly taken into account through $c$, the speed of light in that medium. For gaseous media such as air or argon, the refractive index differs from unity only by some $10^{-4}$. Hence it is often a sufficiently accurate approximation in the discussion of blackbody radiation to assume that $n(\lambda) = 1$ (and thus $c = c(\lambda) = c_0$). In the remainder of this section, this approximation will also be used. It has to be kept in mind, however, that refractive-index effects can no longer be neglected from the outset if relative uncertainties at the $10^{-4}$ level are to be achieved by radiometric measurements.

Figure 5.2 shows examples of the blackbody spectral radiance (in vacuum) for various temperatures in a double-logarithmic graph. These examples illustrate the well-known general features of blackbody radiation: its smooth continuous spectral variation with the steep drop towards short wavelengths

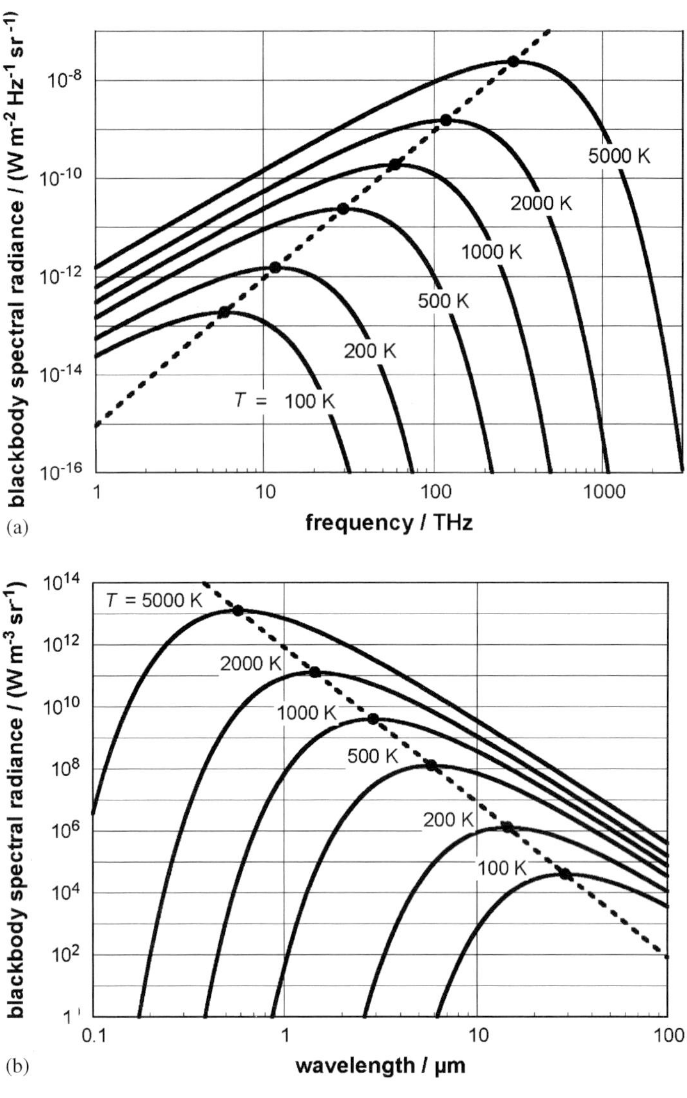


FIG. 5.2. Planck's radiation law for a blackbody (in vacuum): (a) spectral density with respect to frequency, Eq. (5.1); (b) spectral density with respect to wavelength, Eq. (5.2). The broken lines correspond to Wien's displacement law.

and the more gentle decrease towards long wavelengths, the increase at all wavelengths with increasing temperature, and the shift of the wavelength $\lambda_{\max}$, where $L_{\lambda,\mathrm{BB}}(\lambda, T)$ has its maximum as a function of wavelength, towards smaller values for increasing temperature $T$.

The last feature is described quantitatively by what is called Wien's displacement law today, and is a consequence of the more general displacement law (Eq. (5.3)):

$$\lambda_{\max} T = b = c_2/4.965\,114\,23 = 2.897\,77 \times 10^{-3}\,\mathrm{m\,K} \qquad (5.5)$$

Accordingly, $\lambda_{\max}$ is about 500 nm, i.e., well within the visible spectral region, for a temperature of 5800 K, which is approximately the temperature of the Sun's photosphere, and it is larger than about 3.6 μm, i.e., well in the invisible infrared spectral region, for a temperature of 800 K or below. The maximum value of $L_{\lambda,\mathrm{BB}}$ at $\lambda_{\max}$ increases strongly with temperature because it is proportional to $T^5$:

$$L_{\lambda,\mathrm{BB}}(\lambda_{\max}, T) = 4.095\,698 \times 10^{-6}\,(T/\mathrm{K})^5\,\mathrm{W/m^3\,sr^1} \qquad (5.6)$$

In this connection, it should be noted that the frequency $\nu_{\max}$, where $L_{\nu,\mathrm{BB}}(\nu, T)$ has its maximum as a function of $\nu$, is not $c/\lambda_{\max}$, but rather $\nu_{\max} = 5.878\,93 \times 10^{-2}\,\mathrm{THz}\,T/\mathrm{K} = 0.568\,253\,c/\lambda_{\max}$.

From Figure 5.2 it is obvious that the total (spectrally integrated) radiance of a blackbody increases strongly with increasing temperature. This is described by the Stefan–Boltzmann law obtained by integrating Eq. (5.1) or Eq. (5.2) over all frequencies or wavelengths, respectively:

$$L_{\mathrm{BB}}(T) = \sigma T^4/\pi \qquad (5.7)$$

with the Stefan–Boltzmann (exitance) constant $\sigma = 2\pi^5 k^4/(15 c_0^2 h^3) = 5.670\,400 \times 10^{-8}\,\mathrm{W/m^2\,K^4}$. According to Eq. (5.4), the total (i.e., spectrally integrated) exitance is, therefore, $M_{\mathrm{BB}}(T) = \sigma T^4$.

In addition to the total radiance at temperature $T$, the fractional radiance emitted in the wavelength band from $\lambda_1$ to $\lambda_2$ is often required. This can be obtained, for any temperature, as $F(\lambda_2/\lambda_{\max}) - F(\lambda_1/\lambda_{\max})$, with $F$ defined as

$$F(\lambda/\lambda_{\max}) = \pi \int_0^{\lambda} \mathrm{d}\lambda'\, L_{\lambda,BB}(\lambda', T)/(\sigma T^4) \qquad (5.8)$$

A first estimate can be made from the plot of the fractional radiance $F(\lambda/\lambda_{\max})$ shown in Figure 5.3. For more accurate results, the integral in Eq. (5.8) is easily evaluated by numerical integration. It should be noted that about 99.9 % of the total radiance is emitted in the wavelength interval from $0.36\lambda_{\max}$ to $23\lambda_{\max}$. This does not mean that a blackbody cannot be used as a standard of spectral radiance, say, at smaller or larger wavelengths, but spectral filtering will then have to block a large amount of unwanted radiation.

Another quantity useful for the purposes of both radiometry and radiation thermometry is the relative change of the blackbody radiance with temperature, $(\mathrm{d}L_{\lambda,\mathrm{BB}}/\mathrm{d}T)/(L_{\lambda,\mathrm{BB}}/T) = (\mathrm{d}L_{\lambda,\mathrm{BB}}/L_{\lambda,\mathrm{BB}})/(\mathrm{d}T/T)$. This allows us to obtain, on the one hand, the temperature stability required for an

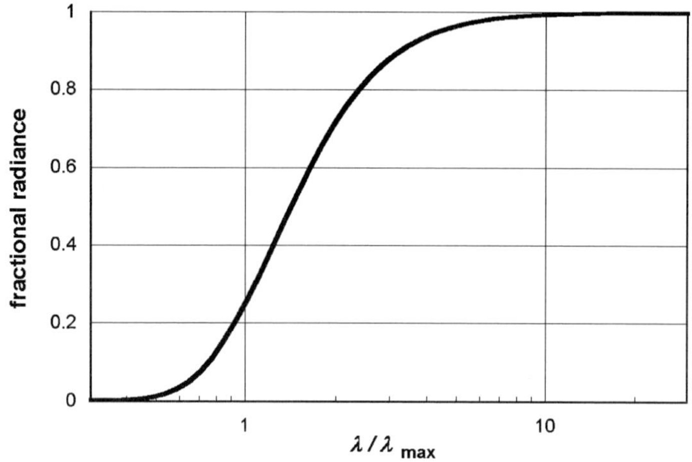


FIG. 5.3. Fractional radiance of a blackbody emitted from wavelengths much shorter than $\lambda_{\max}$ to larger values.

intended stability of the spectral radiance and, on the other, the radiometric accuracy required for an intended accuracy of temperature measurement. Denoting $hc/(\lambda kT)$ by $x$, this quantity is given by

$$(\mathrm{d}L_{\lambda,\mathrm{BB}}/L_{\lambda,\mathrm{BB}})/(\mathrm{d}T/T) = x/[1 - \exp(-x)] \tag{5.9}$$

which is shown in Figure 5.4. From the figure, it is obvious that the sensitivity of radiation thermometry increases steeply toward shorter wavelengths. Conversely, small fluctuations or uncertainties of the blackbody temperature result in large fluctuations or uncertainties of the spectral radiance at short wavelengths. In the wavelength range where Planck's law can be approximated by Wien's law, Eq. (5.10) below, one has $(\mathrm{d}L_{\lambda,\mathrm{BB}}/L_{\lambda,\mathrm{BB}})/(\mathrm{d}T/T) \approx x = hc/(\lambda kT)$. On the other hand, in the long-wavelength limit, where $x$ becomes small, $(\mathrm{d}L_{\lambda,\mathrm{BB}}/L_{\lambda,\mathrm{BB}})/(\mathrm{d}T/T) \approx 1$.

In the limits of either small or large wavelengths, the Planck formula, Eq. (5.2), can be simplified for practical purposes. Even with the computational facilities generally available today, these simplified expressions may be useful in deriving reliable approximate analytic expressions. For small wavelengths, $\lambda \ll c_2/T \approx 5\lambda_{\max}$, the "$-1$" in the denominator of Planck's law can be neglected. This yields Wien's radiation law as the short-wavelength approximation of Planck's law:

$$L_{\lambda,\mathrm{BB,Wien}}(\lambda, T) = (c_{1L}/\lambda^5)\exp[-c_2/(\lambda T)] \tag{5.10}$$

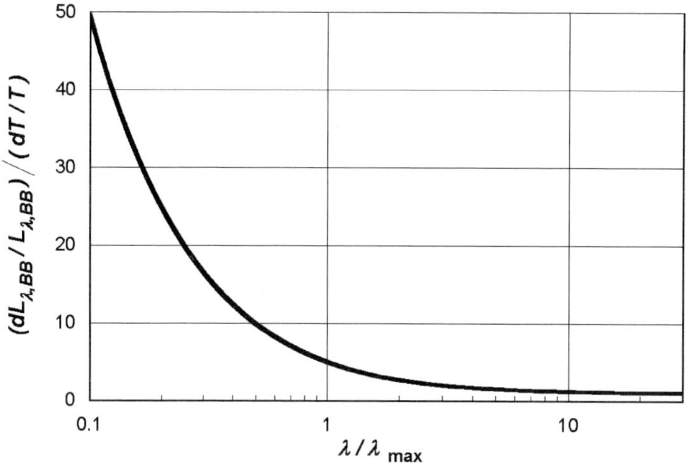


FIG. 5.4. Ratio of the relative change of the spectral radiance and the corresponding relative change of temperature.

The relative error introduced by using Wien's law instead of Planck's law is $-\exp[-c_2/(\lambda T)]$. It is $-1\,\%$ at $\lambda = 1.078\lambda_{\rm max}$, $-0.1\,\%$ at $\lambda = 0.719\lambda_{\rm max}$, $-0.01\,\%$ for $\lambda = 0.539\lambda_{\rm max}$, and $-10^{-5}$ for $\lambda = 0.431\lambda_{\rm max}$.

In the other limit, at large wavelengths the exponential function in the denominator of Eq. (5.2) can be approximated by $1 + c_2/(\lambda T)$. This results in the Rayleigh–Jeans law:

$$L_{\lambda,{\rm BB,R-J}}(\lambda, T) = c_{1L}T/(c_2\lambda^4) \quad \text{or} \quad L_{\nu,{\rm BB,R-J}}(\nu, T) = 2kT\nu^2/c_0^2 \tag{5.11}$$

Obviously, the Rayleigh–Jeans law cannot hold for short wavelengths (or high frequencies), because it predicts the emission of intense short-wavelength radiation such as X-ray radiation by any blackbody, even at low temperature, and thus is afflicted with the "ultraviolet catastrophe". However, at long wavelengths $\lambda \gg \lambda_{\rm max}$, the Rayleigh–Jeans law is the correct approximation of the Planck law. The relative error of this approximation is very nearly $2.5\lambda_{\rm max}/\lambda$ for $\lambda > 250\lambda_{\rm max}$.

When using the Planck formula for blackbody cavity radiation one has to be aware that its derivation is based on several assumptions. One of them is that the wavelength is negligibly small as compared to the linear dimensions of the cavity. This may not be fulfilled for wavelengths in the far infrared [2]. For these wavelengths, however, diffraction effects will usually cause still larger uncertainties.

#### 5.2.1.2 Emissivity of real cavity radiators

For the practical realization of a nearly ideal cavity radiator for the generation of blackbody radiation, two criteria are of decisive importance: temperature uniformity and high effective emissivity.

In order to achieve high temperature uniformity, i.e., small temperature gradients along the cavity wall, the wall should have a thermal conductivity as high as possible. Therefore, modern cavity radiators for high-accuracy radiometry or radiation thermometry are often designed as heatpipes, because heatpipes have a thermal conductivity several hundred times that of copper [13, 14]. For cavity radiators the heatpipe is constructed as a double-walled, hollow tube forming a cylinder with an open end. A small amount of a suitable substance serves as the working fluid inside the tube, where the liquid and gaseous phases of the working fluid are in equilibrium. At the hotter spots, the liquid evaporates, while the gas condenses at the cooler spots. Capillary forces in a wick structure move the condensate back to the evaporator. The high thermal conductance results from the large latent heat of evaporation and leads to very good temperature homogeneity, but the device works only in a limited temperature range for one kind of working fluid.

A cavity to be used as a source of radiation must have an opening or aperture in its wall through which the radiation can escape. This opening disturbs the spectral distribution of radiation as compared to a closed cavity. If the opening is negligibly small in comparison to the total inner surface area, the disturbance is negligible, too, but this condition cannot always be adequately fulfilled. Then it is necessary to determine the actual effective emissivity of the cavity experimentally or theoretically. The emissivity $\varepsilon$ is used to characterize the radiative emission from the surface of a non-black thermal radiator or material [15] with uniform temperature $T$ as compared to a blackbody with the same temperature. It is simply defined as the ratio of the corresponding radiances:

$$\varepsilon(\lambda, T) = L_\lambda(\lambda, T)/L_{\lambda,\mathrm{BB}}(\lambda, T) \tag{5.12}$$

The emissivity cannot become greater than unity. In general, the emissivity depends on the direction of radiation emission, in addition to wavelength and temperature. It is only for Lambertian radiators (such as ideal blackbodies) that $\varepsilon$ is the same for all directions of emission. Sometimes, the wavelength dependence of $\varepsilon$ is negligible at least in the spectral range of interest. The thermal radiator is then called a "gray body".

Since it is difficult to measure high emissivities of cavity radiators with sufficient accuracy, mathematical models are generally used to calculate these emissivities. Various analytical approaches have been developed for this purpose (cf. e.g. [2, 16]). Today, the most widely applied methods rely on Monte Carlo simulations combined with ray tracing [17–22]. These methods are used to calculate the reflectivity or the absorptivity (which equals the emissivity) of the aperture. Photons are virtually inserted into the cavity in random directions, within the defined geometry of the experiment, and the path of these photons is followed inside the cavity until they either are absorbed or leave the cavity. Specular or diffuse reflection of the photons is assumed at the wall. The probability of reflection is given by $1 - \varepsilon_W$, with $\varepsilon_W$ being the emissivity of the wall. The ratio of absorbed to inserted photons yields the emissivity of the cavity. Performing the calculation in this way, temperature gradients along the cavity walls are neglected. However, temperature gradients along the cavity walls can have a significant influence on the effective emissivity, since radiation emitted through the aperture and measured by a detector may originate from areas with different temperatures. Using a modified Monte Carlo ray-tracing method allows this effect to be taken into account, too, by introducing a weight for absorbed photons which is the spectral radiance according to Planck's law with the actual temperature at the place of absorption, multiplied by the emissivity of the wall material [20]. In the end, the (formal) effective emissivity of a non-isothermal cavity is obtained as the ratio of the effective spectral radiance to the Planckian spectral radiance for a reference temperature $T_0$ (usually the temperature of the centre of the cavity bottom):

$$\varepsilon_{\text{eff}}(\lambda) = L_{\lambda,\text{non-isothermal cavity}}(\lambda)/L_{\lambda,\text{BB}}(\lambda, T_0) \tag{5.13}$$

This is not a "true" emissivity since the choice of the reference temperature is arbitrary. As a result, the effective emissivity may exceed unity. Application of this kind of Monte Carlo simulation requires knowledge of the temperature distribution along the cavity walls, which has to be measured separately.

### 5.2.2 Blackbodies as Metrological Standards

For metrological purposes, Planck's law can be exploited in two different ways. On the one hand, a blackbody with known temperature can be used as a radiation-source standard to realize radiometric units. On the other hand, a blackbody can also serve as a temperature standard, if the emitted (spectral or total) radiance is measured with sufficient accuracy. Hence, there is a close connection between blackbody-based radiometry and radiation thermometry.

#### *5.2.2.1 Blackbodies used as temperature standards*

Among the base units of the International Systems of Units (SI), temperature is special because it is not "additive" in the sense that two systems with temperature $T$ would allow realization of the temperature $2T$. Thus, special methods have to be applied to realize the whole temperature scale starting from the temperature of the triple point of water, which is 273.16 K by definition. The measurement of thermodynamic temperatures has to be done with primary thermometers based on well-established fundamental laws, such as the ideal-gas law or the Planck law. However, primary thermometers are not suited for everyday use. Therefore, a "practical" temperature scale was already established at the end of the 19th century and has been continuously improved since then. At present, this practical scale is the International Temperature Scale of 1990 (ITS-90) [23], and the thermometers to be used for the measurement of ITS-90 temperatures $T_{90}$ are platinum resistance thermometers below the freezing point of silver at about 1235 K (down to about 14 K) and radiation thermometers above the silver point. The radiation thermometers are to be used for relative measurements of blackbody radiation to determine $T_{90}$ as a "radiance temperature" from the ratio of the spectral radiances at the unknown temperature $T_{90}$ and the temperature $T_{90}(\mathrm{X})$ of any of the three high-temperature fixed points X (silver, gold or copper freezing point):

$$L_{\lambda,\mathrm{BB}}(\lambda, T_{90})/L_{\lambda,\mathrm{BB}}(\lambda, T_{90}(\mathrm{X})) = \{\exp[c_2/(\lambda T_{90}(\mathrm{X}))] - 1\} / \{\exp[c_2/(\lambda T_{90})] - 1\} \tag{5.14}$$

The value of $c_2$ to be used here is 0.014 388 m K. For temperatures $T_{90}$ distinctly higher than the fixed-point temperatures $T_{90}(\mathrm{X})$, a direct measurement according to Eq. (5.14) may not be advisable in the visible wavelength region, however, because of the large spectral radiance ratio. For a blackbody temperature of 2600 K compared to the gold freezing temperature of 1337 K at a wavelength of 650 nm, for example, this ratio would be about 3000. Therefore, a two-step procedure is common practice for such measurements: first, a stable tungsten strip lamp (cf. Section 5.2.4.1) is calibrated as a transfer standard with radiance temperature of 1800 K, say, by comparison with the gold fixed point, and then this lamp serves as the reference for the high-temperature measurement. In this way, spectral radiance ratios are reduced to 71 and 44 for the first and second step, respectively.

A disadvantage of the ITS-90 procedure results from the comparatively large uncertainty connected with the approximation of the thermodynamical temperature $T$ by $T_{90}(\mathrm{X})$ in Eq. (5.14), because this uncertainty propagates according to $u(T_{90}) = [T_{90}/T_{90}(\mathrm{X})]^2 u(T_{90}(\mathrm{X}))$. For the gold freezing point at 1337 K, $u(T_{90})$ is about 50 mK, resulting in an uncertainty contribution of 300 mK for $T_{90} = 3275$ K. However, owing to recent improvements in detector-based radiometry, "absolute" radiometric measurements of thermodynamic temperatures in the high-temperature range can be done with similar or better accuracy today, using filter radiometers calibrated (directly or indirectly) against cryogenic radiometers [24–33]. Therefore, a future ITS may no longer need to define a practical high-temperature scale, but may refer to absolute radiation thermometry instead.

#### *5.2.2.2 Blackbodies used as radiation standards*

Just as a measurement of its radiance allows a blackbody to be used as primary temperature standard, a measurement of its temperature allows it to be used as radiance standard. If the temperature measurement is done according to the rules of the current International (practical) Temperature Scale, the blackbody radiance standard and the calibrations done with it are traceable to the ITS-90 in a well-defined way. (Of course, the uncertainty of approximating $T$ by $T_{90}$ in Planck's law has to be taken into account in this case.) For a long time, this has been the only way of accurate realization of radiometric units for source-related quantities. With the availability of high-quality silicon photodiodes and the development of high-accuracy cryogenic radiometers during the last two decades, the situation has changed: detector standards with very small uncertainties in the 0.01 % range or below can now be used for direct, absolute radiation thermometry with uncertainties comparable to or even smaller than the uncertainties of approximating thermodynamic temperature by $T_{90}$, at least in the high-temperature range. As a radiation source standard, a blackbody whose temperature is measured in this way is obviously traceable to an absolute radiation detector (and thus to electrical SI units, because the cryogenic radiometer is based on the equivalence of optical and electrical heating power). During the last decade, national metrology laboratories have either switched over to radiometric measurement systems based on absolute detectors, thus anticipating the possible discontinuation of a "practical" high-temperature scale in a future ITS, or have developed measurement facilities providing traceability to both the ITS-90 and to absolute detectors [34–47].

However, even though the cryogenic radiometer is well based on a fundamental principle, the equivalence of optical and electrical (heating) power, it is difficult to make sure that a particular cryogenic radiometer fully realizes this principle (and continues to do so over long periods of time). For example, the absorptivity of its cavity detector may gradually diminish with time or irradiation. Comparisons of different radiometers do provide a means of investigating changes in responsivity or other kinds of deterioration, but common systematic errors or drifts are not detected in this way. Therefore, it has been proposed to link at least one cryogenic radiometer, which can also take part in comparisons, to a fundamental constant such as the Stefan–Boltzmann constant or the Boltzmann constant by measuring the total radiance (or exitance) of a blackbody cavity radiator at the triple-point temperature of water (which is exactly known by the definition of the kelvin) [48]. In the end, this would re-establish the blackbody radiator based on Planck's law as the "highest authority" in radiometry.

As a result of the developments sketched here, radiometry and radiation thermometry have moved closer together again, after they had been drifting gradually apart for nearly a century since Planck's discovery of the radiation law, developing their own designations, etc. Today, absolute radiation thermometry is widely recognized as a major (and most demanding) application of radiometry and, at the same time, as the most important link connecting blackbody-radiator-based radiometry with detector-based radiometry.

### 5.2.3 Types of Modern Blackbodies

Several kinds of blackbodies are used as radiation source standards for the calibration of a great variety of radiation detectors and of various types of other thermal radiators. The temperatures at which these blackbodies are operated cover the range from well below 0 °C to more than 3000 °C today. Many of these blackbodies are variable-temperature blackbodies, but fixed-point-temperature blackbodies are also used for high-accuracy measurements (and, of course, as Planckian reference radiators for ITS-90 radiation temperature measurements). Typical examples of these kinds of blackbodies are presented in some detail in the following.

#### *5.2.3.1 Variable-temperature large-aperture heatpipe blackbodies*

Infrared radiometers are used, e.g., as radiation thermometers, as detectors for remote-sensing, and for thermal imaging. In order to avoid atmospheric absorption due to water vapour and carbon dioxide, these instruments are often equipped with band-limiting filters and work in the spectral ranges around 1.3 μm and 1.6 μm, from 3 μm to 5 μm or from 8 μm to 14 μm. Low-temperature blackbodies are used to calibrate these instruments in terms of radiance (or radiance temperature). Variable-temperature heatpipe blackbodies have proved to be particularly suited for this purpose: first, the whole temperature range from −60 °C to about 1000 °C, which is most interesting for applications, can be covered with four different types of heatpipe cavity radiators, and second, fairly large heatpipe cavities can be manufactured which permit large apertures (as required by radiometers with large fields of view) to be used without corrupting the blackbody emissivity too much. As an example, Table 5.1 gives the characteristic data for the four heatpipe cavity radiators used in the low-temperature calibration facility of

TABLE 5.1. The Heatpipe Blackbodies of the PTB Low-Temperature Calibration Facility

| Heatpipe Working Fluid | Temperature Range (°C) | Cavity Material | Cavity Length (mm) | Cavity Diameter (mm) | Wall Emissivity | Cavity Emissivity and its Standard Uncertainty |
|---|---|---|---|---|---|---|
| $NH_3$ | −60 to 50 | Steel | 525 | 60 | 0.965 | $0.999\,90 \pm 0.000\,06$ |
| $H_2O$ | 50 to 270 | Titanium | 420 | 60 | 0.88 | $0.999\,80 \pm 0.000\,15$ |
| Cs | 270 to 650 | Inconel 600 | 368 | 41 | 0.75 | $0.999\,60 \pm 0.000\,17$ |
| Na | 500 to 962 | Inconel 600 | 368 | 41 | 0.75 | $0.999\,60 \pm 0.000\,17$ |

the Physikalisch-Technische Bundesanstalt (PTB, the German National Metrology Institute) [49–51]. Figure 5.5 shows the set-up of these blackbodies. Other national metrology laboratories have similar facilities [52–63].

The sodium and caesium heatpipe blackbodies (Fig. 5.5a) are of similar design, while the water and ammonia heatpipe blackbodies (Fig. 5.5b) differ from these in some features. All blackbody furnaces meet the requirements of a clean-room environment. The sodium and caesium heatpipe blackbodies have water-cooled stainless-steel housings for heat shielding. They are heated by tubular furnaces with one bifilar heating zone and low-distortion dc power supplies. The temperatures of the heaters are each controlled by a separate sensor with temperature resolution of 0.01 K to a temporal stability of better than 0.1 K. The temperature of the heatpipe cavity is sensed and controlled in a closed loop affecting the set-point of the heater by a standard platinum resistance thermometer (SPRT) inserted into the heatpipe close to the bottom of the cavity. For the ammonia heatpipe, a thermostat circulating a liquid through the cooling elements on top of the heatpipe is used for temperature control, with the temperature typically set to a few degrees below the desired radiation temperature. The cooling element for the water heatpipe is operated with the laboratory water circuit. Additional heating elements are used to reach the desired cavity temperature, which is measured by a SPRT inserted close to the bottom of the heatpipe. This allows all heatpipes to be operated with temperature stability of better than ±10 mK. In order to increase the temperature homogeneity and to avoid condensation or freezing of moisture at low temperatures, the cavities of the water and ammonia heatpipes are purged with dry nitrogen gas. For temperature adaptation, the nitrogen gas flows through a coil structure in the heatpipe before entering the cavity through a ring of small holes in the cavity wall near the aperture of the heatpipe. The large length-to-diameter ratio and the rather high wall emissivities in combination with the excellent temperature homogeneity result in very high emissivities of the heatpipe cavity radiators (Table 5.1).

The overall uncertainty of the radiation temperature and the corresponding relative uncertainty of the spectral radiance, Eq. (5.9), of the heatpipe

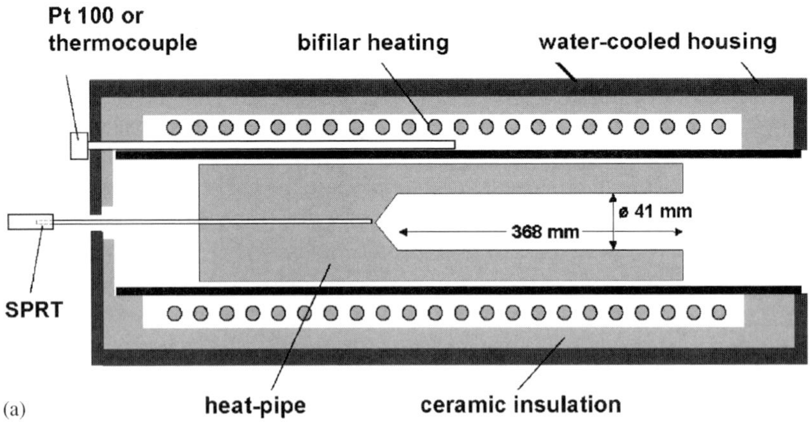


(a)

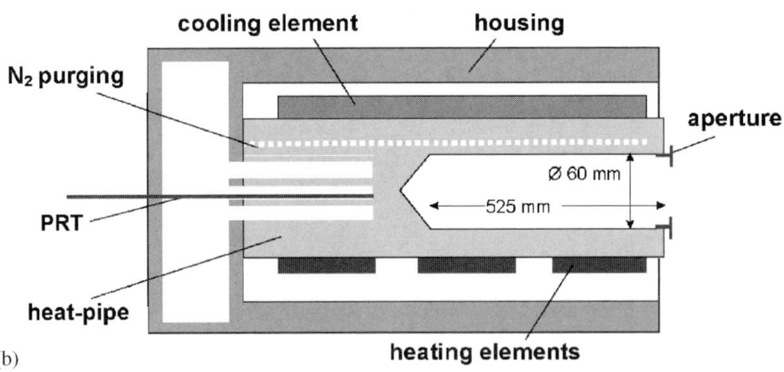


(b)

FIG. 5.5. Schematic cross sections of (a) the sodium and caesium and (b) the water and ammonia heatpipe blackbodies used in the PTB low temperature calibration facility.

blackbodies are given in Table 5.2 for typical wavelengths of infrared radiation thermometers within their typical temperature ranges. Major contributions to the overall uncertainty are

- the uncertainty of the calibration of the SPRT
- the short-term stability of the cavity temperature
- the non-uniformity of the radiation temperature (measured by scanning a radiation thermometer across the blackbody aperture).

Also, the total radiative heat loss through the aperture reduces the surface temperature of the cavity bottom below the temperature inside the heatpipe where the SPRT is located. In order to reduce this effect, the heatpipe

TABLE 5.2. Standard Uncertainties ($k = 1$) of Radiation Temperature, $u(T)$, and Corresponding Relative Standard Uncertainty of the Spectral Radiance, $u(L_\lambda)/L_\lambda$, of the PTB Low-Temperature Calibration Facility. The Uncertainties are Given at Typical Wavelengths of Radiation Thermometers

| Temperature (°C) | Wavelength (μm) | | | | | |
|---|---|---|---|---|---|---|
| | 10 | | 3.9 | | 1.6 | |
| | $u(T)$ (mK) | $u(L_\lambda)/L_\lambda$ | $u(T)$ (mK) | $u(L_\lambda)/L_\lambda$ | $u(T)$ (mK) | $u(L_\lambda)/L_\lambda$ |
| −60 | 35 | 0.11 | 100 | 0.81 | | |
| 0 | 35 | 0.07 | 35 | 0.17 | 35 | 0.42 |
| 50 | 35 | 0.05 | 35 | 0.12 | 35 | 0.30 |
| 100 | 35 | 0.04 | 35 | 0.09 | 35 | 0.23 |
| 200 | 80 | 0.05 | 80 | 0.13 | 80 | 0.32 |
| 300 | 100 | 0.05 | 100 | 0.11 | 100 | 0.27 |
| 400 | 50 | 0.02 | 30 | 0.02 | 20 | 0.04 |
| 600 | 80 | 0.02 | 50 | 0.02 | 30 | 0.04 |
| 800 | 110 | 0.02 | 70 | 0.02 | 60 | 0.05 |
| 960 | 140 | 0.02 | 100 | 0.03 | 80 | 0.05 |

bottoms were made only a few millimetres thick. The resulting temperature drop of a few millikelvin was calculated and corrected, with an uncertainty of half the correction. The isothermal emissivity of the cavities was calculated by Monte Carlo simulation. The remaining uncertainty is mainly due to the uncertainty of the wall emissivity. In order to detect possible temperature gradients along the cavity walls, a high-resolution radiation thermometer [64] was slightly tilted with respect to the cavity axis and scanned along the wall from the aperture to the bottom. From the measured radiation temperature profile, the temperature distribution along the cavity was determined with the help of non-isothermal Monte Carlo simulations. The effect of the temperature gradients on the radiance temperature of the bottom is taken into account as a correction, with an uncertainty of half the correction. For all four heatpipe blackbodies the correction is less than 15 mK, except for the caesium heatpipe below 350 °C, where the heatpipe process gradually starts to fail.

At low blackbody temperatures, room-temperature thermal background radiation (which is also emitted by room-temperature optical elements or the sensor to be calibrated itself) may become a problem. This problem can be avoided, and the thermal environment of instruments for space-based remote sensing can be simulated with high-vacuum chambers cooled down to cryogenic temperatures [65–69].

As discussed above, a blackbody should have a large ratio of the inner surface area to the radiating aperture area in order to avoid degradation of

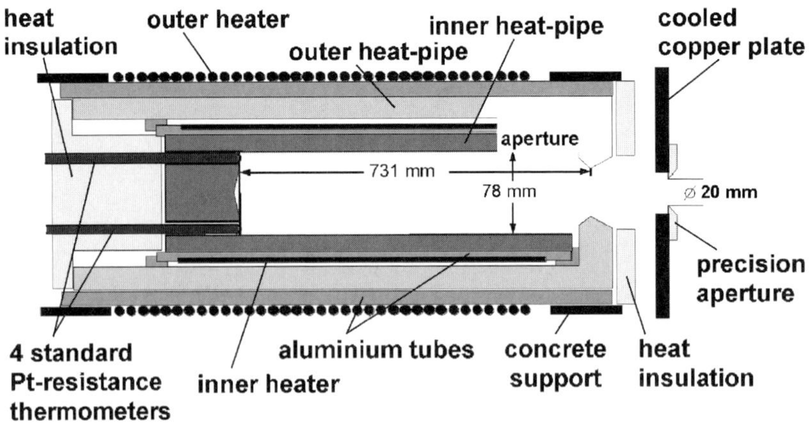


FIG. 5.6. The PTB double-heatpipe large-aperture blackbody.

the emissivity. If the intended use of the blackbody requires a large aperture, this ratio can, in principle, be improved by increasing the cavity length. However, this would impair the temperature homogeneity along the cavity walls, which is also critical for the emissivity. A double-heatpipe design can solve this problem. The PTB has developed such a large-aperture blackbody (LABB), the design of which is shown in Figure 5.6 [70, 71]. With sodium as the heatpipe working fluid, the LABB can be operated in the temperature range from 415 °C to 1000 °C. Four ITS-90-calibrated SPRTs are sensing the temperature of the cavity bottom. One of them is used for temperature control and stabilization. The temperature stability attained is $\pm 2$ mK. The other three SPRTs are used to determine $T_{90}$.

The main feature of the LABB is the outer sodium heatpipe enclosing the radiating inner cavity formed by another sodium heatpipe. This design results in excellent temperature homogeneity. With a 1.2-m-long SPRT, the temperature distribution close to the cylindrical cavity wall was determined by inserting the SPRT into the cavity and measuring the temperature at different immersion depths. The wall temperatures calculated from the measurement results [21] are shown in Figure 5.7.

For temperatures higher than 500 °C, the homogeneity is within 10 mK. Below 480 °C, the temperature inhomogeneity is more pronounced. However, calculating the influence of the temperature inhomogeneity on the effective emissivity of the LABB [21], even the rather large temperature gradient at 415 °C is found to reduce the radiation temperature by only 22 mK in comparison to an isothermal cavity.

The LABB is particularly well suited for measurement of $T_{90} - T$, i.e., deviations of the ITS-90 from the thermodynamic (radiance) temperature

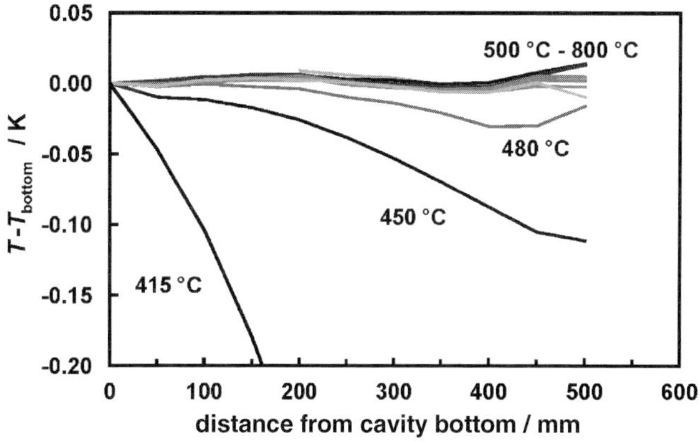


FIG. 5.7. Temperature distribution along the inner cavity wall of the large-aperture blackbody.

scale. The set-up for absolute radiation thermometry of the LABB, i.e., determination of $T$, is shown in Figure 5.8. Its main components (Fig. 5.8a) are a two-aperture assembly with precisely known geometry, and a filter radiometer. The filter-radiometer design used by the PTB is shown in Figure 5.8b; various other designs are used by other laboratories. The spectral-responsivity calibration of the filter radiometer being traceable to a cryogenic radiometer, standard uncertainties of about 20 mK have been obtained with this set-up for $T - T_{90}$ measurements in the temperature range from 450 °C to 500 °C using silicon photodiodes and interference filters with central wavelengths of 800 nm or 900 nm. The corresponding relative standard uncertainty of the LABB spectral radiance is about 0.06 %.

Figure 5.8a shows the "irradiance mode" of absolute radiation thermometry, because the irradiance of aperture A2 (and the corresponding radiant power) can be determined from the radiance of the blackbody. A non-imaging set-up like this, with a filter radiometer calibrated against a cryogenic radiometer, can thus be used to realize the fundamental connection of radiator-related quantities (such as radiant power) to detector responsivities in a radiometric measurement system based on absolute detectors (i.e., cryogenic radiometers). The "radiance mode" of absolute radiation thermometry makes use of an additional optical imaging system between radiator and detector. As an advantage, imaging radiation thermometers can be used with smaller blackbody apertures than non-imaging systems. As a disadvantage, it is more difficult to perform a calibration for spectral responsivity of an imaging system because of the additional optical elements, and to make sure that none of these elements changes noticeably with time.

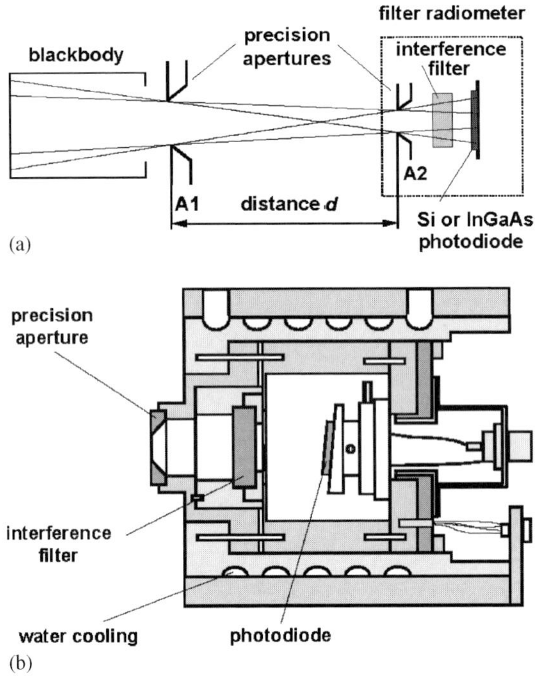


FIG. 5.8. Schematic of (a) the geometry of absolute radiation thermometry of blackbody sources and (b) cross section of an interference filter radiometer used by the PTB.

It should be noted that the uncertainties of the geometrical quantities of a set-up as shown in Figure 5.8a contribute to the total measurement uncertainty and may even become major contributions if all other uncertainty contributions have been improved as much as possible. This is particularly true for the areas of apertures A1 and A2. Because of this, aperture area measurement has been of topical interest during the last years [72–80].

#### *5.2.3.2 Fluid-bath blackbodies*

Heatpipe blackbodies are operated by national metrology institutes in order to achieve small uncertainties for source-related radiometric units and radiation temperatures. However, the production and operation of heatpipe blackbodies is complex and expensive. A simpler and more straightforward solution for temperatures up to about 200°C, which yields sufficient accuracy for many applications, utilizes a copper-cavity radiator immersed in a temperature-regulated bath [81, 82], the temperature of which is measured with a calibrated SPRT. Figure 5.9 gives an example of such a blackbody [49] which can be operated from −20 °C to 200 °C. The inner surface of the cavity is coated with black paint with an emissivity higher than 0.95 in the infrared spectral range. The effective cavity emissivity is about 0.999. As for the ammonia and water heatpipe blackbodies described above, the cavity is purged with dry nitrogen gas adapted to the bath temperature by passing through a copper coil immersed in the liquid.

A simplified bath blackbody radiator has been designed by PTB for the calibration of medical infrared ear thermometers according to the European standard of 2003 (Fig. 5.10). This kind of radiation thermometer determines the human-body temperature by sensing the thermal radiation emitted by the ear canal, which is very nearly a blackbody radiator. For calibration of an infrared ear thermometer, its sensing tip is inserted into the opening of

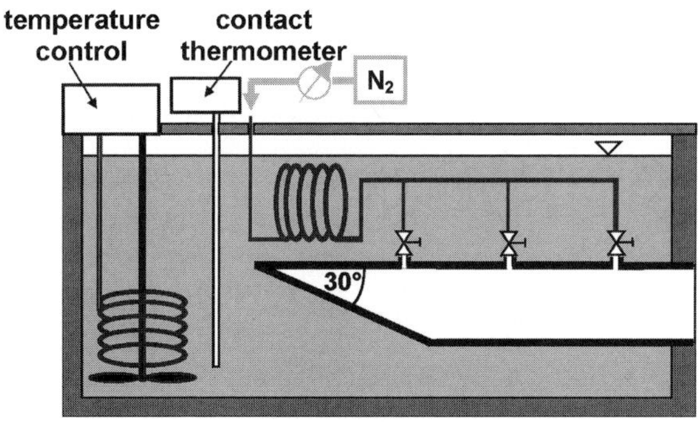


FIG. 5.9. Schematic of a PTB bath blackbody radiator.

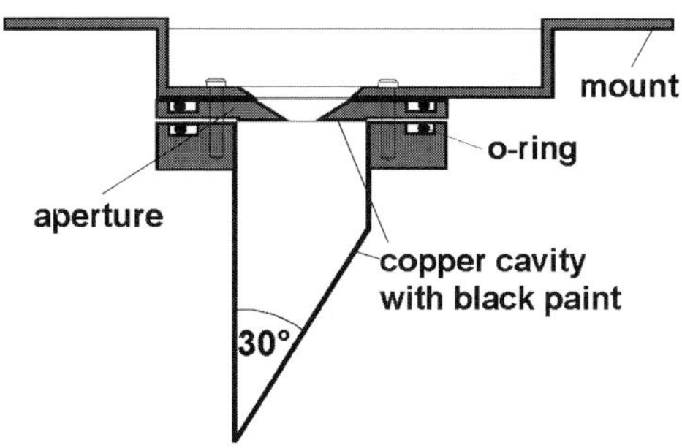


FIG. 5.10. Schematic of the bath blackbody designed by the PTB for the calibration of IR ear thermometers.

the bath blackbody. Nitrogen purging is not required in this case because there is no thermal convection loss through the cavity opening. With the cavity immersed in a 6 L water bath, both temperature stability over 20 minutes and a spatial temperature uniformity within 0.01 °C are achieved. The temperature is measured by a SPRT immersed in the water and located close to the bottom of the copper cavity. This reference bath blackbody radiator is used for calibration of blackbody radiators which are to be employed for infrared-ear-thermometer calibration with a standard uncertainty of 25 mK.

#### *5.2.3.3 Fixed-point blackbodies*

For the variable-temperature blackbodies described in the previous section, temperature has to be determined by a contact or non-contact (radiation) thermometer, and the uncertainty of the temperature determination contributes to the uncertainty of the spectral radiance according to Eq. (5.9). For a blackbody operated at the temperature of one of the ITS-90 fixed points, this thermometer calibration is not necessary—using the ITS-90-prescribed value of $T_{90}$ in Planck's law makes the blackbody radiance (or any quantity realized with its help) uncertain only to the extent of the uncertainty of $T_{90}$ relative to the thermodynamic fixed-point temperature. Therefore, fixed-point blackbodies are used as high-accuracy radiation sources. On the other hand, this type of blackbody can of course be used to determine the thermodynamic temperature $T$ (and thus $T - T_{90}$) of the fixed point by absolute radiation thermometry in a setup as shown in Figure 5.8, for example.

In the temperature range above 0 °C, all of the fixed points of the ITS-90 are metal melting or freezing temperatures. Because of the Stefan–Boltzmann law, Eq. (5.7), the high-temperature fixed points (freezing points of silver, gold, and copper, which are also used as the reference temperatures for ITS-90 relative radiation thermometry) are usually best suited for the operation of blackbodies as radiation sources, but the gallium melting point at a near-ambient temperature of about 30 °C may be preferable for the calibration of remote-sensing or thermal imaging systems. As an example of a fixed-point blackbody, the PTB gold-fixed-point blackbody is shown in Figure 5.11.

Details of the design are described in Reference [83]. The radiating cavity is 97 mm long, with a diameter of 16 mm. The aperture diameter is 3 mm. The cavity with the front aperture is made of graphite and purged with pure argon during operation. The graphite crucible containing the gold ingot is placed in a sodium heatpipe to ensure temperature homogeneity over the cavity length. The single bifilar heating zone is controlled by a

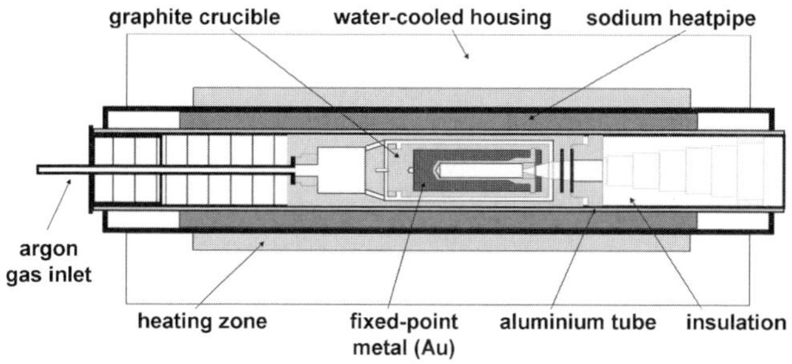


FIG. 5.11. Schematic of the PTB Au fixed point cavity radiator.

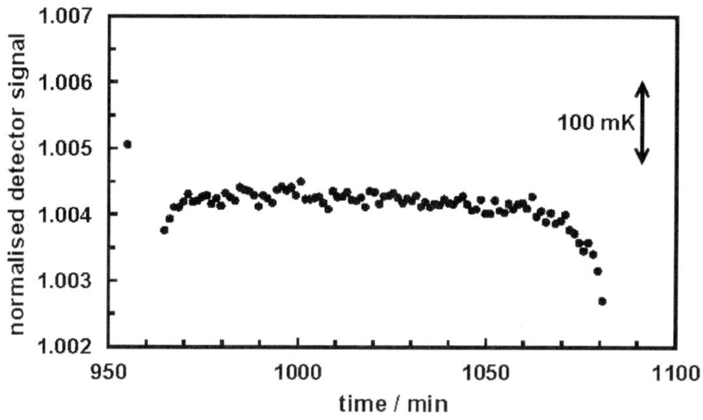


FIG. 5.12. Freezing plateau of the PTB gold fixed-point blackbody. The graph shows the ratio of the detector signals obtained by measuring the gold fixed-point blackbody and a tungsten strip lamp.

microprocessor. The emissivity has been determined as 0.99999 [84] by means of the Monte Carlo ray-tracing simulation technique [21].

The gold fixed-point blackbody emits thermal radiation at the temperature of 1064.18 °C. Owing to the heatpipe housing and the large amount of about 3 kg of gold inside the graphite crucible, the duration of the melting and freezing plateaus is about 90 min with temperature stability of about ±10 mK (Fig. 5.12). This time is sufficient for the calibration of transfer standards of radiation temperature or radiance (e.g. tungsten strip lamps). Due to its excellent performance, this gold fixed-point blackbody was used to determine the thermodynamic temperature of freezing gold as an input for the ITS-90 [83]. Similar gold and silver fixed-point blackbodies are operated by other national laboratories [25, 84, 85].

#### *5.2.3.4 Very-high-temperature blackbodies*

Despite their outstanding accuracy and stability, fixed-point blackbodies are of limited usefulness in radiometry because they can provide thermal radiation at only one particular temperature. Moreover, most of the radiation emitted by the classical high-temperature fixed-point blackbodies is in the infrared (for 1337 K, the gold freezing temperature, $\lambda_{max}$ is about 2.2 μm). According to the results obtained up to now by ongoing world-wide investigations [86–105], metal–carbon and metalcarbide–carbon eutectic mixtures may offer the possibility of introducing additional fixed points, with temperatures ranging from 1400 K to well above 3000 K. At present, however, variable-high-temperature blackbodies (HTBB) [35, 58, 59, 106–113] are usually employed for the realization of high radiation temperatures and of spectral radiance and irradiance. Their temperatures are determined by (relative) radiometric comparison—usually via lamps as secondary standards—with the gold fixed point, say, according to the ITS-90 [47, 106–108] or directly by "absolute" radiation thermometry (cf. Section 5.2.2.2). (As a hybrid intermediate step, a gold-fixed-point blackbody with "absolutely" determined thermodynamic temperature has also been used as the reference for relative measurements [36]).

At the PTB, HTBBs of type BB3200pg [35] are used for this purpose (Fig. 5.13). The cavity of this blackbody consists of pyrolytic graphite rings clamped together with a spring. The cavity is directly Joule-heated with DC as high as 700 A, and can reach temperatures up to 3000 °C. The length of the radiating cavity is about 200 mm, with a diameter of 37 mm.

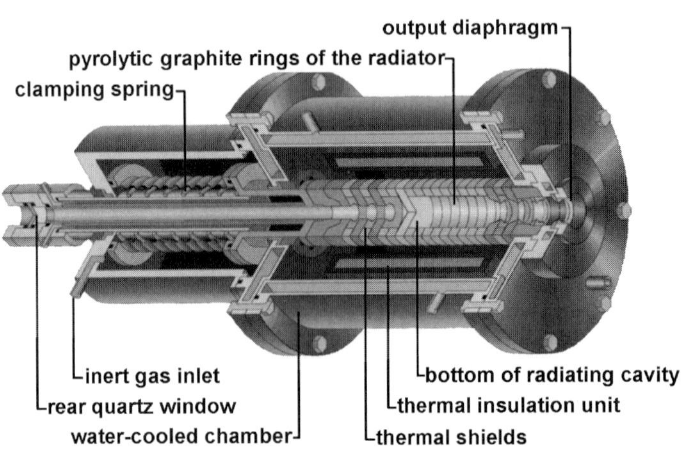


FIG. 5.13. Schematic of the HTBB BB3200pg.

The temperature inhomogeneity of the radiating aperture area is 2 %, and temporal stability of the radiation temperature better than 0.3 K can be achieved. The isothermal emissivity of the HTBB was calculated to be 0.999. In order to determine the influence of the temperature gradients inside the cavity on emissivity, the temperature of the cavity walls was measured using a high-temperature radiation thermometer [113], by tilting its optical axis by 5° with respect to the optical axis of the cavity and moving it along the cavity axis. Temperature profiles measured at different temperatures of the HTBB are shown in Figure 5.14. The temperatures are radiation temperatures which are not the true temperatures at the target locations, but are also influenced by reflected radiation originating in other parts of the cavity. The true local wall temperatures were extracted from the measurements using a Monte Carlo ray-tracing program [21], and employed to determine the effective emissivity of the HTBB (Fig. 5.15). Since the temperature along the cavity wall first increases with increasing distance from the bottom of the cavity, and the temperature of the bottom centre is taken as the reference temperature, the effective cavity emissivity exceeds unity.

Obviously, the influence of the temperature gradients on the effective emissivity is most pronounced at short wavelengths. The variation of the effective emissivity with wavelength cannot be neglected in high-accuracy radiometric measurements. Figure 5.16 shows the temperature error which occurs if the temperature of the HTBB is determined at just the particular

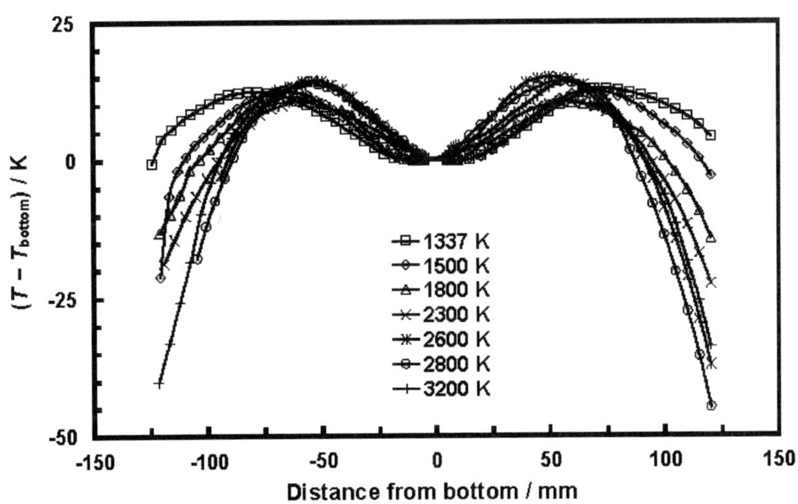


FIG. 5.14. Radiation temperatures of the cavity walls of the HTBB. The measurements in the positive and negative directions are measurements on opposite walls [113].

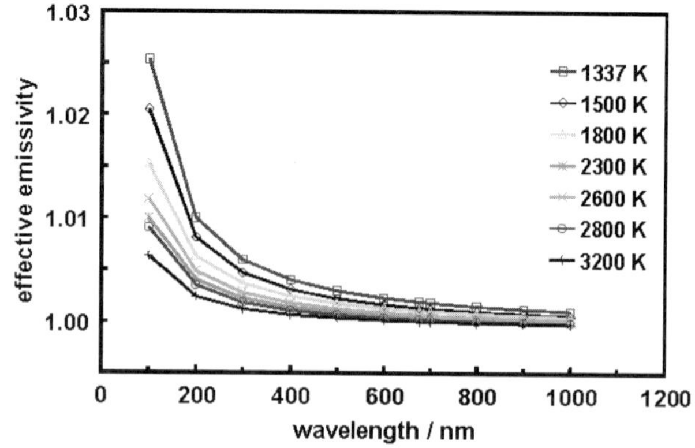


FIG. 5.15. Effective cavity emissivities as functions of wavelength for different blackbody temperatures, calculated from the non-isothermal wall temperatures (Fig. 5.14). The temperature of the cavity bottom was taken as reference temperature [113].

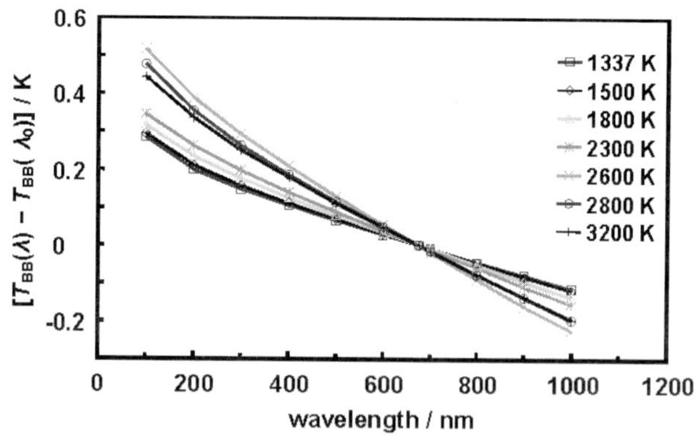


FIG. 5.16. Temperature error when using the HTBB radiation temperature measured at $\lambda_0 = 676\,\text{nm}$ for other wavelengths $\lambda$ as well [113].

wavelength of 676 nm, and the radiator is then used as a primary source of spectral radiance or irradiance or radiation temperature at different wavelengths, assuming the cavity emissivity to be independent of wavelength.

## 5.2.4 Lamps as Secondary and Transfer Standards

### *5.2.4.1 Tungsten strip filament lamps*

Although they are most accurate radiation source standards, fixed-point-temperature and variable-high-temperature blackbodies are not suited as transfer or working standards. Even at many national metrology institutes, they will only be used for calibrations requiring the smallest uncertainties attainable. For the great majority of practical applications, the uncertainties are fully adequate which can be reached with special types of lamps that are more easily transportable and simpler to operate than blackbody radiators.

These lamps are usually characterized by their (spectral) radiance temperature (or radiation temperature) $T_{BB}$, i.e., the blackbody temperature which results in the same spectral radiance as the lamp's spectral radiance $L_\lambda(\lambda)$ at the wavelength $\lambda$ under consideration:

$$L_\lambda(\lambda) = L_{\lambda,BB}(\lambda, T_{BB}) \tag{5.15}$$

Obviously, the radiance temperature will in general depend on $\lambda$. For visual pyrometry, a common "reference wavelength" for $T_{BB}$ is 655 nm. In addition to the emissivity, the radiance temperature provides an alternative way of relating the radiation emission of some radiator to blackbody radiation. In principle, the radiance temperature can also be used for radiators which cannot be described by a temperature $T$, in contrast to the emissivity. For non-black radiators with a true temperature $T$, the radiance temperature $T_{BB}$ will always be lower than $T$ as a result of $\varepsilon \leq 1$.

Well-established, highly stable transfer standards of spectral radiance (and also of radiation temperature) are tungsten strip (or ribbon) filament lamps [2, 99, 114, 115], which are used at radiance temperatures from 600 °C to 2300 °C and over a wavelength range from 220 nm up to 2.5 μm. Up to about 1600 °C, evacuated bulbs can be used. At higher temperatures, a noble-gas filling is required to suppress evaporation of tungsten from the filament. Tungsten strip lamps which serve as transfer standards of spectral radiance are operated at a fixed single current provided by a DC power supply. For lamps calibrated as standards of radiation temperature, the chosen current determines the temperature (Fig. 5.17). For spectral irradiance, 1000 W FEL- or DXW-type quartz halogen tungsten lamps (also available with detector stabilization) have become preferred transfer or working standards, which can even be operated at radiance temperatures up to 3200 °C or 3300 °C and have spectra which are rather similar to blackbody spectra [116–125]. These types of lamps emit useful radiation at wavelengths down to 250 nm or even 200 nm. In the UV spectral range from 190 nm or 200 nm to 400 nm, deuterium ($D_2$) gas-discharge lamps are also used [126–129] as secondary or transfer standards, because their spectral radiant power increases towards shorter wavelengths in this range. However, $D_2$ lamps are less reproducible and have higher ageing rates than tungsten strip or FEL-type lamps. As calibration standards for all these lamps, high-temperature blackbodies are much more suitable than the "classical" ITS-90 fixed-point blackbodies. Plasma sources with still higher

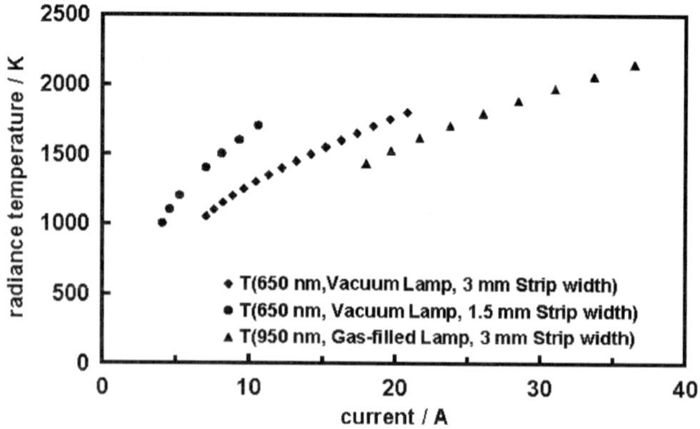


FIG. 5.17. Radiation temperature versus lamp current for different tungsten strip lamps.

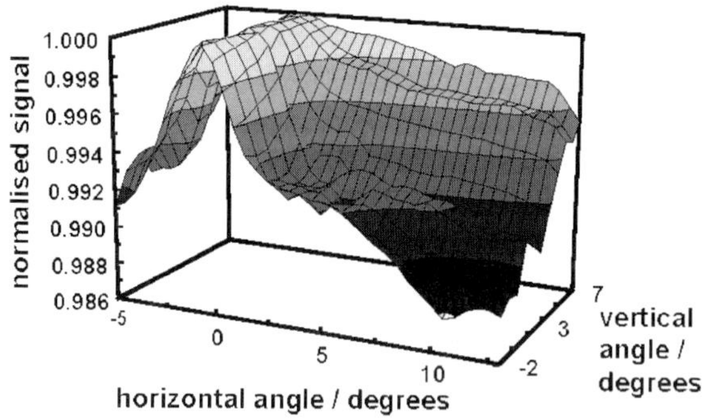


FIG. 5.18. Normalised spectral radiance distribution as a function of the horizontal and vertical emission angle of a tungsten strip lamp at 650 nm.

radiation temperatures such as the argon mini-arc [130] or more powerful wall-stabilized argon arc plasmas [131, 132] are more difficult to handle than the lamps.

Tungsten strip lamps are not Lambertian radiators but exhibit a complex dependence of the emitted radiance on the emission angle (Fig. 5.18). Because of this, precise alignment is required.

A serious limitation to the application of tungsten strip lamps is their small radiating area. With the width of the strip ranging from 1 nm to about 3 mm, tungsten strip lamps can only be used for calibration of spectral-radiance meters (and radiation thermometers) with fields of view smaller than 3 mm in diameter. This limitation can be overcome by the application of integrating-sphere sources. Their advantages are a larger homogeneously radiating area and a Lambertian emission characteristic, while their main disadvantage is their poorer stability in comparison to tungsten strip lamps. In addition, the spectral radiance of integrating-sphere sources is in general distinctly smaller than that of tungsten strip lamps.

#### *5.2.4.2 Calibration of transfer standards*

Tungsten strip lamps and other radiation sources are calibrated for spectral radiance or radiation temperature by comparison to variable-high-temperature blackbodies or to the gold (or silver or copper) fixed-point blackbody using a monochromator for wavelength selection. As an example, Figure 5.19 schematically shows the PTB calibration setup [47]. The gold fixed-point blackbody (Fig. 5.11) and the HTBB (Fig. 5.13) are used for the calibration of source standards of spectral radiance and radiation temperature and of radiation thermometers. Other national metrology institutes or calibration laboratories use similar facilities for this purpose [39, 106, 107, 133, 134].

The detection unit is placed on a translation stage carried in air bearings which can be moved over a distance of 3 m with a resolution of 1 μm and a reproducibility of 10 μm. Radiation is spectrally filtered with a double

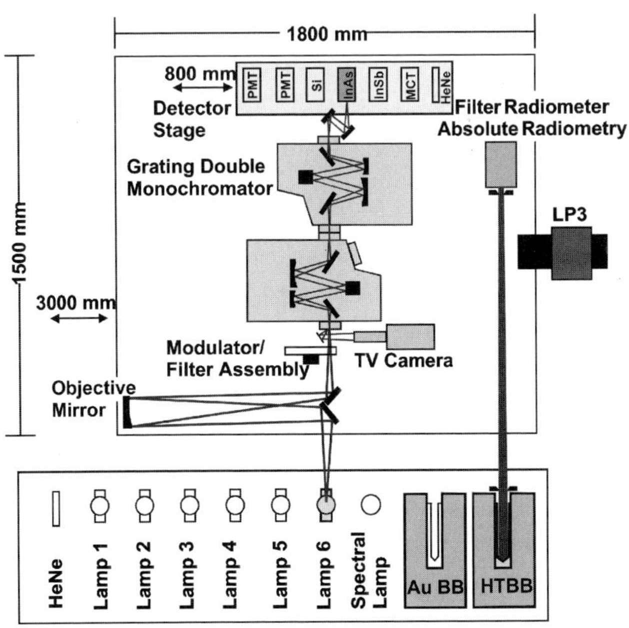


FIG. 5.19. Spectral radiance comparator facility of the PTB.

monochromator consisting of two grating monochromators in additive mode. Filters for order selection and a polarizer are placed in front of the monochromator, so that separate measurements can be performed for the s- and p-components of polarized radiation, e.g. from tungsten strip lamps, as required for the comparison of polarized and unpolarized (blackbody) radiation. For operation over a wavelength range from 200 nm to 15 μm, four automatically exchangeable pairs of gratings are used in the monochromator, and a variety of different detectors, such as Si and InGaAs photodiodes, photomultipliers, InSb and HgCdTe (MCT) photoconductors can be automatically positioned behind the exit slit. An objective mirror forms a one-to-one image of the sources on the entrance slit of the first monochromator. The whole monochromator-detector system acts as a filter radiometer with tunable central wavelength. Therefore, it is often called a spectroradiometer. In addition to the spectroradiometer, filter radiometers and radiation thermometers (such as the LP3 shown in Fig. 5.19 [135]) are operated for temperature measurements at this set-up. Up to six transfer standard lamps can be calibrated at once with this instrumentation. A HeNe laser for alignment and a spectral lamp for wavelength calibration are also installed on the source mount.

The typical set-up for irradiance calibration is essentially the same as for radiance calibration, but standard and test radiators are not imaged onto the monochromator entrance slit. Rather, they illuminate the entrance aperture of an integrating sphere, the exit aperture of which illuminates the monochromator entrance slit [46, 107, 136, 137].

Applying a HTBB as a primary standard source of spectral radiance requires its temperature to be measured with small uncertainty. The PTB facility shown in Figure 5.19 allows the determination of the temperature of the HTBB by two independent methods [47]. With a radiation thermometer (the double monochromator itself might also be used for this purpose), the ITS-90 temperature $T_{90}$ of the HTBB is measured by comparing its spectral radiance with the spectral radiance of the gold-fixed-point blackbody according to Eq. (5.14). Applying interference-filter radiometers and apertures for an absolute irradiance measurement (Fig. 5.8a), the thermodynamic radiation temperature $T$ can be determined, too. This measurement is traceable to a cryogenic radiometer (and thus to electrical SI units) via the calibration of the filter radiometer's spectral transmission. In both measurements, the same part of the blackbody bottom is viewed by the detector systems. Obviously, comparison of the thermodynamic temperature scale with the ITS-90 is also possible in this way.

In order to determine the spectral radiance $L_\lambda(\lambda)$ or, equivalently, the radiance temperature $T_{BB}(\lambda)$ of a tungsten strip lamp operated at its nominal current, the spectroradiometer is set to wavelength $\lambda$. After the temperature $T$ of the HTBB has been measured, the spectroradiometer is moved in front of the HTTB, and its signal $S_{\mathrm{HTTB}}(\lambda)$ is recorded. This measurement provides a calibration for the spectral-radiance responsivity of the spectroradiometer at wavelength $\lambda$. Then, the spectroradiometer is moved in front of the lamp to be calibrated, and the corresponding signal $S_{\mathrm{L}}(\lambda)$ is recorded. Finally repeating the HTBB-temperature measurement allows us to determine (and to correct, if necessary) a possible temperature drift. The basic relation giving the spectral radiance of the lamp in terms of the measured quantities is simply

$$\begin{aligned} L_\lambda(\lambda) &= L_{\lambda,\mathrm{BB}}(\lambda, T_{\mathrm{BB}}) = S_{\mathrm{L}}(\lambda)\varepsilon_{\mathrm{eff}}(\lambda)L_{\lambda,\mathrm{BB}}(\lambda, T)/S_{\mathrm{HTTB}}(\lambda) \\ &= \varepsilon_{\mathrm{eff}}(\lambda)[S_{\mathrm{L}}(\lambda)/S_{\mathrm{HTTB}}(\lambda)](2hc^2/\lambda^5)/\{\exp[hc/(\lambda kT)] - 1\} \end{aligned} \tag{5.16}$$

Here, the slightly non-ideal behaviour of the HTBB is explicitly taken into account by the effective emissivity (Eq. (5.13) and Fig. 5.15). It should be noted that $\varepsilon_{\mathrm{eff}}$ (but usually taken at a wavelength other than $\lambda$) is also implicitly contained as a correction factor in the HTBB temperature.

Obviously, the uncertainties $u(X)$ of all quantities $X$ which appear on the right-hand side of Eq. (5.16) contribute to the uncertainty $u(L_\lambda)$ of the spectral radiance $L_\lambda$ to be determined, according to $u_X(L_\lambda) = (\partial L_\lambda/\partial X)u(X)$. For the PTB facility shown in Figure 5.19 as well as for other facilities, major contributions to $u(L_\lambda)$ come from the uncertainty of the signal ratio $S_{\mathrm{L}}(\lambda)/S_{\mathrm{HTTB}}(\lambda)$ as a result of detector noise and non-linearity in particular at short and long wavelengths, from the usual approximation of $\varepsilon_{\mathrm{eff}}(\lambda)$ by 1, and from the uncertainty of the HTBB temperature $T$. The wavelength uncertainty may also contribute significantly in particular at short wavelengths. Additional less obvious uncertainty contributions result from short-term instabilities of both the HTBB and the lamp to be calibrated, and from the uncertainty of adjusting the lamp current to its nominal value. Eventually, two more subtle effects contribute to the uncertainty of the lamp calibration, which are due to the size of the sources compared and to the spectral responsivity of the spectroradiometer.

The "size-of-source" effect (SSE) [106, 108, 138–140] combines all effects such as scattering or diffraction which cause radiation from outside the geometrical limits to reach the detector of the spectroradiometer. (There is also an inverse process which leads radiation from inside the geometrical limits astray.) Since radiation originating in the vicinity of the geometrical source area will generally be responsible for most of the SSE, there will be a difference for the spectroradiometer signals measured for a tungsten strip lamp with a filament 3 mm wide and a high-temperature blackbody with an aperture of 20 mm diameter, say, even if both of them emit the same spectral radiance from the source area geometrically imaged 1:1 onto a spectroradiometer entrance slit of 0.2 mm width and 1 mm height. There are two main methods for determination (and subsequent correction) of the SSE. An aperture in front of the blackbody can be used to make it look like the tungsten strip lamp. The rise of the signal measured when the aperture is removed indicates the SSE. Alternatively, the geometrical source area as seen by the spectroradiometer can be shaded by a suitable inset in the blackbody aperture and in front of the lamp, and the remaining spectroradiometer signal can be measured. For the set-ups under consideration here, the uncertainty connected with the correction of the SSE is not a major contribution to the total uncertainty.

The spectral responsivity $R$ of the spectroradiometer is another source of uncertainty [46, 106, 107, 141, 142]. Though the bandpass of a double-monochromator system may be made very narrow, the responsivity is not an ideal $\delta$-function, but has a non-vanishing spectral width, mainly determined by the geometrical slit widths, and far-reaching tails towards shorter and longer wavelengths, due to imperfect imaging, diffraction and scattering. Even if they are very nearly vanishing, the long tails may be a particular problem for measurements of blackbody radiation, say, at wavelengths $\lambda$ far away from $\lambda_{\max}$, since the spectral radiance will then be orders of magnitude higher at wavelengths around $\lambda_{\max}$ than at $\lambda$, and may still contribute significantly to the measured signal. Even if such out-of-band radiation is sufficiently suppressed, the signals in Eq. (5.16) are not strictly proportional to the monochromatic spectral radiance $L_\lambda(\lambda)$ at the wavelength $\lambda$ the monochromator is set to, but rather are mathematically expressed as the convolution of the spectral responsivity of the spectroradiometer with the spectral radiance of the source (leaving aside all additional complications arising from the dependence of $R$ on, e.g. the direction or polarization of incoming radiation):

$$S(\lambda) = \int_0^\infty \mathrm{d}\lambda' \, R(\lambda, \lambda - \lambda') L_\lambda(\lambda') \tag{5.17}$$

The integral would only be proportional to $L_\lambda(\lambda)$ for any (continuous) function $L_\lambda(\lambda')$ if $R(\lambda, \lambda' - \lambda)$ was equal to $R_0(\lambda)\delta(\lambda' - \lambda)$, or under special conditions fulfilled by $R$ and/or $L_\lambda$ (e.g., constant $L_\lambda$, or symmetrical $R$ and linear variation of $L_\lambda(\lambda')$ around $\lambda$). In general, however, $L_\lambda(\lambda)$ can only be retrieved from $S(\lambda)$ by deconvolution, which requires sufficiently accurate values of $S(\lambda)$ and $R(\lambda, \lambda' - \lambda)$ to be measured over sufficiently broad wavelength intervals of both $\lambda$ and $\lambda'$. For an accurate determination of $R(\lambda, \lambda' - \lambda)$, the spectroradiometer would have to be set to wavelength $\lambda$, the signal produced by monochromatic radiation with wavelength $\lambda'$ and radiance $L$ would have to be measured, and this would first have to be done in a broad wavelength interval for closely spaced values of $\lambda'$, and then would have to be repeated for all wavelength settings $\lambda$ the spectroradiometer is intended to be used for. Wavelength-tunable lasers illuminating integrating spheres in front of the spectroradiometer are probably the radiation sources most suitable for this purpose, but another spectroradiometer irradiated by a broadband source and combined with a cryogenic radiometer may also provide the required radiation [143].

A more detailed discussion of the calibration of spectroradiometer responsivity is beyond the scope of this contribution. It should be clear, however, that the responsivity calibration of a spectroradiometer is an ambitious undertaking. Moreover, its period of validity is rather uncertain. Therefore, spectroradiometers have not yet been used as absolutely calibrated wavelength-tunable filter radiometers which would no longer need to refer to blackbody standards, and blackbodies are still required as primary source standards of thermal radiation.

## 5.3 Synchrotron Radiation Sources

Synchrotron Radiation (SR) was first observed in 1947. The theory describing its emission is based on classical electrodynamics and can be found in the literature, e.g. in the classical paper of Schwinger [144]. The SR spectrum extends from the infrared to the X-ray region. Its use in many fields such as physics, chemistry, biology, material sciences and elsewhere has expanded rapidly, especially during the 1990s. Currently, more than 60 SR facilities are in operation worldwide, and about 10 new facilities are under construction or being planned. The properties of SR and its applications are described in detail in textbooks [4, 5, 145–147].

The potential for using electron storage rings as almost ideal radiation sources for radiometry from the infrared to the X-ray region and also as calculable radiation sources, i.e., as primary source standards, was recognized as early as 1956 [148]. This option was successfully pioneered during the 1970s and 1980s at only a few facilities. The National Bureau of Standards, which has since been renamed the National Institute of Standards and Technology (NIST), converted its synchrotron SURF into the storage ring SURF II in 1974. SURF II served as a radiometric standard until 1997 when it was upgraded to SURF III [149, 150]. In 1984, the PTB proved the electron storage ring BESSY I to be a primary source standard [151, 152] which was used until its closure in November 1999. At present, the BESSY II storage ring which has been in user operation since January 1999 is being used by the PTB as the European primary source standard from the visible to the X-ray range [153]. SR-based radiometry is also carried out at the TERAS storage ring [154, 155] in Japan and for decades at the VEPP-2M and VEPP-3 facilities [156–158] in Russia (Novosibirsk). Comprehensive studies on the use of SR for radiometry can be found in References [159–161] and the references therein.

### 5.3.1 Description of SR

The spectral radiant intensity of a charged, relativistic particle moving on an arbitrary trajectory can be calculated from classical electrodynamics theory [145–147]. The special case of a relativistic electron moving on a circular arc, i.e., being deflected in a homogenous magnetic field, is treated by the so-called Schwinger equation [144], which allows the spectral radiant intensity of the SR from a bending magnet in an electron storage ring to be calculated from a few storage ring parameters.

In reality, it is more advantageous to work with the spectral radiant power $\Phi_{\mathrm{E}}$ emitted into an aperture of size $a \times b$ placed at a distance $d$ from the radiation source point (see Fig. 5.20). The spectral radiant power as a function of the photon energy $E = hc/\lambda$ can then be written in the general form as

$$\Phi_{\mathrm{E}} = \Phi_{\mathrm{E}}(E; W, B, I, \Sigma_Y, \Psi, d, a, b) \qquad (5.18)$$

where $B$ is the magnetic induction at the radiation source point, $W$ the electron energy, $I$ the stored electron beam current and $\Psi$ the vertical emission angle with respect to the electron orbit plane. The original Schwinger equation describes the spectral energy radiated by one electron on a circular arc. In an electron storage ring, many electrons of number $N$ are stored that

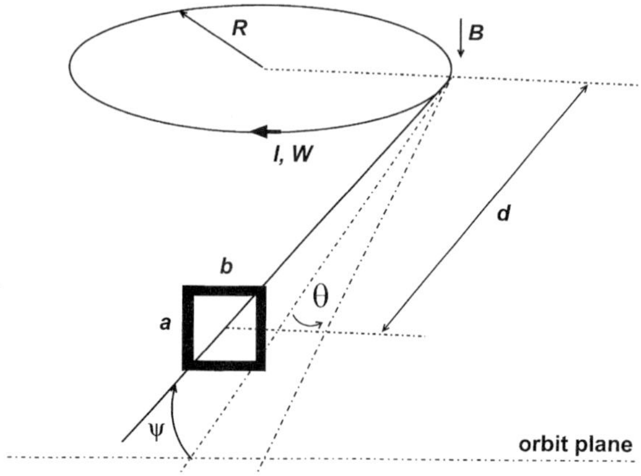


FIG. 5.20. Schematic diagram of the parameters for calculating the spectral radiant power of synchrotron radiation from a bending magnet according to Schwinger.

revolve in the ring with frequency $\nu$. The spectral power radiated by these electrons is then given by multiplying the original Schwinger equation by $\nu N$. The stored electron beam current $I$ is given by $I = e\nu N$. Moreover, the stored electrons have a small deviation from the perfect circular orbit in space as well as in angle. This results in an effective vertical divergence $\Sigma_Y$ which has to be taken into account as a convolution over the vertical emission angle $\Psi$ [162] (see below).

Equation (5.18) is expressed in terms of the magnetic induction $B$ at the radiation source and the electron energy $W$. Alternatively, the bending radius $R$ could be used instead of one of these parameters according to

$$R = W/ecB \tag{5.19}$$

Moreover, for a true circular orbit, as is the case in the single-magnet design of the SURF III electron storage ring, the electron energy $W$ can also be expressed as a function of the magnetic induction $B$ and the revolution frequency $\nu$ as $W = ec^2B/2\pi\nu$.

The derivation of the Schwinger equation can be found in standard textbooks on classical electrodynamics; here, only the result for the spectral radiant power and some basic properties are reviewed.

Equation (5.18) can be written as

$$\begin{aligned}\Phi_{\mathrm{E}} &= \frac{\mathrm{d}\Phi}{\mathrm{d}E}(E) \\ &= \iint_{\text{aperture}} \left[ \int_{-\infty}^{+\infty} \frac{1}{\sqrt{2\pi}\Sigma_Y} [I_{0E}^{\sigma}(\psi'') + I_{0E}^{\pi}(\psi'')] \mathrm{e}^{-\frac{(\psi''-\psi')^2}{2\Sigma_Y^2}} \,\mathrm{d}\psi'' \right] \mathrm{d}\psi \,\mathrm{d}\theta \end{aligned} \tag{5.20}$$

with

$$I_{0E}^{\sigma} = \frac{\mathrm{d}}{\mathrm{d}E}\frac{\mathrm{d}^2}{\mathrm{d}\theta\,\mathrm{d}\psi}\phi^{\sigma} = \frac{2eIR^2}{3\varepsilon_0\gamma^4}\frac{E^2}{(hc)^3}([1+(\gamma\psi)^2]^2 K_{2/3}^2(\xi)) \tag{5.21}$$

and

$$I_{0E}^{\pi} = \frac{\mathrm{d}}{\mathrm{d}E}\frac{\mathrm{d}^2}{\mathrm{d}\theta\,\mathrm{d}\psi}\phi^{\pi} = \frac{2eIR^2}{3\varepsilon_0\gamma^4}\frac{E^2}{(hc)^3}([1+(\gamma\psi)^2](\gamma\psi)^2 K_{1/3}^2(\xi)) \tag{5.22}$$

$$\xi = \frac{2\pi RE}{3hc\gamma^3}(1+(\gamma\psi)^2)^{3/2} = \frac{1}{2}\frac{E}{E_{\mathrm{c}}}(1+(\gamma\psi)^2)^{3/2}, \quad \gamma = \frac{W}{m_0c^2}$$

The $K_{1/3}$ and $K_{2/3}$ are modified Bessel functions of the second kind, which can be expressed in a series expansion as described in Reference [163]. Equations (5.21) and (5.22) describe the $\sigma$ and $\pi$ polarization components of the spectral radiant intensity according to Schwinger for electrons of beam current $I$ and energy $W$, all moving exactly on a circular arc with radius $R$.

In reality, the electron beam in a storage ring has a certain emittance resulting in a Gaussian distribution in space around the orbit as well as in angle with respect to the orbit with standard widths of $\sigma_x$ and $\sigma_y$ in space and $\sigma_{x'}$ and $\sigma_{y'}$ in angle for the horizontal and vertical directions, respectively. The horizontal distribution is of no importance due to the tangential observation of the SR, but in the vertical direction this leads to an effective divergence

$$\Sigma_Y = (\sigma_y^2/d^2 + \sigma_{y'}^2)^{1/2} \tag{5.23}$$

at the location of the aperture stop which is attributed for by a convolution of Eqs. (5.21) and (5.22) before integration over the flux defining aperture, as given in Eq. (5.20). This convolution normally only has little influence besides when dealing with large values of $\Sigma_Y$ or high photon energies. The Gaussian extension of the electron beam makes it unfavorable to work with the radiance of the storage ring source as a radiometric quantity, the spectral radiant power Eq. (5.20) or the spectral radiant intensity (which can be attributed to the expression in the outer brackets of Eq. (5.20)) are easier to handle. In reality, the measurement distance $d$ is much larger than the extension of the electron beam (much less than 1 mm) or the extension of the flux defining aperture and the storage ring can therefore be handled as a point source in most cases.

For a rectangular aperture of size $a \times b$ at distance $d$, the integration in Eq. (5.20) over the horizontal angle $\theta$ in the orbit plane results in a factor $b/d$, in the perpendicular direction the integration over the vertical angle $\psi'$ has to be performed from $(\psi - a/2d)$ up to $(\psi + a/2d)$ with $\psi$ being the observation angle of the centre the aperture with respect to the orbit plane ($a/d$, $b/d$, $\psi \ll 1$). The evaluation of Eq. (5.20) is normally done numerically.

The modification of the spectral radiant power according to Schwinger $\Phi_{\mathrm{E}}^{\mathrm{Schwinger}}$ due to the electron beam emittance is sometimes expressed by a factor $\varepsilon$ according to

$$\begin{aligned}\Phi_{\mathrm{E}} &= \Phi_{\mathrm{E}}(E; W, B, I, \Sigma_Y, \Psi, d, a, b)\\ &= \Phi_{\mathrm{E}}^{\mathrm{Schwinger}}(E; W, B, I, \Psi, d, a, b)(1 + \varepsilon(E; W, B, \Sigma_Y, \Psi, d, a))\end{aligned}$$

The factor $\varepsilon$ is less than about $5 \times 10^{-4}$ for typical calibration geometries at photon energies below 10 keV at the low emittance ring BESSY II but can be of the order of $10^{-2}$ for storage rings with a larger emittance as, e.g. the future Metrology Light Source (MLS) [164].

The SR spectrum extends continuously from the far-infrared through the visible, ultraviolet (UV), vacuum UV (VUV) into the X-ray region and can be classified by the characteristic energy $E_c$ (often called critical energy)

$$E_c = \frac{3\hbar c \gamma^3}{2R} \tag{5.24}$$

In practical units $E_c$ can be evaluated as

$$E_c/\mathrm{eV} = 665.0 B/T\,(W/\mathrm{GeV})^2 = 2218\,(W/\mathrm{GeV})^3/(R/\mathrm{m}) \tag{5.25}$$

For increasing photon energies, the spectral radiant power gradually rises up to a broad maximum around the characteristic energy and thereafter decreases and ultimately vanishes exponentially. One half of the entire, vertically integrated, radiant power per horizontal angular divergence is emitted at photon energies above $E_c$ and the other half below this value. Figure 5.21 shows spectra for the BESSY II, a 7 T superconducting wavelength shifter (WLS) at BESSY II, SURF III and the future MLS electron

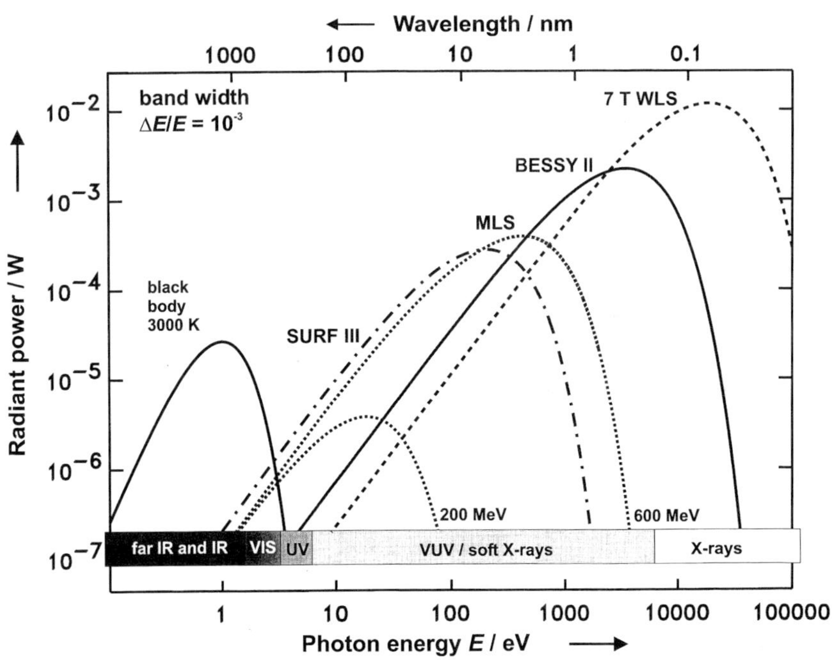


FIG. 5.21. Calculated radiant power for a 3000 K blackbody radiator, SURF III, BESSY II and the future MLS. (The parameters used for the calculation are: BESSY II: $W = 1700$ MeV, $B = 1.3$ T ($B = 7$ T for the WLS); $I = 200$ mA, $d = 20$ m; SURF III: $W = 380$ MeV, $B = 1.51$ T, $I = 500$ mA, $d = 5$ m; MLS: $W = 600$ MeV (200 MeV), $B = 1.3$ T (0.43 T), $I = 200$ mA, $d = 5$ m; black body: $T = 3000$ K, $A_{em} = 78.5\,\mathrm{mm}^2$, $d = 1$ m; all for a flux defining circular aperture of radius 10 mm and a relative spectral bandwidth of $\Delta E/E = 0.1\,\%$.)

storage rings calculated for typical geometries. For comparison, the spectrum of a blackbody radiator is also shown in this figure revealing the potential of electron storage rings for radiometry beyond the UV spectral range.

The angular distribution of the SR is homogeneous in the horizontal direction (orbit plane of the electrons) but shows a narrow distribution in the vertical direction. The vertical divergence depends on the photon energy and decreases with increasing photon energy. In contrast to blackbody radiation, SR is completely polarized. Without taking the electrons emittance into account, the $\sigma$-component, Eq. (5.21), with the electrical field vector parallel to the orbit plane shows a maximum at the orbit plane, whereas the perpendicular $\pi$-component, Eq. (5.22), shifted in phase by 90°, vanishes at the orbit plane. So, in the orbit plane pure linear polarization is present, whereas elliptical polarization of different heliticity is present above and below the orbit plane. Owing to the electron beam emittance and the necessary convolution of $\Sigma_Y$, in reality a small contribution of the $\pi$-component is present also in the orbit plane.

It has already been mentioned that the Schwinger equation is a special case for an electron moving on a circular arc, i.e. in an homogeneous magnetic field. This condition is met for a bending magnet in an electron storage ring with sufficient small field gradients. Nevertheless, the radiation field of an electron moving on an arbitrary trajectory can also be calculated from classical electrodynamics theory. So, the radiation of an electron moving in the magnetic field with stronger gradients, as is the case, e.g. in a superconduction wavelength shifter [165], or in a periodical magnetic field, e.g. of an undulator, can also be calculated, if the magnetic field map of these devices is known [166].

### 5.3.2 Electron Storage Rings Operated as Primary Source Standards

Electron storage rings have some unique properties that make them ideal sources for radiometry:

- The spectrum of bending magnet radiation covers a large spectral range that is not accessible by blackbody radiation. The spectral radiant power of this radiation can be calculated from fundamental electrodynamics relations from a few storage ring parameters and geometrical quantities as described in the previous chapter.
- The spectral radiant power of a storage ring scales linearly with the stored electron beam current, i.e. the number of stored electrons, and thus offers a dynamic range in flux of up to 12 decades. This allows adjusting the spectral radiant power to the problem under study over a wide range.

Unfortunately, since almost all electron storage rings are multi-user facilities, this option can only very seldom be used in special operation shifts.

- Storage rings are clean (operated at ultra-high vacuum), stable and reproducible radiation sources. This, on the other hand, normally requires that the devices under test also have to comply with the ultra high vacuum requirements or windows have to be used which then cause additional uncertainties.

In order to precisely calculate the spectral radiant power, all parameters entering Eq. (5.18) must be known, i.e., in practice must be measured, with high accuracy. This requires a stable and reliable operation of the storage ring and special equipment to be installed into the storage ring as described below. This is—in addition to the necessity of special operation conditions—a reason why, worldwide, only few electron storage rings are used as primary sources. Table 5.3 gives a list of these storage rings, including their electron energies, characteristic photon energies and references for further details. The measurement of the storage ring parameters is presented in detail for SURF III and BESSY II in References [149, 150] and [153], respectively.

In the following, using BESSY II as an example, possible techniques for the measurement of the storage ring parameters with sufficient accuracy are reviewed.

*Measurement of the electron energy $W$.* At BESSY II, the electron energy is measured with two independent and complementary techniques, i.e., by resonant spin depolarization [167] and by Compton backscattering of laser photons [168, 169]. The resonant spin depolarization technique requires a spin-polarized electron beam which takes about 1 hour to build up at BESSY II operated at 1700 MeV. The technique is well established [170] and allows the electron beam energy to be measured with a relative uncertainty

TABLE 5.3. Storage Rings Used as Primary Source Standards, Their Maximal Electron Beam Energy and their Characteristic Photon Energy $E_c$ of Bending Magnet Radiation

| Ring Name (Inst.) | Electron Energy (GeV) | $E_c$ (keV) | Ref. |
|---|---|---|---|
| SURF III (NIST) | 0.4 | 0.17 | [149, 150] |
| MLS* (PTB) | 0.6 | 0.31 | [164] |
| TERAS (NMIJ) | 0.8 | 0.57 | [154, 155] |
| VEPP-2M (INP) | 0.7 | 0.62 | [156, 157] |
| BESSY II (PTB) | 1.7 | 2.5 | [153] |
| VEPP-3 (INP) | 2.0 | 4.8 | [158] |

*Under construction (September 2004).

of better than $5 \times 10^{-5}$. Unfortunately, it is not applicable when BESSY II is operated at a reduced electron energy of 900 MeV in special PTB calibration shifts, since at that electron energy, polarization build-up would take more than 30 h (polarization time scales as $1/\gamma^5$ for a given ring). Therefore, the alternative method of Compton backscattering of laser photons is additionally applied at BESSY II [169], giving a relative uncertainty of better than $10^{-4}$. For 1700 MeV operation of BESSY II, both methods have been applied simultaneously and an excellent agreement of the results was found [169].

*Measurement of the magnetic induction B.* A specially designed vacuum chamber in the bending magnet allows a nuclear magnetic resonance probe to be brought to the source point of the radiation after a beam dump has been performed. The source point lies in a region of the bending magnet with very low field gradients which has been checked by a field mapping of the bending magnet before installation. The relative uncertainty for the determination of the magnetic induction at the radiation source point is better than $10^{-4}$.

*Measurement of the electron beam current I.* BESSY II is operated for special PTB shifts with electron beam currents between 0.2 pA (one stored electron) and normal current of about 250 mA, thus enabling the user to match the photon flux to the sensitivity of the devices to be calibrated over a dynamic range of more than 12 decades [153]. Currents in the upper range, i.e., above 2 mA, are measured with two commercially available DC parametric current transformers. Electron currents in the lower range, i.e., below 40 pA, are determined by counting the number of stored electrons [171–173]. For this, the electrons are gradually kicked out of the storage ring by a mechanical scrapper that can be moved closely to the electron beam while measuring the step-like drop of the SR intensity by cooled photodiodes (see Fig. 5.22). Electron beam currents in the middle range, i.e., from about 40 pA up to 2 mA, are determined by three sets of windowless Si photodiodes with linear response that are illuminated by SR attenuated in intensity by different filters in the upper current range. The calibration factors of these photodiode-filter combinations, which relate the photocurrent to the electron beam current, are determined by comparison with the electron beam current measured at the upper and lower end of the range as described above.

*Measurement of the effective vertical divergence $\Sigma_Y$.* The effective vertical source divergence is very small at BESSY II (3.5 μrad at $d = 30$ m) compared to the vertical opening angle of the SR at the photon energies of interest (e.g. 100 μrad at 8 keV photon energy) and has therefore very little influence on the vertical distribution (see Fig. 5.23). Therefore, it is normally sufficient to rely on the value calculated from the machine parameters or on

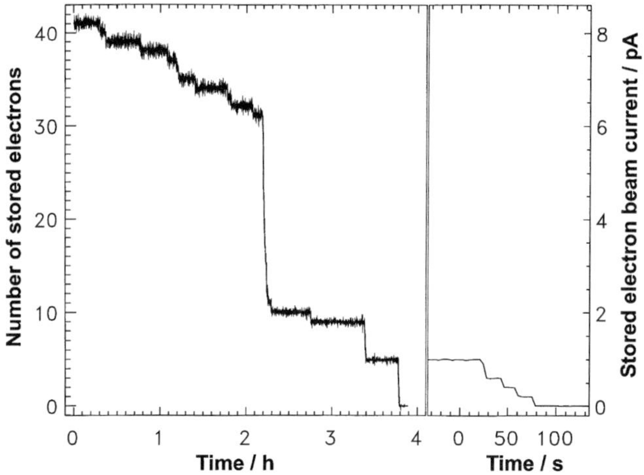


FIG. 5.22. Electron beam current measurement by electron counting at BESSY II. While a scrapper is moved closely to the electron beam to reduce the beam lifetime, the lost electrons are counted by measuring the emitted synchrotron radiation with cooled photodiodes.

beam size measurements which are assumed to have a relative uncertainty of typically about 20 %. The influence of this rather high uncertainty on the uncertainty in the calculation of the spectral photon flux is small, as can be seen in Figure 5.24.

For electron storage rings, where the effective source size has a non-negligible effect—as is the case of SURF III or the future MLS—more accurate determination of the vertical source divergence is needed. This can be done, e.g. by a Bragg polarimeter-monochromator as described in [174].

*Measurement of the distance from the source point and other geometrical quantities*. The distance $d$ to the source point is measured by projecting a fivefold slit into the detection plane [153]. The distance between the slit and the detection plane is precisely known from an interferometric measurement. The distance from the projection plane to the radiation source point at the location of the electron beam can then be calculated from the distance of the projected slits at the detector plane. An accuracy of about 2 mm in the determination of the distance to the radiation source point is reached. Typically, a detector to be calibrated is placed about 30 m from the radiation source point, which gives a relative uncertainty of about $7 \times 10^{-5}$ in the determination of the distance.

The vertical emission angle $\psi$ is normally chosen to be zero (measurement in the orbit plane) by adjusting a detector under test in the vertical plane. A typical uncertainty in positioning is 2 μrad for a calibration at 30 m distance.

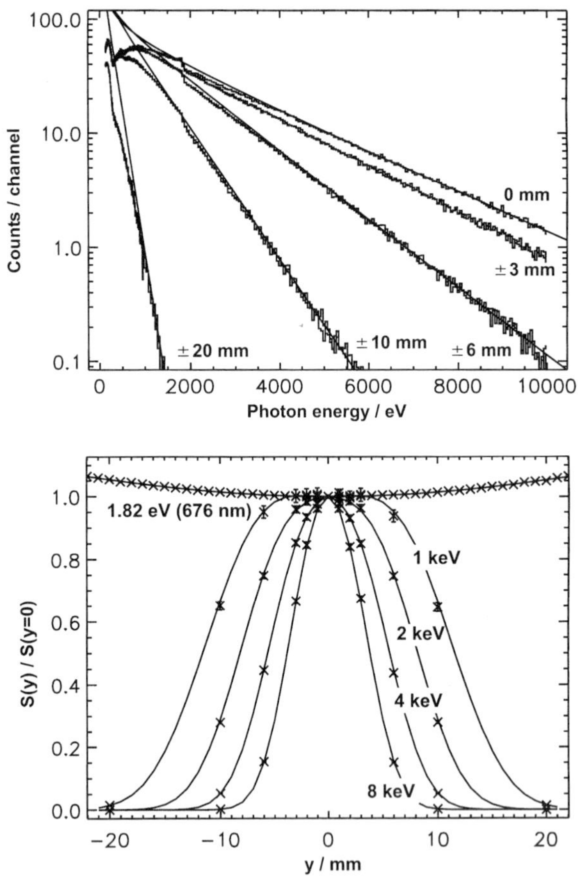


FIG. 5.23. Upper part: spectra measured at BESSY II with a Si(Li) detector with 1 mm × 1 mm entrance aperture at a distance of 30 m from the source point for different vertical offsets from the orbit plane. The solid lines indicate the corresponding calculations according to Schwinger. Lower part: vertical distribution for different photon energies deduced from the measurements shown in the upper part of the figure or with a filter radiometer at 676 nm. The solid lines show the corresponding calculations.

The size $a \times b$ of a flux-defining aperture is normally a detector property and not a property of the primary source standard and therefore not included in the discussion of best measurement capabilities.

*Calculation uncertainty*. The uncertainties in the measurement of the parameters, as summarized for a typical operation of BESSY II in Table 5.4, lead to a relative uncertainty in the calculation of the spectral radiant power

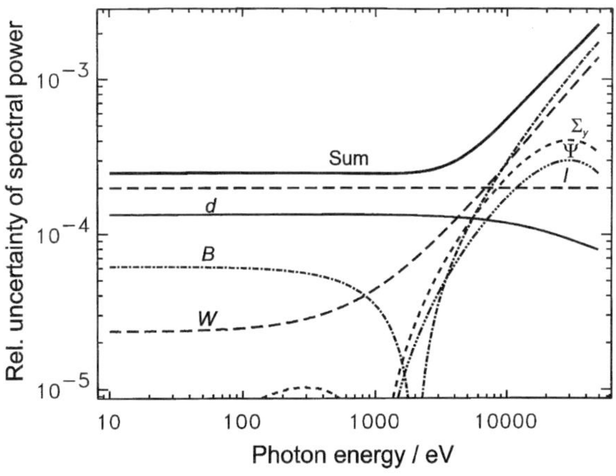


FIG. 5.24. Contributions to the relative uncertainty of the spectral radiant power of BESSY II. The relative uncertainty for the contributing parameters is as listed in Table 5.4 and an aperture size of 5 mm × 5 mm is assumed.

TABLE 5.4. Input Values for the Calculation of the Spectral Photon Flux of BESSY II According to Schwinger, with Their Uncertainties

| Parameter | Value | Rel. Uncertainty |
|---|---|---|
| Electron energy $W$ | 1718.60(6) MeV | $3.5 \times 10^{-5}$ |
| Magnetic induction $B$ | 1.29932(12) T | $1 \times 10^{-4}$ |
| Electron beam current $I$ (example) | 10.000(2) mA | $2 \times 10^{-4}$ |
| Eff. vert. divergence $\Sigma_Y$ | 3.5(7) μrad | 0.2 |
| Vert. emission angle $\psi$ | 0(2) μrad | — |
| Distance $d$ | 30 000(2) mm | $6.7 \times 10^{-5}$ |

as is shown in Figure 5.24. Shown in this figure is the contribution of each parameter, determined numerically from Eq. (5.18) as well as the total relative uncertainty. For low photon energies, the measurement uncertainty of the distance $d$ and the electron beam current $I$ are the limiting factors, whereas for high photon energies the storage ring parameters $W$ and $B$ are limiting. Relative uncertainties in the realization of the spectral radiant power of 0.03 % (for photons below 3 keV) to 0.2 % (for 50 keV photons) are achieved at BESSY II.

Roughly speaking, for photon energies that are high compared to the characteristic energy of an electron storage ring, the relative uncertainty is dominated by the contributions from the electron energy $W$ and magnetic induction $B$ measurement and gradually rises with increasing photon energy. So, for calibrations at high photon energies as compared to the characteristic energy of the given source, it might be advantageous to use another source with a higher characteristic energy, e.g. a superconducting wavelength shifter. At BESSY II, PTB has access to a 7 T superconducting wavelength shifter [175] for this reason.

On the other hand, for calibrations at lower photon energies, where the relative uncertainty in the calculation is almost constant, it is favourable to use a source with a characteristic energy not much higher than the photon energies of interest, since otherwise stray light and higher-order diffraction will limit the achievable accuracy of calibrations of other radiation sources relative to the primary source (see below). Furthermore, the high-energy part of the spectrum causes unwanted heatload and might lead to degradation of optical components. Therefore, e.g. PTB will operate a new electron storage ring, the MLS [164], especially tailored for UV and VUV radiometry at a moderate electron energy adjustable between 200 and 600 MeV. At SURF III the electron energy can also be varied in a wide range from about 180 up to 380 MeV for this purpose [150].

### 5.3.3 Use of Storage Rings as Absolute Radiation Sources

#### *5.3.3.1 Calibration of radiation sources*

Radiation sources are calibrated by comparison of their radiometric properties to the primary source standard by means of suitable monochromator-detector systems. These dedicated systems are mounted in such a way that they can either be illuminated by the primary source standard or by a source under test. No optical components are allowed between the two sources and the systems. The absolute values of the radiometric properties of the source under test are then determined from the ratio of the detector signals measured at the unknown source and the primary source and the calculated, absolute value of the primary source standard.

The specific monochromator-detector system used for the comparison depends on the spectral region of interest and must provide the desired spectral resolution. Ideally, the radiation of the two sources being compared would follow the same optical path in the detector system and illuminate the same area of the optical components in the system. This is hard to accomplish with an electron storage ring as a primary source due to space limitations and therefore leads to limitations in the achievable accuracy due to spatial inhomogenities of the illuminated optics. Also very important is that the system offers a good suppression of stray light and—where applicable—suppression of higher diffraction orders of the monochromator, especially if the spectrum of the electron storage ring includes substantial contributions from photon energies higher than the spectral region of interest for the calibration. If the systems show a polarization-dependent responsivity, the high degree of polarization of the SR also has to be considered. All these aspects enter into the error budget of a calibration and normally are the limiting factors for the achievable accuracy rather than the uncertainty of the realization of the radiometric properties by the primary source itself.

Examples for beamlines and monochromator-detector systems at NIST and at the National Metrology Institute of Japan are given in References [176, 177] and [155], respectively. The PTB beamline for calibrations in the 3 to 30 eV spectral range [178, 179] operated at BESSY II is shown in Figure 5.25. It consists of an aperture defining the solid angle, an imaging mirror, a normal incidence monochromator and a photomultiplier as detector. It can be shifted on rails to either accept the undispersed bending magnet radiation from the BESSY II electron storage ring or from a source to be calibrated. Since, as described before, the SR is strongly polarized in the orbit plane, the whole detector system can be rotated by 90° in order to investigate its polarization dependence. A typical application is the calibration of deuterium lamps [178] that are used as UV secondary standards at other national metrology institutes or at industry, as shown in Figure 5.26. Although the spectral radiant power of the primary source can be realized with a relative uncertainty of 0.03 % in this spectral range, the achievable relative uncertainty in the calibration of the deuterium lamps is typically 2 %, arising from the contributions from the measurement procedure as described above.

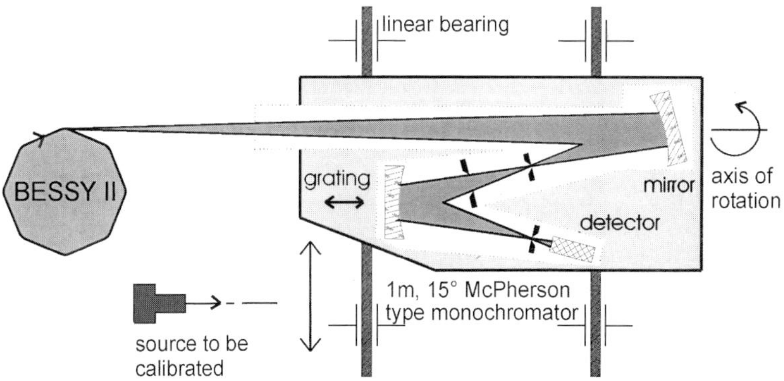


FIG. 5.25. Set-up of the PTB normal-incidence monochromator beamline for source calibration at BESSY II in the spectral range from 3 eV (400 nm) to 30 eV (40 nm). A spherical mirror images the synchrotron radiation source point, placed at 53 m distance, into the entrance slit of a 1 m 15° McPherson-type normal-incidence monochromator. The angular acceptance can be varied by apertures of different size within the range of $(0.1\ \text{mrad})^2$ to $(1\ \text{mrad})^2$. The dispersed radiation is monitored by a photomultiplier tube [179].

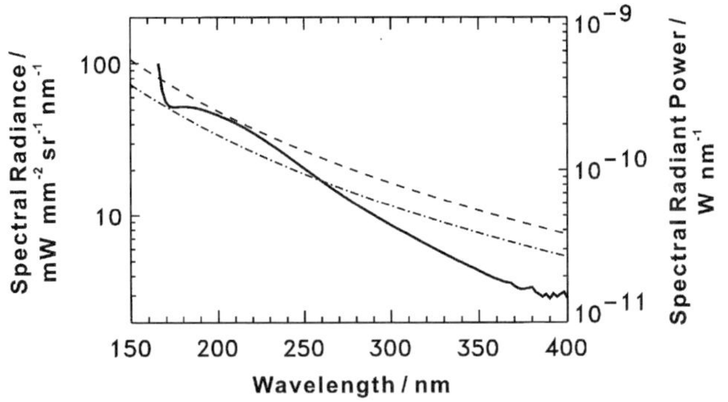


FIG. 5.26. Spectral radiance and spectral radiant power of a deuterium lamp (solid line) compared with the emitted spectral radiant power of BESSY II (dash-dotted line) and former BESSY I (dashed line). The storage rings were operated in a special mode at electron energies of 1700 and 800 MeV and electron beam currents of 0.2 and 0.1 mA, respectively [178].

Sources that have been calibrated in such a way are deuterium lamps [178, 180], hollow cathode sources [181, 182], Penning discharge sources [181, 183], electron-beam excitation sources [181, 184, 185] or electron cyclotron resonance sources [186]. As dedicated sources, special hollow cathode sources have been developed by PTB and calibrated at BESSY I which thereafter were used as a transfer standard for the calibration of the VUV telescopes at the SOHO solar mission [187, 188].

More examples of typical instrumentations, also for other spectral regions, as well as experimental challenges arising from different illumination geometries and polarization properties of the unknown source and an electron storage ring as a primary source standard are given in detail in Reference [189].

#### *5.3.3.2 Calibration of energy-dispersive detector systems*

Energy-dispersive detectors can be calibrated by direct illumination with the calculable, undispersed, continuous spectrum. This calibration scheme has the advantage that the detector can be calibrated over a large spectral range at the one time. Nevertheless, for a precise determination of the detection efficiency, the measured spectrum has to be deconvoluted by the detector response function and possible absorption layers at the detector surface have to be taken into account before being compared to the calculated spectrum of the primary source. Because of the deconvolution of the detector response function and the moderate energy resolution of most energy-dispersive detectors, it is impossible to resolve small band structures in the spectral responsivity by this procedure.

Filter radiometers can be calibrated with this procedure in the IR, VIS and UV spectral region, but most applications of this calibration procedure are in the X-ray spectral region. PTB uses this method to calibrate lithium drifted silicon (Si(Li)) detectors [190, 191], high-purity germanium (HPGe) detectors [192], flow proportional counters [193], CCD detectors [194, 195] or spectrometers consisting, e.g. of a transmission grating and CCD detector [166]. Normally, the storage ring has to be operated with very low electron beam currents, often only a few stored electrons, for these calibrations because of the limited permissible count rate of these devices. The Chandra (NASA) and XMM-Newton (ESA) X-ray astronomy observatories are, e.g. based on calibrations of this type. Figure 5.27 shows an example of the calibration of a Si(Li) detector at BESSY I and BESSY II [196]. In this way, a comparison of the primary sources BESSY I and BESSY II has been made (see below).

As stated above, in principle, also the calculable radiation of undulators can also be used for calibrations. This was demonstrated for the PTB undulator U180 [197] that was used for the calibration of a transmission grating spectrograph used for the characterization of EUV radiation sources [198].

#### *5.3.3.3 Comparisons of primary radiometric standards*

Many efforts have been made to compare the spectral radiant power of storage rings, operated as primary source standards, to that of other established source or detector standards for validation of the radiometric scales or in order to verify the uncertainty budget for the calibration procedure:

- In the region of spectral overlap between SR and blackbody radiation, i.e., in the IR and VIS, these two primary source standards were compared by means of transfers source standards (tungsten halogen lamps) or filter radiometers [152, 199–201].
- The primary SR source standard BESSY I was compared to a primary detector standard (cryogenic electrical-substitution radiometer) by means of filter radiometers [201–203]. Also, the total radiated power of BESSY I was measured with a primary detector standard [204].
- The two primary source standards BESSY I and BESSY II have been compared in the UV and VUV spectral region by means of a deuterium lamp as a transfer standard [178] and in the X-ray region by means of a Si(Li) detector [196].
- The primary source standard BESSY I was compared to radioactive standards in the X-ray spectral region (6.4 keV and 8.0 keV) by means of a Si(Li) detectors [205].

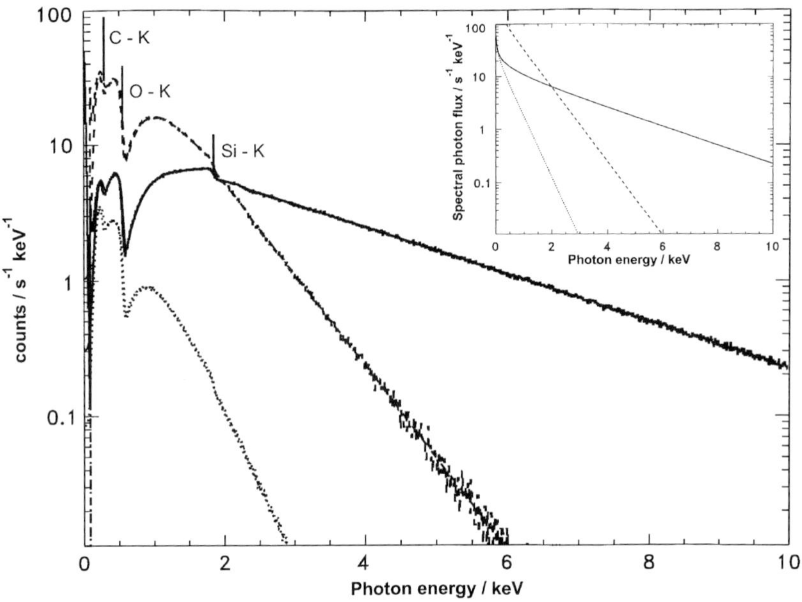


FIG. 5.27. Synchrotron radiation spectra from bending magnets of BESSY I and BESSY II measured with an energy-dispersive Si(Li) detector (left to right: BESSY II operated at 900 MeV, BESSY I operated at 800 MeV and BESSY II operated at 1700 MeV). The measured data are shown in histogram mode, while the smooth lines, which almost coincide with the measurements, are model calculations. The model calculations are derived from the calculated spectra of the storage rings (see small inlay) by taking into account the independently determined detector response function and absorbing layers of the front contact (Au) and window and ice contamination (C, N, O). The main absorption edges of the elements mentioned above are also indicated [196].

For all these different comparisons in the IR, visible, UV and X-ray spectral regions good agreement has always been found. The achieved relative uncertainty was always determined by the procedure of the comparison or by the apparatus used, rather than by the realization of the scale by the primary sources.

### 5.3.4 Use of Storage Rings as Bright Sources from the Visible to the X-Ray Range

Besides their use as primary sources, electron storage rings with appropriate monochromator beamlines are excellent sources of bright monochromatic radiation from the visible to the X-ray region. The beamlines used for radiometry are normally optimized for low stray light and suppression of higher monochromator diffraction orders which is very important for the achievement of low uncertainties. This allows—with the help of a primary detector standard—the calibration of photodetectors that are not energy-dispersive and that cannot be calibrated as described above (Section 5.3.4.1) or a high-accuracy characterization of optical components (Section 5.3.4.2). Some other applications that profit from the well-defined properties of SR are listed at the end of Section 5.3.4.3.

#### *5.3.4.1 Radiometry based on primary detector standards*

For calibration of non-energy-despersive detectors, e.g. photodiodes, the response of the detector to be calibrated to monochromatic radiation is compared to that of a primary detector standard, e.g. a cryogenic electrical-substitution radiometer (ESR). ESRs are the most accurate and well-established primary detector standards in the visible (see Chapter 2). Their utilization in the VUV and soft X-ray region has been pushed by the pioneering work of PTB to adapt them to the low spectral radiant power normally available at monochromator beamlines at electron storage rings as well as to windowless operation in the ultrahigh vacuum needed at these beamlines [206–208]. At its laboratory at BESSY II [209], PTB is presently operating three beamlines suitable for the use of ESRs covering a spectral range from the UV (3 eV) [178, 179] through the VUV [210, 211] up to the X-ray range (10 keV) [212, 213]. At these beamlines, the spectral radiant power, typically in the range μW, can be determined by the use of ESRs with relative uncertainties of about 0.1 % which is much better than the results obtained with double ionization chambers or proportional counters, which served as primary detectors in that spectral range for decades before [214–216]. Also NIST operates dedicated beamlines at SURF III [217, 218] for UV and VUV radiometry with cryogenic ESRs. The National Metrology Institute of Japan has used undulator radiation and a ESR operated at room temperature for UV detector calibration [219].

The use of ESRs led to a significant progress in detector-based UV-, VUV- and X-ray radiometry, e.g. in the calibration of photodiodes commonly used as transfer standards. So the energy-dependent relative uncertainty of the spectral responsivity scale disseminated by PTB—as shown in Figure 5.28—based on cryogenic ESRs from the UV up to 10 keV is typically around 0.5 % [210–213].

With these low uncertainties in the realization of the scale, systematic studies of the homogeneity and stability under irradiation of photodiodes [210, 220, 221] can be performed. Other examples for measurements that are based on the high-accuracy spectral responsivity scale are the determination of the mean energy required for the production of an electron–hole pair in

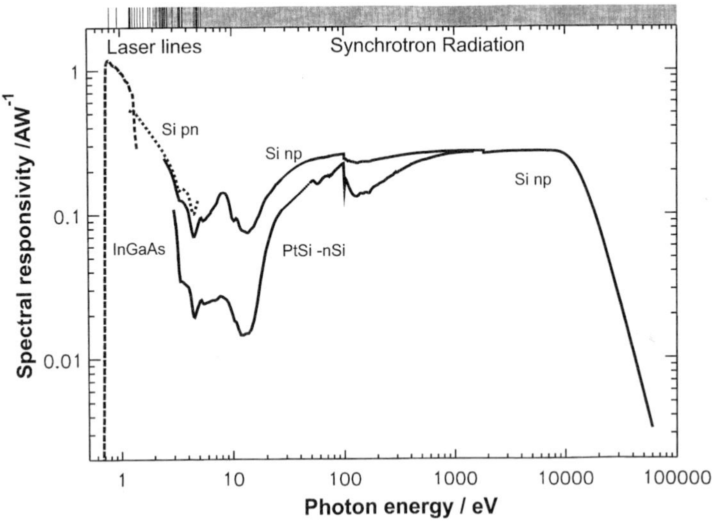


FIG. 5.28. Spectral responsivity scale of PTB-based different types of photodiodes calibrated with cryogenic ESRs.

silicon [222, 223], the determination of electron-impact ionization cross sections of noble gases [224], the quantum efficiency of gold and copper [225, 226] or of cesium iodide photocathodes [227]. Moreover, the high spectral resolution of the monochromatized SR allows resolution of structures close to the absorption edges of the detector materials of energy-dispersive detectors for high-accuracy calibrations [191, 228], that cannot be resolved by calibrations with the calculable, undispersed SR (see Section 5.3.4.2).

### *5.3.4.2 Reflectometry*

For the characterization of optical components like mirrors, gratings or filters using monochromatized SR no primary radiometric standard is needed. Essential is a high spectral purity of the monochromatized radiation. Dedicated experimental stations for the positioning of the optical components with high mechanical precision under UHV conditions (reflectometers) are also needed. Besides at, e.g. the Center for X-Ray Optics at the Advanced Light Source [229], high-accuracy reflectometry is performed at NIST [230] and PTB [211, 231, 232]. Especially the development of EUV lithography (EUVL), which is the leading next-generation lithography technology [233], requires high-accuracy at-wavelength measurements of multilayer-coated mirrors and mask blanks. Today, the peak reflectance of Mo/Si multilayer mirrors at about 13 nm wavelength can be measured with a relative standard uncertainty of less than 0.2 % [211, 229].

Recently, the high-energy transmission gratings of the *Chandra* X-ray observatory have been characterized by SR at four facilities [234]. Another reflectometry application is the investigation of the X-ray transmission characteristics of filters for soft X-ray astronomy [235] or the determination of the thickness of $SiO_2$ films on Si [236].

#### *5.3.4.3 Other applications*

The narrow band and stable monochromatized SR is ideally suited for the determination of detector response functions, e.g. for Si(Li) detectors [191] or other detectors like novel super-conducting tunnel junction detector systems [237].

Besides these main tasks, other example of the use of quantitatively determined SR are the irradiation stability test of optics [220, 238, 239] and detectors [240] or the quantitative X-ray fluorescence analysis [241–243]. Moreover, the ability to alter the intensity of the spectral radiant power by variation of the electron beam current in a controlled manner can be used for linearity testing of detectors, e.g. used at 157 nm [244] or in the EUV [245], over a large dynamic range. A comprehensive description of radiometric tasks at electron storage rings besides their use as primary source standards can be found in Reference [161].

## 5.4 Parametric Down-Conversion-Based Sources

Two additional primary sources for radiometry are based on the correlated photons produced in parametric down-conversion (PDC) [246–248]. One of these is essentially a source of single photons, that is useful for measuring detector efficiency in the photon counting regime (where blackbodies are not appropriate, electron storage rings require a fixed facility, and conventional detector standards are somewhat cumbersome to use). The other source that PDC makes available is effectively a source of spectral radiance, most useful for IR measurements. (This source arises from zero-point fluctuations in the vacuum background.) Measurements made with both of these sources have the unusual characteristics that they are intrinsically absolute (in that they do not rely on any externally calibrated radiometric standards), and they allow for IR calibration using visible detectors and visible optics.

### 5.4.1 Single-Photon Source

Down-conversion occurs when photons from a pump laser are directed into a suitable optically non-linear crystal. A small fraction of the pump laser photons decay into pairs of lower-energy photons under the constraints of energy and momentum conservation (also referred to as phase matching):

$$\omega_{\mathrm{p}} = \omega_1 + \omega_2$$

$$\mathbf{k}_{\mathrm{p}} = \mathbf{k}_1 + \mathbf{k}_2 \tag{5.26}$$

where $\omega_{\mathrm{p}}$ and $\mathbf{k}_{\mathrm{p}}$ are the frequency and wave vector of the pump, and $\omega_i$ and $\mathbf{k}_i$ $(i = 1, 2)$ refer to a pair of down-converted output photons (wavevectors are evaluated within the crystal). Because the photons are created in pairs, the detection of one indicates, with absolute certainty, the existence of the other. Thus, with the addition of a trigger detector, this system can be used as a radiometric source as shown as the dashed box in Figure 5.29. When a photon causes the trigger detector to fire, it indicates that the correlated photon is present in the output port of the source.

This source of individual (and heralded) photons is clearly suitable for calibration of photon counting detectors. The key is to place the device under test (DUT) at the output port of the photon source so that it can collect all the photons correlated to those seen by the trigger detector.

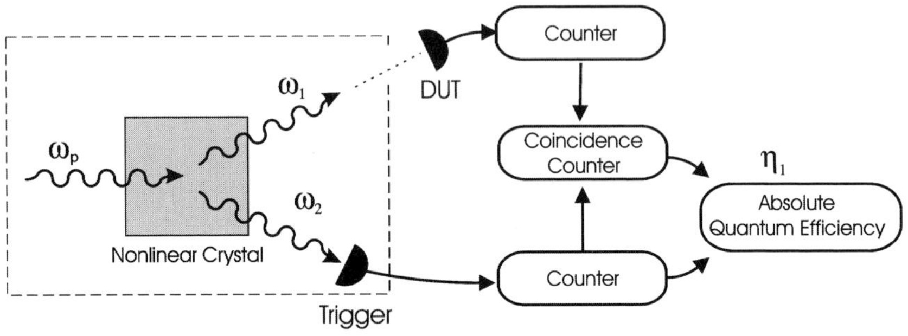


FIG. 5.29. Two photon detection efficiency measurement scheme. Because the photons are always produced in pairs, the detection of a photon by the trigger detector heralds the existence of a photon at the output port of the source. The dashed box containing a pump laser, non-linear crystal, trigger detector and an output port represents a radiometric source of single photons. The determined efficiency, $\eta_1$, is the efficiency of the entire detection path from creation of the photon in the crystal to its detection by the detector.

The detection efficiency is the ratio of the number of coincidence events (where both channels report a detection) to the number of trigger detection events in a given time interval. In terms of the efficiencies of the two detection channels (given by $\eta_1$ and $\eta_2$) the total number of trigger counts, $N_2$, and the total number of coincidence counts, $N_{\mathrm{Coinc}}$, can be written as

$$N_2 = \eta_2 N_{\mathrm{p}}$$

$$N_{\mathrm{Coinc}} = \eta_1 \eta_2 N_{\mathrm{p}} \tag{5.27}$$

where $N_{\mathrm{p}}$ is the total number of down-converted photons emitted into the trigger channel during the counting period. Thus the absolute detection efficiency of channel 1 is simply

$$\eta_1 = \frac{N_{\mathrm{Coinc}}}{N_2} \tag{5.28}$$

A remarkable feature of this result is that $\eta_1$ is independent of the efficiency of the trigger channel. Thus, we can use imperfect detectors to perform an absolute calibration without any comparison to a reference standard [248]. Another useful feature of PDC-based detector calibration is that the spectral selectivity components for the calibration wavelength do not need to be in the DUT optical path, but can be placed in the trigger path. The constraints in Eq. (5.26) can then be used to calculate the range of frequencies that are correlated to the photons incident on the trigger detector. This calculated range can be thought of as a 'virtual' bandpass filter in the DUT path, defining the wavelengths that result in coincidences. This can be of advantage when calibrating a detector in the infrared, where spectral selection may be more convenient in the visible trigger channel.

While this absolute calibration requires no external standard, it is important to note that Eq. (5.28) does not give just the quantum efficiency of the DUT alone ($\eta_{\mathrm{DUT}}$), but rather the detection efficiency of the entire detection channel from where the photons are created within the crystal to where the detections are ultimately recorded (i.e., when the trigger detector senses a photon and the DUT does not detect its twin, Eq. (5.28) does not distinguish whether the photon was actually incident on the detector or was lost upstream in the detection channel.) Any losses within the crystal or in the optical collection system are included in $\eta_1$, along with the efficiency of the detector to be measured. Hence all losses not associated with the DUT must be determined to extract the efficiency of the DUT alone. This is the key to turning this measurement principle into an accurate metrological technique for characterizing detector efficiency.

#### *5.4.1.1 Metrological considerations – DUT channel collection losses and trigger rate*

System losses upstream of the DUT can be classified as either *transmittance* losses due to reflection and absorption or *geometric* losses due to causes such as limiting apertures, finite detector area, or positioning errors. DUT channel transmittance losses may be handled straightforwardly; the transmittances of optical components can be measured conventionally with high accuracy, or in some cases the losses can be calculated with good results (e.g. the losses in the down-conversion crystal). It is even possible to measure the transmission losses in situ using the PDC setup.

The trigger detector channel is also a critical concern, as an accurate determination of the trigger photon rate directly affects the uncertainty of the final detector efficiency (DE) result in Eq. (5.28). Trigger channel considerations differ from DUT channel concerns in that optical losses are not important, but instead dark count, stray light, and deadtime effects must be accounted for. In addition to these factors, it is also important to consider factors related to the suitability of the DUT as a stable device to be measured. These DUT operation parameters including threshold setting, dead time, non-linearities (both reversible and permanent), spatial non-uniformity of the detector response, and other timing related issues will impact the ultimate accuracy of the DUT efficiency determination. For details on these factors and their characterization, see References [249, 250].

#### *5.4.1.2 Comparison of PDC-based calibration to conventional analog calibration of photon-counting detectors*

We now compare the conventional and correlated DE measurement methods. Conventional monochromator-based detector measurements excel in the visible region, where the scale uncertainty is ~0.1 %, while in the NIR the scale uncertainty increases to ~1 % and increases still further in the IR [251]. DE measurement uncertainties approaching these values can only be achieved under optimal conditions, the most critical of which is moderate optical power levels, typically from 10 nW to 1 mW. Thus to use conventional methods to calibrate a photon counting detector, high levels of attenuation are required to avoid saturating the photon counting DUT, which typically has a restricted dynamic range ~ <0.1 pW. The addition of the attenuation components add to the ultimate uncertainty. The two-photon method allows direct calibration without attenuation in the photon counting regime. Uncertainties in the visible region have reached ~0.5 % level with 0.1 % possible and these uncertainties should not necessarily degrade for wavelengths further into in the infrared region. Also, the two-photon method has the advantage that IR spectral selection can be achieved with visible bandpass elements. Overall the two-photon method can achieve comparable uncertainties in the visible and will likely surpass conventional measurements in the NIR and IR. Finally, the fact that it is a "primary standard method" independent of conventional detector standards makes the method a unique and worthwhile addition to the detector metrology toolbox.

### 5.4.2 Radiance Source

In addition to supplying a source of single photons, PDC techniques also provide access to another unique metrological resource—a source of spectral radiance due to a background of zero-point fluctuations in the vacuum field. This source is a primary standard because the spectral radiance of this background can be calculated from fundamental constants (i.e. it needs no external calibration), and it is available everywhere via PDC techniques. This allows spectral radiance measurements to be made by direct comparison to an absolute standard with no intermediary transfer standards—an ideal situation in metrology.

The basic measurement arrangement is that of an optical parametric amplifier, as shown in Figure 5.30. Starting with the spontaneous down-conversion geometry (one photon in, two photons out), the radiance beam to be measured ($R_0$) is directed into the crystal to overlap spatially, angularly, and spectrally with some of the spontaneously down-converted output. This arrangement stimulates the decay of pump photons into that down-converted channel. Because the output photons must be created in pairs, it also enhances the output light in the correlated channel. This arrangement (non-degenerate geometry, i.e. $\omega_1 \neq \omega_2$), allows one to detect an input beam at one wavelength by detecting light at a different wavelength. Thus it is possible to monitor infrared beams with stable, high-quality visible detectors.

This monitoring is turned into an absolute measurement of spectral radiance by considering the origin of the down-converted light produced when

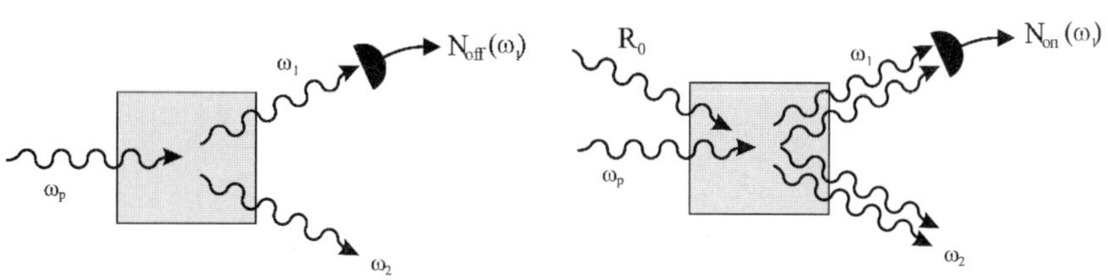


FIG. 5.30. PDC technique for comparing a source of radiance ($R_0$) with the omnipresent standard of spectral radiance provided by background fluctuations.

the beam to be measured is blocked (i.e. when only the pump beam is incident on the crystal). The spontaneous decay of pump photons can be thought of as being stimulated by a one-photon-per-mode background due to zero-point vacuum field fluctuations [246] and radiation reaction [252]. It turns out that this one-photon-per-mode radiance can be written in terms of the wavelength $\lambda$ and fundamental constants as $hc^2/\lambda^5$, which has the units of spectral radiance (i.e. power per unit area, solid angle, and bandwidth). Thus the absolute spectral radiance of an unknown beam can be determined by observing the signal increase in channel 1 when the unknown beam ($R_0$) is put into the crystal to overlap channel 2. The ratio of the increase in the output signal to the spontaneous signal is the radiance of the unknown beam, in units of photons per mode:

$$R_0(\omega_2) = \frac{1}{\varepsilon}\left[\frac{N_{\text{on}}(\omega_1)}{N_{\text{off}}(\omega_1)} - 1\right] \tag{5.29}$$

where $N_{\text{on}}(\omega_1)$ and $N_{\text{off}}(\omega_1)$ are the PDC signals when the source to be measured is on and off, and $\varepsilon$ is the total system efficiency [249]. Because the radiance is specified in terms of a ratio of signals, the detector used to monitor the beam needs no calibration; its only requirement is linearity. As in the DE measurement technique, the spectral band of the radiance measurement is the range of wavelengths correlated to those seen by the detector in channel 1.

Similar to the previous technique, but now on the input side, losses in the optical path between the source-to-be-measured and the PDC crystal are included in the final measurements. These losses must be determined to extract the radiance of the source alone. In addition, the beam to be measured must be made to overlap the volume of the crystal producing the spontaneous signal. Thus we have an overall system efficiency, $\varepsilon$, that is the product of the transmittance of the imaging system and a factor that measures the overlap of $R_0$ with the source of the PDC light. The transmittance factor may be obtained by straightforward measurements and/or calculations. The overlap factor requires more consideration. This factor describes how well the radiation to be measured fills the field modes to which this technique is sensitive. The trick is to design the system with as large an overlap factor as possible. Note that it is not necessary to know how many modes are filled, only how well they are filled by the unknown beam.

The total overlap is determined primarily by multiplying a spatial overlap with an angular overlap. The spatial overlap factor is calculated by integrating the product of pump beam and IR beam spatial profiles within the crystal and normalizing to the integral of the pump beam in the same volume. In practical setups the pump beam often enters the crystal fairly close to normal incidence, while the IR beam enters at a large angle. This results in an astigmatic IR beam within the crystal that must be included in the overlap calculation.

The angular factor is calculated in a similar manner by integrating the product of the angular profile of the IR input beam and the profile of the sensitive IR angular range. The difference for this case is that we determine the sensitive IR angular range by converting from the range of visible angles seen by the detector through the phase matching and energy constraints. This overlap factor also requires a two-dimensional integral, because, while the vertical range of angles seen by the detector is limited solely by the geometric collection limit of the detector (determined by the detector aperture and its distance from the source), the horizontal range may be limited either by the geometry of the detector's collection optics or its spectral limitations. (The spectral filter can limit the range of angles seen though the phase-matching constraint that ties the wavelength of the output to its angle of output [253].)

#### 5.4.2.1 *Comparison to conventional radiance measurements*

Because this method depends on comparing the signal increase due to the unknown beam to the background and the background standard is one photon per mode at all wavelengths, the most accurate comparisons are achieved for sources with similar or larger radiance, i.e., beams with radiances of the order of one-photon-per-mode or more. This makes the technique most suitable for measuring sources in the IR region of the spectrum and with high temperatures where radiance are of this order. So far, sources at temperatures of 2000 to 10,000 K have been measured in the spectral range of 1 to 5 μm with uncertainties on the order of 1 % [254]. These results compare favorably to conventional measurements in this region.

### 5.4.3 History of PDC-Based Sources

The method for using PDC to calibrate detectors has been known for some time. PDC was predicted in 1961 by Louisell et al. [246] and in 1970 [248] the very first experiment to observe coincidences between down-converted photons also included the first detector calibration using a PDC source. The method was not widely disseminated, however, and 7 years later Klyshko [253, 255] independently proposed that PDC could be used to measure detection efficiency. Beginning in the early 1980s, other groups [256–264] continued to develop the technique from demonstration type measurements to more careful metrology efforts. As measurement techniques have been refined the measured values of detection efficiencies have become more precise, and the method has been demonstrated to be a valuable radiometric tool. In particular, with the emergence of the field of quantum information, the need to calibrate photon counting detectors has become more critical.

Even before the advent of PDC, two-photon detector calibration already had a long history of being carried out using atomic cascade [265–267] sources to produce the correlated photons. While successful, the drawback of this type of source is that the large range of solid angles into which the atom can decay results in low source brightness. In comparison to two-photon cascade, the phase matching constraints of PDC restrict the emission of photon pairs to a narrow range of angles, effectively brightening the two-photon source by a factor of $\sim 10^6$. The origin of detector calibration using coincidence techniques can be traced back even further to the early 1900s, when coincidence techniques were already being suggested and used to study particles emitted in nuclear decay [268–270] and other fundamental processes. Even at this early stage it was clear [271] that the coincidence method is ideally suited to single particle counting detector metrology.

The technique for accessing the omnipresent radiance source was proposed in 1977 by David Klyshko [253] at Moscow State University and demonstrated by Penin's group [272] two years later. Penin's Moscow group first demonstrated this method by looking at three different sources: a laser, a fluorescent dye pumped by a laser, and an incandescent lamp. They measured up to a wavelength of 3.9 μm in the infrared with a photomultiplier tube and spectrometer designed for observations in the visible. This achievement highlights the technique's ability to exploit convenient visible components for measurements in the more difficult infrared region. The effective temperatures of the sources and the wavelengths at which they were measured (500,000 K and 70,000 K at 532 nm and 980 K at 3.9 μm) in the early work of Penin's group indicates the range over which this method is most appropriate. The method has been verified to within the expected uncertainty by comparing PDC measurements of a source with those made independently against a blackbody [249, 254].

### 5.4.4 Conclusion

It is useful to point out how the two PDC applications discussed above fit into the world of absolute radiometric standards—i.e., standards whose output or response can be calculated from fundamental physical principles. In Figure 5.31, the existing absolute sources and detectors are shown in gray. Because there are so few, any addition can have a significant impact. The correlated-photon absolute flux source clearly fits in the first column in the figure. The placement of the correlated-photon spectral radiance method is not so clear. As a technique for measuring spectral radiance, it can be regarded as an absolute detector. But alternatively, it is also simply a way of

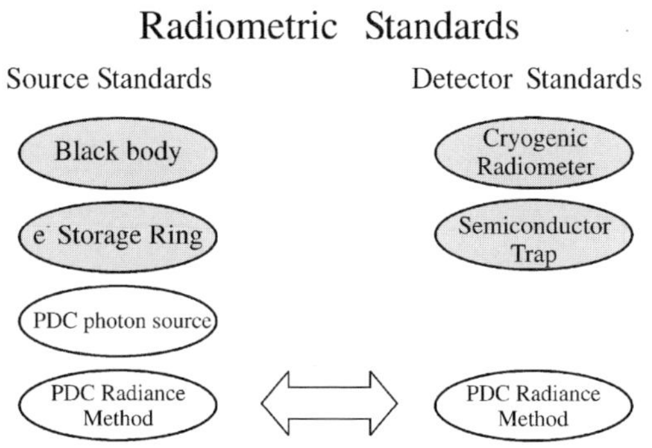


FIG. 5.31. Chart of how the PDC techniques fit into the existing world of radiometric standards. The PDC radiance can be thought of as a source and detector standard so it is shown in both columns.

coupling to an omnipresent absolute spectral-radiance source (the one-photon-per-mode vacuum background) by means of the PDC crystal nonlinearity. In either case, the chart makes it clear that correlated photons offer significant new choices in the world of radiometry.

## References


1. F. Grum and R. J. Becherer, "Optical Radiation Measurements, Volume 1: Radiometry." Academic Press, San Diego, 1979.
2. T. J. Quinn, "Temperature." Academic Press, London, 1990.
3. D. P. DeWitt and G. D. Nutter, "Theory and Practice of Radiation Thermometry." Wiley, New York, 1988.
4. P. J. Duke, "Synchrotron Radiation." Oxford University Press, Oxford, 2000.
5. H. Wiedemann, "Synchrotron Radiation." Springer, Berlin, 2002.
6. H. Kangro, "Vorgeschichte des Planckschen Strahlungsgesetzes." F. Steiner Verlag, Wiesbaden, 1970.
7. D. Cahan, "Meister der Messung," Wiley-VCH, Weinheim, 1992; "An Institute for an Empire," Cambridge University Press, Cambridge, 2004.
8. D. Hoffmann, Schwarze Körper im Labor, *Physikalische Blätter* 56, 43–47 (2000).
9. F. Hengstberger, "Absolute Radiometry," Sect. 1.3.4.3. Academic Press, San Diego, 1989.

*10.* M. Planck, Ueber eine Verbesserung der Wienschen Spektralgleichung, *Verhandlungen der Deutschen Physikalischen Gesellschaft* 2, 202–204 (1900) (also reprinted in Reference [12]).
*11.* M. Planck, Zur Theorie des Gesetzes der Energieverteilung im Normalspektrum, *Verhandlungen der Deutschen Physikalischen Gesellschaft* 2, 237–245 (1900) (also reprinted in Reference [12]).
*12.* M. Planck, "Die Ableitung der Strahlungsgesetze," Ostwalds Klassiker der exakten Wissenschaften, Band 206. Verlag Harri Deutsch, Thun und Frankfurt am Main, 1997.
*13.* G. P. Peterson, "An Introduction to Heat Pipes: Modelling, Testing and Applications." Wiley, New York, 1994.
*14.* O. Brost and M. Groll, Liquid metals heat pipe applications: thermometric calibration tools & heat transfer components for solar-thermal power systems, in "Proc. 9th Int. Heat Pipe Conf." (M. A. Merrigan, Ed.), 110–115. Los Alamos National Laboratory, 1995.
*15.* D. P. DeWitt and J. C. Richmond, "Thermal radiative properties of materials," in Ref. [3], pp. 91–187.
*16.* R. E. Bedford, Calculation of Effective Emissivities of Cavity Sources of Thermal Radiation, in Ref. [3], pp. 653–772.
*17.* A. Ono, Calculation of the directional emissivities of cavities by the Monte Carlo method, *J. Opt. Soc. Am.* 70, 547–554 (1980).
*18.* V. I. Sapritsky and A. V. Prokhorov, Calculation of the effective emissivities of specular-diffuse cavities by the Monte Carlo method, *Metrologia* 29, 9–14 (1992).
*19.* V. I. Sapritsky and A. V. Prokhorov, Spectral effective emissivities of nonisothermal cavities calculated by the Monte Carlo method, *Appl. Opt.* 34, 5645–5652 (1995).
*20.* A. V. Prokhorov, Monte Carlo method in optical radiometry, *Metrologia* 35, 465–471 (1998).
*21.* J. Hartmann, D. R. Taubert, and J. Fischer, Characterization of the double-heatpipe blackbody LABB for use at temperatures below 500°C, in "Proc. TEMPMEKO '99, 7th Int. Symp. on Temperature and Thermal Measurements in Industry and Science" (J. F. Dubbeldam and M. J. de Groot, Eds.), 511–516. IMEKO/NMi VSL, Delft, 1999.
*22.* R. J. Pahl and M. A. Shannon, Analysis of Monte Carlo methods applied to blackbody and lower emissivity cavities, *Appl. Opt.* 41, 691–699 (2002).
*23.* H. Preston-Thomas, The International Temperature Scale of 1990 (ITS-90), *Metrologia* 27, 3–10, 107 (1990).
*24.* H.-J. Jung, On the determination of the thermodynamic temperature of high temperature black bodies via ITS-90 or alternative methods, in

"Proc. TEMPMEKO '96 , 6th Int. Symp. on Temperature and Thermal Measurements in Industry and Science" (P. Marcaroni, Ed.), 235–244. Levrotto & Bella, Torino, 1996.

25. K. D. Mielenz, R. D. Saunders, and J. B. Shumaker, Spectroradiometric determination of the freezing temperature of gold, *J. Res. Natl. Inst. Stand. Technol.* 95, 49–67 (1990).
26. Fischer, Fu Lei, and M. Stock, Present state of the determination of thermodynamic temperatures near the freezing point of silver by absolute cryoradiometry, *Metrologia* 28, 243–246 (1991).
27. N. J. Harrison, N. P. Fox, P. Sperfeld, J. Metzdorf, B. B. Khlevnoy, R. I. Stolyarevskaya, V. B. Khromchenko, S. N. Mekhontsev, V. I. Shapoval, M. F. Zelener, and V. I. Sapritsky, International comparison of radiation-temperature measurements with filtered detectors over the temperature range 1380 K to 3100 K, *Metrologia* 35, 283–288 (1998).
28. H. W. Yoon and C. E. Gibson, Determination of radiance temperatures using detectors calibrated for absolute spectral power response, in "Proc. TEMPMEKO '99, 7th Int. Symp. on Temperature and Thermal Measurements in Industry and Science" (J. F. Dubbeldam and M. J. de Groot, Eds.), 737–742. IMEKO/NMi VSL, Delft, 1999.
29. H. W. Yoon, P. Sperfeld, S. Galal Yousef, and J. Metzdorf, NIST-PTB measurements of the radiometric temperatures of a high-temperature black body using filter radiometers, *Metrologia* 37, 377–380 (2000).
30. H. W. Yoon and C. E. Gibson, Comparison of the absolute detector-based spectral radiance assignment with the current NIST-assigned spectral radiance of tungsten strip lamps, *Metrologia* 37, 429–432 (2000).
31. B. B. Khlevnoy, N. J. Harrison, L. J. Rogers, D. F. Pollard, N. P. Fox, P. Sperfeld, J. Fischer, R. Friedrich, J. Metzdorf, J. Seidel, M. L. Samoylov, R. I. Stolyarevskaya, V. B. Khromchenko, S. A. Ogarev, and V. I. Sapritsky, Intercomparison of radiation temperature measurements over the temperature range from 1600 K to 3300 K, *Metrologia* 40, S39–S44 (2003).
32. H. W. Yoon, C. E. Gibson, and J. L. Gardner, Spectral radiance comparisons of two blackbodies with temperatures determined using absolute detectors and ITS-90 techniques, *AIP Conf. Proc.* 684, 601–606 (2003).
33. K. Anhalt, J. Hartmann, J. Hollandt, G. Machin, D. Lowe, H. Mc Evoy, F. Sakuma, and L. Ma, Comparison of the high-temperature scales of the NMIJ and the NPL with the scale of the PTB in the temperature range from 1000°C up to 2900°C, in "Proc. TEMPMEKO 2004,

9th Int. Symp. on Temperature and Thermal Measurements in Industry and Science" (D. Zvizdic, Ed.), 1063–1068. Laboratory for Process Measurement, Faculty of Mechanical Engineering and Naval Architecture, Zagreb, 2005.

*34.* V. I. Sapritsky, National primary radiometric standards of the USSR, *Metrologia* 27, 53–60 (1990).

*35.* V. I. Sapritsky, Black-body radiometry, *Metrologia* 32, 411–417 (1995/1996).

*36* K. D. Mielenz, R. D. Saunders, A. C. Parr, and J. J. Hsia, The 1990 NIST scales of thermal radiometry, *J. Res. Natl. Inst. Stand. Technol.* 95, 621–629 (1990).

*37.* B. C. Johnson, C. L. Cromer, R. D. Saunders, G. Eppeldauer, J. Fowler, V. I. Sapritsky, and G. Dezsi, A method of realizing spectral irradiance based on an absolute cryogenic radiometer, *Metrologia* 30, 309–315 (1993).

*38.* A. C. Parr, "A National Measurement System for Radiometry, Photometry, and Pyrometry Based upon Absolute Detectors," NIST Technical Note 1421, U.S. Government Printing Office, Washington, DC, 1996. (Available via http://physics.nist.gov/Pubs/TN1421/contents.html

*39.* A. C. Parr, The candela and photometric and radiometric measurements, *J. Res. Natl. Inst. Stand. Technol.* 106, 151–186 (2001).

*40.* M. White, N. P. Fox, V. E. Ralph, and N. J. Harrison, The characterisation of a high temperature blackbody as the basis for the NPL spectral irradiance scale, *Metrologia* 32, 431–434 (1995/1996).

*41.* T. J. Quinn and J. E. Martin, Total radiation measurements of thermodynamic temperature, *Metrologia* 33, 375–381 (1996).

*42.* N. P. Fox, Radiometry with cryogenic radiometers and semiconductor photodiodes, *Metrologia* 32, 535–543 (1995/1996).

*43.* N. P. Fox, Primary radiometric quantities and units, *Metrologia* 37, 507–513 (2000).

*44.* J. Metzdorf, Network and traceability of the radiometric and photometric standards at the PTB, *Metrologia* 30, 403–408 (1993).

*45.* P. Sperfeld, K.-H. Raatz, B. Nawo, W. Möller, and J. Metzdorf, Spectral-irradiance scale based on radiometric black-body temperature measurements, *Metrologia* 32, 435–439 (1995/1996).

*46.* P. Sperfeld, "Entwicklung einer empfängergestützten spektralen Bestrahlungsstärkeskala," Ph.D. Thesis, TU Braunschweig, 1999 (available via http://www.biblio.tu-bs.de/ediss/data/19990628a/19990628a.html.

*47.* R. Friedrich and J. Fischer, New spectral radiance scale from 220 nm to 2500 nm, *Metrologia* 37, 539–542 (2000).

*48.* J. E. Martin and P. R. Haycocks, Design considerations for the construction of an absolute radiation detector at the NPL, *Metrologia* 35, 229–233 (1998).

*49.* J. Fischer, Developments in infrared radiation thermometry, in "Proc. TEMPMEKO '99, 7th Int. Symp. on Temperature and Thermal Measurements in Industry and Science" (J. F. Dubbeldam and M. J. de Groot, Eds.), 27–34. IMEKO/NMi VSL, Delft, 1999.

*50.* J. Hartmann and J. Fischer, Radiator standards for accurate IR calibrations in remote sensing based on heatpipe blackbodies, *Proc. SPIE* 3821, 395–403 (1999).

*51.* J. Hartmann, B. Gutschwager, J. Fischer, and J. Hollandt, Calibration of thermal cameras for temperature measurements using blackbody radiation in the temperature range −60°C up to 3000°C, in "Proc. of Infrared Sensors & Systems 2002," 119–124. AMA Service GmbH, Wunstorf, Germany, 2002.

*52.* B. W. Mangum, G. T. Furukawa, K. G. Kreider, C. W. Meyer, D. C. Ripple, G. F. Strouse, W. L. Tew, M. R. Moldover, B. C. Johnson, H. W. Yoon, C. E. Gibson, and R. D. Saunders, The Kelvin and temperature measurements, *J. Res. Natl. Inst. Stand. Technol.* 106, 105–149 (2001).

*53.* B. Chu, H. C. McEvoy, and J. W. Andrews, The NPL reference sources of blackbody radiation, *Meas. Sci. Technol.* 5, 12–19 (1994).

*54.* B. Chu and G. Machin, A low-temperature blackbody reference source to −40°C, *Meas. Sci. Technol.* 10, 1–6 (1999).

*55.* G. Machin and B. Chu, High quality blackbody sources for infra-red thermometry and thermography between −40°C and 1000°C, *Imaging Sci.* 48, 15–22 (2000).

*56.* E. W. M. van der Ham, M. Battuello, P. Bloembergen, R. Bosma, S. Clausen, O. Enouf, E. Filipe, J. Fischer, B. Gutschwager, T. Hirvonen, J. U. Holtoug, J. Irvasson, G. Machin, H. McEvoy, J. Pérez, T. Ricolfi, P. Ridoux, M. Sadli, V. Schmitt, C. Staniewicz, O. Struss, and T. Weckström, Intercomparison of local temperature scales with transfer radiation thermometers between –50°C and 300°C, in "Proc. TEMPMEKO 2001, 8th Int. Symp. on Temperature and Thermal Measurements in Industry and Science" (B. Fellmuth, J. Seidel, and G. Scholze, Eds.), 831–838. VDI Verlag, Berlin, 2002.

*57.* M. Battuello, F. Girard, T. Ricolfi, M. Sadli, P. Ridoux, O. Enouf, J. Pérez, V. Chimenti, T. Weckström, O. Struss, E. Filipe, N. Machado, E. van der Ham, G. Machin, H. Mc Evoy, B. Gutschwager, J. Fischer, V. Schmidt, S. Clausen, J. Ivarsson, S. Uğur, and A. Diril, The European project TRIRAT: arrangements for and results of the comparison of local temperature scales with transfer infrared thermometers between 150°C and 962°C, *AIP Conf. Proc.* 684, 903–908 (2003).

*58.* V. I. Sapritsky, B. B. Khlevnoy, V. B. Khromchenko, B. E. Lisiansky, S. N. Mekhontsev, U. A. Melenevsky, S. P. Morozova, A. V. Prokhorov, L. N. Samoilov, V. I. Shapoval, K. A. Sudarev, and M. F. Zelener, Precision blackbody sources for radiometric standards, *Appl. Opt.* 36, 5403–5408 (1997).

*59.* V. I. Sapritsky, B. B. Khlevnoy, V. B. Khromchenko, S. A. Ogarev, S. P. Morozova, B. E. Lisiansky, M. L. Samoylov, V. I. Shapoval, and K. A. Sudarev, Blackbody sources for the range 100 K to 3500 K for precision measurements in radiometry and radiation thermometry, *AIP Conf. Proc.* 684, 619–624 (2003).

*60.* J. Ishii and A. Ono, Low-temperature infrared radiation thermometry at NMIJ, *AIP Conf. Proc.* 684, 657–662 (2003).

*61.* K. D. Hill and D. J. Woods, The NRC blackbody-based radiation thermometer calibration facility, *AIP Conf. Proc.* 684, 669–674 (2003).

*62.* A. Diril, H. Nasibov, and S. Uğur, UME radiation thermometer calibration facilities below the freezing point of silver (961.78°C), *AIP Conf. Proc.* 684, 663–668 (2003).

*63.* Chen Yinghang, Liu Yaping, Li Yongqian, Jin Xiuying, and Song Hengxue, A medium temperature radiation calibration facility using a new design of heatpipe blackbody as a standard source, *Meas. Sci. Technol.* 12, 491–494 (2001).

*64.* B. Gutschwager and J. Fischer, An InGaAs radiation thermometer with an accurate reference function as transfer standard from 150°C to 960°C, in "Proc. TEMPMEKO '99, 7th Int. Symp. on Temperature and Thermal Measurements in Industry and Science" (J. F. Dubbeldam and M. J. de Groot, Eds.), 567–572. IMEKO/NMi VSL, Delft, 1999.

*65.* R. Datla, M. Croarkin, and A. C. Parr, Cryogenic blackbody calibrations at the National Institute of Standards and Technology low background infrared calibration facility, *J. Res. Natl. Inst. Stand. Technol.* 99, 77–87 (1994).

*66.* S. R. Lorentz, S. C. Ebner, J. H. Walker, and R. U. Datla, NIST low-background infrared spectral calibration facility, *Metrologia* 32, 621–624 (1995/1996).

*67.* J. B. Fowler, B. C. Johnson, J. P. Rice, and S. R. Lorentz, The new cryogenic vacuum chamber and blackbody source for infrared calibrations at NIST's FARCAL facility, *Metrologia* 35, 323–327 (1998).

*68.* V. S. Ivanov, B. E. Lisiansky, S. P. Morozova, V. I. Sapritsky, U. A. Melenevsky, L. Y. Xi, and L. Pei, Medium-background radiometric facility for calibration of sources or sensors, *Metrologia* 37, 599–602 (2000).

69. A. C. Carter, T. M. Jung, A. Smith, S. R Lorentz, and R. Datla, Improved broadband blackbody calibrations at NIST for low-background infrared applications, *Metrologia* 40, S1–S4 (2003).
70. M. Stock, J. Fischer, R. Friedrich, H. J. Jung, and B. Wende, The double-heatpipe black body: A high-accuracy standard source of spectral irradiance for measurements of T-$T_{90}$, *Metrologia* 32, 441–444 (1995/1996).
71. J. Fischer, J. Seidel, and B. Wende, The double-heatpipe black body: A radiance and irradiance standard for accurate infrared calibrations in remote sensing, *Metrologia* 35, 441–445 (1998).
72. E. Ikonen, P. Toivanen, and A. Lassila, A new optical method for high-accuracy determination of aperture area, *Metrologia* 35, 369–372 (1998).
73. J. E. Martin, N. P. Fox, N. J. Harrison, B. Shipp, and M. Anklin, Determination and comparisons of aperture areas using geometric and radiometric techniques, *Metrologia* 35, 461–464 (1998).
74. J. B. Fowler, R. S. Durvasula, and A. C. Parr, High-accuracy aperture-area measurement facilities at the National Institute of Standards and Technology, *Metrologia* 35, 497–500 (1998).
75. J. B. Fowler, R. D. Saunders, and A. C. Parr, Summary of high-accuracy aperture-area measurement capabilities at the NIST, *Metrologia* 37, 621–623 (2000).
76. M. Stock and R. Goebel, Practical aspects of aperture-area measurements by superposition of Gaussian laser beams, *Metrologia* 37, 633–636 (2000).
77. J. Hartmann, J. Fischer, and J. Seidel, A non-contact technique providing improved accuracy in area measurements of radiometric apertures, *Metrologia* 37, 637–640 (2000).
78. J. Fowler and M. Litorja, Geometric area measurements of circular apertures for radiometry at NIST, *Metrologia* 40, S9–S12 (2003).
79. A. W. Smith, A. C. Carter, S. R. Lorentz, T. M. Jung, and R. U. Datla, Radiometrically deducing aperture sizes, *Metrologia* 40, S13–S16 (2003).
80. M. Stock and R. Goebel, Influence of the beam shape on aperture measurements with the laser beam scanning technique, *Metrologia* 40, S208–S211 (2003).
81. J. B. Fowler, A third generation water bath based blackbody source, *J. Res. Natl. Inst. Stand. Technol.* 100, 591–599 (1995).
82. J. B. Fowler, An oil-bath-based 293 K to 473 K blackbody source, *J. Res. Natl. Inst. Stand. Technol.* 101, 629–637 (1996).

83. J. Fischer and H. Jung, Determination of the thermodynamic temperatures of the freezing points of silver and gold by near-infrared pyrometry, *Metrologia* 26, 245–252 (1989).
84. H. McEvoy, G. Machin, R. Friedrich, J. Hartmann, and J. Hollandt, Comparison of the new NPL primary standard Ag fixed-point blackbody source with the primary standard fixed-point of PTB, *AIP Conf. Proc.* 684, 909–914 (2003).
85. J. E. Martin, H. C. McEvoy, J. Fischer, and M. Stock, Comparison of NPL and PTB silver and gold point black bodies using an absolute spectral radiometer, in "Temperature, Its Measurement and Control in Science and Industry," Vol. 6 (J. F. Schooley, Ed.), 59–62. Am. Inst. of Phys., New York. 1992.
86. Y. Yamada, H. Sakate, F. Sakuma, and A. Ono, Radiometric observation of melting and freezing plateaus for a series of metal-carbon eutectic points in the range 1330°C to 1950°C, *Metrologia* 36, 207–209 (1999).
87. Y. Yamada, H. Sakate, F. Sakuma, and A. Ono, A possibility of practical high temperature fixed-points above the copper point, in "Proc. TEMPMEKO '99, 7th Int. Symp. on Temperature and Thermal Measurements in Industry and Science" (J. F. Dubbeldam and M. J. de Groot, Eds.), 535–540. IMEKO/NMi VSL, Delft, 1999.
88. N. Sasajima, Y. Yamada, B. M. Zailani, K. Fan, and A. Ono, Melting and freezing behavior of metal-carbon eutectic fixed-point blackbodies, in "Proc. TEMPMEKO 2001, 8th Int. Symp. on Temperature and Thermal Measurements in Industry and Science" (B. Fellmuth, J. Seidel, and G. Scholze, Eds.), 501–506. VDI Verlag, Berlin, 2002.
89. M. Sadli, G. Machin, D. Lowe, J. Hartmann, and R. Morice, Realisation and comparison of metal-carbon eutectic points for radiation thermometry applications and W-Re thermocouple calibration, in "Proc. TEMPMEKO 2001, 8th Int. Symp. on Temperature and Thermal Measurements in Industry and Science" (B. Fellmuth, J. Seidel, and G. Scholze, Eds.), 507–512. VDI Verlag, Berlin, 2002.
90. B. Khlevnoy, V. Sapritsky, M. Samoylov, and Y. Yamada, Ir-C and Re-C fixed-point blackbodies' spectral radiance reproducibility at 650 nm, in "Proc. TEMPMEKO 2001, 8th Int. Symp. on Temperature and Thermal Measurements in Industry and Science" (B. Fellmuth, J. Seidel, and G. Scholze, Eds.), 513–518. VDI Verlag, Berlin, 2002.
91. D. Lowe and G. Machin, Development of metal-carbon eutectic blackbody cavities to 2500°C at NPL, in "Proc. TEMPMEKO 2001, 8th Int. Symp. on Temperature and Thermal Measurements in Industry and Science" (B. Fellmuth, J. Seidel, and G. Scholze, Eds.), 519–524. VDI Verlag, Berlin, 2002.

*92.* B. B. Khlevnoy, V. Khromchenko, M. Samoylov, V. Sapritzky, N. Harrison, P. Sperfeld, and J. Fischer, Determination of the temperatures of metal-carbon eutectic fixed-points by different detectors from VINIIOFI, NPL and PTB, in "Proc. TEMPMEKO 2001, 8th Int. Symp. on Temperature and Thermal Measurements in Industry and Science" (B. Fellmuth, J. Seidel, and G. Scholze, Eds.), 845–850. VDI Verlag, Berlin, 2002.

*93.* G. Machin, Y. Yamada, D. Lowe, N. Sasajima, F. Sakuma, and Fan Kai, A comparison of metal-carbon eutectic blackbody cavities of rhenium, iridium, platinum and palladium between NPL and NMIJ, in "Proc. TEMPMEKO 2001, 8th Int. Symp. on Temperature and Thermal Measurements in Industry and Science" (B. Fellmuth, J. Seidel, and G. Scholze, Eds.), 851–856. VDI Verlag, Berlin, 2002.

*94.* V. I. Sapritsky, S. A. Ogarev, B. B. Khlevnoy, M. L. Samoylov, and V. B. Khromchenko, Development of metal–carbon high-temperature fixed-point blackbodies for precision photometry and radiometry, *Metrologia* 40, S128–S131 (2003).

*95.* G. Machin, G. Beynon, F. Edler, S. Fourrez, J. Hartmann, D. Lowe, R. Morice, M. Sadli, and M. Villamanan, HIMERT: A Pan-European project for the development of metal-carbon eutectics as temperature standards, *AIP Conf. Proc.* 684, 285–290 (2003).

*96.* M. Sadli, M. Fanjeaux, and G. Bonnier, Construction and implementation of a set of metal-carbon eutectic fixed points, *AIP Conf. Proc.* 684, 267–272 (2003).

*97.* V. I. Sapritsky, B. B. Khlevnoy, V. B. Khromchenko, S. A. Ogarev, M. L. Samoylov, and Yu. A. Pikalev, High temperature fixed-point blackbodies based on metal-carbon eutectics for precision measurements in radiometry, photometry and radiation thermometry, *AIP Conf. Proc.* 684, 273–278 (2003).

*98.* N. Sasajima, Y. Yamada, and F. Sakuma, Investigation of fixed points exceeding 2500°C using metal carbide-carbon eutectics, *AIP Conf. Proc.* 684, 279–284 (2003).

*99.* P. Bloembergen, Y. Yamada, N. Yamamoto, and J. Hartmann, Realizing the high-temperature part of a future ITS with the aid of eutectic metal-carbon fixed points, *AIP Conf. Proc.* 684, 291–296 (2003).

*100.* G. Machin, Y. Yamada, D. Lowe, N. Sasajima, K. Anhalt, J. Hartmann, R. Goebel, and H. McEvoy, A comparison of high temperature fixed points of Pt-C and Re-C constructed by BIPM, NMIJ and NPL, in "Proc. TEMPMEKO 2004, 9th Int. Symp. on Temperature and Thermal Measurements in Industry and Science" (D. Zvizdic, Ed.), 1049–1056. Laboratory for Process Measurement, Faculty of Mechanical Engineering and Naval Architecture, Zagreb, 2005.

*101.* J. Hartmann, K. Anhalt, P. Sperfeld, J. Hollandt, M. Sakharov, B. Khlevnoy, Yu. Pikalev, S. Ogarev, and V. Sapritsky, Thermodynamic temperature measurements of the melting curves of Re-C, TiC-C and ZrC-C Eutectics, in "Proc. TEMPMEKO 2004, 9th Int. Symp. on Temperature and Thermal Measurements in Industry and Science" (D. Zvizdic, Ed.), 189–194. Laboratory for Process Measurement, Faculty of Mechanical Engineering and Naval Architecture, Zagreb, 2005.

*102.* G. Machin, C. E. Gibson, D. Lowe, D. W. Allen, and H. W. Yoon, A comparison of ITS-90 and detector-based scales between NPL and NIST using metal-carbon eutectics, in "Proc. TEMPMEKO 2004, 9th Int. Symp. on Temperature and Thermal Measurements in Industry and Science" (D. Zvizdic, Ed.), 1057–1062. Laboratory for Process Measurement, Faculty of Mechanical Engineering and Naval Architecture, Zagreb, 2005.

*103.* M. K. Sakharov, B. B. Khlevnoy, V. I. Sapritsky, M. L. Samoylov, and S. A. Ogarev, Development and investigation of high temperature fixed point based on TiC-C eutectic, in "Proc. TEMPMEKO 2004, 9th Int. Symp. on Temperature and Thermal Measurements in Industry and Science" (D. Zvizdic, Ed.), 319–324. Laboratory for Process Measurement, Faculty of Mechanical Engineering and Naval Architecture, Zagreb, 2005.

*104.* M. Sadli, J. Fischer, Y. Yamada, V. Sapritsky, D. Lowe, and G. Machin, Review of metal carbon eutectic temperatures: proposal for new ITS-90 secondary points, in "Proc. TEMPMEKO 2004, 9th Int. Symp. on Temperature and Thermal Measurements in Industry and Science" (D. Zvizdic, Ed.), 341–348. Laboratory for Process Measurement, Faculty of Mechanical Engineering and Naval Architecture, Zagreb, 2005.

*105.* D. Lowe and G. Machin, Development of metal-carbon eutectic based high-temperature fixed-points for reproducibility studies, in "Proc. TEMPMEKO 2004, 9th Int. Symp. on Temperature and Thermal Measurements in Industry and Science" (D. Zvizdic, Ed.), 177–182. Laboratory for Process Measurement, Faculty of Mechanical Engineering and Naval Architecture, Zagreb, 2005.

*106.* J. H. Walker, R. D. Saunders, and A. T. Hattenburg, "Spectral Radiance Calibrations," p. 26. NBS Special Publication 250-1. U.S. Government Printing Office, Washington, DC, 1987.

*107.* J. H. Walker, R.D. Saunders, J. K. Jackson, and D. A. McSparron, "Spectral Irradiance Calibrations," p. 37. NBS Special Publication 250-20. U.S. Government Printing Office, Washington, DC, 1987.

*108.* C. E. Gibson, B. K. Tsai, and A. C. Parr, "Radiance Temperature Calibrations," p. 56. NIST Special Publication 250-43. U.S. Government Printing Office, Washington, DC, 1998.

*109.* P. Sperfeld, J. Metzdorf, N. J. Harrison, N. P. Fox, B. B. Khlevnoy, V. B. Khromchenko, S. N. Mekhontsev, V. I. Shapoval, M. F. Zelener, and V. I. Sapritsky, Investigation of high-temperature black body BB3200, *Metrologia* 35, 419–422 (1998).

*110.* S. G. Yousef, P. Sperfeld, and J. Metzdorf, Measurement and calculation of the emissivity of a high-temperature black body, *Metrologia* 37, 365–368 (2000).

*111.* E. R. Woolliams, N. J. Harrison, and N. P. Fox, Preliminary results of the investigation of a 3500 K black body, *Metrologia* 37, 501–504 (2000).

*112.* H. W. Yoon, C. W. Gibson, and B. C. Johnson, The determination of emissivity of the variable-temperature blackbody used in the dissemination of the US national scale of radiance temperature, in "Proc. TEMPMEKO 2001, 8th Int. Symp. on Temperature and Thermal Measurements in Industry and Science" (B. Fellmuth, J. Seidel, and G. Scholze, Eds.), 221–226. VDI Verlag, Berlin, 2002.

*113.* J. Hartmann, R. Friedrich, S. Schiller, and J. Fischer, Non-isothermal temperature distribution and resulting emissivity corrections for the high temperature blackbody BB3200, in "Proc. TEMPMEKO 2001, 8th Int. Symp. on Temperature and Thermal Measurements in Industry and Science" (B. Fellmuth, J. Seidel, and G. Scholze, Eds.), 227–232. VDI Verlag, Berlin, 2002.

*114.* P. B. Coates, Tungsten ribbon lamps, in Ref. [3], pp. 773–819.

*115.* R. D. Saunders, C. E. Gibson, K. D. Mielenz, V. I. Sapritsky, K. A. Sudarev, B. B. Khlevnoy, S. N. Mekhontsev, and G. D. Harchenko, Results of a NIST/VNIIOFI comparison of spectral-radiance measurements, *Metrologia* 32, 449–453 (1995/1996).

*116.* Y. Ohno and J. K. Jackson, Characterization of modified FEL quartz-halogen lamps for photometric standards, *Metrologia* 32, 693–696 (1995/1996).

*117.* J. Metzdorf, A. Sperling, S. Winter, K.-H. Raatz, and W. Möller, A new FEL-type quartz-halogen lamp as an improved standard of spectral irradiance, *Metrologia* 35, 423–426 (1998).

*118.* L. K. Huang, R. P. Cebula, and E. Hilsenrath, New procedure for interpolating NIST FEL lamp irradiances, *Metrologia* 35, 381–386 (1998).

*119.* A. Sperling and V. Bentlage, A stabilized transfer-standard system for spectral irradiance, *Metrologia* 35, 437–440 (1998).

*120.* N. P. Fox, C. J. Chunnilall, and M. G. White, Detector-based transfer standards for improved accuracy in spectral irradiance and radiance measurements, *Metrologia* 35, 555–561 (1998).

*121.* K. D. Stock, K.-H. Raatz, P. Sperfeld, J. Metzdorf, T. Kübarsepp, P. Kärhä, E. Ikonen, and L. Liedquist, Detector-stabilized FEL lamps as transfer standards in an international comparison of spectral irradiance, *Metrologia* 37, 441–444 (2000).

*122.* N. J. Harrison, E. R. Woolliams, and N. P. Fox, Evaluation of spectral irradiance transfer standards, *Metrologia* 37, 453–456 (2000).

*123.* S. A. Windsor, N. J. Harrison, and N. P. Fox, The NPL detector-stabilized irradiance source, *Metrologia* 37, 473–476 (2000).

*124.* J. Gröbner, D. Rembges, A. F. Bais, M. Blumthaler, T. Cabot, W. Josefsson, T. Koskela, T. M. Thorseth, A. R. Webb, and U. Wester, Quality Assurance of Reference Standards From Nine European Solar-Ultraviolet Monitoring Laboratories, *Appl. Opt.* 41, 4278–4282 (2002).

*125.* L. Ylianttila, K. Jokela, and P. Kärhä, Ageing of DXW-lamps, *Metrologia* 40, S120–S123 (2003).

*126.* R. D. Saunders, W. R. Ott, and J. M. Bridges, Spectral irradiance standard for the ultraviolet: The deuterium lamp, *Appl. Opt.* 17, 593–600 (1978).

*127.* J. Hollandt, U. Becker, W. Paustian, M. Richter, and G. Ulm, New developments in the radiance calibration of deuterium lamps in the UV and VUV spectral range at the PTB, *Metrologia* 37, 563–566 (2000).

*128.* R. P. Lambe, R. Saunders, C. Gibson, J. Hollandt, and E. Tegeler, A CCPR international comparison of spectral radiance measurements in the air-ultraviolet, *Metrologia* 37, 51–54 (2000).

*129.* P. Sperfeld, K. D. Stock, K.-H. Raatz, B. Nawo, and J. Metzdorf, Characterization and use of deuterium lamps as transfer standards of spectral irradiance, *Metrologia* 40, S111–S114 (2003).

*130.* J. M. Bridges and W. R. Ott, Vacuum ultraviolet radiometry. 3: The argon mini-arc as a new secondary standard of spectral radiance, *Appl. Opt.* 16, 367–376 (1977).

*131.* J. M. Bridges and A. L. Migdall, Characterization of argon arc source in the infrared, *Metrologia* 32, 625–628 (1995/1996).

*132.* K. Grützmacher, Wall-stabilized arc-plasma source for radiometric applications in the range 200 nm to 10 μm, *Metrologia* 37, 465–468 (2000).

*133.* National Physical Laboratory (UK), "Spectral Radiance & Irradiance Primary Standard," available via http://www.npl.co.uk/optical_radiation/instrumentation/leaflets/spectral_radiance_irradiance.pdf Spectral Radiance and Irradiance Primary Scales, http://www.npl.co.uk/optical_radiation/primary_standards/srips.html

*134.* S.-N. Park, J.-C. Seo, D.-J. Shin, and I.-W. Lee, Realization of a spectral radiance scale at KRISS, *Metrologia* 40, S196–S199 (2003).

*135.* J. Fischer, G. Neuer, E. Schreiber, and R. Thomas, Metrological characterisation of a new transfer-standard radiation thermometer, in "Proc. TEMPMEKO 2001, 8th Int. Symp. on Temperature and Thermal Measurements in Industry and Science" (B. Fellmuth, J. Seidel, and G. Scholze, Eds.), 801–806. VDI Verlag, Berlin, 2002.

*136.* H. W. Yoon, C. E. Gibson, and P. Y. Barnes, Realization of the National Institute of Standards and Technology detector-based spectral irradiance scale, *Appl. Opt.* 41, 5879–5890 (2002);
The realization of the NIST detector-based spectral irradiance scale, *Metrologia* 40, S172–S176 (2003).

*137.* H. W. Yoon, J. E. Proctor, and C. E. Gibson, FASCAL 2: A new NIST facility for the calibration of the spectral irradiance of sources, *Metrologia* 40, S30–S34 (2003).

*138.* G. Machin and M. Ibrahim, SSE and temperature uncertainty I: High temperature systems, in "Proc. TEMPMEKO '99, 7th Int. Symp. on Temperature and Thermal Measurements in Industry and Science" (J. F. Dubbeldam and M. J. de Groot, Eds.), 681–686. IMEKO/NMi VSL, Delft, 1999.

*139.* G. Machin and M. Ibrahim, SSE and temperature uncertainty II: Low temperature systems, in "Proc. TEMPMEKO '99, 7th Int. Symp. on Temperature and Thermal Measurements in Industry and Science" (J. F. Dubbeldam and M. J. de Groot, Eds.), 687–692. IMEKO/NMi VSL, Delft, 1999.

*140.* G. Machin and R. Sergienko, A Comparative Study of Size of Source Effect (SSE) Determination Techniques, in "Proc. TEMPMEKO 2001, 8th Int. Symp. on Temperature and Thermal Measurements in Industry and Science" (B. Fellmuth, J. Seidel, and G. Scholze, Eds.), 155–160. VDI Verlag, Berlin, 2002.

*141.* H. J. Kostkowski, The relative spectral responsivity and slit-scattering function of a spectroradiometer, Natl. Bur. Stand. U.S. Tech. Note 910-4, 1979, in "Self-Study Manual on Optical Radiation Measurements" (F. E. Nicodemus, Ed.). NBS Technical Notes 910-1 through 910-8, 1976–1985.

*142.* R. D. Saunders and J. R. Shumaker, Apparatus function of a prism-grating double monochromator, *Appl. Opt.* 25, 3710–3714 (1986).

*143.* E. W. M. van der Ham, H. C. D. Bos, and C. A. Schrama, Primary realization of a spectral irradiance scale employing monochromator-based cryogenic radiometry between 200 nm and 20 μm, *Metrologia* 40, S177–S180 (2003).

*144.* J. Schwinger, *Phys. Rev.* 75, 1912 (1949).

*145.* "Handbook on Synchrotron Radiation," Vol. 1a to 4, North-Holland, Amsterdam, 1983 to 1991.
*146.* G. Margaritondo, "Introduction to Synchrotron Radiation." Oxford University Press, New York, 1988.
*147.* S. L. Hulbert and G. P. Williams, Synchrotron radiation sources, in "Vacuum Ultraviolet Spectroscopy I" (J. A. R. Samson and D. L. Ederer, Eds.). Academic Press, San Diego, 1998.
*148.* D. H. Tomboulian and P. L. Hartman, *Phys. Rev.* 102, 1423 (1956).
*149.* D. L. Ederer, E. B. Saloman, S. C. Ebner, and R. P. Madden, *J. Res. Natl. Bur. Stand.* 79A, 761 (1975).
*150.* U. Arp, R. Friedman, M. L. Furst, S. Makar, and P.-S. Shaw, *Metrologia* 37, 357 (2000).
*151.* F. Riehle and B. Wende, *Opt. Lett.* 10, 365 (1985).
*152.* F. Riehle and B. Wende, *Metrologia* 22, 75 (1986).
*153.* R. Thornagel, R. Klein, and G. Ulm, *Metrologia* 38, 385 (2001).
*154.* H. Suzuki, *Nucl. Instrum. Meth.* 228, 201 (1984).
*155.* T. Zama and I. Saito, *Metrologia* 40, S115 (2003).
*156.* E. S. Gluskin, E. M. Trakhenberg, I. G. Feldman, and V. A. Kochubei, *Space Sci. Instrum.* 5, 129 (1980).
*157.* A. N. Subbotin, V. A. Chernov, V. V. Gaganov, A. V. Kalutsky, N. V. Kovalenko, A. K. Krasnov, K. E. Kuper, A. G. Legkodmymov, A. D. Nikolenko, I. N. Nesterenko, V. F. Pindyurin, and V. N. Romaev, *Nucl. Instrum. Meth. A* 470, 452 (2001).
*158.* A. N. Subbotin, V. V. Gaganov, A. V. Kalutsky, V. F. Pindyurin, V. P. Nazmov, A. D. Nikolenko, and A. K. Krasnov, *Metrologia* 37, 497 (2000).
*159.* G. Ulm and B. Wende, in "Röntgen Centennial" (A. Haase, G. Landwehr, and E. Umbach, Eds.), p. 81. World Scientific, Singapore, 1997.
*160.* G. Ulm and B. Wende, *Rev. Sci. Instrum.* 66, 2244 (1995).
*161.* G. Ulm, *Metrologia* 40, S101 (2003).
*162.* F. Riehle, E. Tegeler, and B. Wende, *Proc. SPIE* 733, 80 (1986).
*163.* V. O. Kostroun, *Nucl. Instrum. Meth.* 172, 371 (1980).
*164.* R. Klein, G. Ulm, M. Abo-Bakr, P. Budz, K. Bürkmann-Gehrlein, D. Krämer, J. Rahn, and G. Wüstefeld, in "Proc. of EPAC 2004," Lucerne, Switzerland, 2004, p. 2290; http://accelconf.web.cern.ch/accelconf/e04
*165.* R. Thornagel, R. Fliegauf, R. Klein, F. Scholze, and G. Ulm, *Rev. Sci. Instrum.* 67, 653 (1996).
*166.* K. Molter and G. Ulm, *Rev. Sci. Instrum.* 63, 1296 (1992).
*167.* P. Kuske, R. Goergen, R. Klein, R. Thornagel, and G. Ulm, in "Proc. of EPAC 2000," Vienna, Austria, 2000, p. 1771; http://accelconf.web.cern.ch/accelconf/e00

*168.* R. Klein, T. Mayer, P. Kuske, R. Thornagel, and G. Ulm, *Nucl. Instrum. Meth. A* 384, 293 (1997).

*169.* R. Klein, P. Kuske, R. Thornagel, G. Brandt, R. Görgen, and G. Ulm, *Nucl. Instrum. Meth. A* 486, 541 (2002).

*170.* A. Lysenko, I. Koop, A. Polunin, E. Pozdeev, V. Ptitsin, and Yu. Shatunov, *Nucl. Instrum. Meth. A* 359, 419 (1995).

*171.* A. R. Schaefer, L. R. Hughey, and J. B. Fowler, *Metrologia* 19, 131 (1984).

*172.* F. Riehle, S. Bernstorff, R. Fröhling, and F. P. Wolf, *Nucl. Instrum. Meth. A* 268, 262 (1988).

*173.* G. Ulm, W. Hänsel-Ziegler, S. Bernstorff, and F. P. Wolf, *Rev. Sci. Instrum.* 60, 1752 (1989).

*174.* F. Riehle, *Nucl. Instrum. Meth. A* 246, 385 (1986).

*175.* W. Görner, M. P. Hentschel, B. R. Müller, H. Riesemeier, M. Krumrey, G. Ulm, W. Diete, U. Klein, and R. Frahm, *Nucl. Instrum. Meth. A* 467–468, 703 (2001).

*176.* P.-S. Shaw, D. Shear, R. J. Stamilo, U. Arp, H. W. Yoon, R. D. Saunders, A. C. Parr, and K. R. Lykke, *Rev. Sci. Instrum.* 73, 1576 (2002).

*177.* P.-S. Shaw, U. Arp, H. W. Yoon, R. D. Saunders, A. C. Parr, and K. R. Lykke, *Metrologia* 40, S124 (2003).

*178.* M. Richter, J. Hollandt, U. Kroth, W. Paustian, H. Rabus, R. Thornagel, and G. Ulm, *Metrologia* 40, S107 (2003).

*179.* M. Richter, J. Hollandt, U. Kroth, W. Paustian, H. Rabus, R. Thornagel, and G. Ulm, *Nucl. Instrum. Meth. A* 467–468, 605 (2001).

*180.* R. P. Lambe, R. Saunders, C. Gibson, J. Hollandt, and E. Tegeler, *Metrologia* 37, 51 (2000).

*181.* M. Kühne, Hollow Cathode, Penning, and Electron-Beam Excitation Sources, in "Vacuum Ultraviolet Spectroscopy I" (J. A. R. Samson and D. L. Ederer, Eds.), p. 65. Academic Press, San Diego, 1998.

*182.* J. Hollandt, M. Kühne, and B. Wende, *Appl. Opt.* 33, 68 (1994).

*183.* A. Heise, J. Hollandt, R. Kling, M. Kock, and M. Kühne, *Appl. Opt.* 33, 5111 (1994).

*184.* W. Jans, B. Möbus, M. Kühne, G. Ulm, A. Werner, and K.-H. Schartner, *Phys. Rev. A* 55, 1890 (1997).

*185.* A. McPherson, N. Rouze, W. B. Westerveld, and J. S. Risley, *Appl. Opt.* 25, 298 (1986).

*186.* P. Grübling, J. Hollandt, and G. Ulm, *Rev. Sci. Instrum.* 71, 1200 (2000).

*187.* J. Hollandt, U. Schühle, W. Paustian, W. Curdt, M. Kühne, B. Wende, and K. Wilhelm, *Appl. Opt.* 35, 5125 (1996).

*188.* R. A. Harrison, B. J. Kent, E. C. Sawyer, J. Hollandt, M. Kühne, W. Paustian, B. Wende, and M. C. E. Huber, *Metrologia* 32, 647 (1995).

*189.* M. Kühne, Radiometric Characterization of VUV Sources, in "Vacuum Ultraviolet Spectroscopy I" (J. A. R. Samson and D. L. Ederer, Eds.), p. 119. Academic Press, San Diego, 1998.
*190.* P. Müller, F. Riehle, E. Tegeler, and B. Wende, *Nucl. Instrum. Meth. A* 246, 569 (1986).
*191.* F. Scholze and M. Procop, *X-Ray Spectrom.* 30, 69 (2001).
*192.* S. Kraft, F. Scholze, R. Thornagel, G. Ulm, W. C. McDermott, and E. M. Kellogg, *Proc. SPIE* 3114, 101 (1997).
*193.* J. M. Auerhammer, G. Brandt, F. Scholze, R. Thornagel, G. Ulm, B. J. Wargelin, W. C. Mc Dermott, T. J. Norton, I. N. Evans, and E. M. Kellogg, *Proc. SPIE* 3444, 19 (1998).
*194.* M. W. Bautz, M. J. Pivovaroff, S. E. Kissel, G. Y. Prigozhin, T. Isobe, S. E. Jones, G. R. Ricker, R. Thornagel, S. Kraft, F. Scholze, and G. Ulm, *Proc. SPIE* 4012, 53 (2000).
*195.* R. Hartmann, G. Hartner, U. G. Briel, K. Dennerl, F. Haberl, L. Strüder, J. Trümper, E. Bihler, E. Kendziorra, J.-F. Hochedez, E. Jourdain, P. Dhez, Ph. Salvetat, J. Auerhammer, D. Schmitz, F. Scholze, and G. Ulm, *Proc. SPIE* 3765, 703 (1999).
*196.* F. Scholze, R. Thornagel, and G. Ulm, *Metrologia* 38, 391 (2001).
*197.* R. Klein, J. Bahrdt, D. Herzog, and G. Ulm, *J. Synchrotron Rad.* 5, 451 (1998).
*198.* U. Vogt, H. Stiel, I. Will, P. V. Nickles, W. Sandner, M. Wieland, and T. Wilhein, *Proc. SPIE* 4343, 535 (2001).
*199.* A. R. Schaefer, R. D. Saunders, and L. R. Hughey, *Opt. Eng.* 25, 892 (1986).
*200.* H. J. Kostkowski, J. L. Lean, R. D. Saunders, and L. R. Huhgey, *Appl. Opt.* 25, 3297 (1986).
*201.* M. Stock, J. Fischer, R. Friedrich, H. J. Jung, R. Thornagel, G. Ulm, and B. Wende, *Metrologia* 30, 439 (1993).
*202.* N. P. Fox, P. J. Key, F. Riehle, and B. Wende, *Appl. Opt.* 25, 2409 (1986).
*203.* R. Thornagel, J. Fischer, R. Friedrich, M. Stock, G. Ulm, and B. Wende, *Metrologia* 32, 459 (1995).
*204.* H. Rabus, R. Klein, F. Scholze, R. Thornagel, and G. Ulm, *Metrologia* 39, 381 (2002).
*205.* D. Arnold and G. Ulm, *Nucl. Instrum. Meth. A* 339, 43 (1994).
*206.* A. Lau-Främbs, U. Kroth, H. Rabus, E. Tegeler, G. Ulm, and B. Wende, *Metrologia* 32, 571 (1995/1996).
*207.* H. Rabus, F. Scholze, R. Thornagel, and G. Ulm, *Nucl. Instrum. Meth. A* 377, 209 (1996).
*208.* H. Rabus, V. Persch, and G. Ulm, *Appl. Opt.* 36, 5421 (1997).

*209.* G. Ulm, B. Beckhoff, R. Klein, M. Krumrey, H. Rabus, and R. Thornagel, *Proc. SPIE* 3444, 610 (1998).

*210.* F. Scholze, G. Brandt, P. Müller, B. Meyer, F. Scholz, J. Tümmler, K. Vogel, and G. Ulm, *Proc. SPIE* 4688, 680 (2002).

*211.* F. Scholze, J. Tümmler, and G. Ulm, *Metrologia* 40, S224 (2003).

*212.* M. Krumrey and G. Ulm, *Nucl. Instrum. Meth. A* 467–468, 1175 (2001).

*213.* M. Krumrey, L. Büermann, M. Hoffmann, P. Müller, F. Scholze, and G. Ulm, *AIP Conf. Proc.* 861, 861 (2004).

*214.* J. A. R. Samson, *J. Opt. Soc. Am.* 54, 6 (1964).

*215.* L. R. Canfield and N. Swanson, *J. Res. Natl. Bur. Stand.* 92, 97 (1987).

*216.* T. Saito and H. Onuki, *Metrologia* 32, 525 (1995/1996).

*217.* P.-S. Shaw, K. R. Lykke, R. Gupta, T. R. O'Brian, U. Arp, H. H. White, T. B. Lucatorto, J. L. Dehmer, and A. C. Parr, *Appl. Opt.* 38, 18 (1999).

*218.* P.-S. Shaw, T. C. Larason, R. Gupta, S. W. Brown, R. E. Vest, and K. R. Lykke, *Rev. Sci. Instrum.* 72, 2242 (2001).

*219.* T. Saito, I. Saito, T. Yamada, T. Zama, and H. Onuki, *J. Electr. Spectr. Relat. Phenom.* 80, 397 (1996).

*220.* F. Scholze, B. Beckhoff, G. Brandt, R. Fliegauf, A. Gottwald, R. Klein, B. Meyer, U. Schwarz, R. Thornagel, J. Tümmler, K. Vogel, J. Weser, and G. Ulm, *Proc. SPIE* 4344, 402 (2001).

*221.* R. Korde, C. Prince, D. Cunningham, R. Vest, and E. Gullikson, *Metrologia* 40, S145 (2003).

*222.* F. Scholze, H. Rabus, and G. Ulm, *J. Appl. Phys.* 84, 2926 (1998).

*223.* F. Scholze, H. Henneken, P. Kuschnerus, H. Rabus, M. Richter, and G. Ulm, *Nucl. Instrum. Meth. A* 439, 208 (2000).

*224.* A. A. Sorokin, L. A. Shmaenok, S. V. Bobashev, B. Möbus, M. Richter, and G. Ulm, *Phys. Rev. A* 61, 022723 (2000).

*225.* H. Henneken, F. Scholze, M. Krumrey, and G. Ulm, *Metrologia* 37, 485 (2000).

*226.* H. Henneken, F. Scholze, and G. Ulm, *J. Appl. Phys.* 87, 257 (2000).

*227.* H. Rabus, U. Kroth, M. Richter, G. Ulm, J. Friese, R. Gernhäuser, A. Kastenmüller, P. Maier-Komor, and K. Zeitelhack, *Nucl. Instrum. Meth. A* 438, 94 (1999).

*228.* S. Serej, E. Kellogg, R. Edgar, F. Scholze, and G. Ulm, *Proc. SPIE* 3765, 777 (1999).

*229.* E. M. Gullikson, S. Mrowka, and B. B. Kaufmann, *Proc. SPIE* 4343, 363 (2001).

*230.* C. Tarrio, S. Grantham, M. B. Squires, R. E. Vest, and T. B. Lucatorto, *J. Res. Natl. Inst. Stand. Technol.* 108, 267 (2003).

*231.* F. Scholze, B. Beckhoff, G. Brandt, R. Fliegauf, R. Klein, B. Meyer, D. Rost, D. Schmitz, M. Veldkamp, J. Weser, G. Ulm, E. Louis, A. E. Yakshin, S. Oestreich, and F. Bijkerk, *Proc. SPIE* 4146, 72 (2000).

*232.* J. Tümmler, F. Scholze, G. Brandt, B. Meyer, F. Scholz, K. Vogel, G. Ulm, M. Poier, U. Klein, and W. Diete, *Proc. SPIE* 4688, 338 (2002).

*233.* H. Meiling, V. Banine, P. Kürz, and N. Harned, *Proc. SPIE* 5374, 31 (2004).

*234.* K. A. Flanagan, T. H. Markert, J. E. Davis, M. L. Schattenburg, R. L. Blake, F. Scholze, P. Bulicke, R. Fliegauf, S. Kraft, G. Ulm, and E. M. Gullikson, *Proc. SPIE* 4140, 559 (2000).

*235.* M. Bavdaz, A. Peacock, R. den Hartog, A. Poelaert, P. Underwood, V.-P. Viitanen, D. Fuchs, P. Bulicke, S. Kraft, F. Scholze, G. Ulm, and A. C. Wright, *Proc. SPIE* 2808, 301 (1996).

*236.* M. Krumrey, M. Hoffmann, G. Ulm, K. Hasche, and P. Thomsen-Schmidt, *Thin Solid Films* 459, 241 (2004).

*237.* S. Bechstein, B. Beckhoff, R. Fliegauf, J. Weser, and G. Ulm, *Spectrochim. Acta B* 59, 215 (2004).

*238.* H. Meiling, B. Mertens, F. Stietz, M. Wedowski, R. Klein, R. Kurt, E. Louis, and A. Yakshin, *Proc. SPIE* 4506, 93 (2001).

*239.* R. Klein, A. Gottwald, F. Scholze, R. Thornagel, J. Tümmler, G. Ulm, M. Wedowski, F. Stietz, B. Mertens, N. Koster, and J. van Elp, *Proc. SPIE* 4506, 105 (2001).

*240.* F. Scholze, R. Klein, and T. Bock, *Appl. Opt.* 42, 5621 (2003).

*241.* B. Beckhoff, R. Fliegauf, G. Ulm, G. Pepponi, C. Streli, P. Wobrauschek, L. Fabry, and S. Pahlke, *Spectrochim. Acta B* 56, 2073 (2001).

*242.* C. Streli, P. Wobrauschek, B. Beckhoff, G. Ulm, L. Fabry, and S. Pahlke, *X-ray Spectrom.* 30, 24 (2001).

*243.* B. Beckhoff and G. Ulm, *Adv. X-ray Anal.* 44, 349 (2001).

*244.* M. Richter, U. Kroth, A. Gottwald, C. Gerth, K. Tiedtke, T. Saito, I. Tassy, and K. Vogler, *Appl. Opt.* 41, 7167 (2002).

*245.* F. Scholze, R. Klein, and R. Müller, *Proc. SPIE* 5374, 926 (2004).

*246.* W. H. Louisell, A. Yariv, and A. E. Siegman, *Phys. Rev.* 124, 1646–1654 (1961).

*247.* F. Zernike and J. E. Midwinter, "Applied Nonlinear Optics." Wiley, New York, 1973.

*248.* D. C. Burnham and D. L. Weinberg, *Phys. Rev. Letts.* 25, 84–87 (1970).

*249.* A. Migdall, R. Datla, A. Sergienko, J. S. Orszak, and Y. H. Shih, *Appl. Opt.* 37, 3455–3463 (1998).

*250.* A. L. Migdall, *IEEE Trans. Instrum. Meas.* 50, 478–481 (2001).

*251.* T. Larason, S. Bruce, and A. Parr, "Spectroradiometric Detector Measurement: Part I—Ultraviolet Detectors and Part II—Visible to Near-Infrared Detectors." NIST Special Pub. 250-41, 1998.
*252.* C. Cohen-Tannoudji, J. Dupont-Roc, and G. Grynberg, "Atom-Photon Interactions," p. 321. Wiley, New York, 1992.
*253.* D. N. Klyshko, *Sov. J. Quantum Electr.* 7, 591–594 (1977).
*254.* E. Dauler, A. Migdall, N. Boeuf, R. Datla, A. Muller, and A. Sergienko, *Metrologia* 35, 295 (1998).
*255.* D. N. Klyshko, *Sov. J. Quantum Electr.* 10, 1112–1116 (1981).
*256.* A. A. Malygin, A. N. Penin, and A. V. Sergienko, *Sov. J. Quantum Electr.* 10, 939–941 (1981).
*257.* S. R. Bowman, Y. H. Shih, and C. O. Alley, The use of geiger mode avalanche photodiodes for precise laser ranging at very low light levels: An experimental evaluation, *Proc. SPIE* 663, 24–29 (1986).
*258.* J. G. Rarity, K. D. Ridley, and P. R. Tapster, *Appl. Opt.* 26, 4616–4619 (1987).
*259.* A. N. Penin and A. V. Sergienko, *Appl. Opt.* 30, 3582–3588 (1991).
*260.* V. M. Ginzburg, N. Keratishvili, E. L. Korzhenevich, G. V. Lunev, A. N. Penin, and V. Sapritsky, *Opt. Eng.* 32, 2911–2916 (1993).
*261.* P. G. Kwait, A. M. Steinberg, R. Y. Chiao, P. H. Eberhard, and M. D. Petroff, *Appl. Opt.* 33, 1844–1853 (1994).
*262.* A. Migdall, R. Datla, A. Sergienko, J. S. Orszak, and Y. H. Shih, *Metrologia* 32, 479–483 (1995).
*263.* G. Brida, S. Castelletto, I. P. Degiovanni, C. Novero, and M. L. Rastello, *Metrologia* 37, 625–628 (2000).
*264.* G. Brida, S. Castelletto, I. P. Degiovanni, M. Genovese, C. Novero, and M. L. Rastello, *Metrologia* 37, 629–632 (2000).
*265.* N. Feather and J. V. Dunworth, Absorption and coincidence experiments on the radiations from the radioactive sodium, $Na^{24}$, *Proc. Cambridge Phil. Soc.* 34, 442–449 (1938).
*266.* E. Brannen, F. R. Hunt, R. H. Adlington, and R. W. Nicholls, *Nature* 175, 810–811 (1955).
*267.* F. Cristofori, P. Fenici, G. E. Frigerio, N. Molho, and P. G. Sona, *Phys. Lett.* 6, 171–172 (1963).
*268.* H. Geiger and E. Marsden, *Phys. Zeitschr.* 11, 7–11 (1910).
*269.* N. Feather and J. V. Dunworth, Note on the production of positron-electron pairs during the passage of $\beta$-particles through matter, *Proc. Cambridge Phil. Soc.* 34, 435–441 (1938).
*270.* N. Feather and J. V. Dunworth, A further study of the problem of nuclear isomerism: The application of the method of coincidence counting to the investigation of the $\gamma$-rays emitted by uranium z and

the radioactive silver $Ag^{106}$, *Proc. Roy. Soc. London* 168, 566–584 (1938).

*271.* J. V. Dunworth, The application of the method of coincidence counting to experiments in nuclear physics, *R. S. L. (Royal Society of London)* 11, 167–180 (1940).

*272.* G. Kitaeva, A. N. Penin, V. V. Fadeev, and Yu. A. Uanait, *Sov. Phys. Dokl.* 24, 564 (1979).

# 6. UNCERTAINTY ESTIMATES IN RADIOMETRY

J. L. Gardner
National Measurement Institute, Lindfield, Australia

## 6.1 Introduction

No measurement is complete without an estimate of its uncertainty. The relatively recent introduction of the ISO Guide to the Expression of Uncertainty of Measurement (GUM) [1] has led to the application of a common framework across all fields of metrology for the method by which uncertainties are estimated and expressed. This applies both to parameters influencing the measurement of a particular quantity and to any resultant formed by combining a number of directly measured quantities. The role of the metrology expert in a given field is to know the dominant factors (*influence quantities*) that affect a reported quantity (the *measurand*), and the method of estimating the likely range of those factors about their values (the *uncertainties*). The GUM then shows how to propagate those uncertainties to that of the measurand, and how to express that uncertainty so that it in turn can be propagated in subsequent combinations using the measurand. This chapter deals with some uncertainty components found in the field of radiometry only to show the method of propagation—it does not cover the range of components to be considered in a particular application.

As discussed in Chapter 1, many radiometric quantities of interest are formed by integration, usually over a geometric range of angles or area elements or over more fundamental spectral measurements. Quantities in photometry and colorimetry are often broadband and calculated by integration of spectral measurements over a range of wavelengths. The spectral values from which these integrals are formed are often assumed to have independent (*random*) uncertainties, but this is usually not true because factors common to all wavelengths (*systematic* factors) affect all values in a related manner, that is, they are *correlated.* Correlations complicate the propagation of uncertainty through the integration process. Correlations of spectral quantities are most likely to be strong in the national laboratories providing primary reference standards or secondary working standards.

Uncertainty *components* are the individual factors affecting a measurement. Many systematic effects may be treated by component, fully correlated within a measurement. Spectral values (spectral irradiance, spectral responsivity) at different wavelengths are generally partially correlated



EXPERIMENTAL METHODS IN THE PHYSICAL SCIENCES, vol. 41
ISSN 1079-4042 DOI: 10.1016/S1079-4042(05)41006-1

because both random and systematic effects contribute to their uncertainties. Treating these as fully correlated when evaluating spectral integrals tends to overestimate uncertainties of the integrals and so mask other systematic effects that may be present when separate results are compared. Proper treatment of partial correlations can lead to an improvement in measurement understanding and practice.

At the highest levels of accuracy, systematic effects dominate an uncertainty budget. In many cases, negligible uncertainty is added by a direct transfer from a reference artifact under similar conditions to those used to calibrate it—the uncertainty of the prior calibration often dominates.

In this chapter, we cover the basic principles of uncertainty propagation and apply the methods to various operations and examples encountered in radiometry. We then cover guidelines from the GUM for the estimation of uncertainties of influence quantities and the expression of uncertainty.

## 6.2 Propagation of Uncertainty

To propagate means to pass along. Uncertainty propagation is a passing along of uncertainty of one or more measured or estimated values to the uncertainty of another quantity calculated from those values through a relationship which is the measurement equation. The calculated quantity in turn may be an input value to another measurement equation, with uncertainties passed along through more stages. Curve-fitting and interpolation, covered in Sections 6.4.2 and 6.4.3, respectively, are examples of passing uncertainties from one set of parameters to another, and then continuing the passing-along process to other calculated values.

Spectral measurements in radiometry are examples where the uncertainties of some measured or estimated parameters are passed along to a whole family of values at different wavelengths. For systematic components in a measurement system, each member of the family then has an uncertainty such that the values of the family members move up or down within their uncertainty range in a coordinated, or correlated, way as the value of the contributing component moves up or down within its uncertainty range. When we then combine the spectral family members in a measurement equation to form another parameter, usually as an integral over wavelength, these correlated effects can have a significant effect on the resultant uncertainty.

Section 6.2.1 begins by covering the basic relationships and definition of terms for uncertainty propagation. Covariance is defined in Section 6.2.3 and its propagation covered in detail in subsequent subsections, because it is so important to the understanding of systematic effects in spectral radiometry.

Appendix B contains a detailed example of a calibration method where covariance significantly alters the uncertainty of the result.

### 6.2.1 Basic Relationship

Most quantities in radiometry are not measured directly, but are the result of a calculation, which may be as simple as subtracting a background value, or as complex as calculating correlated color-temperature from a series of spectral measurements. Many calculations are made in smaller steps, producing intermediate values as various corrections are made. Uncertainties are propagated through each calculation in the consistent manner described in the GUM. The standard uncertainty in a quantity $y$ formed by combining $N$ measured quantities $x_i$ through some functional relationship $y = f(x_1, x_2, \ldots, x_N)$—the measurement equation—is given by

$$u^2(y) = \sum_{i=1}^{N} \left(\frac{\partial f}{\partial x_i}\right)^2 u^2(x_i) + \sum_{i=1}^{N} \sum_{j \neq i=1}^{N} \frac{\partial f}{\partial x_i} \frac{\partial f}{\partial x_j} u(x_i, x_j) \tag{6.1}$$

or

$$u^2(y) = \sum_{i=1}^{N} \left(\frac{\partial f}{\partial x_i}\right)^2 u^2(x_i) + 2 \sum_{i=1}^{N-1} \sum_{j=i+1}^{N} \frac{\partial f}{\partial x_i} \frac{\partial f}{\partial x_j} u(x_i, x_j) \tag{6.2}$$

where $u(x_i)$ is the standard uncertainty in $x_i$ and $u(x_i, x_j) = u(x_j, x_i)$ is the covariance of $x_i$ and $x_j$. (We also note that $u^2(x_i)$ is the variance of $x_i$.) This is a first-order approximation that applies in most circumstances. The term *standard uncertainty* is based on the assumption that the measured values, particularly after combining a number of effects, follow a normal probability distribution (the familiar bell-shaped curve) about the true value, or are taken to be so distributed. It is equivalent to the standard deviation of a quantity that is normally distributed. The assumption of a normal distribution is generally made in radiometry, particularly for the systematic effects whose uncertainty may be too difficult or time-consuming to estimate by repeat measurements.

It is often useful to describe correlations in terms of correlation coefficients. These are normalized covariances, defined by

$$r(x_1, x_2) = \frac{u(x_1, x_2)}{u(x_1)u(x_2)} \tag{6.3}$$

Rewriting Eq. (6.2) in terms of correlation coefficients gives

$$u^2(y) = \sum_{i=1}^{N} \left(\frac{\partial f}{\partial x_i}\right)^2 u^2(x_i) + 2 \sum_{i=1}^{N-1} \sum_{j=i+1}^{N} \frac{\partial f}{\partial x_i} \frac{\partial f}{\partial x_j} r(x_i, x_j) u(x_i) u(x_j) \tag{6.4}$$

There are two special cases of uncertainty propagation. The first is that of independent variables, i.e., their covariances are zero. Equations (6.1–6.4) then reduce to

$$u^2(y) = \sum_{i=1}^{N} \left(\frac{\partial f}{\partial x_i}\right)^2 u^2(x_i) \tag{6.5}$$

the familiar "sum-of-squares" commonly applied. Sum-of-squares is equivalent to adding variances. The second is that of completely correlated variables, for which $u(x_i, x_j) = \pm\, u(x_i)u(x_j)$, or $r(x_i, x_j) = \pm 1$. Where the sign of the correlation coefficient is the same for all pairs, Eq. (6.4) reduces to

$$u^2(y) = \left(\sum_{i=1}^{N} \frac{\partial f}{\partial x_i}\, u(x_i)\right)^2$$

or

$$u(y) = \left|\sum_{i=1}^{N} \frac{\partial f}{\partial x_i}\, u(x_i)\right| \tag{6.6}$$

In these cases, the standard uncertainty of the combination is found as a linear sum. The modulus is included because uncertainties are taken to be positive.

If the correlation coefficient is constant for all pairs, it can be removed outside the second term of Eq. (6.4), which then reduces to

$$u^2(y) = (1-r) \sum_{i=1}^{N} \left(\frac{\partial f}{\partial x_i}\right)^2 u^2(x_i) + r\left(\sum_{i=1}^{N} \frac{\partial f}{\partial x_i}\, u(x_i)\right)^2 \tag{6.7}$$

### 6.2.2 Sensitivity Coefficients

The derivatives $\partial f/\partial x_i$ in Eq. (6.1) are *sensitivity coefficients* for the dependence of $y$ on the various measured parameters $x_i$, reflecting the weight that a parameter has in the combined result. Sensitivity coefficients may be determined in a number of ways.

The relationship $y = f(x_1, x_2, \ldots, x_N)$ may be differentiated directly, especially for simple functions. A particularly useful technique in complex cases is that of implicit differentiation [2, 3]. As an example, consider the determination of the temperature of a blackbody radiator by measuring the photocurrent from a filtered detector whose spectral responsivity has been measured as a function of wavelength, under known geometric conditions [3]. The photocurrent $i$ can be expressed in the form

$$i = k \sum_{n} S_n L_n(T) \tag{6.8}$$

where $L_n$ is the spectral radiance of a blackbody at wavelength $\lambda_n$ for temperature $T$, $k$ a collection of various terms and $S_n$ the spectral responsivity of the detector, measured at regularly spaced wavelengths $\lambda_n$. The sensitivity of the temperature to each of the spectral values $S_n$ can be found by implicit differentiation [4] of the zero-valued function

$$f = i/k - \sum_n S_n L_n(T) \tag{6.9}$$

as

$$\frac{\partial T}{\partial S_n} = -\frac{\partial f}{\partial S_n} \bigg/ \frac{\partial f}{\partial T} = L_n(T) \bigg/ \frac{\partial f}{\partial T} \tag{6.10}$$

where

$$\frac{\partial f}{\partial T} = -\sum_n S_n \frac{\partial L_n(T)}{\partial T} \tag{6.11}$$

This method is also applied to find the sensitivity of the temperature with respect to each of the measurements forming the quantity $k$.

Numerical differentiation is useful in complex cases. The measurement system can be modeled in terms of a given parameter (e.g. the temperature of the blackbody in the previous example) and sensitivity determined by varying the parameter in a given range about the measured value.

Finally, direct measurement of the sensitivity may be made. For example, we can measure the response of a photodetector at different temperatures, and then calculate the sensitivity as the ratio of the responsivity difference to the temperature difference. We can then apply the sensitivity both to correct measurements made where temperature has varied from reported conditions and to estimate the uncertainty of that correction from the range of temperatures encountered during a particular response measurement.

### 6.2.3 Covariance

Fully correlated components have a correlation coefficient of $+1$ or $-1$. Uncorrelated components have a correlation coefficient of zero. The covariance (and hence correlation coefficient) of two quantities is non-zero if they in turn depend on one or more common parameters. The two quantities vary in a related manner when a common parameter varies. If $x_1$ and $x_2$ are our quantities and $p$ is a common parameter, the covariance of $x_1$ and $x_2$ due to $p$ is given by [1, Eq. F.2]

$$u(x_1, x_2) = \frac{\partial x_1}{\partial p} \frac{\partial x_2}{\partial p} u^2(p) \tag{6.12}$$

If our variables depend on two common parameters $p_1$ and $p_2$ which may be correlated ($u(p_1, p_2) \neq 0$), the covariance becomes

$$u(x_1, x_2) = \frac{\partial x_1}{\partial p_1}\frac{\partial x_2}{\partial p_1}u^2(p_1) + \frac{\partial x_1}{\partial p_2}\frac{\partial x_2}{\partial p_2}u^2(p_2) + \left(\frac{\partial x_1}{\partial p_1}\frac{\partial x_2}{\partial p_2} + \frac{\partial x_2}{\partial p_1}\frac{\partial x_1}{\partial p_2}\right)u(p_1, p_2) \tag{6.13}$$

or, in matrix form,

$$u(x_1, x_2) = \begin{bmatrix} \dfrac{\partial x_1}{\partial p_1} & \dfrac{\partial x_1}{\partial p_2} \end{bmatrix} \begin{bmatrix} u^2(p_1) & u(p_1, p_2) \\ u(p_1, p_2) & u^2(p_2) \end{bmatrix} \begin{bmatrix} \dfrac{\partial x_2}{\partial p_1} \\ \dfrac{\partial x_2}{\partial p_2} \end{bmatrix} \tag{6.14}$$

Complex correlations between multiple variables are best expressed in matrix terms, as shown in the next section. Note that if $x_1 = f(p_1)$ and $x_2 = f(p_2)$, as in the example given in Section 6.4.1, two of the sensitivity coefficients are zero and Eq. (6.14) reduces to

$$u(x_1, x_2) = \frac{\partial x_1}{\partial p_1}\frac{\partial x_2}{\partial p_2}u(p_1, p_2) \tag{6.15}$$

Correlation coefficients are formed by applying the normalization of Eq. (6.3).

Uncertainty propagation is more complicated where the measured quantities are correlated. If we consider the complete measurement equation, we can often cancel terms that would otherwise introduce correlation. A common example is the calibration of a meter used to measure ratios of photocurrents. Suppose we measure the voltage output of two photodetectors with two amplifiers of current-to-voltage gains $g_1$ and $g_2$. The ratio of the photocurrents is given by

$$i_{\text{ratio}} = \frac{V_1}{V_2}\frac{g_2}{g_1} \tag{6.16}$$

If we propagate uncertainties from this equation, we need to include correlations between the readings of the voltmeter (through a common calibration factor), between the gain values (through a likely common calibration resistor) and between the gain values and the voltage values (if the same voltmeter was used to calibrate the gains). Each voltage value is a product of reading $R$ and calibration factor $k$, common if the same range is used or the meter is linear across ranges. Similarly, if the same system was used to calibrate the gains of the amplifiers, the gain ratio $g_R$ is independent of the voltage readings. Hence $i_{\text{ratio}} = g_R R_1/R_2$, where the readings are now random and uncorrelated. Many such simplifications are implicitly present in good radiometric measurement practice.

### 6.2.4 Matrix Form of Uncertainty Propagation

Given that $u^2(x_i) \equiv u(x_i, x_i)$, Eq. (6.1) can be expressed as

$$u^2(y) = \mathbf{f}_x \mathbf{U}_x \mathbf{f}_x^{\mathbf{T}} \tag{6.17}$$

where

$$\mathbf{f}_x = \left( \frac{\partial f}{\partial x_1} \frac{\partial f}{\partial x_2} \cdots \frac{\partial f}{\partial x_n} \right) \tag{6.18}$$

is a row vector of sensitivity coefficients (superscript **T** indicates the transpose) and $\mathbf{U}_x = (u(x_i, x_j))$ is the symmetric $N \times N$ variance–covariance matrix with variance of the input quantities in diagonal elements, and the covariance between input quantities elsewhere. We can propagate uncertainty from input quantities, held in the variance–covariance matrix, to any output quantity formed by combining them if we know the sensitivity factors. The covariance of any two quantities $y_1, y_2$ that depend on the same set of input quantities is found by including two sensitivity vectors in Eq. (6.17) ([1, Eq. H.9], written in terms of covariance),

$$u(y_1, y_2) = \mathbf{f}_x \mathbf{U}_x \mathbf{g}_x^{\mathbf{T}} \tag{6.19}$$

where $y_1 = f(x_1, x_2, \ldots, x_N)$ and $y_2 = g(x_1, x_2, \ldots, x_N)$ have different functional dependencies on the input quantities. In the remaining sections, no distinction will be made between variance and covariance. Variance is taken as the covariance of a quantity with itself and Eq. (6.19) is used to calculate uncertainty as the square root of the variance found by setting the two sensitivity vectors to be the same.

For uncertainty propagation through a series of multiplications it is sometimes convenient to use the relative variance–covariance matrix, one containing values of relative variance and relative covariance.

### 6.2.5 Relative Uncertainty

Relative uncertainty is the ratio of uncertainty to the absolute value of a quantity,

$$u_{\text{rel}}(x) = \left| \frac{u(x)}{x} \right| \tag{6.20}$$

It is often expressed as a percentage, although this can lead to confusion where a value is itself expressed as a percentage (a common occurrence in radiometry). Results are often calculated as a series of linear multiplications and divisions. In such cases where the terms are uncorrelated, it is readily shown from Eq. (6.1) that the square of the relative uncertainty in the result is found by adding the square of the relative uncertainties of each of the terms. In non-linear combinations where a quantity appears to a particular power, each relative uncertainty is multiplied by that power. As an example, consider the determination of the luminous intensity $I_V$ of a point source by measuring its illuminance $E_V$ at a distance $d$, using the relationship $I_V = E_V/d^2$. By calculating sensitivity coefficients and propagating the uncertainty, we can show that

$$u_{rel}^2(I_V) = u_{rel}^2(E_V) + [2u_{rel}(d)]^2 \tag{6.21}$$

Most uncertainties in radiometry are quoted in relative terms. Note that relative uncertainty has little meaning for a near-zero quantity, such as spectral response at wavelengths far removed from the central wavelength of a bandpass-filtered detector.

### 6.2.6 Relative Covariance

Relative covariance is given by

$$u_{rel}(x_1, x_2) = \frac{u(x_1, x_2)}{x_1 x_2} \tag{6.22}$$

It is analogous to the relative uncertainty of Eq. (6.20), or more correctly, relative variance. The correlation coefficient of $x_1$ and $x_2$, Eq. (6.3), can be given from the relative covariance as

$$r(x_1, x_2) = \frac{u_{rel}(x_1, x_2)}{u_{rel}(x_1) u_{rel}(x_2)} \tag{6.23}$$

### 6.2.7 Propagation by Component

Equations (6.5) and (6.6) can often be used to advantage, especially when combining spectral data at different wavelengths. This occurs if we separately consider each independent effect on a measurement, propagating the uncertainty completely through the final required quantity. In Table 6.1 we list the uncertainty components that arise from various independent effects, each influencing the uncertainty of our quantities $x_i$ and subsequently the uncertainty of their combination through the measurement equation $y = f(x_1, x_2, \ldots, x_n)$. Because the effects are independent, the combined uncertainty in each quantity $x_i$ is given as a quadratic sum, shown in the right column. Similarly, once we have found the uncertainties in $y$ caused by each particular effect, we can combine them in a quadratic sum, as shown in the last row. However, if we propagate the uncertainties for all the effects

TABLE 6.1. Uncertainties of Independent Effects (Columns) on Quantities (Rows) that are Subsequently Combined through the Measurement Equation $y = f(x_1, x_2, \ldots, x_n)$

| Quantity | Effect 1 | Effect 2 | Effect 3 | | Effect $k$ | Combined |
|---|---|---|---|---|---|---|
| $x_1$ | $u_{1,1}$ | $u_{1,2}$ | $u_{1,3}$ | $\cdots$ | $u_{1,k}$ | $u^2(x_1) = \sum_{i=1}^{k} u_{1,i}^2$ |
| $x_2$ | $u_{2,1}$ | $u_{2,2}$ | $u_{2,3}$ | $\cdots$ | $u_{2,k}$ | $u^2(x_2) = \sum_{i=1}^{k} u_{2,i}^2$ |
| $\vdots$ | | | | | | |
| $x_n$ | $u_{n,1}$ | $u_{n,2}$ | $u_{n,3}$ | $\cdots$ | $u_{n,k}$ | $u^2(x_n) = \sum_{i=1}^{k} u_{n,i}^2$ |
| Combined | $u_1(y)$ | $u_2(y)$ | $u_3(y)$ | | $u_k(y)$ | $u^2(y) = \sum_{i=1}^{k} u_i^2(y)$ |

through each value $x_i$, these $x_i$ values in general will be partially correlated because some of the independent effects will be systematic over the set and some will be random. Then the full double summation of Eq. (6.2) will be required to form the final uncertainty in $y$.

The calculation method for the combined uncertainty through the measurement relation for a particular effect depends on whether the resultant uncertainties in the quantities $x_i$ for that effect are uncorrelated, partially correlated, or completely correlated.

- Partially correlated effects must be treated using the full sums of Eq. (6.1). Spectral reference values at different wavelengths are generally partially correlated through the means of establishing the primary standards. Interpolated spectral values will be partially correlated to each other and to the original values. Neither of these cases can be calculated with the simpler Eqs. (6.5) or (6.6), unless the correlation coefficient $r$ is constant. Then both equations can be used, with variances combined in the ratios $(1-r)$ and $r$, respectively, as shown in Eq. (6.7).
- Uncorrelated effects, such as the random noise of an amplifier, are combined using Eq. (6.5).
- Fully correlated effects arise when a common factor influences all values. If effect $k$ of Table 6.1 influences all values of $x_i$, the magnitude of the covariance $u(x_i, x_j)$ is given by $u_{k,i}u_{k,j}$. This follows from Eq. (6.12) for a single influence parameter. If more than one independent parameter influences the uncertainty we can treat them as separate effects. As we are considering effect $k$ alone to be contributing to the uncertainty, the total uncertainty in $x_i$ is $u_{k,i}$ and that in $x_j$ is $u_{k,j}$. Hence we have

$|u(x_i, x_j)| = u(x_i)u(x_j)$, which is the condition for complete correlation and possible application of Eq. (6.6) to find the combined uncertainty for this effect. Note that for combinations of many quantities $x_i$ we must consider the sign of the correlation, not only its magnitude. We shall see in Section 6.4.5 that there is a special class of spectral combinations, important in radiometry, where the correlation coefficients can take values $+1$ and $-1$ for different pairs but we can still use a linear sum provided we consider the uncertainty to carry the sign of the sensitivity coefficient.

Most individual effects will be either uncorrelated or fully correlated over the measurement set $x_i$ and Eqs. (6.5) and (6.6) are used to propagate their uncertainty contribution through the measurement relation.

One clear advantage of propagation by component is that the contribution of each effect to the uncertainty of the final desired quantity is delineated. Then efforts can be made to improve the measurement process to reduce the uncertainties of the dominant components. Those effects that dominate the uncertainty of the spectral quantity itself at the different wavelengths are not necessarily those that dominate a combination of those spectral values. A simple example is that in colorimetry, where factors that contribute to the uncertainty of the absolute value of the spectral irradiance of a source do not contribute to the uncertainty of $(x, y)$ chromaticity values.

### 6.2.8 Propagation of Covariance

Covariance of two quantities may change as we add/subtract a common offset or multiply/divide them by a common factor. If the factors are not constant, but carry uncertainty, quantities which were originally uncorrelated become correlated by these operations. These changes are analogous to the propagation of uncertainty through the same operations. We will consider the following operations on two quantities $x_1$ and $x_2$ with covariance $u(x_1, x_2)$ and forming new quantities $y_1$ and $y_2$. All results follow from Eq. (6.13). The addition or multiplication quantities $c$ are assumed not correlated to the original quantities ($u(c, x_1) = u(c, x_2) = 0$).

(a) Addition of a constant

$$y_1 = x_1 + c, \qquad y_2 = x_2 + c$$
$$u^2(y_1) = u^2(x_1), \quad u^2(y_2) = u^2(x_2), \quad u(y_1, y_2) = u(x_1, x_2)$$

Uncertainty and covariance remain unchanged.

(b) Addition of a quantity

$$y_1 = x_1 + c, \qquad y_2 = x_2 + c$$
$$u^2(y_1) = u^2(x_1) + u^2(c), \quad u^2(y_2) = u^2(x_2) + u^2(c), \quad u(y_1, y_2) = u(x_1, x_2) + u^2(c)$$

The increase in covariance is the same as that of the variance of each of the quantities.

(c) Addition of correlated quantities

$$y_1 = x_1 + c_1, \qquad y_2 = x_2 + c_2$$
$$u^2(y_1) = u^2(x_1) + u^2(c_1), \quad u^2(y_2) = u^2(x_2) + u^2(c_2), \quad u(y_1, y_2) = u(x_1, x_2) + u(c_1, c_2)$$

The covariance of the new values is also increased, but by a different amount compared to the last example.

(d) Multiply/divide by a constant

$$y_1 = cx_1, \qquad y_2 = cx_2$$
$$u(y_1) = cu(x_1), \quad u(y_2) = cu(x_2), \quad u(y_1, y_2) = c^2 u(x_1, x_2)$$

The relative covariance and relative uncertainties remain unchanged.

(e) Multiply/divide by a quantity. Such scaling of quantities by various calibration factors is common in radiometry.

$$y_1 = cx_1, \qquad y_2 = cx_2$$
$$u_{\text{rel}}^2(y_1) = u_{\text{rel}}^2(x_1) + u_{\text{rel}}^2(c), \quad u_{\text{rel}}^2(y_2) = u_{\text{rel}}^2(x_2) + u_{\text{rel}}^2(c), \quad u(y_1, y_2) = c^2 u(x_1, x_2) + x_1 x_2 u^2(c)$$

The relative covariance is $u_{\text{rel}}(y_1, y_2) = u_{\text{rel}}(x_1, x_2) + u_{\text{rel}}^2(c)$.

(f) Multiply/divide by correlated quantities

$$y_1 = c_1 x_1, \qquad y_2 = c_2 x_2$$
$$u_{\text{rel}}^2(y_1) = u_{\text{rel}}^2(x_1) + u_{\text{rel}}^2(c_1), \quad u_{\text{rel}}^2(y_2) = u_{\text{rel}}^2(x_2) + u_{\text{rel}}^2(c_2), \quad u(y_1, y_2) = c_1 c_2 u(x_1, x_2) + x_1 x_2 u(c_1, c_2)$$

The relative covariance is $u_{\text{rel}}(y_1, y_2) = u_{\text{rel}}(x_1, x_2) + u_{\text{rel}}(c_1, c_2)$.

From these examples we see that relative covariance propagates in the same way as relative variance. However, covariance can be positive or negative and hence the relative covariance between two quantities can decrease as well as increase in a given operation.

## 6.3 Uncertainty in a Single Quantity

Many radiometric quantities are reported as a single value that is unlikely to be combined with others. Response of a photometer is one such example. In such cases the uncertainty must be estimated, but possible correlations with other quantities is not of interest. (This statement can mislead, as those providing a measurement may not know its intended use. For example, the response values of two photometers calibrated separately and then used as transfer standards in an inter-laboratory comparison are correlated through common elements in the calibration process.) In this section, we deal with quantities formed by combining a small number of measured or estimated components. Each component can be treated as uncorrelated, fully correlated, or partially correlated to the others, and their uncertainties are propagated through the measurement equation to obtain the standard uncertainty of the combination.

### 6.3.1 Ratio Example

As an example consider the ratio $t$ of two quantities $y_1$ and $y_2$, formed by measuring signals $x_1$, $x_2$ and background signals $b_1$, $b_2$, all dominated by random effects. The ratio is

$$t = \frac{y_1}{y_2} = \frac{x_1 - b_1}{x_2 - b_2} \tag{6.24}$$

and, as the readings are uncorrelated, the relative uncertainty in the ratio is given by

$$u_{\mathrm{rel}}^2(t) = \frac{u^2(x_1) + u^2(b_1)}{y_1^2} + \frac{u^2(x_2) + u^2(b_2)}{y_2^2} \tag{6.25}$$

Now consider the effect of a constant offset $e$ in the background (uncorrelated to the other measurements), possibly caused by reflection of a room light from the rear of the shutter used to block the main signal beams. The signals now become $y_1 = x_1 - b_1 - e$, $y_2 = x_2 - b_2 - e$. If $e$ is significant a correction must be made, and the uncertainty of that correction included in the analysis. We will consider the case $e \approx 0$ and treat the uncertainty analysis with three separate methods.

#### *6.3.1.1 Method 1: Propagation through the expanded measurement equation*

The expanded measurement equation is now

$$t = \frac{x_1 - b_1 - e}{x_2 - b_2 - e} \tag{6.26}$$

The sensitivity coefficient $\partial t/\partial e$ is $((1/y_2) - (y_1/y_2^2))$. The readings are uncorrelated; using Eq. (6.5) and dividing by $t^2$ we can show

$$u_{\mathrm{rel}}^2(t) = \frac{u^2(x_1) + u^2(b_1)}{y_1^2} + \frac{u^2(x_2) + u^2(b_2)}{y_2^2} + \left(\frac{1}{y_1} - \frac{1}{y_2}\right)^2 u^2(e) \tag{6.27}$$

#### *6.3.1.2 Method 2: Addition of a correlated component*

The relative uncertainty in the ratio $t$ from the signal components $x_1$, $x_2$, $b_1$, and $b_2$ remains that shown in Eq. (6.25). The offset $e$ introduces a correlation between $y_1$ and $y_2$, with a covariance

$$u(y_1, y_2) = \frac{\partial y_1}{\partial e}\frac{\partial y_2}{\partial e}\,u^2(e) = u^2(e) \tag{6.28}$$

If $e$ was the only term contributing to the uncertainties of $y_1$ and $y_2$, and hence of the ratio, the correlation coefficient of $y_1$ and $y_2$ would be

$$r(y_1, y_2) = \frac{u(y_1, y_2)}{u(y_1)u(y_2)} = \frac{u^2(e)}{u(e)u(e)} = 1 \tag{6.29}$$

i.e., $y_1$ and $y_2$ are fully correlated for this effect and we can directly determine the uncertainty as a linear sum. The sensitivity coefficients for $t$ are

$$\frac{\partial t}{\partial y_1} = \frac{1}{y_2}, \qquad \frac{\partial t}{\partial y_2} = -\frac{y_1}{y_2^2} \tag{6.30}$$

and the resultant uncertainty (for this component alone) from Eq. (6.6) is

$$u(t)|_e = \left(\frac{1}{y_2} - \frac{y_1}{y_2^2}\right)u(e) \tag{6.31}$$

This expands to

$$u_{\text{rel}}^2(t)|_e = \left(\frac{1}{y_1} - \frac{1}{y_2}\right)^2 u^2(e) \tag{6.32}$$

identical to the term in Eq. (6.27) and adds to Eq. (6.25). While a relatively simple exercise here where only two terms are combined, this separation of systematic effects into separate components can be a powerful tool in more complicated spectral combinations.

#### *6.3.1.3 Method 3: Partial correlation*

Partially correlated quantities require the covariance to be estimated. As shown in the last section, the covariance between $y_1$ and $y_2$ is $u^2(e)$. From Eq. (6.2) and converting to relative uncertainty, we have

$$u_{\text{rel}}^2(t) = \frac{u^2(x_1) + u^2(b_1) + u^2(e)}{y_1^2} + \frac{u^2(x_2) + u^2(b_2) + u^2(e)}{y_2^2} + 2\left(\frac{1}{y_2}\right)\left(\frac{-y_1}{y_2^2}\right)\left(\frac{y_2}{y_1}\right)^2 u^2(e) \tag{6.33}$$

Collecting the terms in $u^2(e)$ yields

$$\left(\frac{1}{y_1^2}+\frac{1}{y_2^2}-\frac{2}{y_1 y_2}\right)u^2(e)=\left(\frac{1}{y_1}-\frac{1}{y_2}\right)^2 u^2(e) \tag{6.34}$$

again equal to the additional term of Eq. (6.27).

Addition of a correlated component is the simplest way of treating the extra effect.

As another example of partial correlation, consider that of an average value formed by measuring a device against two reference artifacts calibrated against a common reference. We can write the two values as

$$y_1 = kt_1, \qquad y_2 = kt_2 \tag{6.35}$$

with $y_1 \approx y_2$ and uncertainties $u(y_1) \approx u(y_2) \approx u(y)$. Here $k$ is the value of the common reference artifact with uncertainty $u(k)$, and $t_1, t_2$ are the total (uncorrelated) transfer ratios with uncertainties $u(t_1) \approx u(t_2) \approx u(t)$. The covariance of $y_1$ and $y_2$ is

$$u(y_1, y_2) = \frac{\partial y_1}{\partial k}\frac{\partial y_2}{\partial k}\, u^2(k) = t^2 u^2(k) \tag{6.36}$$

It follows that the relative uncertainty in the mean $m$ is

$$u_{\text{rel}}^2(m) = \tfrac{1}{2}\left(u_{\text{rel}}^2(y) + u_{\text{rel}}^2(k)\right) \tag{6.37}$$

If the relative uncertainty in the common reference is dominant, $u_{\text{rel}}(y) \approx u_{\text{rel}}(k)$ and the uncertainty is unchanged by averaging over the two measurements. This example can also be treated by averaging the two transfer uncertainties $y_1/k$ and $y_2/k$ as uncorrelated, then recombining that result with the uncertainty of the common reference.

### 6.3.2 Propagation of Uncertainty through Combining Fully Correlated Quantities

Propagation of uncertainty is simpler when combining fully correlated quantities. Method 2 above shows the advantage of dealing with systematic effects by component, arising because the quantities being combined are fully correlated. Here we treat operations commonly applied to derive a quantity from measurements containing a common systematic component, where the measurements are fully correlated with a correlation coefficient $r = \pm 1$. Most systematic effects, treated singly, have a correlation coefficient of $+1$, but values of $-1$ also occur. In all cases we consider the quantity $y$ formed by combining $x_1$ and $x_2$, with $u(x_1, x_2) = ru(x_1)u(x_2)$.

(a) Addition of fully correlated quantities

$$y = x_1 + x_2 \tag{6.38}$$

Sensitivity coefficients for the dependence of $y$ on $x_1$ and $x_2$ are both 1, and hence from Eq. (6.2),

$$u^2(y) = u^2(x_1) + u^2(x_2) + 2ru(x_1)u(x_2) \tag{6.39}$$

or

$$u(y) = |u(x_1) + ru(x_2)| \tag{6.40}$$

(b) Subtraction of fully correlated quantities

$$y = x_1 - x_2 \tag{6.41}$$

The sensitivity coefficients now become +1, −1 respectively, with the result

$$u(y) = |u(x_1) - ru(x_2)| \tag{6.42}$$

These results are important for estimating the uncertainty contribution of systematic components in linear spectral sums, as shown in Section 6.4.5. Note that in contrast to propagation of random uncertainty components, uncertainty can decrease depending on the sign of the correlation.

(c) Multiplication of fully correlated quantities

$$y = x_1 x_2 \tag{6.43}$$

Sensitivity coefficients for the dependence of $y$ on $x_1$ and $x_2$ are $x_2$ and $x_1$, respectively. Hence

$$u^2(y) = x_2^2 u^2(x_1) + x_1^2 u^2(x_2) + 2rx_1x_2u(x_1)u(x_2) \tag{6.44}$$

or

$$u_{\text{rel}}(y) = |u_{\text{rel}}(x_1) + ru_{\text{rel}}(x_2)| \tag{6.45}$$

(d) Division of fully correlated quantities

$$y = \frac{x_1}{x_2} \tag{6.46}$$

Sensitivity coefficients for the dependence of $y$ on $x_1$ and $x_2$ are $1/x_2$ and $-x_1/x_2^2$, respectively, and it follows that

$$u_{\text{rel}}(y) = |u_{\text{rel}}(x_1) - ru_{\text{rel}}(x_2)| \tag{6.47}$$

Again, relative uncertainty in a product or ratio can decrease depending on the sign of the correlation. Most systematic errors in radiometry (offsets and scaling factors) have positive correlation for quantities being combined, usually as ratios, and uncertainties in the ratio decrease. Where the systematic factor is one of a scaling, i.e., has a constant relative uncertainty, its contribution to the uncertainty of a ratio is zero.

## 6.4 Uncertainty Across a Spectrum of Quantities

The treatment of uncertainty across a spectrum of quantities is in principle identical to that for a single quantity. We treat this as a separate topic because the spectral integrals encountered in radiometry—photometric response, the tristimulus values of colorimetry, response of a filtered detector to a broadband source of radiation are examples—can combine a large number of similar quantities measured at different wavelengths. Reference spectral responsivity standards are derived at a limited number of wavelengths then fitted to a model or interpolated to provide reference values at a greater number of wavelengths. The measured values will have common systematic errors in measurements made at different wavelengths. Reference irradiance standards may be derived in a similar manner. Reference spectral radiance and irradiance standards are often obtained from a single value of the temperature of a blackbody radiator. In all of these cases, the reference quantities obtained at different wavelengths are at least partially correlated. Uncertainty values propagated from these reference standards should include these correlations.

### 6.4.1 Common Systematic Components

Suppose we make a number of measurements under similar conditions. The example we will take is that of the spectral responsivity of a silicon photodiode or trap detector using a cryogenic radiometer at a number of laser wavelengths. Each of these measurements contains a number of uncertainty components. Table 6.2 lists some components which may be present and identifies those common components that we take as systematic at all wavelengths from those which are specific to a particular wavelength. By systematic, we mean that not only is the value of the uncertainty common at all wavelengths, it is fully correlated. All results are affected in the same way if the cavity absorbance is in error, for example. A higher value of absorbance, within its probability distribution, means that all measured power levels would be lower, independent of wavelength. Power stability, for example, may be significantly different if lasers of different power levels are available—the assigned uncertainty of the component is not common to all wavelengths—but its variation will be random between wavelengths. We may also decide, having determined the window transmittance at each

TABLE 6.2. Typical Uncertainty Components for Calibration of the Responsivity of a Silicon Photodetector Against a Cryogenic Radiometer Reference. Components Common to Measured Values at All Wavelengths (Systematic Components) are Identified

| Component | Common? |
|---|---|
| Cavity absorbance | Yes |
| Window transmittance | |
| Electrical power in the cavity | Yes |
| Repeatability with beam position | Yes |
| Power stability (measurement noise) | |
| Detector amplifier gain | Yes |

wavelength, that its value was constant within an acceptable uncertainty and flag it as a common component. In taking the window transmittance to be independent at each wavelength we have also assumed that systematic effects in its measurement at different wavelengths are zero. Each of the components in Table 6.2 is taken to be independent of the others in determining the responsivity at a particular wavelength.

The components common to all wavelengths introduce correlations between responsivity values at different wavelengths. We can determine the resultant covariance most simply by collecting the correlated and non-correlated terms at each wavelength. Most systematic errors in radiometry lead to a positive correlation between values in a data set. (One exception is a systematic wavelength offset in spectral data—this is dealt with in Section 6.5.3.) Provided all correlations are positive and offsets are negligible, we write the spectral responsivity at the $i$th wavelength in the form $S_i = a_i b$, where $b$ is a collection of the measurements and corrections for components common to all wavelengths and $a_i$ is a collection of those that are not. Equation (6.5) is used to determine the uncertainty of each of these collections. The $a_i$ and $b$ components are independent and so $u_{\mathrm{rel}}^2(S_i) = u_{\mathrm{rel}}^2(a_i) + u_{\mathrm{rel}}^2(b)$. From Eq. (6.12) we have the covariance of spectral responsivity values for wavelengths $i$ and $j$ as $u(S_i, S_j) = a_i a_j u^2(b)$ and hence $u_{\mathrm{rel}}(S_i, S_j) = u_{\mathrm{rel}}^2(b)$ is constant (but the correlation coefficient is not). This is applied when propagating uncertainty through combinations of values of spectral responsivity at different wavelengths. The correlation coefficient $r(S_i, S_j) = u_{\mathrm{rel}}^2(b)/u_{\mathrm{rel}}(S_i)u_{\mathrm{rel}}(S_j)$ tends to the value $+1$ as the common components dominate.

In the primary laboratories, spectral responsivity values may be combined with external loss (reflectance or transmittance) measurements to determine the spectral quantum efficiency of silicon photodiodes, either directly as the internal quantum efficiency or as the quantum defect. These quantities are then fitted to a model so that spectral responsivity values can be generated at wavelengths other than those used for direct measurement with the cryogenic radiometer [5]. Proper fitting requires the correlation of the fitted quantities to be taken into account [6]. The correlation of quantum defect or internal quantum efficiency values is obtained from that of the spectral responsivity values using the methods of Section 6.2.8. If we take the values of spectral responsivity $S_i$ to be those corrected for radiation lost by reflection, the internal quantum efficiency $n_i$ and quantum defect $q_i$ at wavelength $\lambda_i$ are given by

$$n_i = 1 - q_i = S_i \frac{hc}{\lambda_i} \tag{6.48}$$

where $h$ is Planck's constant and $c$ the velocity of light. The covariance (Section 6.2.3) between these quantities at the $i$th and $j$th wavelengths is then

$$u(n_i, n_j) = u(q_i, q_j) = \frac{(hc)^2}{\lambda_i \lambda_j} u(S_i, S_j) \tag{6.49}$$

### 6.4.2 Fitting

Fitting is a process to estimate the parameters of a function that represents the relationship between a limited set of input quantities, one dependent and the other independent (usually representing wavelength in radiometry). The function is then used to calculate the dependent value at other independent values. Uncertainty propagation through Eq. (6.17) is applied to each of these stages. The parameters are correlated because they depend on a common set of input measurements. Values calculated using the fitted function are correlated because they depend on a common set of parameters.

Suppose that we have a set of $N$ values $y_i$ each dependent on its wavelength $\lambda_i$, and a set of $P$ parameters $a_j$, where the values $y_i$ are assumed to be represented by the function

$$y_i' = f(a_1 \ldots a_P; \lambda_i) \tag{6.50}$$

Estimates of the values of the parameters $a_j$ are found by least-squares minimization of

$$\chi^2 = \sum_i (y_i' - y_i)^2/(N - P) \tag{6.51}$$

Once the parameter estimates are found, it is possible to calculate the sensitivity coefficients for the dependence of each parameter $a_j$ on each of the input quantities $y_i$. Hence the variance–covariance matrix $\mathbf{U}_a = (u(a_i, a_j))$ of the parameters (correlated through dependence on the common set of input quantities) is calculated through Eq. (6.19).

The minimization is improved by weighting the terms of Eq. (6.51) by the inverse of the variance of the $y_i$ values. A complication here is that these input quantities may be correlated through systematic errors in the measurement process, or through a common dependence on other quantities. Woeger [7] has described a technique that propagates uncertainties including input correlations, with Eq. (6.51) redefined as

$$\chi^2 = \mathbf{M}^{\mathrm{T}}\mathbf{U}_y^{-1}\mathbf{M}/(N - P) \tag{6.52}$$

where $\mathbf{M}$ is a vector of the differences between calculated $y_i'$ and measured $y_i$ values (the terms summed in the numerator of Eq. (6.51)).

Using the fitted parameters, we can now calculate the value for the quantity $y_k$ (we have dropped the prime for convenience) at wavelength $\lambda_k$ through Eq. (6.50) and calculate its standard uncertainty through applying Eq. (6.17),

$$u^2(y_k) = \left(\frac{\partial f}{\partial a_i}|_{\lambda_k}\right)\mathbf{U}_{\mathrm{a}}\left(\frac{\partial f}{\partial a_i}|_{\lambda_k}\right)^{\mathrm{T}} \tag{6.53}$$

Equation (6.53) gives the uncertainty propagated from the input, independent of the quality of the fit. The "goodness" of the fit is represented by the value of $\chi^2$ of Eq. (6.52), in which the $(N-P)$ term is the number of degrees of freedom. The parameter $\chi^2$ for a weighted fit (with or without the covariance terms included) is a measure of the average ratio of the variance of the fitted value to the variance of its input value for $\lambda_k = \lambda_i$; for a set of data well described by the chosen fitting function, $\chi^2 \approx 1$ [8]. We can include uncertainty due to the choice of fitting function by expanding the variance calculated in Eq. (6.53) as

$$u^2(y_k) = (1 + \chi^2)\left(\frac{\partial f}{\partial a_i}|_{\lambda_k}\right)\mathbf{U}_{\mathrm{a}}\left(\frac{\partial f}{\partial a_i}|_{\lambda_k}\right)^{\mathrm{T}} \tag{6.54}$$

This expansion overestimates the variance where the function is a good representation of the data ($\chi^2 \approx 1$), but it does partly address the problem of a poor fit. It however is an averaging process and does not take into account any obvious structure, or correlation, between the fitted and calculated points over the wavelength range. The expanded uncertainty is not a substitute for a proper fit function. Note that for an unweighted fit, $\chi^2$ has dimensions equivalent to variance, and in such a case its value is added directly to the propagated variance to account for the goodness-of-fit; where the input uncertainty values are not known and the model is assumed to be accurate, $\chi^2$ itself is taken as a measure of the variance of the input points [8].

The covariance of any two calculated values $y_k$, $y_m$ at wavelengths $\lambda_k$, $\lambda_m$, propagated from that of the parameters, is given from Eq. (6.19):

$$u(y_k, y_m) = \left(\frac{\partial f}{\partial a_i}|_{\lambda_k}\right)\mathbf{U}_\mathrm{a}\left(\frac{\partial f}{\partial a_i}|_{\lambda_m}\right)^\mathrm{T} \tag{6.55}$$

In expanding the uncertainty to account for the goodness-of-fit in Eq. (6.54), we have effectively added correlated quantities with value zero to the calculated values $y_k$ and $y_m$. It follows from Section 6.2.8 that we must also expand the covariance as

$$u(y_k, y_m) = (1 + \chi^2)\left(\frac{\partial f}{\partial a_i}|_{\lambda_k}\right)\mathbf{U}_\mathrm{a}\left(\frac{\partial f}{\partial a_i}|_{\lambda_m}\right)^\mathrm{T} \tag{6.56}$$

and the correlations of the propagated fit are maintained. Equations (6.54) and (6.56) provide the complete variance-covariance matrix for values of a spectral distribution obtained by fitting.

Values of the parameters may be little different if correlations between the input values are included or not in the fitting. However, the correlations between the parameters may be significantly different, and this in turn influences both the uncertainties and the correlations of calculated values [6, 9].

The covariance matrix $\mathbf{U}_y$ in Eq. (6.52) may be ill-conditioned, especially for data points with highly correlated uncertainties. Rounding errors then can accumulate in routines used to calculate the inverse matrix, which is used to calculate uncertainties in the parameters [7]. Hence caution should be applied when propagating uncertainties through the fitting process by component, for systematic effects.

### 6.4.3 Interpolation

Where data are available at sufficient points to define a distribution known (or suspected) to be smooth, direct interpolation is an alternative to fitting. This is discussed in detail elsewhere [10]. Interpolation provides a mathematical means to estimate new values from a set of input values. The parameters of the interpolation formula within a given range depend on a common set of input values, and are thus correlated. Sensitivity coefficients for the dependence of the new values on the input values can be estimated either numerically or by differentiation of the relationship defining them in terms of the original values. The interpolated values in a given range depend on the common interpolation parameters for the range, and hence are correlated. Uncertainties of the new values, and correlations between them, can be propagated through Eq. (6.19).

Interpolation is often used when a limited set of data are available to be convolved with some specified function and integrated. Interpolation is used to improve the accuracy of the integral when the function changes significantly in the region between the measured data points. Uncertainty propagation through this process requires that the correlations introduced by the fitting process be properly taken into account. A conclusion of Reference [10] is that uncertainties are correctly estimated by propagating them through the integration using the function values interpolated to the points where the data are measured.

### 6.4.4 Propagation of Uncertainty through a Spectral Transfer

Once reference standards are established at different wavelengths, spectral comparison is used to transfer values to secondary or working standards, and from those along a chain leading to the final calibration of a device whose spectral properties are to be measured. The spectral comparison occurs wavelength by wavelength. In this section, we wish to estimate the uncertainties and covariances of the measured spectral data, with contributions from both the reference values and the transfer process. This information is not necessary if the measurement is the first stage of a calculation combining the spectral values—then it is simpler to calculate the effect of correlations in the reference standard and the effect of random and systematic effects of the transfer for the combination by component, as in Section 6.2.7. However, a calibrated artifact at one level may be the reference artifact for the next lower level, and in such a case we need to know the uncertainties and correlations of the spectral data after the transfer.

In general terms the transfer process can be written as

$$Q_i = t_i R_i \tag{6.57}$$

where $Q_i$ represents the transferred spectral quantity at the $i$th wavelength, $R_i$ the reference value and $t_i$ the transfer ratio. The transfer itself is of the form

$$t_i = y_1/y_2 \tag{6.58}$$

where at each wavelength, $y_1$ and $y_2$ are background-corrected signals for the artifact under test and the reference artifact, respectively. Individual independent effects contributing to the uncertainty of the transfer are modeled and their uncertainties combined (including correlations between the two signal channels if necessary). The uncertainty of $Q_i$ is then found by combining the reference and transfer uncertainty values by sum-of-squares (adding variances), assuming that the respective measurement systems are independent.

Where both the reference and transfer values are significantly greater than zero, the combined uncertainty at a given wavelength is found by adding relative uncertainties by sum-of-squares. However, filtered detectors such as photometers have spectral regions where the transfer ratio is approximately zero. Here the uncertainty must be propagated directly, as relative uncertainty has no meaning. It follows from Eq. (6.58) that

$$u^2(t_i) = \left(\frac{1}{y_2}\right)^2 u^2(y_1) + \left(\frac{y_1}{y_2^2}\right)^2 u^2(y_2) - \frac{y_1}{y_2^3}\, u(y_1, y_2) \approx \frac{u^2(y_1)}{y_2^2} \tag{6.59}$$

for $y_1 \approx 0$. The uncertainty is then found from

$$u^2(Q_i) = R_i^2 u^2(t_i) + t_i^2 u^2(R_i) \tag{6.60}$$

These last two expressions can of course be applied at all wavelengths, instead of using relative uncertainties.

The end result is that the general form of the uncertainty at a given wavelength is

$$u(Q_i) = a + bQ_i \tag{6.61}$$

where the constant term $a$ dominates at low values of $Q_i$ and the scaling term $b$ dominates at high values of $Q_i$. Both $a$ and $b$ may be wavelength-dependent, although in practice they can be taken as constant over wide spectral ranges. Reference artifacts are chosen to have strong values over the wavelength region of interest, and the offset term $a$ is often negligible; uncertainties for reference artifacts are usually given as relative to the reference value.

If the spectral data are to be used in a subsequent combination, we need to know the correlations present between values at different wavelengths. Correlations arise from those present in the reference data, and from systematic components present in the measurement system which correlate the transfer ratio values at different wavelengths.

Correlations in the reference data are often not stated explicitly for data obtained from another laboratory. At a minimum, the total uncertainty (or alternatively the total random component) and the total systematic component at each wavelength should be stated. Usually relative uncertainty values only are required for reference spectral data, although different values may apply in different wavelength regions. Correlations in the reference data are then determined as in Section 6.4.1, assuming that the correlations for the systematic components are all positive.

Correlations in the transfer are calculated by component. Individual uncertainty components of the transfer are either uncorrelated or fully correlated. The covariance of random components is zero. The covariance of the transfer ratios between wavelengths for the fully correlated effects is found by multiplying their uncertainties, taking into account the sign of the correlation as described in Section 6.4.5. The total covariance for a given wavelength pair is then found by summing over all components contributing to the uncertainty of the transfer.

Typically, spectral data are obtained at a number of wavelengths, and it is then convenient to form the variance–covariance matrices $\mathbf{T}$ of the transfer ratios and $\mathbf{R}$ of the reference data.

If the result of the transfer is non-zero at all wavelengths, such as generating a lower-level reference standard or a comparison of similar detector types, for example, we form these in relative form as $\mathbf{T}_{\text{rel}} = [u_{\text{rel}}(t_i, t_j)]$, with $u_{\text{rel}}(t_i, t_i) = u^2_{\text{rel}}(t_i)$, for the transfer ratios and $\mathbf{R}_{\text{rel}} = [u_{\text{rel}}(R_i, R_j)]$ for the spectral reference data. The matrix of relative variance–covariance values for the transferred quantities is then given by $\mathbf{Q}_{\text{rel}} = \mathbf{T}_{\text{rel}} + \mathbf{R}_{\text{rel}}$—this follows from Eq. (6.57). The uncertainties and covariances of the transferred spectral quantities are found by multiplying the relative values by the values of the quantities.

For those artifacts for which the transfer ratio is approximately zero for parts of the spectrum, individual covariance values are calculated as

$$u^2(Q_i) = R_i^2 u^2(t_i) + t_i^2 u^2(R_i) \tag{6.62}$$

where $u^2(t_i)$ is given by Eq. (6.59).

It is then useful to calculate the matrix of correlation coefficients for the spectral data, as often these are approximately constant and hence correlations in subsequent calculations are more easily applied, as shown in Eq. (6.7).

### 6.4.5 Complete Correlation of Systematic Components in Spectral Combinations

Spectral combinations in radiometry often involve similar terms, varying with wavelength but having the same form. Specific examples are given in the following sections. Our general combination is $y = f(x_1, x_2, \ldots, x_N)$, where $x_i$ is a set of spectral irradiance or spectral responsivity data. We consider each independent effect contributing to the uncertainty in the spectral data separately, describing it by the influence quantity $p$, possibly wavelength dependent but uncorrelated to any other. The uncertainty in the spectral data at the $i$th and $j$th wavelengths is then given by

$$u(x_i) = \left|\frac{\partial x_i}{\partial p}\right| u(p), \quad u(x_j) = \left|\frac{\partial x_j}{\partial p}\right| u(p) \tag{6.63}$$

For those effects which are random (uncorrelated) at different wavelengths, the uncertainty contribution for this effect is propagated through Eq. (6.5). For those effects which apply to the whole system of measurements of the spectral data, the covariance between wavelength pairs is given by

$$u(x_i, x_j) = \frac{\partial x_i}{\partial p}\frac{\partial x_j}{\partial p}\, u^2(p) \tag{6.64}$$

Because we are treating this independent uncertainty component on its own as if it was the only contributor, Eqs. (6.63) and (6.64) represent the total uncertainties and covariance; hence $|u(x_i, x_j)| = u(x_i)u(x_j)$, which is the condition for complete correlation.

If the correlation coefficient has the same sign for all wavelength pairs, we can use Eq. (6.6) to propagate uncertainties through our measurement relation. Where the sign of the correlation changes through the spectrum, we must consider the sign of the product $(\partial f/\partial x_i)(\partial f/\partial x_j)u(x_i, x_j)$ of Eq. (6.1). If the sign of $(\partial f/\partial x_i)(\partial f/\partial x_j)$ is the same as that of $u(x_i, x_j)$, we can also use Eq. (6.6).

Spectral sums are important in radiometry. They appear in the integrated response of a photometer or any other filtered radiometer to broadband radiation, and in the tristimulus values upon which all color-coordinates are based. They are all of the form

$$y = \sum_{i=1}^{N} c_i x_i \tag{6.65}$$

where $c_i$ is usually (but not necessarily) a positive value (e.g. color matching function, $V(\lambda)$ photometer response, source irradiance distribution) and $x_i$ is a spectral quantity such as spectral irradiance or spectral responsivity. Using the sign function sgn( ), for our influence parameter $p$ we have

$$u(x_i, x_j) = \operatorname{sgn}\left(\frac{\partial x_i}{\partial p}\right)\left|\frac{\partial x_i}{\partial p}\right| \operatorname{sgn}\left(\frac{\partial x_j}{\partial p}\right)\left|\frac{\partial x_j}{\partial p}\right| u^2(p) \tag{6.66}$$

From Eq. (6.2), the variance of $y$ is

$$u^2(y)|_p = \sum_{i=1}^{N}\left(c_i \operatorname{sgn}\left(\frac{\partial x_i}{\partial p}\right)\left|\frac{\partial x_i}{\partial p}\right|\right)^2 u^2(p) + 2\sum_{i=1}^{N-1}\sum_{j=i+1}^{N} c_i c_j \operatorname{sgn}\left(\frac{\partial x_i}{\partial p}\right) \times \left|\frac{\partial x_i}{\partial p}\right| \operatorname{sgn}\left(\frac{\partial x_j}{\partial p}\right)\left|\frac{\partial x_j}{\partial p}\right| u^2(p) \tag{6.67}$$

(where we have added the sign term inside the square of the first term) or

$$u^2(y)|_p = \left( \sum_{i=1}^{N} c_i \,\mathrm{sgn}\left(\frac{\partial x_i}{\partial p}\right) \left|\frac{\partial x_i}{\partial p}\right| \right)^2 u^2(p) \tag{6.68}$$

If we treat the uncertainty of the spectral quantity as carrying the sign of the sensitivity coefficient for the effect we are considering,

$$u_s(x_i) = \frac{\partial x_i}{\partial p}\, u(p) \tag{6.69}$$

we then have

$$u(y)|_p = \left| \sum_{i=1}^{N} c_i u_s(x_i) \right| \tag{6.70}$$

Signed uncertainties are especially useful when calculating the correlated systematic uncertainty components of color quantities, where the variance–covariance matrix is easily calculated for the tristimulus values and uncertainties are propagated through the three sensitivity factors of the color quantity to each of the tristimulus values [11]. In this case, the covariance of two tristimulus values is simply the product of two sums inside the modulus sign of Eq. (6.70), calculated with different coefficient sets $c_i$.

## 6.5 Examples of Spectral Combinations

Examples illustrating uncertainty propagation through combinations of spectral measurements, applying the principles of the GUM, can be found in the literature for color [11, 12], distribution temperature [7] and correlated color temperature [2, 13]. Here a few specific examples are given for particular uncertainty components to illustrate the methods. All use the practical example of the calibration of the response of a photometer to a source of Commission Internationale de l'Eclairage (CIE) Illuminant A, by a direct measurement of the spectral responsivity against a silicon photodiode reference detector. The spectral energy distribution of the source is defined, and carries no uncertainty, unlike practical sources which approximate Illuminant A and so require a further correction (which in turn adds uncertainty to the estimated response). Any photometer has a spectral responsivity which is approximately that of $V(\lambda)$. We can use the $V(\lambda)$ distribution to model our sources of uncertainty and know that the results obtained will apply to real photometers.

The photometer spectral response is shown in Figure 6.1, along with that of an ideal photodetector. A silicon trap detector, or a silicon photodiode

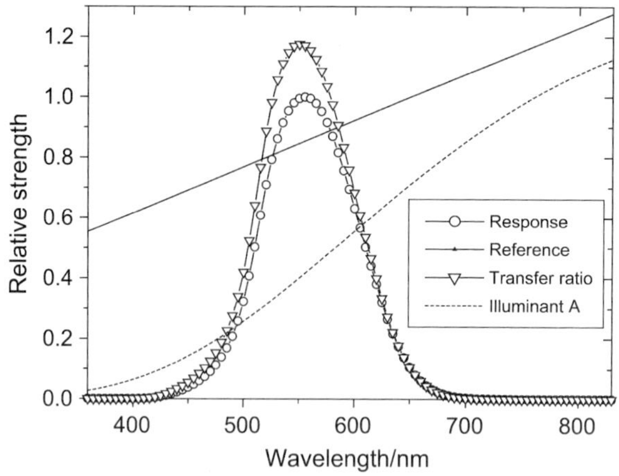


FIG. 6.1. Transfer calibration of a photometer against a silicon photodiode detector. Also shown is the spectral source distribution of CIE Illuminant A.

with thin oxide coating has approximately such an ideal relative response through the visible wavelength region. Absolute scaling of these spectra is not important for this exercise. The transfer ratio then has the shape shown. Data are assumed to have been acquired at 5 nm intervals over the wavelength range 360–830 nm.

The quantity we require is the integrated response of the photometer to CIE Illuminant A:

$$R_V = \sum_n V_n S_n^{\mathrm{A}} = \sum_n t_n R_n S_n^{\mathrm{A}} \tag{6.71}$$

where $V_n$ is the spectral response of the photometer, $t_n$ the transfer ratio, $R_n$ the spectral responsivity of the reference detector and $S_n^{\mathrm{A}}$ the spectral irradiance of Illuminant A, all at the $n$th wavelength. We wish to estimate the contribution of different components to the uncertainty of the response.

The reference spectral responsivity values in general are partially correlated and hence we calculate their contribution to $R_V$ by the matrix multiplication in Eq. (6.17), where **U** is the covariance matrix for the responsivity and the elements of the sensitivity matrix **f** are $\partial R_V / \partial R_n = t_n S_n^{\mathrm{A}}$. Table 6.3 shows the relative uncertainty in the photometric response to Illuminant A for a relative uncertainty of 0.1% and as a function of the correlation coefficient in the spectral responsivity values of the reference detector. The effect of averaging random data is that at low correlations the uncertainty contribution from that of the reference values is negligible, but at high correlations it approaches that of the spectral values themselves.

Table 6.3. Uncertainty in Photometer Response due to 0.1% Relative Uncertainty in the Reference Spectral Responsivity Uncertainty Correlated at Different Levels

| Correlation coefficient | Photometer relative uncertainty (%) |
|---|---|
| 0 | 0.019 |
| 0.3 | 0.057 |
| 0.8 | 0.090 |

The uncertainty in the transfer ratio in practice depends on the signal levels through the spectrum, which in turn depend on the efficiency of the wavelength selector and the spectral irradiance of the lamp. The effect of random noise in the relative value of the transfer is calculated through Eq. (6.5), with sensitivity coefficients $\partial R_V/\partial t_n = R_n S_n^{\mathrm{A}}$. The relative uncertainty in the response is 0.002% for a transfer uncertainty fixed at the level of 0.01% of the transfer value at 555 nm.This negligible value arises from the averaging of the random values over the 95 values in the spectrum.

Note that the noise in the transfer ratio may appear random but in fact be systematic and strongly correlated between wavelengths. This is more likely to be true if the transfer device is an array spectrometer which can record a spectrum in a time small compared to a drift in gain or background, for example. Such correlation has also been seen with a stepped grating system, where the detector was a photomultiplier driven from a DC–DC voltage converter, with the low-voltage DC supply drifting. These correlations can be detected by recording a number of transfer spectra and then calculating not only the standard deviation of the transfer but also the correlations between values at different wavelengths.

Now we will consider various systematic effects that might have affected our transfer measurement.

### 6.5.1 Constant Offset in the Transfer Ratio

Our measurement equation now becomes

$$R_V = \sum_n (t_n + t_0) R_n S_n^{\mathrm{A}} \tag{6.72}$$

where $t_0 \approx 0$ is a constant error in the transfer ratio. This is the same at all wavelengths, uncorrelated to the other parameters and so we can estimate its contribution to the total uncertainty independently. Strictly we should treat this effect in two parts, as separate offsets in the photometer and reference signals. In many cases, the effect of an offset in the sample signals (our photometer in this case) dominates that in the reference signals and it is that case we treat here. Repeating Section 6.4.5 explicitly for this specific example, we have variances

$$u^2(t_n + t_0) = \left(\frac{\partial(t_n + t_0)}{\partial t_0}\right)^2 u^2(t_0) = u^2(t_0) \tag{6.73}$$

and covariances

$$u(t_n + t_0, t_m + t_0) = \frac{\partial(t_n + t_0)}{\partial t_0} \frac{\partial(t_m + t_0)}{\partial t_0} u^2(t_0) = u^2(t_0) \tag{6.74}$$

and hence the correlation coefficient $r(t_n + t_0, t_m + t_0) = +1$ for all pairs. The sensitivity coefficient of $R_V$ for $t_0$ is

$$\frac{\partial R_V}{\partial t_0} = R_n S_n^{\mathrm{A}} \tag{6.75}$$

It then follows that $u(R_V)|_{t_0}$, the contribution of uncertainty in $R_V$ from component $t_0$ alone, is

$$u(R_V)|_{t_0} = \sum R_n S_n^{\mathrm{A}} u(t_0) \tag{6.76}$$

If the background offset is $10^{-4}$ of that at the peak, the relative uncertainty in $R_V$ is 0.058%, which is significant. A similar offset but random between wavelengths contributes only 0.007% to the uncertainty. Note that Eq. (6.76) can be used to estimate the uncertainty that follows from extending the wavelength range of the spectral transfer to 900 nm, in an attempt to define any response in the long-wavelength tail of the photometer. While the mean response in the tail may be measured as zero, its uncertainty contribution to the integral is 0.028%, and increases as the range is extended. It is a better practice to use cut-on filters, e.g. Schott glass RG830, to estimate the integrated response at long wavelengths.

### 6.5.2 Gain Drift in an Amplifier

Suppose that the gain of an amplifier used in the recording path for the photometer drifts in a linear fashion as the wavelength range is stepped during a spectral transfer, but the amplifier used in the reference path is stable. The measurement equation (6.71) becomes

$$R_V = \sum_n t_n(1 + k(\lambda_i - \lambda_1))R_n S_n^{\mathrm{A}} \tag{6.77}$$

where $k$ is the relative change in the gain during the measurement. The sensitivity coefficient of $R_V$ for $k$ at the $n$th wavelength is

$$\frac{\partial R_V}{\partial k}|_n = t_n(\lambda_i - \lambda_1)R_n S_n^{\mathrm{A}} \tag{6.78}$$

Because the component is fully correlated between wavelengths, the contribution of uncertainty in $R_V$ is

$$u(R_V)|_{t_0} = \left|\sum (\lambda_i - \lambda_1)V_n S_n^{\mathrm{A}}\right| u(k) \tag{6.79}$$

For a gain drift with an uncertainty of 0.1% as we recorded results from 360 to 830 nm, the uncertainty in the response integral is 0.045%. This in fact is the change in gain in moving from 360 to 570 nm at an even rate, 570 nm being the peak of the convolved spectral of Illuminant A and $V(\lambda)$. Note that the uncertainty is dependent on the wavelength at which we know the gain to be correct, taken as 360 nm in this example. In practice, gain drifts are corrected by interpolating between repeat reference measurements, but the uncertainty of the correction remains wavelength-dependent.

### 6.5.3 Wavelength Offset

Now consider transfer calibration of the spectral responsivity of a detector where the wavelength setting of a monochromator has a fixed offset $\Delta$ from the true value at which the reference detector responsivity $R$ is known and the photometer responsivity $V$ is required. At each wavelength $\lambda$, the transfer term of Eq. (6.71) can be written as

$$t_{\lambda+\Delta} = \frac{S_{\lambda+\Delta}V_{\lambda+\Delta}}{S_{\lambda+\Delta}R_{\lambda+\Delta}} \tag{6.80}$$

(ignoring amplifier gain differences), where $S_\lambda$ is the spectral power incident on the detector.

To first-order,

$$t_{n,\lambda+\Delta} = t_n\left(1 + \left[\frac{1}{V_n}\frac{\mathrm{d}V_n}{\mathrm{d}\lambda} - \frac{1}{R_n}\frac{\mathrm{d}R_n}{\mathrm{d}\lambda}\right]\Delta\right) \tag{6.81}$$

This is equivalent to saying that the wavelength offset is fully correlated between the test and reference measurement cycles. The relative uncertainty of the transfer, due to this component is

$$u_{\mathrm{rel}}(t_n) = \left|\frac{1}{V_n}\frac{\mathrm{d}V_n}{\mathrm{d}\lambda} - \frac{1}{R_n}\frac{\mathrm{d}R_n}{\mathrm{d}\lambda}\right| u(\Delta) \tag{6.82}$$

The covariance of values of the transfer at wavelengths denoted by $n$ and $m$,

$$u(t_{n+\Delta}, t_{m+\Delta}) = \frac{\partial t_{n+\Delta}}{\partial \Delta}\frac{\partial t_{m+\Delta}}{\partial \Delta}\, u^2(\Delta) \tag{6.83}$$

is now positive or negative depending on the relative slopes of the response functions. If we are considering this uncertainty component alone, the uncertainties for the transfer are fully correlated but the correlation coefficient between wavelengths may be +1 or −1. However, we note that the measurement equation (6.71) is of the class discussed in Section 6.4.5, with $V(\lambda)$ as the sensitivity coefficient at a given wavelength. Hence the total uncertainty contribution is found as the linear sum

$$u(R_V)|_\Delta = \left|\sum V_n S_n^{\mathrm{A}} \left(\frac{1}{V_n}\frac{\mathrm{d}V_n}{\mathrm{d}\lambda} - \frac{1}{R_n}\frac{\mathrm{d}R_n}{\mathrm{d}\lambda}\right)\right| u(\Delta) \tag{6.84}$$

depending on the signs of the relative slopes in Eq. (6.81).

An uncertainty of 0.2 nm in the wavelength offset caused by uneven illumination of the grating of a monochromator—1/20 of the bandwidth for a typical measurement—produces an uncertainty of 0.17% in the photometer response to Illuminant A. By comparison, a random uncertainty of 0.03 nm—typical for a scanning system calibrated at high resolution with a number of well-defined spectral lines—produces an uncertainty of 0.01%.

We can understand the change in sign of the correlation coefficient from Figure 6.1. At wavelengths below the peak of the photometer spectral response, a positive offset in wavelength leads to overestimating the spectral response; at wavelengths above the peak where the relative slopes have different signs, the spectral response is underestimated. These two effects compensate each other in the sum.

## 6.6 Type A and B Uncertainties in Radiometry

In the sections above we have concentrated on the propagation of uncertainty assuming that we know the uncertainty of the measurement components. An important part of any measurement is the estimation of those uncertainties. The GUM defines type A uncertainties as those evaluated by statistical means and type B as those evaluated by other means. Systematic effects are most often estimated by experience, although repeat measurements and a type A analysis may have been made to gain that experience. Such uncertainties are sometimes classed as "pooled type A" components, difficult and time consuming to evaluate but unlikely to change from measurement to measurement unless there is a major change in the system. Positioning of a discharge lamp that may not fully illuminate the grating of a monochromator and so generate a wavelength offset error is one example.

For type A uncertainties we assume a normal distribution and calculate the sample variance and mean from the repeat measurements. We need to be clear here on the difference between the variance of the sample and the variance of the mean. An estimate of the sample variance is given by

$$s^2(x_i) = \frac{1}{N-1} \sum_{i=1}^{N} (x_i - \bar{x})^2 \tag{6.85}$$

where $x_i$ are our $N$ sample readings and $\bar{x}$ is the mean,

$$\bar{x} = \frac{1}{N} \sum_{i=1}^{N} x_i \tag{6.86}$$

The standard uncertainty for the sample is taken as the square root of the sample variance. An estimate of the variance of the mean is given by

$$s^2(\bar{x}) = \frac{s^2(x_i)}{N} \tag{6.87}$$

This is the variance we would expect to obtain if we repeated our $N$ measurements many times and then computed the variance of the individual means. If we determine a type A component by repetition each time the experiment is performed, we can use the standard deviation of the mean as the uncertainty. However, if we have made our $N$ repeated measurements for one application, then use the results in a later uncertainty calculation, we must use the sample standard deviation. It is the variability than we might expect to encounter for any single sample [1, Section 4.2.4].

Systematic effects are most often type B, estimated through experience with the measuring system. They may have been evaluated as type A and then reapplied in other measurements, as discussed in the last paragraph. Most often the component is defined as falling within estimated limits. If the range of the limits is $\pm a$ and the value is judged to have equal probability anywhere in the range, the equivalent standard uncertainty is $u(x_i) = a/\sqrt{3}$. An example is the uncertainty caused by the limited resolution of a digital instrument, where $2a$ is equal to the value of the least significant digit. If the probability is greatest at the center of the range and zero at the limits, the equivalent standard uncertainty is $u(x_i) = a/\sqrt{6}$. The GUM gives other examples of equivalent standard uncertainties where the range is estimated to cover a particular fraction of the measurements. The same result can be obtained by expanding the limits to encompass all likely readings.

One component that is difficult to estimate is the long-term stability of an artifact. At best this can cover only careful use, and the estimate is very much based on experience with similar artifacts. New artifacts must be checked on a regular basis until sufficient information is available to estimate changes with time. For spectral values such as irradiance or responsivity, drift rates are likely to be wavelength-dependent. They may be able to be modeled based on experience, or by a change in operating temperature of a lamp, for example, and will most likely be strongly correlated between wavelengths. For spectral irradiance lamps, drift may be the most significant uncertainty component [6]; its magnitude can only be reduced by more frequent calibration.

## 6.7 Expression of Uncertainty

In the preceding sections we have discussed the ways by which uncertainties are estimated for given components, and then combined. All are in terms of standard uncertainties, equivalent to standard deviations if all were to be obtained by statistical means. The GUM also recommends a standard method for the expression of uncertainty, based on a particular level of confidence in the final result. The level of confidence most often adopted for modern measurement is 95%, meaning that the value of the measurand has a 95% chance of falling within the range of the quoted uncertainty if the measurement was repeated. The multiplier of the standard uncertainty to determine a given level of confidence is termed the coverage factor; it in turn depends on the degrees of freedom of the measurement.

For type A measurements, the degrees of freedom are the $(N-1)$ factor shown in Eq. (6.85). Uncertainties obtained from a small sample size are systematically underestimated. Student's $t$ [1, Section G.3] estimates the coverage factor to achieve a particular level of confidence for a limited data set. In practice, a sample size of $>30$ repeats is sufficient for the scaling factor of 1.96 to be sufficiently close to its limit of 2.04 for an infinite sample size, for a 95% level of confidence. It is commonly taken as 2.0 for infinite degrees of freedom, although strictly this provides a coverage factor of 95.45%. (We need to remember that we are *estimating* uncertainty, and that this is an inexact process!) Very small sample sizes are common in radiometry—e.g. two or three different photometers may be characterized and differences between their illuminance response be ascribed to an unknown systematic effect. It is then more realistic to estimate an uncertainty for this effect from the range of the differences and to generally assume infinite degrees of freedom, rather than apply a type A analysis.

To calculate the coverage factor to apply to the combined standard uncertainty, we must estimate the effective degrees of freedom $\nu_{\text{eff}}$ that apply to the combination process. This is given by the Welch–Satterwaithe equation [1, Section G.4],

$$\nu_{\text{eff}} = \frac{u_c^4(y)}{\sum_{i=1}^{k} u_i^4(y)/\nu_i} \tag{6.88}$$

where $u_c(y)$ is the standard uncertainty of the combination, $u_i(y)$ the contribution to the standard uncertainty of the $i$th component and $\nu_i$ the number of degrees of freedom of that component. A complete expression of uncertainty for a measurand then quotes the combined standard uncertainty scaled by Student's $t$ for the effective degrees of freedom, the level of confidence of the result, and either the effective degrees of freedom or the coverage factor to achieve that level of confidence. All of this information is required if the measured result is to be part of a further uncertainty calculation, since it is the standard uncertainty and its effective degrees of freedom that will be used there.

Radiometric measurements are usually dominated by systematic uncertainty components, and it is typical to assume that the effective degrees of freedom are infinite and the coverage factor to achieve a 95% level of confidence is 2.

In some cases it is convenient to provide an expanded uncertainty to cover an offset known to be present in a measurement, but difficult to determine in a particular case. This should be included in the measurement equation, the result corrected and the uncertainty of the offset included in the uncertainty calculation. Any such offset treated as an uncertainty is independent of the coverage factor or effective degrees of freedom; it should be added as a linear sum to the expanded uncertainty and identified as such for subsequent uncertainty calculations.

## 6.8 Bayesian Concepts

The method of estimating and propagating uncertainty that is discussed above is classed as frequentist—it is based on sampling theory and probability distributions estimated each time we make a measurement. Bayesian statistics is based on the thought that there is other prior information that we can also take into account. The analysis of uncertainty by Bayesian methods is quite different to that of sampling theory, although for uncorrelated random distributions and no prior information the outcome is the same, the normal distribution. Coverage factors and degrees of freedom relate to sampling and are not relevant to Bayesian statistics. The probability distribution itself is computed and used to estimate uncertainty for a given confidence interval. A detailed critique of the GUM in Bayesian terms is given in Reference [14].

We might apply the concept of prior information to the interpolation of a blackbody reference spectrum. We know that if we actually measure the spectral irradiance at a wavelength between two others, we can reasonably expect that the uncertainty we would obtain would be that interpolated from the existing readings. We might also reasonably expect that the correlation between this point and other interpolated or existing points could also be interpolated from that of the measured values. However, any alternative treatment must be consistently applied. If we were to interpolate the uncertainty without considering any correlations, the uncertainty of any spectral irradiance integral could be driven to zero by merely interpolating many times. Even if we apply prior knowledge to this simple interpolation idea, then apply frequentist statistics, the implication is that only correlated uncertainty components remain because the uncertainty due to random effects would be driven to zero merely by repeated interpolation.

The assumption of a normal distribution is likely to persevere in radiometry, as a means of dealing with the systematic components that dominate uncertainties and for which detailed statistical analysis is not practical. However, general interest in Bayesian analysis is rising and we can expect to see application to radiometric measurements and uncertainty analyses.

## References


*1.* "Guide to the Expression of Uncertainty in Measurement," International Organization for Standardization, Geneva, 1993.

*2.* J. Fontecha, J. Campos, A. Corrons, and A. Pons, An analytical method for estimating correlated colour temperature uncertainty, *Metrologia* 39, 531–536 (2002).

*3.* H. W. Yoon, C. E. Gibson, and J. L. Gardner, Spectral radiance comparison of two blackbodies with temperatures determined using absolute detectors and ITS-90 techniques, *Temp. Its Meas. Control Sci. Ind.* 7, 601–606 (2003).

*4.* E. W. Weisstein, "Implicit Differentiation," http://mathworld.wolfram.com/ImplicitDifferentiation.html

*5.* T. R. Gentile, J. M. Houston, and C. L. Cromer, Realization of a scale of absolute spectral response using the National Institute of Standards and Technology high-accuracy cryogenic radiometer, *Appl. Opt.* 35, 4392–4403 (1996).

*6.* J. L. Gardner, Uncertainty propagation for NIST visible spectral standards, *J. Res. Natl. Stand. Technol.* 109, 305–318 (2004).

*7.* W. Woeger, Uncertainties in models with more than one output quantity, "Proceedings of the CIE Expert Symposium 2001," 12–17. CIE, Vienna, 2001.

*8.* P. R. Bevington and D. K. Robinson, "Data Reduction and Error Analysis for the Physical Sciences," 2nd edition, p. 147. McGraw-Hill, MA, 1992.

*9.* J. L. Gardner, Correlations in primary spectral standards, *Metrologia* 40, S167–S171 (2003).

*10.* J. L. Gardner, Uncertainties in interpolated spectral data, *J. Res. Natl. Stand. Technol.* 108, 69–78 (2003).
*11.* J. L. Gardner, "Uncertainties in colour measurements," Report to Korean Research Institute for Standards and Science, 2003.
*12.* E. A. Early and M. E. Nadal, Uncertainty analysis for reflectance colorimetry, *Color Res. Appl.* 29, 205–216 (2004).
*13.* J. L. Gardner, Correlated colour temperature—uncertainty and estimation, *Metrologia* 37, 381–384 (2000).
*14.* R. Kacker and A. Jones, On use of Bayesian statistics to make the *Guide to the Expression of Uncertainty in Measurement* consistent, *Metrologia* 40, 235–248 (2003).

# 7. PHOTOMETRY*


Yoshi Ohno

National Institute of Standards and Technology, Gaithersburg, Maryland, USA


## 7.1 Introduction

The aim of photometry is to measure light in such a way that the results correlate with human vision. Traffic signals and computer displays, for example, are meant for human eyes, and therefore, must be evaluated based on the spectral responsivity of the average human eyes. While radiometry covers all spectral regions from ultraviolet to infrared, photometry deals with only the spectral region from 360 to 830 nm (the visible region) where human eyes are sensitive. Photometry is essential for evaluation of light sources and objects used for lighting, signaling, displays, and other applications where light is seen by the human eye.

In earlier times, real human eyes were used as detectors in photometry. The intensity of a test light source placed at a varied distance was compared with that of a standard light source at a fixed distance by visual observation. The distance of the test light source was adjusted so that the two light sources would look equally bright. The intensity of the test light source was given from the intensity of the standard source and the ratio of the distances squared. Such a quantity for the intensity of light in one direction at that time was called candle power (luminous intensity in present terminology), which was the first photometric quantity defined. Until about 1940, such visual comparison measurement techniques were predominant in photometry [1].

In modern photometric practice, measurements are made with photodetectors. This is referred to as physical photometry. Physical photometry uses either optical radiation detectors constructed to mimic the spectral response of the eye, or spectroradiometry coupled with appropriate calculations for weighting by the spectral response of the eye. Typical photometric units include the lumen (luminous flux), the candela (luminous intensity), the lux (illuminance), and the candela per square meter (luminance). These photometric units correspond to radiometric units: watt (radiant flux), watt per steradian (radiant intensity), watt per square meter (irradiance), and watt per square meter per steradian (radiance) (see Chapter 1, Table 1.2).

*Contribution of the National Institute of Standards and Technology.



EXPERIMENTAL METHODS IN THE PHYSICAL SCIENCES, vol. 41
ISSN 1079-4042 DOI: 10.1016/S1079-4042(05)41007-3
Published by Elsevier Inc.

Photometric quantities are spectrally integrated radiometric quantities weighted by the human eye response.

Similar to photometry, measurement of color of light sources and objects also deals with broadband measurement of the visible radiation and is referred to as *colorimetry*. Colorimetry is ascribed to measurement of light spectra weighted by three standardized spectral weighting functions, one of which is identical to the standardized human eye response used in photometry.

This chapter describes the state-of-the-art of modern photometry and also describes a part of colorimetry that is essential to the evaluation of photometric quantities. The terminology used in this chapter follows international standards and recommendations [2–4].

## 7.2 Basis of Physical Photometry

### 7.2.1 Visual Response

In order to achieve the aim of photometry, one must take into account the characteristics of human vision. The relative spectral responsivity of the human eye was first defined by the Commission Internationale de l'Éclairage (CIE), (the International Commission on Illumination), in 1924 [5]. It is called the *spectral luminous efficiency for photopic vision*, with a symbol $V(\lambda)$, defined in the domain from 360 to 830 nm, and is normalized to unity at its peak, 555 nm (Fig. 7.1). This model has gained wide acceptance. The values were republished by CIE in 1983 [6], and adopted by Comité International des Poids et Mesures (CIPM), (the International Committee on Weights and

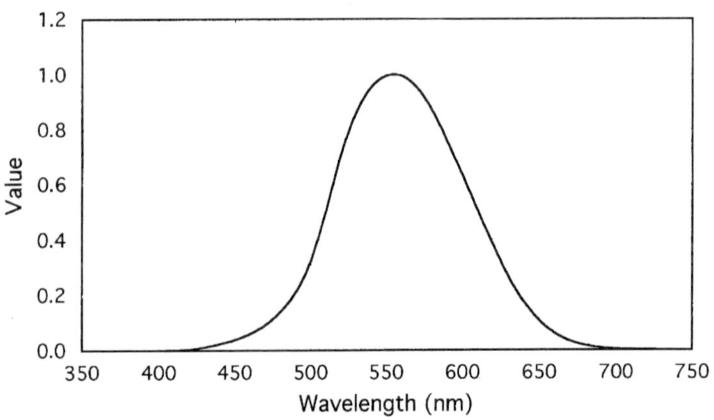


FIG. 7.1. CIE $V(\lambda)$ function.

Measures), in 1983 [7] to supplement the 1979 definition of the candela. The tabulated values of the function at 1 nm increments are available in References [6–8]. In most cases, the region from 380 to 780 nm suffices for calculation with negligible errors since the value of the $V(\lambda)$ function falls below $10^{-4}$ outside this region.

As specified in the definition of the candela (see Section 7.2.2) in 1979 [9] and a supplemental document from CIPM in 1983 [7], a photometric quantity $X_v$ is now defined in relation to the corresponding radiometric quantity $X_{e,\lambda}$ by the equation

$$X_v = K_m \int_{360\ \text{nm}}^{830\ \text{nm}} X_{e,\lambda} V(\lambda)\, d\lambda \tag{7.1}$$

The constant, $K_m$, relates the photometric quantities and radiometric quantities, and is called the *maximum spectral luminous efficacy of radiation for photopic vision*. The value of $K_m$ is given in the 1979 definition of candela, which defines the spectral luminous efficacy of radiation at the frequency $540 \times 10^{12}$ Hz (at the wavelength 555.016 nm in standard air) to be 683 lm/W. Note that this is not exactly at the peak of $V(\lambda)$ at 555 nm. The exact value of $K_m$ is calculated as $683 \times V(555.000\ \text{nm})/V(555.016\ \text{nm}) = 683.002$ lm/W [6]. $K_m$ is normally rounded to 683 lm/W, with negligible error for all practical applications.

It should be noted that the $V(\lambda)$ function is defined for the *CIE standard photometric observer for photopic vision*, which assumes additivity of sensation and a 2° field of view at relatively high luminance levels (higher than $\sim$1 cd/m$^2$). The human vision in this level is called *photopic* vision. The spectral responsivity of human eyes deviates significantly at very low levels of luminance (at luminance levels $<\sim 10^{-3}$ cd/m$^2$) when the rods in the eyes are the dominant receptors. This type of vision is called *scotopic* vision. Its spectral responsivity, peaking at 507 nm, is designated as $V'(\lambda)$, and was defined by CIE in 1951 [10], recognized by CIPM in 1976 [11], and republished by CIPM in 1983 [7]. The human vision in the region between photopic vision and scotopic vision is called *mesopic vision*. While there has been active research in this area [12], there is not yet an internationally accepted spectral luminous efficiency function for the mesopic region. In current practice, almost all photometric quantities are given in terms of photopic vision, even at low light levels. Quantities in scotopic vision are seldom used except for special calculations for research purposes.

### 7.2.2 Photometric Base Unit, the Candela

The history of photometric standards dates back to the early 19th century, when the intensity of light sources was measured by comparison with a standard candle using visual bar photometers [1]. At that time, the flame of a candle was used as a unit of luminous intensity that was called *the candle*. The old name for luminous intensity "candle power" came from this origin. Standard candles were gradually superseded by flame standards of oil lamps, and in 1920, the unit of luminous intensity, recognized as *the international candle*, was adopted by the CIE. In 1948, it was adopted by the Conférence Générale des Poids et Mesures (CGPM), (the General Conference on Weights and Measures) with a new name "candela" defined as the luminous intensity of a platinum blackbody at its freezing temperature under specified geometry [13]. Although the 1948 definition served to establish the uniformity of photometric measurements throughout the world, most national laboratories found it difficult and expensive to maintain a platinum freezing point blackbody and hence a consensus was developed to change the definition of the candela to the one based solely upon the measurement of optical power and remove the reliance for the definition of the candela upon the performance of high temperature sources. In 1979 the CGPM adopted a recommendation from the CIPM and redefined the candela in terms of a specific amount of optical power from a source which could be measured by an appropriate detector.

The 1979 candela is defined as follows:

> The candela is the luminous intensity, in a given direction, of a source that emits monochromatic radiation of frequency $540 \times 10^{12}$ hertz and that has a radiant intensity in that direction of (1/683) watt per steradian.

The value of $K_m$ (683 lm/W) was determined based on measurements by several national laboratories in such a way that consistency was maintained with the prior unit. Technical details on this redefinition of the candela are reported in References [14, 15]. This 1979 redefinition of the candela has enabled the derivation of the photometric units from the radiometric units using various techniques.

## 7.3 Quantities and Units in Photometry

In 1960, the Système International (SI) was established, and the candela became one of the seven SI base units [16]. For further details on the SI (see References [16–19]). Several quantities and units, defined in different geometries, are used in photometry and radiometry. Table 1.2 in Chapter 1 lists the photometric quantities and units, and their corresponding radiometric quantities and units. While the candela is the SI base unit, the luminous flux (lumen) is perhaps the most fundamental photometric quantity, as the other photometric quantities are defined in terms of lumen with an appropriate geometric unit. The official definitions of these photometric quantities are available in Reference [2].

## 7.4 Luminous Intensity Standards and Measurements

### 7.4.1 Detector-Based Realization of the Candela

During the previous definition of the candela from 1948 to 1979, a platinum-point blackbody was used to realize the candela. Now the candela is most often realized based on the absolute responsivity of detectors as provided in the 1979 redefinition. In this method, referred to as the detector-based candela, calibrated detectors provide the illuminance unit and the candela is deduced from the measured illuminance and the distance from the source to the photometer.

After the 1979 definition of the candela, many national laboratories started realizing the candela based on the absolute responsivity of detectors. Until the early 1980s, electrical substitution radiometers (ESRs) operating at room temperature were predominantly used [20, 21]. Later, the silicon photodiode self-calibration technique [22, 23], using 100% quantum efficient silicon photodiodes in a trap configuration, was often used for the realization of the candela [24–26]. From the late 1980s, absolute cryogenic radiometers started to be employed as primary radiometric standards for national laboratories. Typical cryogenic radiometers are cooled to $\approx 5\,\mathrm{K}$ by liquid helium, and work on the principle of electrical substitution [27, 28]. Cryogenic radiometers, having relative uncertainties of the order of $10^{-4}$, are now the most accurate means for establishing radiometric scales, and are now commonly used to establish spectral responsivity scale, and thus to realize the candela [29–31]. More detail on cryogenic radiometers can be found in Chapter 2.

The principles of the detector-based realization of the candela are described below. A standard photometer, consisting of a silicon photodiode, a $V(\lambda)$-correction filter, and a precision aperture, is depicted in Figure 7.2. The absolute spectral power responsivity $s(\lambda)$ (in A/W) of the whole photometer (as an average over the aperture area) is calibrated against the spectral responsivity scale based on an absolute cryogenic radiometer. The area of the aperture $A$ is calibrated by using a dimension measuring instrument. The illuminance responsivity $s_{\mathrm{v}}$ (in A/lx) of the photometer is then obtained by

$$s_{\mathrm{v}} = \frac{A \int_{\lambda} S(\lambda)\, s(\lambda)\, \mathrm{d}\lambda}{K_{\mathrm{m}} \int_{\lambda} S(\lambda) V(\lambda)\, \mathrm{d}\lambda} \tag{7.2}$$

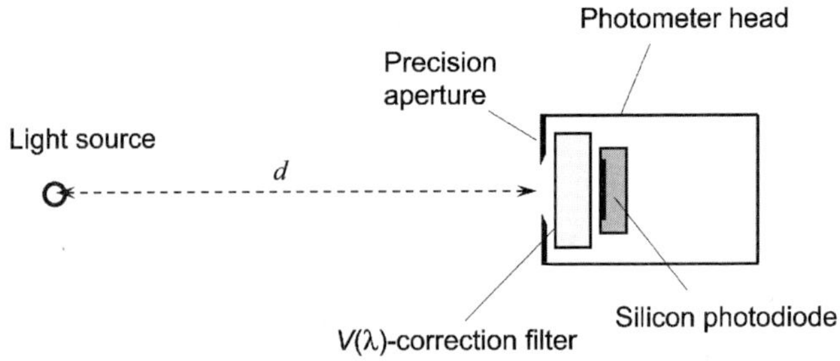


FIG. 7.2. Geometry for the detector-based candela realization.

where $S(\lambda)$ is the relative spectral power distribution of the light to be measured, $V(\lambda)$ the spectral luminous efficiency function, and $K_m$ the maximum spectral luminous efficacy (683 lm/W). Planckian radiation at 2856 K (CIE Standard Illuminant A [32]) is normally used for $S(\lambda)$.

When a light source having spectral distribution $S(\lambda)$ is measured with the photometer at a distance $d$, the illuminance $E_v$ [lx] at the plane of the photometer aperture (reference plane) is given by

$$E_v = \frac{y}{s_v} \tag{7.3}$$

where $y$ is the output current of the photometer. This establishes the illuminance unit, lux. With the distance $d$ (measured from the light source to the reference plane of the photometer) accurately known, the luminous intensity $I_v$ [in cd] of the source is given, according to the inverse square law, by

$$I_v = E_v\, d^2 \frac{1}{\Omega_0} = \frac{y}{s_v} \frac{d^2}{\Omega_0} \tag{7.4}$$

where $y$ is the output current of the photometer, and $\Omega_0$ is the unit solid angle ( = 1 sr). This establishes the luminous intensity unit, the candela.

Note that, as shown in Eqs. (7.2) and (7.4), the photometer by itself does not establish the luminous intensity unit, but it does with knowledge of the relative spectral distribution of the light source and distance measurement.

For a practical application of the principles, the spectral responsivity is often separated into the absolute and relative terms, and Eq. (7.2) can be expressed as

$$s_v = \frac{A\, s(555) \int_\lambda S(\lambda)\, s_{rel}(\lambda)\, d\lambda}{K_m \int_\lambda S(\lambda) V(\lambda)\, d\lambda} \tag{7.5}$$

where $s(555)$ is the absolute spectral responsivity at 555 nm, and $s_{rel}(\lambda)$ the relative spectral responsivity normalized at 555 nm. In this equation, if $s_{rel}(\lambda)$ is perfectly matched to $V(\lambda)$, the two integral terms would be cancelled out, and $s_v$ would be independent of $S(\lambda)$. This implies that, if $s_{rel}(\lambda)$ is close to $V(\lambda)$, $s_v$ will not be sensitive to small differences in $S(\lambda)$.

Nevertheless, since the relative spectral responsivity of photometers cannot be perfectly matched to $V(\lambda)$, the illuminance responsivity $s_v$ is more or less dependent on $S(\lambda)$. Therefore, when $s_v$ is specified, the source spectrum $S(\lambda)$ should also be specified. It is a common practice among national laboratories to use a CIE Source A lamp [33] (a practical realization of CIE Standard Illuminant A having a relative spectral distribution of Planckian radiation at a correlated color temperature of 2856 K) for calibration of standard photometers. The illuminance responsivity of a standard photometer for CIE Standard Illuminant A, denoted as $s_v^*$, is given as

$$s_v^* = \frac{\int_\lambda S_A(\lambda)\, s(\lambda)\, d\lambda}{K_m \int_\lambda S_A(\lambda) V(\lambda)\, d\lambda} \tag{7.6}$$

where $S_A(\lambda)$ is the relative spectral distribution of CIE Standard Illuminant A.

When a light source other than CIE Source A, having spectral distribution $S(\lambda)$, is measured, the illuminance $E_v$ is measured from the photometer signal $y$ by introducing a correction factor $F^*$ as follows

$$E_v = \frac{y}{s_v^*} F^* \tag{7.7}$$

with

$$F^* = \frac{\int_\lambda S(\lambda) V(\lambda)\, d\lambda \int_\lambda S_A(\lambda)\, s_{rel}(\lambda)\, d\lambda}{\int_\lambda S(\lambda)\, s_{rel}(\lambda)\, d\lambda \int_\lambda S_A(\lambda) V(\lambda)\, d\lambda} \tag{7.8}$$

This factor $F^*$ is called a *spectral mismatch correction factor*. When different types of lamps are measured, $F^*$ for each type of lamp can be calculated as a list. The photometer can be recalibrated only with a CIE Source A (for $s_v^*$), as the relative spectral responsivity of photometers is fairly stable over a long period of time.

The calibration chain for the detector-based realization of the candela used at NIST is shown in Figure 7.3 as an example. An absolute cryogenic radiometer [30] at the top of the chain provides the primary radiometric scale. Silicon trap detectors are calibrated at several laser wavelengths on the cryogenic radiometer and used to establish the spectral responsivity scale. The scale is transferred to a monochromator-based Spectral Comparator Facility [34], where reference photodiodes are calibrated against the trap detectors and used for routine calibrations. The absolute spectral responsivity [A/W] of the reference photometers is calibrated on the Spectral Comparator Facility. Since the photometer aperture is underfilled in this measurement, corrections are made for the spatial nonuniformity of spectral

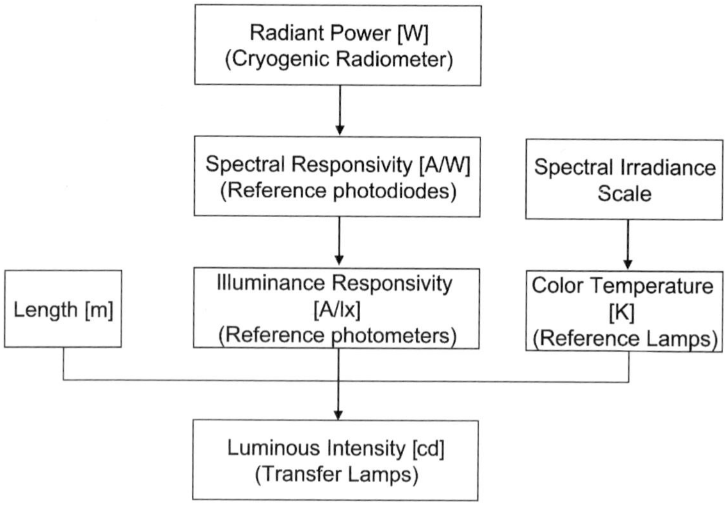


FIG. 7.3. Realization and maintenance of the photometric units at NIST.

responsivity over the aperture area. The illuminance responsivity [A/lx] of each photometer is then calculated using Eq. (7.2). Then, the luminous intensity of a transfer lamp (test lamp) is calibrated based on the illuminance measurement by the standard photometers according to Eq. (7.4). No luminous intensity standard lamps are maintained, and instead, the standard photometers are used to maintain the units of illuminance and luminous intensity. Further details of the realization of the candela at NIST are described in Reference [35].

### 7.4.2 Photometers as Reference and Transfer Standards

In the traditional photometric practice, the luminous intensity units are maintained on a group of standard lamps. Since lamps age with use, the operating time of standard lamps has to be strictly limited. Therefore, the scale on the primary standard lamps is transferred to secondary standard lamps, then to working standard lamps that are used for routine calibrations. The uncertainty of calibration increases when the scale is transferred from one lamp to another. As presented in the previous section, photometers can be used to establish the photometric units, and also to maintain the units.

The use of standard photometers utilizing high quality silicon photodiodes and filters provides many advantages over standard lamps. The photometers do not age through use, they are robust against mechanical shocks—thus easy to transport, and they have excellent short-term stability

(0.01%) and a large linearity range (14 orders of magnitude is reported [36]). The long-term stability of photometers varies depending on the $V(\lambda)$ filter. Selected photometers exhibit a stability of better than 0.1% over a period of 1 year, while poor ones show changes over 1% in 1 year. The long-term stability of photometers must be evaluated before they can be used as standards. Lamps are generally more stable than photometers for long-term storage over many years, but lamps age as they are used and are generally not as reproducible as photometers due to difficulty in alignment and other characteristics. Selected photometers can be used as low-uncertainty photometric standards with periodic calibrations.

Typical constructions of standard photometers are shown in Figure 7.4, showing two types commonly used—diffuser type and non-diffuser type. An important requirement of a standard photometer is that its reference plane is clearly defined, and thus a limiting aperture (or some structure that works as a limiting aperture) is required. In both types shown in Figures 7.4(a) and (b), the reference plane is the front surface of the aperture.

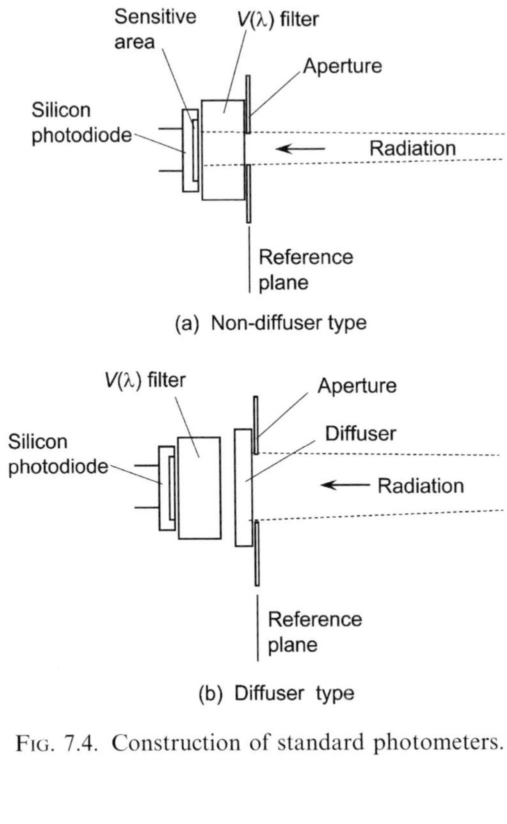


FIG. 7.4. Construction of standard photometers.

At national laboratories, standard photometers are used to realize the units, whereas in industry, they are used only to maintain and transfer the units. Therefore, the requirements of standard photometers used by national laboratories and in industry are different. Non-diffuser photometers have a uniform responsivity over the aperture area, and they are preferred for realization of the candela (as seen in Fig. 7.2) because the absolute spectral responsivity is normally measured in a power mode using a narrow beam. Thus, the spatial nonuniformity over the light-receiving area is critical for realization with low uncertainty. In addition, a standard photometer does not need cosine correction over a large angular range because it is used with a standard lamp of a limited size, at a sufficient distance, and placed on the optical axis of the photometer. A narrow acceptance angle of non-diffuser photometers actually is advantageous to reduce the effect of ambient stray light. A disadvantage of the non-diffuser photometer is that the aperture on top of the $V(\lambda)$ filter should never be touched, and it is difficult to clean the filter surface without damaging the aperture edges. Compressed air is normally used to blow off dust, and the photometers should be kept in dry air environment.

Diffuser-type photometers can be constructed in such a way that the aperture and the diffuser surface are flush so the diffuser surface can be easily cleaned by soft dry cloth. This type is much easier to handle, e.g., in industrial laboratories. The diffuser type also has a wider acceptance angle and, thus it can be used to measure extended sources at shorter distances. The diffuser also acts to thermally isolate the $V(\lambda)$-correction filter from the incident radiation. As the $V(\lambda)$ filter is not exposed to direct illumination, this type of photometer may be used for higher illuminance levels than the non-diffuser type. Care should be taken, however, because the reflectance of the diffuser is fairly high and diffusive, and the reflection from the diffuser can cause some stray light to be reflected back from any nearby objects. Note that the diffuser type photometer should not be confused with a cosine-corrected photometer head of illuminance meters. Adding a diffuser gives only approximate cosine response in a limited angle range. A cosine-corrected photometer head has a geometrically structured acceptance surface for good cosine response in the incident angle range of up to nearly $\pm 90°$, which is not necessary for standard photometers.

It is also well known that the illuminance responsivity of a photometer changes with temperature. While the temperature coefficient of typical silicon photodiodes in the visible region is insignificant, the transmittance of the $V(\lambda)$ filter tends to change significantly with temperature. To avoid such effects, modern standard photometers are equipped with temperature stabilization or a temperature monitor to allow for corrections. A temperature-controlled photometer incorporates a temperature sensor and a heater or a thermoelectric cooler to maintain the temperature of the detector-filter package at a constant temperature (e.g. within $\pm 0.1^\circ C$). A temperature-monitored photometer incorporates a temperature sensor, which is thermally connected to the detector-filter package, and its signal is used first to determine the temperature coefficient of the photometer head, then used to correct for the temperature difference during measurements. Examples of temperature-controlled and temperature-monitored photometers are found in Reference [37]. Such standard photometers are also commercially available.

### 7.4.3 Maintenance of the Luminous Intensity Unit

The candela is annually realized at NIST and maintained via a group of eight standard photometers (NIST reference photometers). The stability of these photometers over the past several years is shown in Figure 7.5. Five of the photometers have shown long-term stability within 0.4% over a 7-year period and better than 0.1% per year. These five photometers are used in the routine calibrations of luminous intensity and illuminance. A few other photometers showed much poorer stability and are not used to maintain the scale (though they are still used in the realization). These photometers employ a $V(\lambda)$ filter from a different manufacture than the others, and tend to suffer from contamination by moisture.

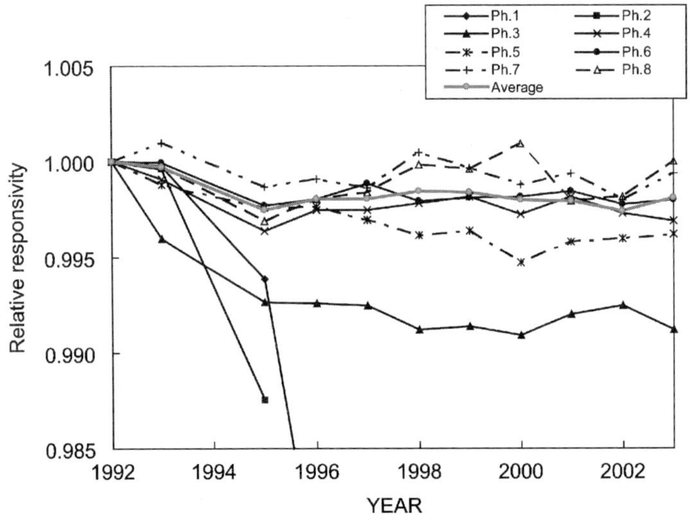


FIG. 7.5. Stability of the NIST reference photometers (data normalized to 1992).

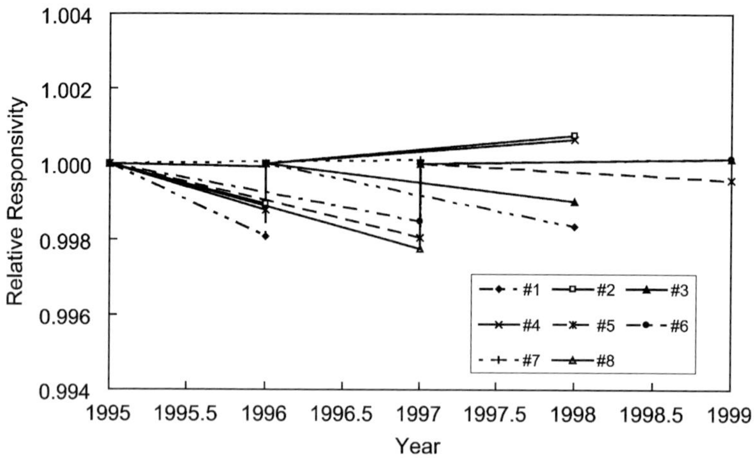


FIG. 7.6. Calibration history of eight standard photometers from a customer.

The detector-based method has also been introduced in industry. Several companies in the United States now use standard photometers as their corporate reference standards for photometry. These photometers are annually calibrated by NIST. As an example, Figure 7.6 shows the calibration history of eight standard photometers from a customer. In this case, the photometers were divided into two groups since 1996, and each group of four photometers is calibrated in a 2-year cycle, overlapping 1 year with each other. The data are normalized to 1 after recalibration. These data show the stability of the photometers (against NIST scale) to be within 0.2% between calibration cycles and that the average scale of each group maintains the traceability to the NIST scale within $\approx$0.1%. Some other photometers are not performing as well as this case.

### 7.4.4 Application of Detector-Based Methods for Illuminance Calibration

Calibrations of illuminance meters are often performed over a large range of illuminance levels and often require a variety of standard lamps of different power levels and a long photometric bench. Use of standard photometers for calibration of illuminance meters is particularly advantageous in that many of the requirements for the standard sources are eliminated or eased. The only critical requirements are that the lamps must have spectral power distribution close to that of CIE Standard Illuminant A, a sufficient short-term stability, and a spatially uniform field of illuminance. The luminous intensity of the lamp need not be known, and neither long-term stability nor reproducibility is critical because the illuminance is determined by the standard photometers at the time of each use. The alignment and distance settings of the lamp are not critical, either.

To perform illuminance meter calibrations in a large range, for example, additional components for attenuation (e.g., neutral density filters or transmitting diffusers) can be inserted in the light path to lower the illuminance level. Alternatively, an integrating sphere source with a variable aperture can be used as a variable illuminance source. To increase the illuminance level, a convex, achromatic lens can be used in front of the lamp. In these cases, it should be ensured that the spectral distribution and spatial non-uniformity of illuminance are acceptable for the required calibration uncertainty. The accuracy of the matching of spectral distribution to CIE Standard Illuminant A is normally not very critical. For example, deviations of distribution temperature up to 50 K from Illuminant A would not cause notable errors when the illuminance meter has a reasonable spectral responsivity match to $V(\lambda)$—e.g., with its $f_1'$ value [38] of 3% or less.

As a good example of utilization of the detector-based method, a high illuminance calibration facility was developed at NIST. With traditional luminous intensity lamp standards, the illuminance scale is available up to a level of $\approx$5000 lx. A much higher illuminance level is often required for calibration of illuminance meters. To meet such needs, a high illuminance calibration facility as shown in Figure 7.7 was developed [39]. The facility utilizes a commercial solar simulator source employing a 1000 W xenon arc lamp with an optical feedback control and originally provides $\approx$300 klx of illumination of xenon spectra (6500 K). The source is also combined with a set of selected color glass filters that modify its spectral power distribution to

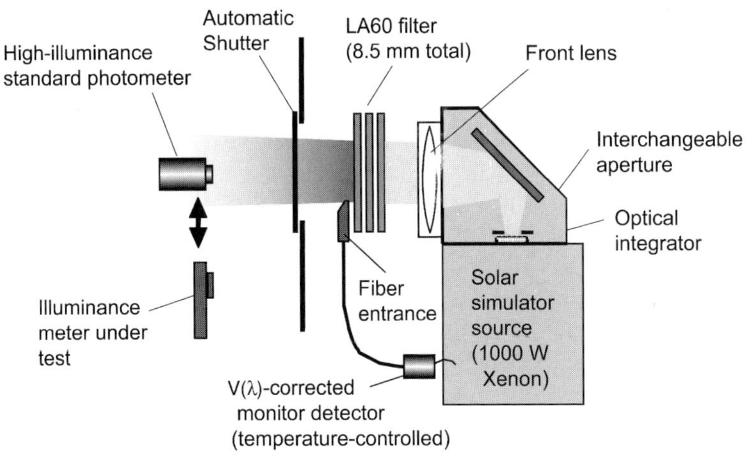


FIG. 7.7. Configuration of the high illuminance calibration facility.

approximate CIE Source A (2840–2860 K) at an illuminance level of ≈75 klx. The illuminance level can be varied without changing the color temperature significantly and without changing the distance. The illuminance scale is provided by a set of high-illuminance standard photometers that are calibrated against the NIST reference photometers (used to maintain the candela) and have been verified for linear response up to 100 klx. This is also an example utilizing the large linearity range of detector standards.

### 7.4.5 Application of Detector-Based Methods for Luminous Intensity Calibration

Utilizing the wide linearity range of standard photometers, the luminous intensity of lamps in a large range of wattage can be directly calibrated. A need for maintaining a number of working standard lamps is eliminated. When measuring luminous intensity of lamps using a standard photometer, care should be taken to ensure that stray light is controlled to a negligible level because, unlike the conventional lamp-based substitution method, the effects of stray light will not be cancelled out with the detector-based method for luminous intensity measurement.

## 7.5 Luminous Flux Standards and Measurements

Lumen is commonly realized by national laboratories using goniophotometers, which require a large dark room and a costly, high-precision positioning mechanism. It also takes long hours of operation for a goniophotometer to take data at many points (e.g., 2592 points for $5^\circ \times 5^\circ$ scan); and, yet, the photometer head scans only a small portion of the total spherical area (typically less than 5% with a point-by-point measurement, and less than 20% with a spiral, continuous scan [40]). Rigorous uncertainty analyses are required for lamps having structured angular intensity distributions. As the burning time of the lamps should be kept to a minimum, the scanning intervals and the measurement time are always compromised. Integrating spheres, on the other hand, provide instantaneous and continuous spatial integration (almost 100% coverage) over the entire spherical area, which is a great benefit over goniophotometers. However, they could not be used for absolute measurement of luminous flux mainly due to their spatial nonuniformities, which were not well understood. Integrating spheres were used only as a relative measurement device. While detector-based methods were already applied to illuminance and luminous intensity measurements with many benefits, it was not possible to apply detector-based methods to total luminous flux measurements until a new method described in the next section was developed.

### 7.5.1 Absolute Integrating Sphere Method

A new method for realization of the lumen using an integrating sphere (referred to as the *Absolute Integrating-Sphere Method*) was recently developed at NIST. The feasibility of this method was first studied through computer simulations [41] and a preliminary experiment [42]. With this method, the total flux of a lamp inside the sphere is calibrated against the known amount of flux introduced into the sphere from an external source through a calibrated aperture. The key element of this method is the correction for the spatial nonuniformity of the integrating sphere. A theory and experimental procedure using a scanning beam were developed to allow for this correction. A NIST luminous flux unit was realized using this method in 1995 with the relative expanded uncertainty ($k = 2$) of 0.53% [43].

Figure 7.8 shows the concept design for the Absolute Integrating-Sphere Method. The flux from the external source is introduced through a calibrated aperture placed in front of the opening. The internal source, a lamp to be calibrated, is mounted in the center of the sphere. The external source and the internal source are operated alternately. When the external flux is

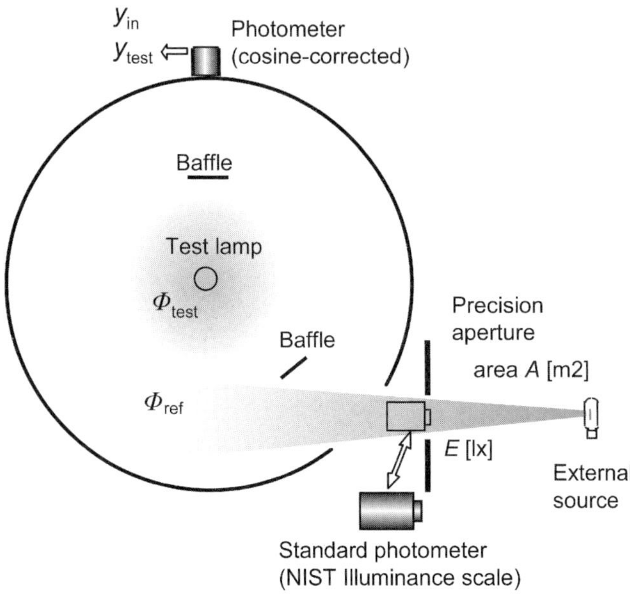


FIG. 7.8. Concept design for the Absolute Integrating-sphere Method.

introduced, the internal source is not operated but remains in the sphere. The principles of this method are given below.

The flux $\Phi_{\text{ext}}$ [lm][1] from the external source is given by

$$\Phi_{\text{ext}} = E \cdot A \tag{7.9}$$

where $E$ [lx] is the average illuminance from the external source over the limiting aperture of known area $A$. Then, the total luminous flux $\Phi_{\text{int}}$ of the internal source is obtained by comparison to the luminous flux introduced from the external source as given by

$$\Phi_{\text{int}} = c_{\text{f}}\, \Phi_{\text{ext}} \frac{y_{\text{int}}}{y_{\text{ext}}} \tag{7.10}$$

where $y_{\text{int}}$ is the detector signal for the internal source, and $y_{\text{ext}}$ that for the flux from the external source. The quantity $c_{\text{f}}$ is a correction factor for various non-ideal behaviors of the integrating sphere, and given by

$$c_{\text{f}} = \frac{F^*_{\text{int}}}{F^*_{\text{ext}}} \frac{k_{\text{s,int}}}{k_{\text{s,ext}}} \frac{\rho_{45}}{\rho_0} \tag{7.11}$$

where $F^*_{\text{int}}$ and $F^*_{\text{ext}}$ are the spectral mismatch correction factors of the integrating sphere system for the internal source and for the external source, respectively, against CIE Standard Illuminant A. $k_{\text{s,int}}$ and $k_{\text{s,ext}}$ are the spatial nonuniformity correction factors of the integrating sphere system for the internal source and for the external source, respectively, against an isotropic point source. These factors are obtained by scanning a narrow beam source inside the integrating sphere. The details of such spatial mapping measurements are described in the next section. The quantities $\rho_0$ and $\rho_{45}$ are the diffuse reflectances of the sphere coating at 0° and 45° incidence, respectively. This correction is necessary because the light from the external source is incident at 45° while the light from the internal source is incident normally. A self-absorption correction is not necessary as the test source stays in the sphere when the sphere is calibrated with the external source.

The absolute integrating sphere method was introduced or experimented with, at a few other national laboratories [44–47]. One of them established its luminous flux unit officially using this method [47]. There is also an interest from industry to apply the detector-based measurements of luminous flux, and procedures simplified for use in industrial laboratories are proposed [48].

[1]As an aid to the reader, the appropriate SI unit in which a quantity should be expressed is indicated in brackets when the quantity is first introduced.

### 7.5.2 Correction for Spatial Nonuniformity Errors

The responsivity of the integrating sphere is not uniform over the sphere wall due to baffles and other structures inside the sphere and also due to nonuniform reflectance of the sphere wall. A method was developed to measure the spatial nonuniformity of the sphere by using a scanning beam source.

The spatial responsivity distribution function (SRDF) of the sphere, $K(\theta, \phi)$, is defined as the sphere response for the same amount of flux incident on a point $(\theta, \phi)$ of the sphere wall or on a baffle surface, relative to the response at the origin, $K(0, 0)$.

The SRDF of a real integrating sphere depends not only on the theoretical effect of the baffle but also on the uneven thickness of coating, contamination of its surface (particularly the lower hemisphere), the gap between the two hemispheres, and other structures such as the auxiliary lamp and the lamp holder. For correction purposes, the SRDF must be obtained by actual measurements.

The SRDF $K(\theta, \phi)$ of an integrating sphere can be obtained by measuring the detector signals while rotating a beam spot inside the sphere. The rotating lamp must be insensitive to its burning position. $K(\theta, \phi)$ is then normalized to $K^*(\theta, \phi)$ for the sphere response to an ideal point source as defined by

$$K^*(\theta, \phi) = \frac{4\pi K(\theta, \phi)}{\int_{\phi=0}^{2\pi} \int_{\theta=0}^{\pi} K(\theta, \phi) \sin\theta \, \mathrm{d}\theta \, \mathrm{d}\phi} \tag{7.12}$$

From $K^*(\theta, \phi)$, the spatial nonuniformity correction factor $k_{\mathrm{s,ext}}$ for the external source with respect to an isotropic point source is given by

$$k_{\mathrm{s,ext}} = \frac{1}{K^*(\theta_{\mathrm{e}}, \phi_{\mathrm{e}})} \tag{7.13}$$

where $(\theta_{\mathrm{e}}, \phi_{\mathrm{e}})$ is the location of the center of the beam spot of the external source. The spatial correction factor $k_{\mathrm{s,int}}$ for the internal source with respect to a point source is given by

$$k_{\mathrm{s,int}} = \frac{\int_{\phi=0}^{2\pi} \int_{\theta=0}^{\pi} I_{\mathrm{rel}}(\theta, \phi) \sin\theta \, \mathrm{d}\theta \, \mathrm{d}\phi}{\int_{\phi=0}^{2\pi} \int_{\theta=0}^{\pi} I_{\mathrm{rel}}(\theta, \phi) K^*(\theta, \phi) \sin\theta \, \mathrm{d}\theta \, \mathrm{d}\phi} \tag{7.14}$$

where $I_{\mathrm{rel}}(\theta, \phi)$ is the relative luminous intensity distribution of the internal source. The factor $k_{\mathrm{s,int}}$ can be assumed to be unity for most luminous flux standard lamps, which have a relatively uniform spatial distribution, if the sphere is designed and fabricated appropriately. The correction factor $k_{\mathrm{s,int}}$ needs to be evaluated for directional light sources such as light-emitting

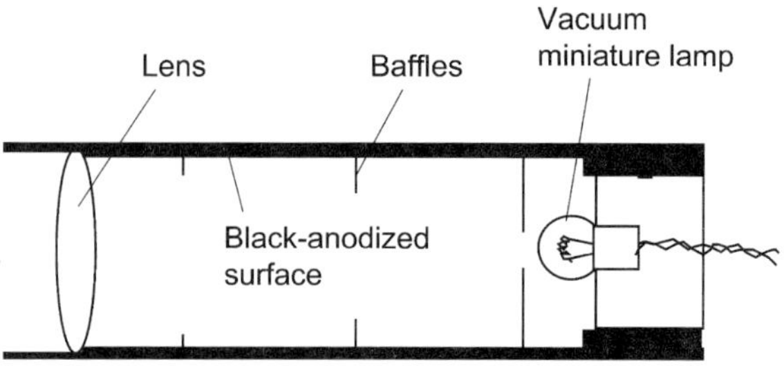


FIG. 7.9. Construction of the beam source for the sphere scanner.

diodes (LEDs). The factor $k_{s,ext}$, on the other hand, is a critical component of the correction. See Reference [49] for further details on these correction factors.

Figure 7.9 shows a construction of a beam scanner for spatial nonuniformity measurements used at NIST. The NIST integrating sphere is equipped with computer-controlled rotation stages on the top and the bottom of the sphere, which can rotate the lamp holder horizontally. Then, at the lamp socket, another small rotation stage is mounted and rotates the beam source vertically. Thus, the beam is scanned over the $4\pi$ solid angle. The beam source consists of a vacuum miniature lamp and a lens as shown in the figure. A vacuum lamp is used to make the source insensitive to its burning position (gas-filled lamps are burning-position sensitive). However, the luminous flux from the beam source using a vacuum lamp is very low ($\approx 0.2$ lm). White LEDs are promising for this purpose, and a development is already reported [45]. A new beam source using a high power white LED is also being developed at NIST for this purpose.

Figure 7.10 shows an example of the mapping data of the NIST 2.5 m integrating sphere. In this figure, $\theta = 0°$ at the top and $\theta = 180°$ at the bottom of the sphere, and $\phi = 0°/360°$ is the plane where the photometer head is located. Various structures in the sphere are seen in the data. A large drop in the center of the figure is the effect of the shadow of the baffle. The two grooves at $\phi = 70°$ and $250°$ are the hemisphere joints. It is also observed that the responses in the upper hemisphere appear slightly lower than in the lower hemisphere (probably due to more spray falling in the lower half when it was coated). The overall uniformity of this integrating sphere, however, is considered excellent. From the SRDF data, the spatial nonuniformity correction factor for the external beam $k_{s,ext}$ was determined to be 0.9992.

This correction technique for the spatial nonuniformity of integrating spheres is useful not only for the Absolute Integrating Sphere Method but also for conventional substitution methods where test lamps having various

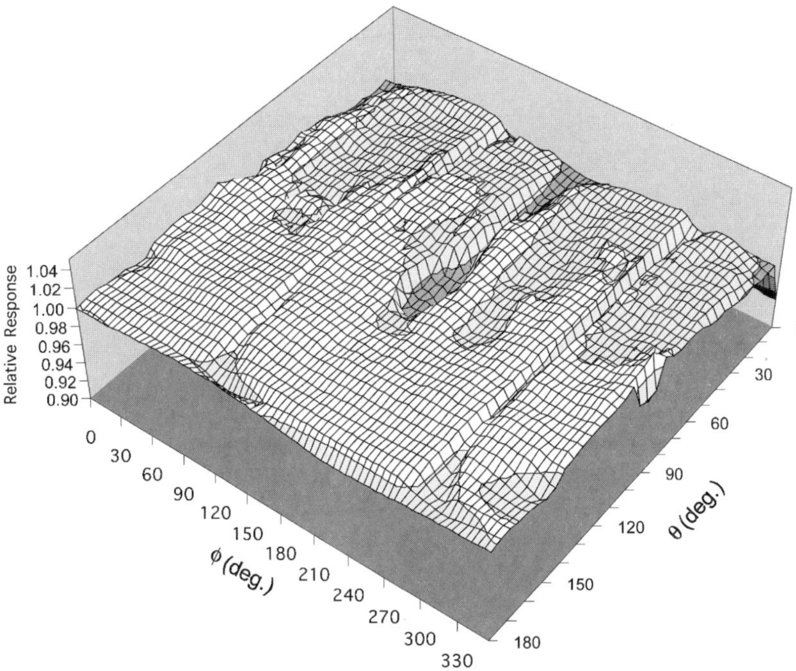


FIG. 7.10. The mapping of the NIST 2.5 m integrating sphere responsivity (normalized SRDF).

different intensity distributions are measured against one type of standard lamp. The errors or uncertainties in luminous flux measurements of lamps having various different intensity distributions in an integrating sphere were not clearly understood. In collaboration with a lamp manufacturer, a series of computer simulations were performed for several different types of incandescent and discharge lamps under various conditions (reflectance of the coating, etc.) of an integrating sphere, and the estimated errors were reported [50].

### 7.5.3 The Detector-Based Luminous Flux Calibration

At the time of the first implementation of the Absolute Integrating Sphere Method at NIST in 1995, the facility with a 2 m integrating sphere was not automated, and only the primary standard lamps were calibrated. The routine calibrations of luminous flux were still based on substitution with the standard lamps. A new, 2.5 m integrating sphere system as shown in Figure 7.11 was built at NIST in 1997; it is automated so that the Absolute

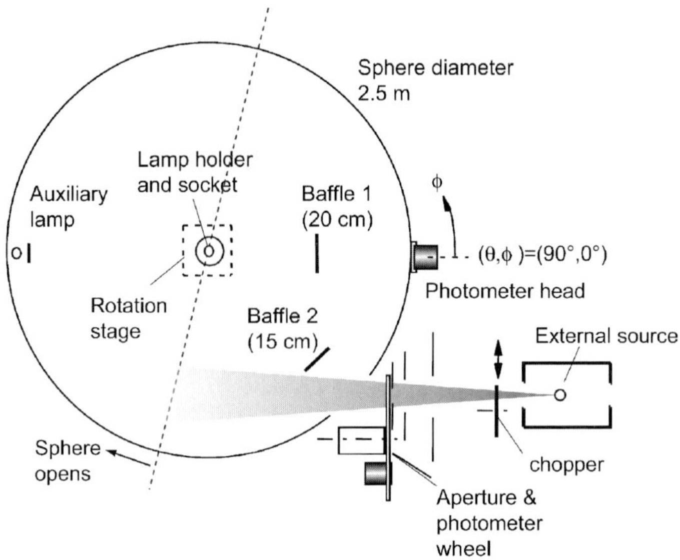


FIG. 7.11. Arrangement of the NIST 2.5 m integrating sphere for the detector-based total luminous flux calibration.

Integrating Sphere Method is used not only for realization of the lumen but also for routine calibration of each test lamp [49, 51]. The sphere is designed similar to the original concept design described in Section 7.5.1, with a small modification of geometry for instrumentation convenience. This system allows for calibration of test lamps based on the illuminance standard photometers, with no need for luminous flux standard lamps, and has enabled a direct, detector-based calibration of luminous flux for the first time.

The sphere system is equipped with an aperture/photometer wheel at the sphere opening. The wheel is computer controlled and has four positions. A precision aperture (50 mm diameter) is mounted in one position, and another position works as a shutter to block the incoming beam. The wheel is placed as close to the sphere opening as possible to minimize diffraction losses. The other two positions are used to mount the standard photometers to measure the illuminance at the center of the aperture. These standard photometers are the temperature-controlled type and annually calibrated against the NIST illuminance unit, and have long-term stability of better than 0.1% per year. The illuminance distribution over the aperture area was measured by spatially scanning a cosine corrected photometer to determine the ratio of the average illuminance to the aperture center illuminance.

An external source, a 1000 W FEL lamp, is operated throughout a measurement session. When a test lamp is mounted in the sphere (not yet turned on), the wheel is set to one of the photometers to measure the illuminance of the external source, and then the aperture is set to introduce the flux into the sphere. The photometer signal is measured to calibrate the sphere responsivity under this condition. The wheel is turned to close the shutter for the external source, and then the test lamp is turned on, stabilized, and its luminous flux is measured based on the sphere responsivity. In this manner, the sphere is calibrated immediately before (or after) each test lamp is measured, taking into account such factors as the self-absorption, long-term drift of the sphere responsivity, and sphere responsivity variations due to mechanical repeatability of the sphere closure. The sphere is coated with a barium sulfate-based coating having a reflectance of approximately 98% in the visible region since the spatial uniformity of the sphere responsivity is very critical in this method.

### 7.5.4 AC/DC Technique for the Integrating Sphere Calibration

The flux level of typical incandescent lamps ($\approx 10^3$ lm) and that of the external beam used in the Absolute Integrating Sphere Method differ by a few orders of magnitude. The measurement results are therefore liable to errors due to a possible effect of heat by the test lamp on the sphere coating and/or to the nonlinearity of the detector system. A technique (referred to as the *AC/DC technique*) has been developed under a collaboration between NIST and International Bureau of Weights and Measures (BIPM), Sevres, France, to measure the integrating sphere characteristics while the test lamp is operating inside the sphere [46].

The AC/DC technique is used for an integrating sphere system with an external source. A chopper is inserted in the beam path of the external source, and the introduced light is chopped at 90 Hz. When the internal lamp is turned on in the sphere, the AC signal from the chopped external beam is superimposed on the DC signal from the internal lamp. As the AC signal is very small (typically $10^{-3}$ of the DC signal), a lock-in amplifier is used to separate and measure the AC signal with a sufficient signal-to-noise ratio. The AC signal is monitored simultaneously with the DC signal when the internal lamp is turned on and off. The AC signal should stay constant if the sphere responsivity is constant. Any changes of the AC signal when turning on the internal lamp indicate a change of the sphere responsivity. The cause of the change can be a thermal effect by the internal lamp on the sphere coating or on the photometer head (which appear as a gradual change), the nonlinearity of the photometer head (which appears as a sudden change), and/or the change of the self-absorption of the internal lamp (which is not well known).

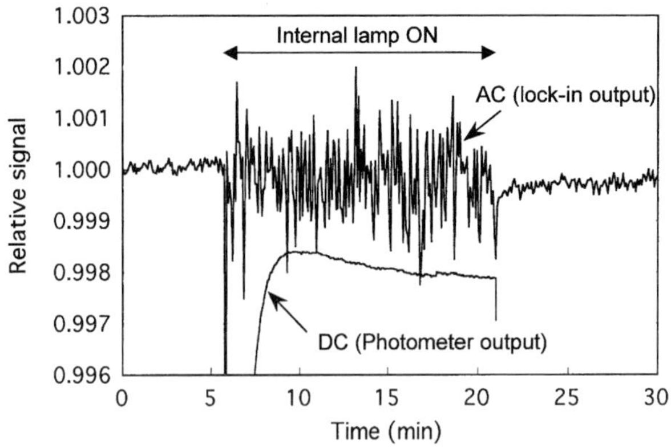


FIG. 7.12. The result of the AC/DC measurement of the NIST integrating sphere with a 1000 W tungsten lamp operated in the sphere.

Figure 7.12 shows an example of the result of the AC/DC measurement of the NIST 2.5 m integrating sphere with a 1000 W tungsten halogen lamp ($\approx$25,000 lm) operated in the sphere. Although the noise of the AC signal is high when the internal lamp is on, the average level of the AC signal stays nearly constant. No obvious sudden change of the AC signal is observed when the lamp is turned on, which validates the linearity of the photometer head. A slight gradual change (0.03–0.04%) is observed over the 15 min burning time, which is probably due to some effect of heat, but is at a negligible level in most cases. This technique can also be used to measure the change of the self-absorption of a discharge lamp when the lamp is turned on or off, which was not possible to evaluate before. Such results have not yet been reported.

## 7.6 Detector-Based Methods for Other Photometric Quantities

### 7.6.1 Detector-Based Luminance Scale

Luminance units are traditionally established using a white reflectance standard or a transmitting diffuser illuminated by a luminous intensity standard lamp. The determination of the luminance factor of the material includes comparison of the incident and outgoing illuminances, which differ by 3–4 orders of magnitude, and which makes precise calibration difficult. The uncertainty is also limited by that of the standard lamp used.

A luminance scale can be established by using an illuminance standard photometer and an integrating sphere source, with less uncertainty and

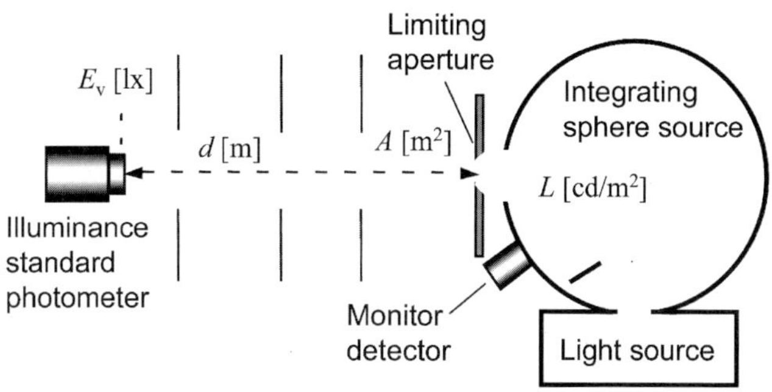


FIG. 7.13. Configuration for luminance unit realization at NIST.

difficulty than the traditional method using a diffuse reflectance or transmittance standard. As an example, Figure 7.13 shows the geometry and the principles of the realization of a luminance scale used at NIST.

A limiting aperture with known area $A$ [m$^2$] is mounted in front of the opening of the integrating sphere source. The illuminance standard photometer measures the illuminance $E_v$ [lx] at distance $d$ [m] from the aperture reference plane. The average luminance $L_v$ [cd/m$^2$] over the aperture plane is given by

$$L_v = k_G \frac{E_v d^2}{A} \tag{7.15}$$

where $k_G$ is a geometrical correction factor determined by the radius $r_s$ of the source aperture, the radius $r_d$ of the detector aperture, and the distance $d$, as given by

$$k_G = \frac{d^2}{2r_s^2 r_d^2}\left[d^2 + r_d^2 + r_s^2 - \sqrt{(d^2 + r_d^2 + r_s^2)^2 - 4r_d^2 r_s^2}\right] \tag{7.16}$$

The aperture should be placed close to the sphere opening in order to reduce the diffraction loss [52] to a negligible level caused by the aperture. Geometrical factors in various geometries are found [53].

The sphere source is normally operated at a correlated color temperature of 2856 K. If the source is operated at a different correlated color temperature, a spectral mismatch correction (see Section 7.4.1) should be applied to the illuminance standard photometer.

### 7.6.2 Photometric Unit for Flashing Lights

The unit for luminous exposure, lux second (lx s), is realized at NIST using flashing-light standard photometers [54]. As one of the derivation methods employed in this scheme, a flashing-light standard photometer

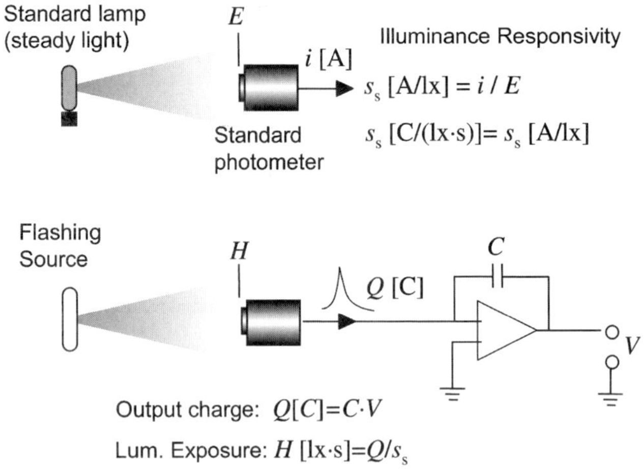


FIG. 7.14. Derivation of the luminous exposure unit.

(a silicon photodiode combined with a $V(\lambda)$ filter and a front aperture) is calibrated for illuminance responsivity $s_s$ [A/lx] with steady light against NIST reference photometers, and then, the same value holds for the responsivity in coulomb/(lx s), [C/(lx s)], since coulomb is ampere second. With a current integrator using a calibrated capacitor as shown in Figure 7.14, the photometer output current is integrated with the capacitance $C$ [F]. The electric charge $Q$ [C] in the capacitor is related to the capacitance and the output voltage $V$ [V] by the formula

$$Q = CV \tag{7.17}$$

The luminous exposure $H_v$ [lx s] incident on the photometer is then determined by

$$H_v = Q/s_s \tag{7.18}$$

From Eqs. (7.17) and (7.18), the responsivity $s_f$ [V/(lx s)] of the photometer including the current integrator is given by

$$s_f = s_s/C \tag{7.19}$$

From the output voltage of the current integrator, the luminous exposure $H_v$ is obtained by

$$H_v = V/s_f \tag{7.20}$$

The subscripts s and f in the quantity symbols represent steady light and flashing light, respectively.

### 7.6.3 LED Intensity Measurement

For measurement of LEDs, CIE recommends that test LEDs be calibrated against standard LED, having spectral power distributions and geometrical characteristics as close to the test LEDs as possible [55], whereby no corrections are needed with such substitution measurements (source-based method). While this method would assure simple and accurate measurements in industry, many standard LEDs of different colors and types would be required, and it is not realistic.

As an alternative approach, assuming that the approximate relative spectral distributions of LEDs are known, a detector-based method can be employed where the photometer is simply calibrated against CIE Illuminant A and measures LEDs with spectral mismatch correction factors. Such approach is taken by NIST [56]. Two LED standard photometers equipped with a $1\,\text{cm}^2$ circular aperture have been developed for calibration of Averaged LED Intensity defined by CIE [55]. The measurement geometry, shown in Figure 7.15, is standardized due to the fact that most LEDs are not point sources and luminous intensity values vary depending on the distance used. Since these standard photometers are used at short distances (10 and 31.6 cm), care should be taken when calibrating them. Errors may occur if these photometers are calibrated in a normal far field conditions (e.g., lamp-to-photometer distance $\approx 3\,\text{m}$) due to near field effects. Errors in the position of reference plane will also be critical at such short distances. At NIST, these LED photometers were calibrated (against the primary standard photometers holding the illuminance scale) using an integrating sphere source with a 6 mm aperture placed at exactly the same distances (10 and 31.6 cm) [56]. More recently, a tunable-laser based facility [57] was used to measure absolute spectral irradiance responsivity of the LED photometers at the same distances [58]. This technique using tunable lasers producing near-Lambertian monochromatic irradiance provides accurate

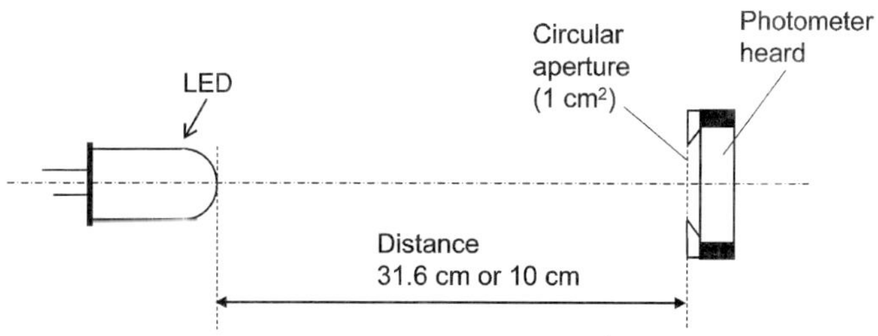


Fig. 7.15. Geometry for CIE Averaged LED Intensity. The distance 31.6 or 10 cm is selected for CIE Condition A or B.

measurement of the spectral responsivity as well as illuminance responsivity in the near field.

## 7.7 Color Temperature Standards and Measurements

*Color temperature* is a quantity related to the spectrum of a light source, and is important in photometry, in that photometric measurements often require knowledge of the spectrum of the light source. For example, the luminous intensity unit, the candela, is realized using a calibrated photometer, but only with knowledge of the spectrum of the source $S(\lambda)$ as found in Eq. (7.2). In this section, color temperature and a few related color quantities are discussed as an important part of photometry. For definitions and measurements of many other color quantities, references on colorimetry, [33, 59] for example, should be consulted.

Color temperature is officially defined as "the temperature of a Planckian radiator whose radiation has the same chromaticity as that of a given stimulus" [2]. However, real light sources other than blackbodies, almost never produce exactly "the same" chromaticity as Planckian radiation. Therefore, rigorously speaking, color temperature applies only to theoretical Planckian radiation and blackbodies. In common photometric practice, however, color temperature is often used to describe the spectrum (and thus the operating point) of incandescent lamps. The use of color temperature for normal incandescent lamps is acceptable since their chromaticity coordinates are very close to Planckian locus (normally, within the perceivable color difference) and their spectral distribution can be approximated by Planckian radiation with negligible errors in many photometric applications. Color temperature, in this case, describes the approximate spectral distribution of the source. It is calculated using the same formula as correlated color temperature (described next).

*Correlated color temperature* (CCT) is a quantity to describe the color of light stimulus compared with that of Planckian radiation, and can be used for sources whose spectral power distribution is dissimilar to that of Planckian radiation; such as discharge lamps. Correlated color temperature is defined as "the temperature of the Planckian radiator whose perceived color most closely resembles that of a given stimulus at the same brightness and under specified viewing conditions" [2]. In the current recommendation [33], correlated color temperature is determined from the chromaticity coordinate of the point on the Planckian locus that is at the closest distance from that of the light source in question on the CIE 1960 $u,v$ diagram (now obsolete). CCT can be used for all white light sources including incandescent lamps, having chromaticities within a certain distance from the Planckian locus (see Reference [33] for the details). It should be noted that CCT describes only the color and not the spectral distribution of the source.

Another similar quantity to note is *distribution temperature*, which is defined as "the temperature of the Planckian radiator whose relative spectral distribution is the same or nearly the same as that of the radiation considered in the spectral range of interest" [2]. Distribution temperature is determined by a least square fit of Planckian radiation to the given spectral distribution by changing the temperature and absolute scale of the Planck's equations. Distribution temperature can be used for sources having a relative spectral distribution that deviates within $\pm 10\%$ from the fitted Planckian curve in the defined spectral range (400 to 750 nm). Typical incandescent lamps satisfy this requirement. See Reference [60] for the official definition and other details of distribution temperature.

Distribution temperature is used to specify the color (and thus the operating point) of incandescent lamps; thus, both color temperature and distribution temperature serve the same purpose. Color temperature is widely used in the USA and United Kingdom whereas distribution temperature is widely used in other European countries and Asia. If the relative spectral power distribution of the given radiation is identical to that of the Planckian radiator, the values of color temperature, distribution temperature, and CCT will be all the same. The differences between distribution temperature and CCT for typical incandescent lamps are within only a few kelvins.

### 7.7.1 Realization and Maintenance of the Color Temperature Scale

The color temperature (or CCT) of incandescent lamps is calculated simply from the relative spectral power distribution of the lamp. Therefore, the color temperature scale is derived directly from the spectral irradiance scale. The color temperature scale is derived at NIST from the NIST spectral irradiance scale [61], which is based on the International Temperature Scale of 1990 [62]. The scale realization chain for the spectral irradiance scale and the color temperature scale is shown in Figure 7.16. Two 1000 W FEL type quartz halogen lamps are maintained as the NIST color temperature primary working standard lamps in the range of 2000–3200 K. These lamps have demonstrated stability of operation in this color temperature range [63]. The spectral irradiance of these lamps is calibrated periodically (based on burning hours) against the spectral irradiance scale at 2000, 2300, 2600, 2856, and 3200 K. The correlated color temperatures of these lamps are computed from the spectral irradiance values at 5 nm intervals according to the definition given in Reference [33].

The color temperature of a test lamp is determined with a spectroradiometer that is calibrated against these primary working standard lamps

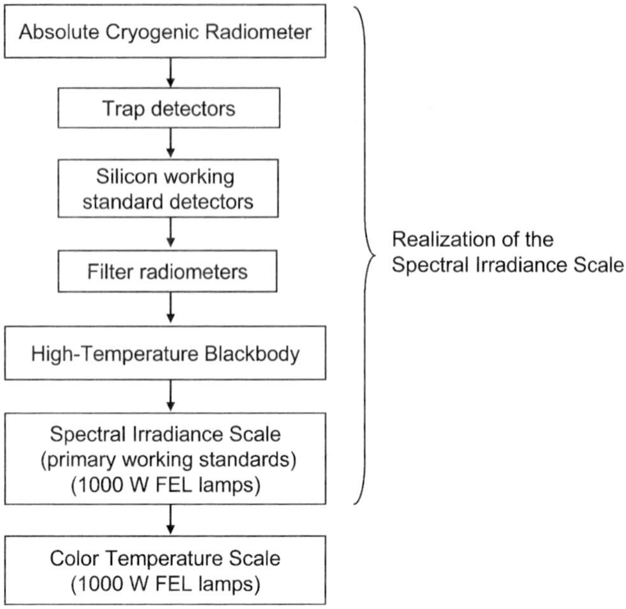


FIG. 7.16. Realization of the NIST spectral irradiance scale and the color temperature scale.

operated at a color temperature closest to that of the test lamp being measured. By calibrating the spectroradiometer using a similar spectrum as that of the test source, the errors of the spectroradiometer such as stray light can be minimized. This allows the use of a single-grating instrument such as an array spectroradiometer, which could otherwise have significant stray light errors.

### 7.7.2 Detector-Based Calibration of the Color Temperature Scale

The calibration of color temperature described above is based on blackbody radiation. The current uncertainty of the color temperature calibration at NIST is 8 K at 2856 K [35]. It is also possible to derive the color temperature scale based on the absolute responsivity of detectors, as was done in the realization of the candela. It is proposed that the uncertainty of the color temperature scale can be reduced by employing the detector-based method, as the uncertainty of the spectral irradiance responsivity calibration is improved [64]. The detector-based calibration of color temperature is achieved by using a tristimulus colorimeter whose spectral responsivity is absolutely calibrated. The principles are summarized below.

Assume a tristimulus colorimeter having three detector channels with absolute spectral irradiance responsivity $s_x(\lambda)$, $s_y(\lambda)$, $s_z(\lambda)$ that are approximately matched to the CIE color matching functions, $\bar{x}(\lambda)$, $\bar{y}(\lambda)$, $\bar{z}(\lambda)$. The responsivity of each detector channel to each tristimulus value for CIE Standard Illuminant A is given by

$$
\begin{aligned}
s_x^* &= \frac{\int_\lambda S_A(\lambda)\, s_x(\lambda)\, d\lambda}{k \int_\lambda S_A(\lambda)\, \bar{x}(\lambda)\, d\lambda} \\
s_y^* &= \frac{\int_\lambda S_A(\lambda)\, s_y(\lambda)\, d\lambda}{k \int_\lambda S_A(\lambda)\, \bar{y}(\lambda)\, d\lambda} \\
s_z^* &= \frac{\int_\lambda S_A(\lambda)\, s_z(\lambda)\, d\lambda}{k \int_\lambda S_A(\lambda)\, \bar{z}(\lambda)\, d\lambda}
\end{aligned}
\tag{7.21}
$$

where $k$ is a normalizing constant and $S_A(\lambda)$ is the spectral distribution of CIE Standard Illuminant A. When a test source (incandescent lamp with unknown color temperature) is measured with this colorimeter, the tentative tristimulus values of the test source are obtained by

$$
\begin{aligned}
X_1 &= j_x / s_x^* \\
Y_1 &= j_y / s_y^* \\
Z_1 &= j_z / s_z^*
\end{aligned}
\tag{7.22}
$$

where $j_x$, $j_y$, and $j_z$ are the output signals from each detector channel. From these tristimulus values, tentative chromaticity coordinates $(x_1, y_1)$, and then the tentative correlated color temperature $T_1$ is calculated. These values are tentative because they include spectral mismatch errors (caused by the difference between the relative spectral responsivity of the detector channels and the color matching functions, and by the difference between the spectral distribution of the source and that of CIE Standard Illuminant A). Then, the relative spectral distribution $S_{T_1}(\lambda)$ of Planckian radiation at temperature $T_1$ is obtained, and the responsivity in Eq. (7.21) is recalculated as

$$
\begin{aligned}
s_{x,T_1} &= \frac{\int_\lambda S_{T_1}(\lambda)\, s_x(\lambda)\, d\lambda}{k \int_\lambda S_{T_1}(\lambda)\, \bar{x}(\lambda)\, d\lambda} \\
s_{y,T_1} &= \frac{\int_\lambda S_{T_1}(\lambda)\, s_y(\lambda)\, d\lambda}{k \int_\lambda S_{T_1}(\lambda)\, \bar{y}(\lambda)\, d\lambda} \\
s_{z,T_1} &= \frac{\int_\lambda S_{T_1}(\lambda)\, s_z(\lambda)\, d\lambda}{k \int_\lambda S_{T_1}(\lambda)\, \bar{z}(\lambda)\, d\lambda}
\end{aligned}
\tag{7.23}
$$

By using these revised responsivities, the tristimulus values are recalculated, and the second value of color temperature $T_2$ is obtained. With this iteration, spectral mismatch errors will be mostly removed. Another iteration to obtain $T_3$ will be sufficient to converge the result to required accuracy.

Even after sufficient iterations, a small spectral mismatch error remains due to the small deviation of the spectral distribution of the test source from the Planckian radiation at the same color temperature. If the colorimeter has well-matched spectral responsivities ($f'_1 < 3\%$ for all channels), the residual theoretical error for normal incandescent lamps (operated at 2800–3200 K) may be practically negligible ($<0.1$ K).

The uncertainty of the color temperature values determined using this method is dominated by the uncertainty of measurements of the absolute responsivity $s_x(\lambda)$, $s_y(\lambda)$, $s_z(\lambda)$ of the colorimeter channels. However, fully correlated uncertainty components (systematic errors) affecting the relative values of all the channels are cancelled out due to the fact that the color temperature is determined by the ratios of $X$, $Y$, and $Z$ to their sum. For approximate estimation, a 0.1% relative error in one of tristimulus values leads to a $\approx 5$ K error in measured color temperature at around 2850 K. The measurement of spectral irradiance responsivity of colorimeter heads at a 0.1% level of uncertainty will be a considerable challenge, and requires well-designed colorimeter heads and a low uncertainty spectral irradiance responsivity calibration facility. The spectral responsivity curves, especially for the z channel, are fairly steep, and a narrowband spectral measurement is required. A laser-based spectral responsivity calibration facility [57] is promising for low-uncertainty calibration of colorimeter heads. An attempt to derive a color temperature scale using such a facility is in progress [65].

## 7.8 International Intercomparisons of Photometric Units

International intercomparisons of photometric units as well as radiometric units are occasionally conducted by CCPR to evaluate and ensure the agreement of photometric units disseminated by different countries. Such comparisons originally started for scientific purposes to determine the best estimate of the SI units. Since the Mutual Recognition Arrangement (MRA) was signed in 1999 [66], these intercomparisons are given a new objective, which is to establish the equivalence of the units disseminated by different countries and to mutually recognize their calibration, in order to facilitate international commerce and trade. After the MRA, essential quantities required for periodic intercomparisons were chosen and these comparisons are named "Key Comparisons." Comparisons of other quantities are conducted as "Supplementary Comparisons" with lower priorities. Also, these comparisons are limited to participation by CCPR member countries. To cover many other countries, similar comparisons are conducted within Regional Metrology Organizations (RMOs) such as EUROMET (for Europe), SIM (Americas), and APMP (for Asia and Pacific countries). The results of these regional comparisons are linked to the results of Key Comparisons.

In photometry, three Key Comparisons were conducted most recently in 1998; CCPR K3.a luminous responsivity, CCPR K3.b Luminous intensity, and CCPR K4 Luminous flux. While these comparisons are conducted for the purpose of the MRA, the results of these comparisons provide us with information on the current, state-of-the-art uncertainties of photometric measurements worldwide. Figures 7.17–7.19 show the summary of the results of these comparisons. K3.a compared the luminous intensity using incandescent lamps as transfer standards. K3.b compared the luminous responsivity of photometer heads, thus this compared the illuminance unit using detectors. Therefore, the results of K3.a and K3.b should be, theoretically, in reciprocal relationship. K3.b was the first photometric comparison conducted using detectors as transfer standards with an expectation that a better agreement of results would be obtained using detectors than lamps. K4 compared the total luminous flux of incandescent standard lamps. For both luminous intensity and luminous flux, the overall results are

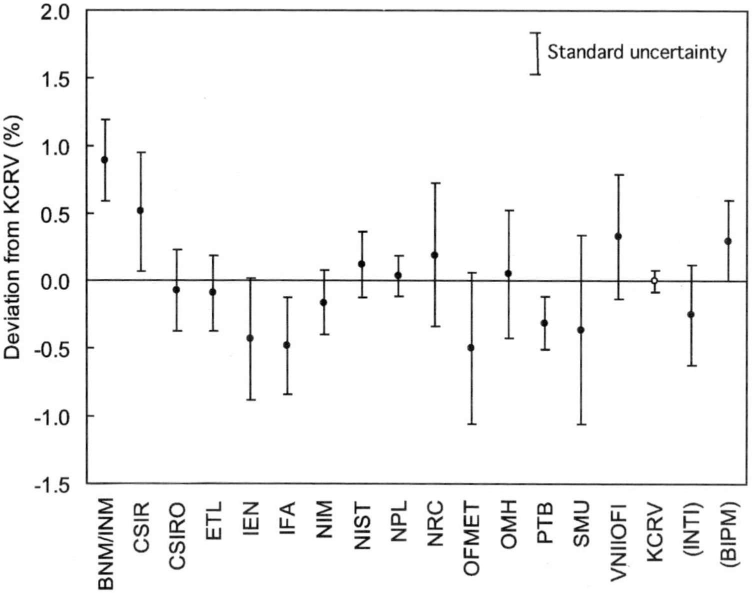


FIG. 7.17. Results of CCPR K3.a Luminous Intensity.

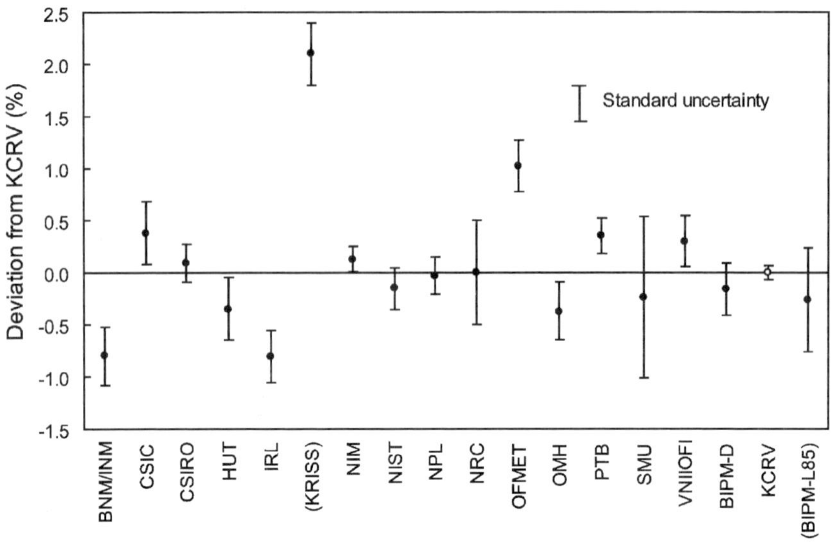


FIG. 7.18. Results of CCPR K3.b Luminous Responsivity.

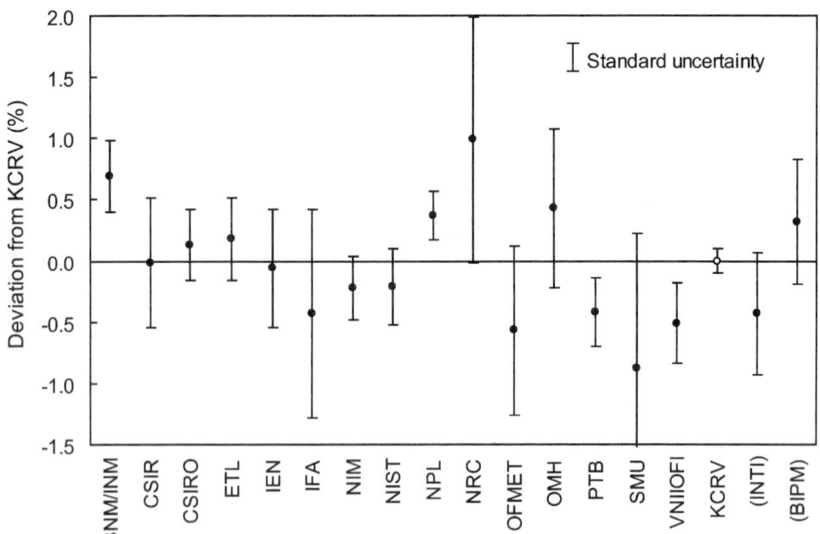


FIG. 7.19. Results of CCPR K4 Luminous Flux.

mostly within ±0.5% except a few outliers. The agreement of overall results slightly improved from the previous comparisons in 1985 though not as much as had been expected. It is noted, however, that the agreement among several major national laboratories is within a few tenths of a percent in both luminous intensity and flux, and is a considerable improvement from 1985. For further details of these comparisons, see the final reports of the comparisons (Reference [67] for K3.a and K4, and Reference [68] for K3.b).

## 7.9 Future Prospects in Photometry

As discussed in this chapter, significant improvements in photometry have been made over the last decade at the national laboratory level utilizing detector-based methods. As demonstrated in the international intercomparisons, the state-of-the-art uncertainties of photometric calibrations among major national laboratories are around 0.5% ($k = 2$). This level of uncertainty in photometry is much larger than that in radiometry where uncertainties of one order of magnitude smaller are achieved. The difficulties in realizing photometric units are in photometry they are for broadband radiation in irradiance geometry (as opposed to monochromatic radiation and in a beam geometry in radiometry). A new technique that is expected to fill this gap will be the tunable laser-based facility for spectral irradiance responsivity calibration [57]. Similar facilities are being developed at other national laboratories. It is expected that the uncertainties for photometric base unit will be reduced to a half of the current level by utilizing such new facilities. This level of uncertainty will probably suffice for the current needs of the photometric base unit at a national laboratory level.

The purpose of realizing photometric units by national laboratories is to disseminate the units with the lowest uncertainties possible and ensure that measurements in industry are done with required accuracy. The uncertainties achieved (as demonstrated in the intercomparisons) and standards provided by the national laboratories are mostly for incandescent lamps, though some effort is made to provide calibrations for other sources such as LEDs. The situation in industry is that all kinds of light sources are measured under various conditions. For example, uncertainties ($k = 2$) of $\approx 5\%$ are estimated for measurement of various types of discharge lamp measured in lamp factories [69]. In the case of LED measurements, discrepancies of results as much as 30% among different industrial laboratories were reported [70]. Such a large gap in measurement uncertainty between the national standards and industrial measurements may be one of the important aspects in photometry that needs to be addressed.

One of the reasons for the large gap may be that large errors are involved when transferring the scale from incandescent lamp standards to various different sources having dissimilar characteristics. Necessary corrections tend not to be applied in industry due to the cost of such operations or lack of knowledge. One way to assist industry in this respect has been to provide standards and calibrations for more varieties of artifacts to allow them to perform strict substitution measurements. Such efforts are being made, e.g., for calibration of LEDs and flashing lights as discussed in Sections 7.6.2 and 7.6.3. However, such an extension of calibration services by national laboratories is limited and cannot cover all applications. It is necessary that measurement techniques used in the industry be improved. Further research is needed for simpler methods or calibration schemes that can reduce uncertainty and can be easily implemented in the industry environment. For example, methods used by national laboratories can be adapted for easier implementation by industry [71].

Another reason for the large variation in measurement in industry may be due to differences in measurement conditions used. This was experienced, e.g., in LED intensity measurements, where the luminous intensity values varied significantly at different distances used. This was addressed by the CIE and standardized procedures were developed as discussed in Section 7.6.3. Similar problems now exist in measurement of luminous flux of LEDs, which is being addressed by a CIE technical committee. A large variation in measurement of retroreflective materials has been also due to different measurement geometries used [72]. Further standardizations of measurement methods and education will play important roles in future improvements in photometry.

The largest difficulty in photometric measurements of various light sources is probably due to spectral mismatch errors. Spectral mismatch errors are inevitable for all photometers ($V(\lambda)$-corrected detectors). While procedures are available for the corrections (as given in Eq. (7.8)), they are often not followed in industry. The spectrum of the source is needed for this correction, which necessitates the use of a spectroradiometer. The spectral mismatch errors are particularly serious in the LED measurements, and the requirements for spectral responsivity match of photometers are of great concern. To minimize spectral mismatch errors, strict substitution is generally recommended. For example, Reference [55] recommends that photometers to measure LEDs be calibrated against standard LEDs of the same color (spectrum) as the LED under test. Such strict substitution methods, however, are not realistic for sources like LEDs because there are so many different colors of LEDs produced, and it would cost too much for industry to obtain or maintain many standards.

A better way to deal with this problem may be to use a spectroradiometer. Photometric quantities can be measured theoretically with no spectral mismatch errors using a spectroradiometer. This method is becoming realistic as array spectroradiometers are becoming increasingly common and affordable; they provide the same speed as photometers and their quality has been significantly improved recently. Spectroradiometers also provide a benefit of measuring a photometric quantity and colorimetric quantities at the same time. Integrating spheres using a spectroradiometer as a detector are increasingly used by the lamp manufacturers and the LED industry, but measurement variation is not improving due to the lack of standards. Such systems need to be calibrated spectrally, against a spectral radiant flux standard. The need for total spectral radiant flux standards has been stressed in the optical radiation measurement community [73]. In response, a new traceability chain for luminous intensity and luminous flux measurements will be established using such spectral standards, in addition to conventional standards for photometric units. The industry should also be guided toward such a new way of photometric measurements. The concept of strict substitution, however, should still be followed for geometrical aspects. Standard lamps of different sizes and power levels will be needed in industry. For LED applications, it would be ideal if spectral radiant flux standards in the form of a broadband LED that has emission in the entire visible region can be developed. Even when measurements are done spectrally, photometrically calibrated standards such as LEDs of different colors and some discharge lamps will still be useful as check standards to validate uncertainties of measurements of particular type of sources at laboratories in the industry.

This chapter covered only physical photometry, i.e., measurement of photometric quantities as SI units. If we consider the real purpose of photometry (to evaluate radiation as human eyes perceive), there are many issues to be addressed in the future. For example, measurement in the mesopic region is becoming important in roadway and outdoor lighting, and standardization of the spectral luminous efficiency function(s) in the mesopic region is urgently needed. In the area of measurements of flashing lights, such as those for aircraft anti-collision lights and emergency warning lights, several different formulae are used to specify effective intensity (unit: candela, the measure of conspicuity of flashing signals), and standardization in this area is also urgently required [74]. Also, it is often questioned whether the 80-year-old spectral luminous efficiency function, $V(\lambda)$, should continue to be used as the basis of photometric units. For example, the function $V_M(\lambda)$, corrected by Judd, is believed to be more accurate than $V(\lambda)$ and has been recognized by the CIE [75], but it has not been adopted for use in metrology due to the long history of the use of $V(\lambda)$ and due to the small changes. The need for spectral luminous efficiency function for 10° fields is discussed [76] and work is in progress to recognize $V_{10}(\lambda)$ in the CIE.

When such a function is officially established, there will be a question as to whether such a new function should be adopted for metrology use and how it can be implemented without causing confusion.

## References


1. J. W. T. Walsh, "Photometry." Constable & Company, London, 1953.
2. "International Lighting Vocabulary," CIE 17.4/IEC 50(845) (1987).
3. "International Vocabulary of Basic and General Terms in Metrology," BIPM, IEC, IFCC, ISO, IUPAC, IUPAP, and OIML (1994).
4. "Quantities and Units: ISO Standards Handbook," IEC, Switzerland, 1993.
5. CIE Compte Rendu, p. 67, (1924).
6. "The Basis of Physical Photometry," CIE 18.2 (1983).
7. "Principles Governing Photometry." BIPM, F-92310 Sèvres, France, 1983.
8. CIE D001: Disc Version of CIE Photometric and Colorimetric Data (tables from Publ. 18.2, 86, ISO 10526/CIE S005 and ISO/CIE 10527) (1988).
9. "Comptes Rendus des Séances de la 16e Conférence Générale des Poids et Mesures (CGPM)." BIPM, F-92310 Sèvres, France, 1979.
10. CIE Compte Rendu, Vol. 3, Table II, 37–39 (1951).
11. CIPM Procés-Verbaux 44, 4 (1976).
12. "Mesopic photometry: History, Special Problems and Practical Solutions," CIE Publication No. 81 (1989).
13. 9e CGPM Compte Rendu, 54 (1948).
14. W. R. Blevin and B. Steiner, Redefinition of the candela and the lumen, *Metrologia* 11, 97–104 (1975).
15. W. R. Blevin, The candela and the watt, *CIE Proceedings P-79-02*, Kyoto, Japan, 1979.
16. "Le Système International d'Unité (SI), The International System of Units (SI)." BIPM, Sèvres, France, 1991.
17. B. N. Taylor, "Guide for the Use of the International System of Units (SI)." Natl. Inst. Stand. Technol. Spec. Publ. 811, US Government Printing Office, Washington, DC, 1995.
18. B. N. Taylor, Ed., "Interpretation of the SI for the United States and Metric Conversion Policy for Federal Agencies." Natl. Inst. Stand. Technol. Spec. Publ. 814, 1991.
19. "SI Units and Recommendations for the Use of their Multiples and of Certain other Units." ISO, Geneva, Switzerland, 1992.

*20.* C. Carreras and A. Corrons, Absolute spectroradiometric and photometric scales based on an electrically calibrated pyroelectric radiometer, *Appl. Opt.* 20, 1174–1177 (1981).
*21.* L. P. Boivin, A. A. Gaertner, and D. S. Gignac, Realization of the new candela (1979) at NRC, *Metrologia* 24, 139–152 (1987).
*22.* E. F. Zalewski and J. Geist, Silicon photodiode absolute spectral response self-calibration, *Appl. Opt.* 19, 1214–1216 (1980).
*23.* "Determination of the Spectral Responsivity of Optical Radiation Detectors," CIE 64 (1984).
*24.* J. L. Gardner, Radiometric and photometric standards in Australia, *CIE Proceedings 22nd Session* 1-1, Melbourne 1991, Div.2, 5–8 (1991).
*25.* C. L. Cromer, G. Eppeldauer, J. E. Hardis, T. C. Larason, and A. C. Parr, National institute of standards and technology detector-based photometric scale, *Appl. Opt.* 32, 2936–2948 (1993).
*26.* Y. Ohno, Silicon photodiode self-calibration using white light for photometric standards—Theoretical analysis, *Appl. Opt.* 31, 466–470 (1992).
*27.* J. E. Martin, N. P. Fox, and P. J. Key, A cryogenic radiometer for absolute radiometric measurements, *Metrologia* 21, 147–155 (1985).
*28.* T. R. Gentile, J. M. Houston, J. E. Hardis, C. L. Cromer, and A. C. Parr, The NIST high-accuracy cryogenic radiometer, *Appl. Opt.* 35, 1056–1068 (1996).
*29.* T. M. Goodman and P. J. Key, The NPL radiometric realization of the candela, *Metrologia* 25, 29–40 (1988).
*30.* C. L. Cromer, G. Eppeldauer, J. E. Hardis, T. C. Larason, Y. Ohno, and A. C. Parr, The NIST detector-based luminous intensity scale, *J. R. NIST* 101(2), 109–131 (1996).
*31.* W. Erb and G. Sauter, PTB network for realization and maintenance of the candela, *Metrologia* 34, 115–124 (1997).
*32.* "Colorimetry – Part 2: CIE Standard Illuminants," CIE DS 014-2.2/E (2004).
*33.* "Colorimetry," 3rd edition, CIE Publication 15, 2004.
*34.* T. C. Larason, S. S. Bruce, and A. C. Parr, Spectroradiometric Detector Measurements, NIST Special Publication 250-41, 1998.
*35.* Y. Ohno, Photometric Calibrations, NIST Special Publication 250-37, 1997.
*36.* G. Eppeldauer and J. E. Hardis, Fourteen-decade photocurrent measurements with large-area silicon photodiodes at room temperature, *Appl. Opt.* 30, 3091–3099 (1991).
*37.* G. Eppeldauer, Temperature monitored/controlled silicon photodiodes for standardization, *Proc. SPIE* 1479, 71–77 (1991).
*38.* "Methods of Characterizing Illuminance Meters and Luminance Meters," CIE 69 (1987).

39. Y. Ohno, High illuminance calibration facility and procedures, *J. IES* 27, 132–140 (1998).
40. "Measurements of Luminous Flux," CIE 84 (1987).
41. Y. Ohno, Integrating sphere simulation—application to total flux scale realization, *Appl. Opt.* 33, 2637–2647 (1994).
42. Y. Ohno, New method for realizing a total luminous flux scale using an integrating sphere with an external source, *J. IES* 24, 106–115 (1995).
43. Y. Ohno, Realization of NIST 1995 luminous flux scale using integrating sphere method, *J. IES* 25(1), 13–22 (1996).
44. M. L. Rastello, E. Miraldi, and P. Pisoni, Luminous-flux measurements by an absolute integrating sphere, *Appl. Opt.* 35, 4385–4391 (1996).
45. K. Lahti, J. Hovila, P. Toivanen, E. Vahala, I. Tittonen, and E. Ikonen, Realisation of the luminous-flux unit using a LED scanner for the absolute integrating sphere method, *Metrologia* 37, 595–598 (2000).
46. Y. Ohno, R. Köhler, and M. Stock, An ac/dc technique for the absolute integrating sphere method, *Metrologia* 37, 583–586 (2000).
47. J. Hovila, P. Toivanen, and E. Ikonen, Realization of the unit of luminous flux at the HUT using the absolute integrating-sphere method, *Metrologia* 41, 407–413 (2004).
48. Y. Ohno, R. Bergman, Detector-referenced integrating sphere photometry for industry, *Proceedings of 2002 IESNA Annual Conference*, Salt Lake City, UT, 311–316 (2002).
49. Y. Ohno, Detector-based luminous flux calibration using the absolute integrating-sphere method, *Metrologia* 35, 473–478 (1998).
50. Y. Ohno and R. O. Daubach, Integrating sphere simulation on spatial nonuniformity errors in luminous flux measurement, *J. IES* 30, 105–115 (2001).
51. Y. Ohno, and Y. Zong, Detector-based integrating sphere photometry, *Proceedings of 24th Session of the CIE*, Vol. 1, Part 1, Warsaw, Poland, 155–160 (1999).
52. W. R. Blevin, Diffraction losses in radiometry and photometry, *Metrologia* 7, 39–44 (1970).
53. "Lighting Handbook," 9th edition, Chapter 9, "Lighting Calculations," Illuminating Engineering Society of North America (2000).
54. Y. Ohno and Y. Zong, Establishment of the NIST flashing-light photometric unit, *Proceedings of SPIE* 3140, 2–11 (1997).
55. "Measurement of LEDs," CIE Publ. 127 (1997).
56. C. C. Miller, and Y. Ohno, Luminous intensity measurements of light emitting diodes at NIST, *Proceedings of 2nd CIE Expert Symposium on LED Measurement*, May 11–12, 2001, Gaithersburg, Maryland, USA, 28–32 (2001).

*57.* S. W. Brown, G. P. Eppeldauer, and K. R. Lykke, NIST facility for spectral irradiance and radiance response calibrations with a uniform source, *Metrologia* 37, 579 (2000).

*58.* C. C. Miller, Y. Zong, and Y. Ohno, LED photometric calibrations at the National Institute of Standards and Technology and future measurement needs of LEDs, *Proceedings of SPIE Fourth International Conference on Solid State lighting*, August 2004, Denver, CO, 69–79 (2004).

*59.* G. Wysescky, and W. S. Stiles, "Color Science, Concepts and Methods, Quantitative Data and Formulae," 2nd edition, Wiley, New York.

*60.* "CIE Collection in Photometry and Radiometry," CIE Publication, 114(4) (1994).

*61.* H. W. Yoon, C. E. Gibson, and P. Y. Barnes, Realization of the National Institute of Standards and Technology detector-based spectral irradiance scale, *Appl. Opt.* 41, 5879–5890 (2002).

*62.* K. D. Mielenz, R. D. Saunders, A. C. Parr, J. J. Hsia, The new international temperature scale of 1990 and its effect on radiometric photometric and colorimetric measurements and standards, *Proceedings of CIE 22nd Session – Division 2*, Melbourne, Australia, 65–68 (1991).

*63.* Y. Ohno and J. K. Jackson, Characterization of modified FEL quartz-halogen lamps for photometric standards, *Metrologia* 32, 693–696 (1996).

*64.* G. Eppeldauer, Spectral response based calibration method of tristimulus colorimeters, *J. Res. NIST* 103, 615–619 (1998).

*65.* G. P. Eppeldauer, S. W. Brown, C. C. Miller, and K. R. Lykke, Improved accuracy photometric and tristimulus-color scales based on spectral irradiance responsivity, *Proceedings of 25th Session of the CIE*, San Diego, CA, Vol. 1, p. D2-30–D2-33 (2003).

*66.* Mutual Recognition of national measurement standards and of calibration and measurement certificates issued by national metrology institutes, Paris, 14 October 1999, Comité International des Pois et Measures (International Bureau of Weights and Measures).

*67.* G. Sauter, D. Lindner, and M. Lindemann, CCPR Key Comparisons K3a of Luminous Intensity and K4 of Luminous Flux with Lamps and Transfer Standards, PTB-Report, Physikalisch Technische Bundesanstalt, December 1999.

*68.* Key Comparison Data Base, available at http://kcdb.bipm.fr/

*69.* R. S. Bergman and Y. Ohno, The art and science of lamp photometry, *Proceedings of Ninth International Symposium on the Science and Technology of Light Sources (LS:9)*, August 2001, Ithaca, NY, 165–176 (2001).

*70.* K. Suzuki, et al, Round Robin LED Photometry Test in Japan, *Proceedings of 2nd CIE Expert Symposium on LED Measurement*, May 2001, Gaithersburg, Maryland, 11–13 (2001).

*71.* Y. Ohno and R. Bergman, Detector-referenced integrating sphere photometry for industry, *J. IES* 32(2), 21–26 (2003).

*72.* ASTM E810-1994, Standard Test method for Coefficient of Retroreflection of Retroreflective Sheeting (1994).

*73.* "Pressing Problems and Projected National Needs in Optical Radiation Measurements," U.S. Council on Optical Radiation Measurements (CORM), 7th Report (2001).

*74.* Y. Ohno, Physical measurement of flashing lights – now and then, *Proceedings of CIE Symposium'02*, Veszprem, Hungary, CIE x025:2003, 31–36 (2003).

*75.* "CIE 1988 2° Spectral Luminous Efficiency Function for Photopic Vision," CIE Publication 86 (1990).

*76.* J. Schanda, L. Morren, M. Rea, L. Rositani-Ronchi, and P. Walraven, Does lighting need more photopic luminous efficiency functions?, *Lighting Res. Technol.* 34, 69–78 (2002).

# 8. LASER RADIOMETRY

Gordon W. Day
National Institute of Standards and Technology (retired), Boulder, Colorado, USA

## 8.1 Properties of Laser Radiation

Following the invention of the ruby laser in 1960 [1], it quickly became apparent that contemporary radiometric techniques were not adequate to characterize the intense output of lasers. Early ruby lasers produced single or low repetition rate pulses (at 694.3 nm), with a duration of about a millisecond and energy of as much as 1 J/pulse. The pulse waveforms of ruby lasers and other early solid state lasers were generally complex, and commonly described as having a "spiking" behavior [2] (Fig. 8.1), with individual spikes having durations of the order of a microsecond. Soon, other solid-state lasers were developed, providing other wavelengths, shorter pulse duration, and higher peak power. Today, a wide range of pulsed lasers is available, with wavelengths from the deep ultraviolet to the mid-infrared, pulse durations as brief as the femtosecond range [3] and peak power levels approaching a terawatt (Fig. 8.2). It is not only necessary to determine power and energy at the highest levels, but also for many applications at the lowest levels of detectability.

The power levels and modulation characteristics of continuous-wave (cw) and high-duty-cycle pulse lasers also pose challenges. Continuous-wave lasers used in manufacturing applications often provide several kilowatts of optical power, and the output power of quasi-cw lasers used in military

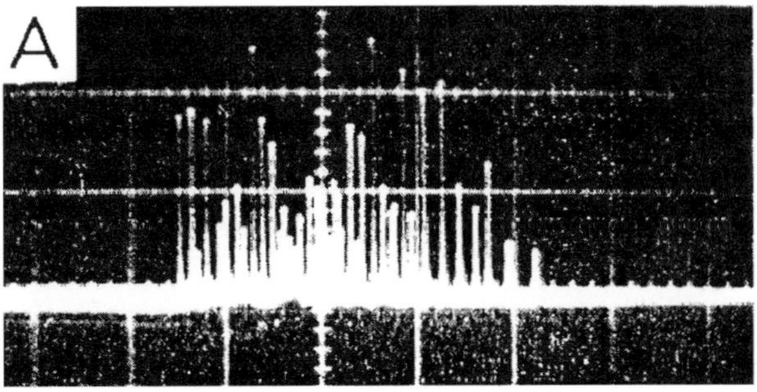


FIG. 8.1. Glass laser pulse with spiking (20 μs/div). © IEEE



EXPERIMENTAL METHODS IN THE PHYSICAL SCIENCES, vol. 41
ISSN 1079-4042 DOI: 10.1016/S1079-4042(05)41008-5

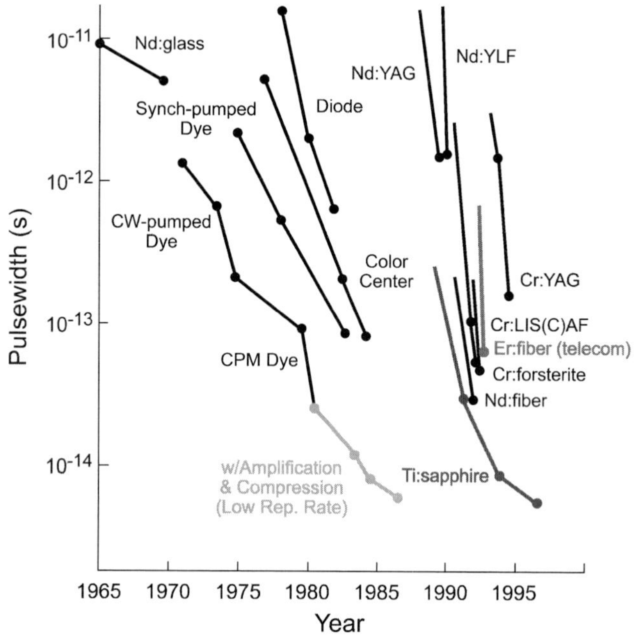


FIG. 8.2. Progress in the reduction of laser pulse duration for various lasers.

applications may easily exceed 100 kW. High-speed digital and analog modulation of lower power lasers used in communications and other applications in information technology often requires measurements of optical waveforms modulated at rates of 100 GHz or greater.

The definitive evidence for laser oscillation (commonly called "lasing") in early experiments was the dramatic narrowing of the optical emission spectra, more precisely known as an increase in temporal coherence and frequently characterized by the coherence length,

$$l_c \approx \frac{c}{\Delta\nu} = \frac{\lambda^2}{\Delta\lambda} \tag{8.1}$$

where $c$ is the speed of light, $\Delta\nu$ the spectral width of the output expressed in frequency, $\lambda$ the wavelength, and $\Delta\lambda$ the spectral width expressed in wavelength. The coherence length is the approximate path difference in the direction of propagation over which interference can be observed. For laser radiation, it can be quite large, especially in comparison to the dimensions of optical components or measurement systems. In a multimode semiconductor laser it may be less than a millimeter. But in common 633 nm He–Ne lasers, which may oscillate in one or more modes within their inherent gain bandwidth of about 1 GHz, the coherence length is not less than about 30 cm and can be much larger. In a stabilized laser with a linewidth of the order of 1 kHz, it could be, at least in principle, hundreds of kilometers.

When working with highly coherent sources or coherent detection methods, interference associated with reflections within and between components can dramatically affect the transmittance of light through a system. These effects are sensitive to small changes in the ratio of optical path length to wavelength, and thus to temperature and wavelength changes. For a plane wave incident on an etalon, the transmittance is given by

$$T = \frac{1}{\left[1 + \frac{4R}{(1-R)^2} \sin^2\left(\frac{2\pi nl}{\lambda}\right)\right]} \tag{8.2}$$

where $n$ is the refractive index, $l$ the thickness, $\lambda$ the vacuum wavelength, and $R = (n-1)^2/(n+1)^2$ the reflectance of the first surface. Figure 8.3 shows the variation of this function with temperature for a 1 cm thick etalon with typical glass parameters ($n = 1.5$, temperature coefficient of path length $\sim 1 \times 10^{-5}$).

Laser light is generally well polarized (perfectly coherent light is completely polarized), though the polarization state may not be linear and may vary with time, even within the duration of a pulse. Except at normal incidence, the reflection and transmission characteristics of most optical elements are polarization dependent (note the familiar plots of the Fresnel equations in Fig. 8.4) and birefringence in many optical elements modifies polarization states during propagation. It is thus particularly important to consider polarization effects in the design of systems for laser radiometry.

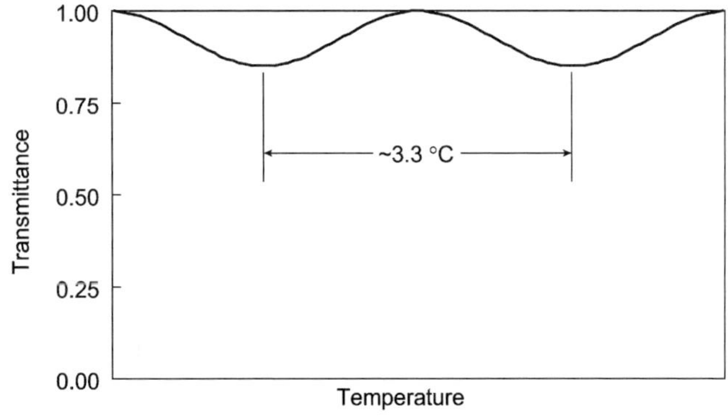


FIG. 8.3. Transmittance of a 1-cm-thick glass etalon versus temperature.

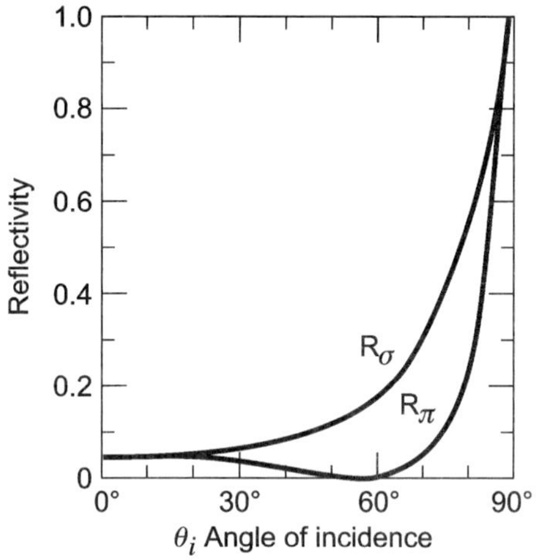


FIG. 8.4. Plots of the Fresnel equations for reflection at a dielectric interface.

The spatial characteristics of lasers must also be considered. In the 1960s, most detectors used in radiometry were designed for spatially uniform radiation, at least over the dimensions of the detector. In contrast, laser radiation is usually described as a "collimated beam," and in fact its spatial profile consists of one or more transverse modes of the laser resonator. For many laser resonators, the fundamental transverse mode has a Gaussian irradiance profile and is stable. But lasers designed for higher power or energy sometimes use more complex resonator designs and often have spatial beam profiles that are the superposition of many higher-order modes, and the mode patterns may not be stable with time, even within the duration of a pulse. When such a source is characterized with, for example, a thermopile that consists of an array of thermocouples, the result depends on the details of the beam shape. To further complicate the problem, it is sometimes necessary to make radiometric measurements of laser radiation that, though spatially limited, is highly divergent, for example, the outputs of some laser diodes or the radiation from an optical fiber. In these cases, it can be a challenge to capture all the radiation, and to be certain that the detector response is independent of the angle of incidence.

Conventional radiometric techniques of the early 1960s could have been adapted to address most of these difficulties, but the early developers of lasers generally chose to approach their measurement needs by developing new types of detectors–detectors that could withstand the radiation levels directly, that did not require many ancillary components whose transmittance might be affected by interference or polarization changes, and that provided a response that was uniform spatially and in direction. Most of the new detectors developed were thermal—the absorption of laser radiation raised the temperature of the absorbing element, and the change in temperature was measured. They were calibrated by comparing the radiation-induced temperature rise with that caused by the dissipation of a known amount of electrical power or energy.

Many concepts and designs for electrically calibrated laser detectors were explored in the 1960s and 1970s and are described in review articles of that era [4].

## 8.2 Primary Standards

The idea of making radiometric measurements by comparing absorbed optical energy to dissipated electrical energy with a thermal detector did not originate with the laser pioneers. It had been conceived in the 1890s, apparently independently by Knut Ångström (son of Anders Ångström) at the University of Uppsala in Sweden, and Ferdinand Kurlbaum, at the Physikalisch–Technische Bundesanstalt (PTB), in Germany [5]. Optical–electrical comparisons were widely used in the early part of the 20th century for solar measurements and later for millimeter-wave measurements. Beginning around 1930, optical–electrical comparison techniques based on the work of Callendar [6] became the basis for traceable optical radiometry at the National Physical Laboratory (NPL) [7, 8]. Sometimes the technique is called "absolute radiometry," although that name seems misleading, since the function of the thermal detector is only that of a comparator, providing traceability to electrical quantities which could, and can, be much more accurately measured than optical quantities.

But electrically calibrated thermal detectors were just what were needed to characterize the output of high-power lasers, and began to be used shortly after the demonstration of the ruby laser. Most of the early examples were called calorimeters and were designed to withstand and absorb the full output of the laser, avoiding problems associated with attenuators and other optical components. Two key factors in determining the accuracy were thus assuring that all of the energy was absorbed and that electrical and optical inputs produced equivalent heating.

In what is apparently the first use of this approach with lasers, Li and Sims [9] used a thin carbon cone as both the absorber and the electrical heater. Other work on cone-shaped detectors followed [10, 11]. An alternative approach was the use of liquid to absorb the laser energy, apparently

FIG. 8.5. First NBS primary standard for laser energy measurements.

first explored by Damon and Flynn [12] and refined by Donald Jennings [13] at the National Bureau of Standards (NBS).[1]

The Jennings calorimeter (Fig. 8.5) used a solution of $CuSO_4$ in a 3-mm-deep cell to absorb 99.9% of the energy of ruby laser pulses up to 15 J (not including Fresnel reflections at the window) and, in the case of Q-switched pulses, with peak power levels up to 150–200 MW. Jennings found that electrical calibrations and first-principles calibrations, based on the thermal and physical properties of the materials in the calorimeter, agreed to 0.3%, and that different calorimeters agreed to about 0.7%. He estimated the total uncertainty at 1%.

Efforts to validate the uncertainty of the Jennings calorimeter by comparison with detectors traceable to standard sources were not very successful [14], largely because of the dramatically different characteristics of the sources for which they were designed. This led to a decision at NBS to expand work on electrical calibration techniques specifically for laser radiometry. The Jennings calorimeter became the basis for the first traceable laser energy calibrations conducted at NBS in 1967.

Similar decisions were made at NPL, the Electro-Technical Laboratory (ETL) in Japan (now NMIJ/AIST), and at PTB. NPL developed electrically calibrated cone calorimeters and other electrically calibrated designs, primarily for pulsed lasers [15]. ETL developed a "microcalorimeter" as a laser

[1]In this chapter, references to the National Bureau of Standards (NBS) are generally associated with work done before 1988, and references to National Institute of Standards and Technology (NIST) are associated with work done thereafter.

standard for Japan [16]. PTB also focused on an electrically calibrated cone design when its program in laser radiometry began in 1973 [17]. Comparisons among these four National Measurement Institutes (NMIs)—NBS, NPL, ETL, and PTB—began in 1975, generally with agreement of between 0.5 and 1% [18]. Continuing comparisons, some of which are cited below, assure the consistency of measurements performed at these and other NMIs.

At NBS, Dale West and Kenneth Churney set the direction for future NBS/NIST work on laser radiometry with theoretical work on the application of isoperibol calorimetry to laser measurements [19]. This was followed by the design of the first isoperibol laser calorimeter [20], which became known as the C-series calorimeter. With improvements, it is still used at NIST as a standard for cw laser measurements at levels in the milliwatt range.

The presently used version of the C-series calorimeter is shown in Figures 8.6 and 8.7. Light enters through a window, slightly wedged to avoid interference effects, and is absorbed in a thin-walled copper cylinder, closed at the distal end with a planar surface at an angle to the axis and coated on the inside with a highly absorbing black paint. About 97% of the light entering the cylinder is absorbed at the end surface and most of the remainder on the cylinder walls. Heater elements are placed near the end surface and a thermopile is incorporated into the supporting structure. The region surrounding the cylinder is evacuated.

It is important to remember that calorimeters are energy-measuring instruments, not power-measuring instruments. When measuring the output of pulsed lasers, they provide a measure of the energy per pulse or the total energy of a series of pulses. Determining the peak power of a pulse thus

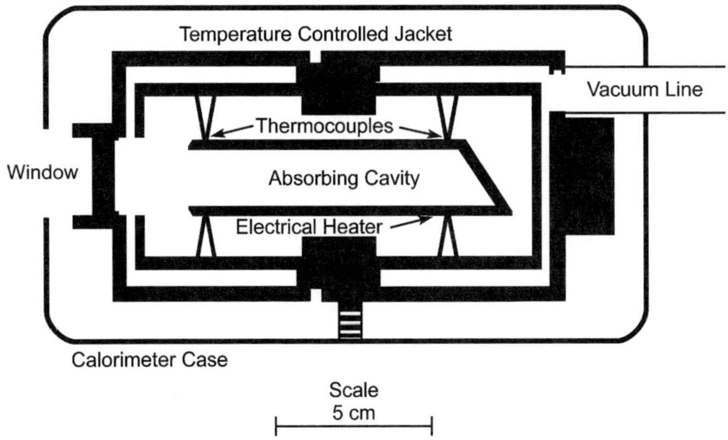


FIG. 8.6. Construction of C-series calorimeter.

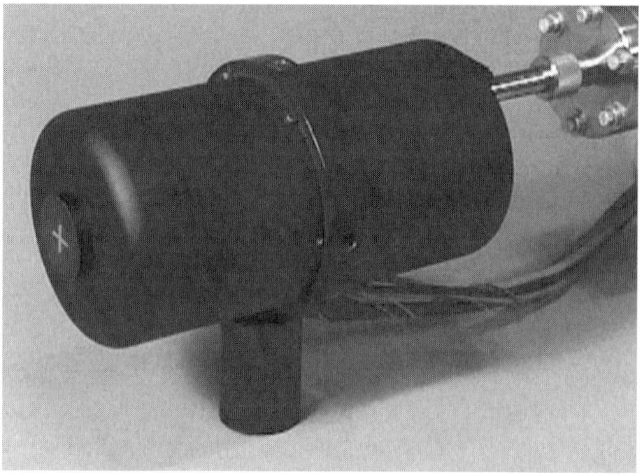

FIG. 8.7. Photograph of C-series calorimeter.

depends on the ability to accurately determine the pulse shape (see Section 8.8) and the number of pulses absorbed.

When measuring the output of cw lasers, calorimeters determine the total energy applied during an interval usually determined by an external shutter [21]. In this case, knowing the time interval accurately, and assuring that the turn-on and turn-off times are either negligibly short or accurately known, are very important.

Figure 8.8 shows data from a calorimeter in which energy has been injected for a period of about 2 min. At times sufficiently distant from the injection period (sometimes called the rating period), the output voltage of the thermopile that measures the temperature of the calorimeter decays exponentially with time as

$$V(t) = (V_0 - V_\infty)e^{-\varepsilon t} + V_\infty \tag{8.3}$$

where $V_0$ is the voltage at an arbitrarily chosen time, $V_\infty$ the voltage at very long times after the energy input, and $\varepsilon$ a parameter known as the cooling constant. $V(t_1)$ and $V(t_2)$ are voltages at arbitrarily chosen times, before and after energy injection, during periods when the voltage is decaying as the simple exponential function of time shown above. The energy absorbed by the calorimeter during the injection period is equal to the sum of the change in internal energy and the energy transferred from the system to the environment, over the chosen interval from $t_1$ to $t_2$. Mathematically, this is

$$E = K[V(t_2) - V(t_1)] + \varepsilon \int_{t_1}^{t_2} [V(t) - V_\infty]\,dt \tag{8.4}$$

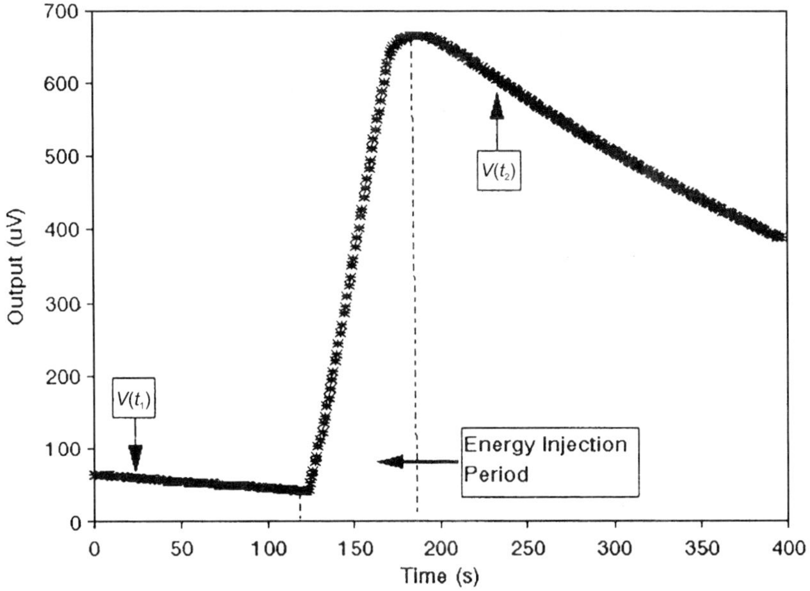


FIG. 8.8. Output voltage from the thermopile of a laser calorimeter.

where the first term is the change in internal energy and the second term is the energy transferred to the environment. The parameters $\varepsilon$ and $V_\infty$ are determined by fitting the cooling data to an exponential function. The parameter $K$, which is generally known as the calibration factor, is determined by injecting a known amount of electrical energy. This is the basic mechanism by which the measurement of optical energy becomes traceable to electrical quantities. The fundamental uncertainty in that comparison is the equivalence between the temperature changes resulting from the two types of heating, commonly known as the in-equivalence or non-equivalence error.

The energy incident on the calorimeter is the energy absorbed and the energy reflected by the window (if any) and any light not absorbed, that is, light reflected from the trapping structure. These quantities must be determined independently, but generally need only infrequent evaluation.

Since they were first used for traceable calibrations in the early 1970s, the C-series calorimeters have been refined, and associated measurement procedures have been improved. Uncertainty estimates are ongoing, but experimental and theoretical analyses currently lead to an inherent uncertainty in the measurement of optical (as opposed to electrical) energy in the neighborhood of 0.25% (95% confidence) [22]. Comparisons with other NMIs have yielded variations in the range of 0.2–0.5% [23]. Calibration of instruments against the C-series calorimeter generally yields an uncertainty of about 1% (95% confidence) [24] depending on the quality of the instrument calibrated.

It is generally believed that it will be difficult to reduce the inherent uncertainty of an instrument such as the C-series calorimeter to below 0.1%, because of the difficulty in establishing and evaluating the optical–electrical in-equivalence. An important alternative approach is to cool an absorbing structure (typically an angle-truncated metallic cylinder similar to that used in the C-series) to cryogenic temperatures, where greater thermal diffusivity leads to better equivalence between optical and electrical power.

Unfortunately, greater thermal diffusivity is accompanied by non-linearity in the heat capacity of the materials in the absorbing structure. Thus, there is a very limited temperature range over which the optical–electrical comparison can be made. Typically, the desired temperature is set through electrical dissipation and, when optical radiation is applied, the decrease in electrical dissipation necessary to achieve the same temperature is measured. Instruments of this sort, which compare power rather than energy, are more properly called radiometers than calorimeters; hence the common name, "cryogenic radiometer." More details on the design, operation, and performance of cryogenic radiometers can be found elsewhere in this volume [25].

Cryogenic radiometers are now widely available, and are among the tools used by most NMIs to provide traceable measurements of optical power [26]. For laser measurements where they can be used directly (cw lasers at power levels between about 1 and 100 μW), they provide the lowest uncertainty comparisons between electrical and optical power presently available—at power levels around 100 μW, NIST provides direct calibrations of laser power meters against its Laser Optimized Cryogenic Radiometer (LOCR) with uncertainties around 0.02% [27].

A comparison between a C-series calorimeter and a cryogenic radiometer [28] yielded differences of 0.04–0.06%. This was significantly better than would have been expected from the inherent uncertainty of the calorimeter and uncertainty in the comparison procedure, and suggests that the inherent uncertainty of the C-series calorimeter may be smaller than can be independently demonstrated. Another recent study of calibrations traceable through the C-series, the LOCR, and a monochromator-based system [29] yielded consistent results.

For cw lasers that provide power levels up to 1 kW, which are commonly used for cutting, welding, and other large machine tool applications, modifications to the basic isoperibol calorimeter are required to avoid damage to the absorbing surface and improve optical–electrical equivalence. Work on this problem at NIST resulted in the development of the K-series calorimeter [30] (Fig. 8.9). The most noticeable difference from the C-series is the

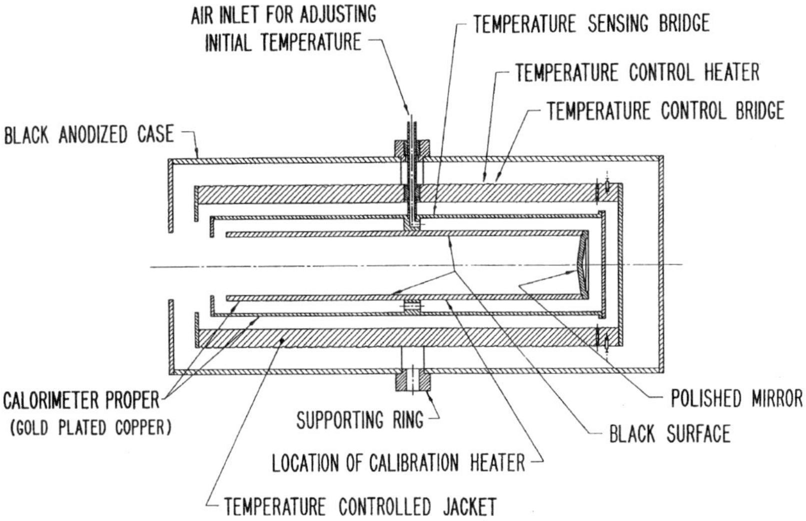


FIG. 8.9. Diagram of K-series calorimeter.

replacement of the angled absorber at the distal end of the absorbing cavity with a convex polished reflector, which serves to spread the incident light over the cylindrical wall and keep the irradiance below ~200 W/cm$^2$, which is about the damage threshold of black paint. Another significant difference is that the K-series operates at atmospheric pressure, and does require a window.

The K-series calorimeter is typically operated at cw power levels of 5 W to 1 kW, with input timed to inject between 300 J and 3 kJ. Presently, the inherent uncertainty of the calorimeter is estimated to be 0.85% (95% confidence) [31] and generally instruments can be calibrated against the K-Series calorimeter with an uncertainty near 1%, depending on the quality of the instrument being calibrated. Recent comparisons between the K-series calorimeter and the standards used at PTB (presently the only other NMI that provides calibrations in this power range) yielded agreement in the range of 0.5–0.7% over power levels between 80 and 550 W [32].

For still higher power cw lasers, including those used in military applications, the problems of avoiding optical damage and removal of optical energy are still more difficult. Figure 8.10 shows the absorber design of a calorimeter developed at NIST for energy levels between about 30 kJ and 10 MJ, at power levels between 300 W and 100 kW [33]. This instrument, known as the BB calorimeter, spreads the input power using a series of specular reflectors, a diffuser, and absorbers, all water-cooled. Temperature rise in the water is monitored and compared to that for a similar level of electrical power

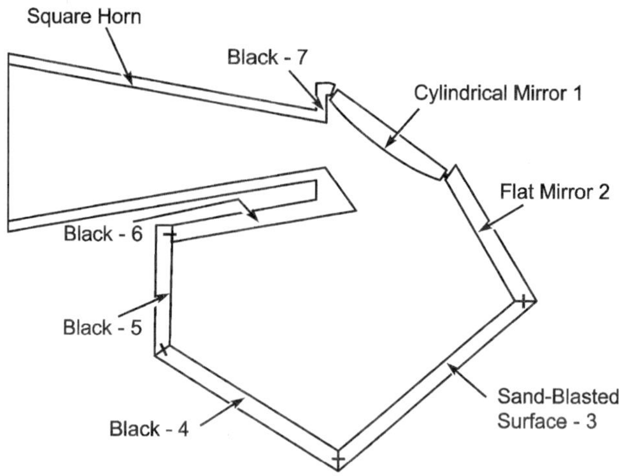


FIG. 8.10. BB calorimeter design.

dissipated in the water. Analysis and measurements suggest that the inherent uncertainty of the BB calorimeter is between 2 and 3% (95% confidence).

The C- and K-series calorimeters can, in principle, be used with lasers that emit short pulses, but paint and other materials that absorb at their surfaces are typically damaged by laser pulses at energy densities well below 1 $J/cm^2$. The Jennings calorimeter addressed this problem by using a liquid absorber designed to absorb the energy in a volume of material rather than at a surface. Later, calorimeters using solid volume absorbers were developed.

Apparently the first of these was developed by Stuart Gunn at Lawrence Livermore Laboratory [34], for measuring the output of $Nd^{3+}$:YAG lasers operating at 1.06 μm and $CO_2$ lasers operating at 10.6 μm. Pulse durations were in the range of 0.1–1 ns with energy densities up to a few joules per square centimeter and peak power densities up to $10^{10}$ $W/cm^2$. Work on calorimeters with volume absorbers followed at NBS with the development of the Q-series calorimeter [35] (Fig. 8.11), which remains the US national standard for such measurements. It incorporates two pieces of absorbing glass (Schott NG-10), one covering the distal end of the absorbing structure (in this case, having a square cross-section), and the other positioned on the top wall of the structure to absorb the light reflected from the first absorber. Early versions of the Q-series calorimeter were estimated to have an inherent uncertainty of about 2%; current estimates are about 0.75% (95% confidence).

As an alternative to volume absorbers one may use absorbing materials that will withstand the laser energy without damage, most of which have a relatively high reflectivity, in a structure that will nonetheless collect nearly

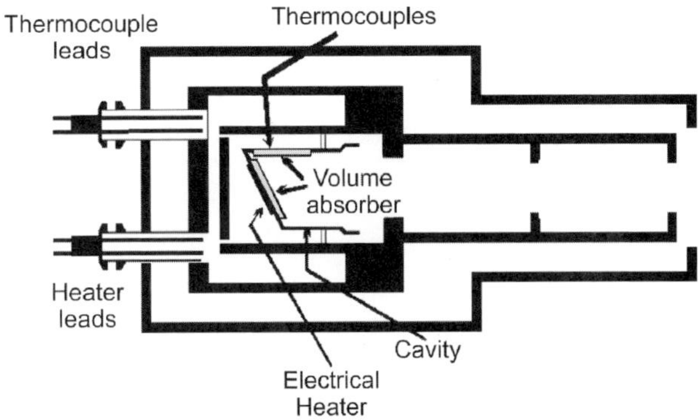


FIG. 8.11. Diagram of Q-series calorimeter.

all of the light. This concept was used by Julian Edwards at NPL, whose early work on carbon cone calorimeters [36] was followed by the development of a nickel (reflectivity ~20%) cone calorimeter that was used at NPL in the 1970s as a primary standard for pulsed laser measurements [37]. During their development, both the Q-series and NPL-Cone calorimeters were compared to the NBS C-series calorimeters, with agreement of 1% or better [35, 37].

The basic design of the Q-series calorimeter can be adapted to almost any wavelength range for which suitable volume absorbers can be identified. In the mid-1990s, a version known as the Q-UV, using Schott UG-11 glass as the absorber, was developed for the measurement of the pulsed output of a krypton fluoride excimer laser operating at 248 nm [38]. This was followed by the Q-DUV calorimeter, using a custom-formulated glass, for measurements of the argon fluoride excimer laser operating at 193 nm [39]. The uncertainties of these calorimeters are similar to those of the Q-series, about 0.75% (95% confidence).

Appropriate volume absorber materials suitable for use with the 157 nm $F_2$ laser are not readily available, however. This led to the design of the Q-VUV calorimeter [40], which has an entirely new absorber configuration illustrated in Figure 8.12. The cavity is formed from four flat pieces of SiC, tilted to each other and joined by thermally conducting vacuum grade epoxy. The SiC surfaces are highly polished for specular reflection and the configuration is designed to achieve up to 15 reflections. With a reflection coefficient of 0.55 or smaller (reflection depends on angle), 99.99% of the light is absorbed in the cavity. The electrical heater for calibration is placed at the distal end of the cavity, on the back side of the surface which initially receives the light.

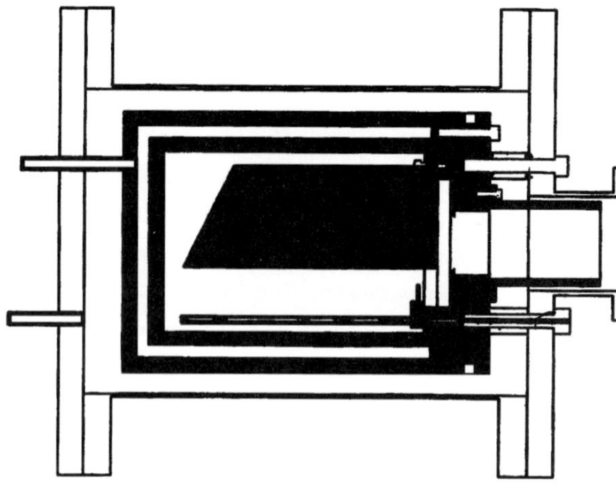

FIG. 8.12. Diagram of Q-VUV Calorimeter.

The evaluation of uncertainties in all of the UV calorimeters continues, and no comparisons between NMIs at these wavelengths have yet been completed. Recent comparisons among the Q-UV, Q-DUV, and Q-VUV calorimeters, where comparisons can be made, yielded agreement close to 0.1%, though the uncertainties in the measurements were significantly higher [41].

## 8.3 Transfer Standards

With some exceptions, the principal use of the primary standards described above is to calibrate other radiometric instruments, thereby disseminating traceable measurements and perhaps expanding the range of parameters (power or energy level, temporal characteristics, spatial characteristics, wavelength, etc.) for which traceable measurements can be made. Ideally, the instruments calibrated, often called transfer standards, should be precise, stable under variable environmental conditions, independent of optical parameters (beam profile, divergence, polarization, and wavelength, etc.), fast enough to resolve temporal characteristics, and linear over a wide range of power and energy levels. In laser radiometry, spatial uniformity, independence on direction of incidence, speed of response, and linearity are typically more critical than in broadband radiometry. Transfer standards are often less complex and easier to use than primary standards and thus may become important parts of calibration systems.

Thermal detectors, sometimes with electrical calibration used as an adjunct to calibration against primary standards, are often suitable transfer standards. Thermopiles and arrays of bolometers, often used as transfer standards for spatially extended sources, can sometimes be used with lasers, but spatial variations in responsivity across the active surface can be a problem with small laser beams (Fig. 8.13a). Pyroelectric detectors [42] can provide more spatially uniform responsivity (Fig. 8.13b), though, they too have limitations.

The output current of a pyroelectric detector is proportional to the time derivative of the temperature of the material. Thus, if pyroelectric detectors are to be used with cw radiation, the radiation must be modulated, preferably with 100% modulation depth such as can be accomplished with a

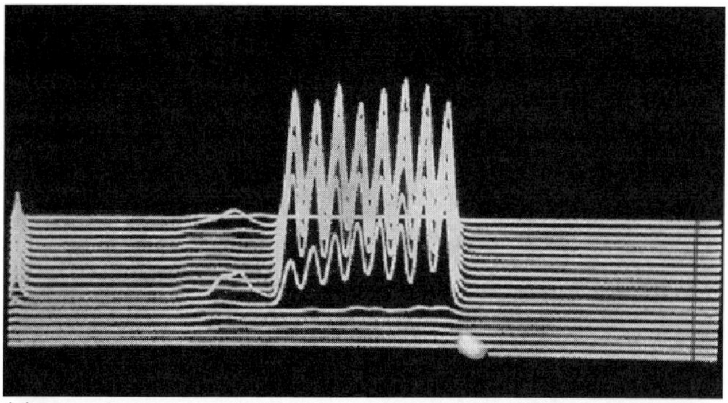

(a)

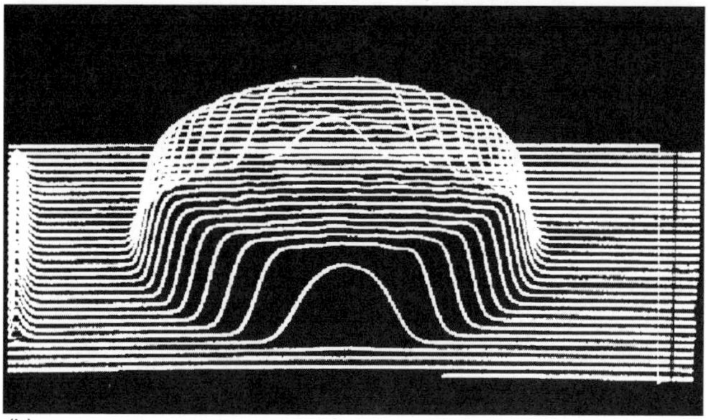

(b)

FIG. 8.13. (a) Raster scan of a 64 element thermopile with a focused laser beam. (Source: R. J. Phelan, Jr., NIST.) (b) Raster scan of a 1-cm-diameter pyroelectric detector with the same laser beam diameter. (Source: R. J. Phelan, Jr., NIST.)

chopper. Achieving a large temperature derivative requires a small thermal mass, so pyroelectric detector elements are usually very thin, but they can have a large area so many configurations are possible. Spatial uniformity depends on uniformity in materials characteristics and in thickness. Lithium tantalite, polished to a thickness of the order of 10 μm is perhaps the most widely used material for pyroelectric detectors, but thin (~6 to 12 μm) polyvinylidene fluoride films [43], have also been used successfully. A problem with pyroelectric detectors is that all pyroelectric materials are also piezoelectric, and thus also respond to acoustic disturbances.

One of the most successful applications of pyroelectric detectors to laser radiometry has been the Electrically Calibrated Pyroelectric Radiometer (ECPR), developed in the 1970s at NBS [44, 45]. The basic concept is that, since the pyroelectric detector uses surface electrodes, a conducting absorber such as gold-black can serve as electrode, absorber, and electrical heater. Equivalence between absorbed optical power and dissipated electrical power is good because they occur in the same thin film (Fig. 8.14). Further, if the optical and electrical signals are modulated with the same waveform, 180° out of phase, the instrument can become a null-sensing radiometer (Fig. 8.15).

Although it was originally hoped that the ECPR could become a primary standard, noise and uniformity issues have limited its accuracy. It has nonetheless proved to be very useful as a stable and versatile transfer standard and has been produced commercially for nearly 30 years.

One of the most valuable applications of transfer standards is to extend the spectral range of traceable measurements. In early work, it was common to rely on the spectral independence of black coatings on thermal detectors but, in spite of efforts to improve the spectral absorptivity of various coatings, variations are almost always significant and sometimes very large [46–48].

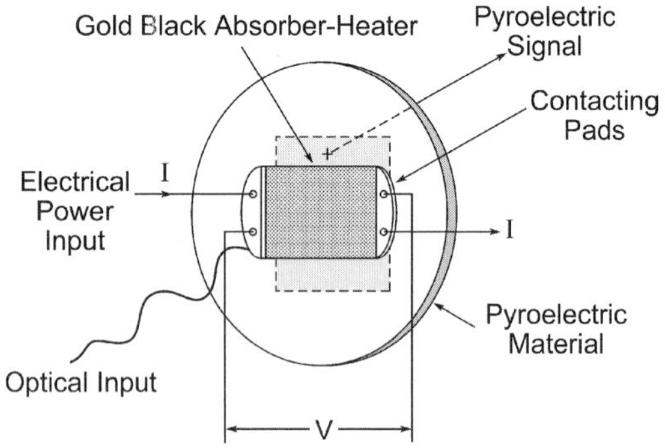


FIG. 8.14. The sensing element in an ECPR designed for laser power measurements.

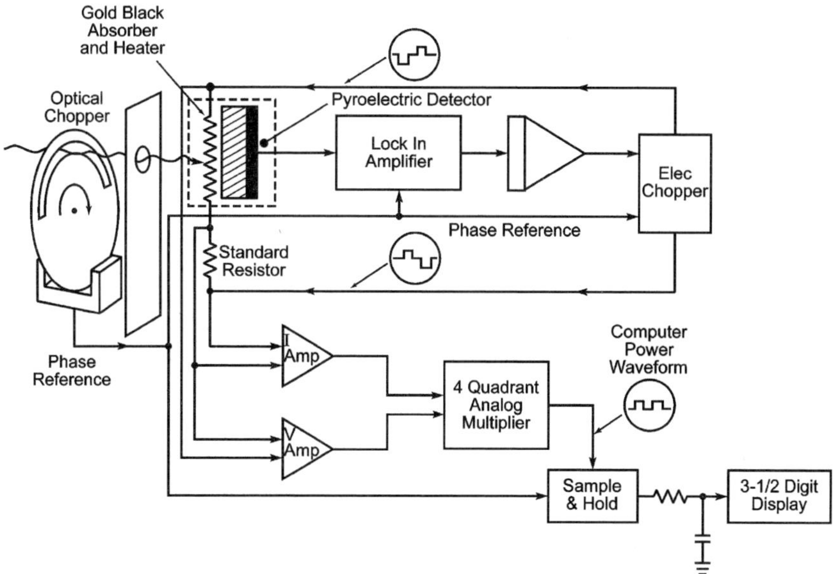


FIG. 8.15. The ECPR null radiometer system for laser radiometry.

To assure that all of the radiation is absorbed, one can also use a cavity structure such as those used in the primary standards described above. Such devices typically have a relatively large thermal mass, however, and it is therefore difficult to design them to have a response time fast enough for chopped radiation. One of the most successful attempts was a cone detector developed by W. L. 'Lum' Eisenman and colleagues [49], of the US Navy, which could be used with radiation chopped at 1 Hz. From its development in the 1960s through much of the 1970s, the Eisenman cone was widely regarded as the best infrared spectral responsivity standard available in the US.

For much more convenient and accurate measurements using higher chopping frequencies, one can use faster detectors incorporated into reflective traps that direct light that is not absorbed back onto the active area. One of the first examples of this approach consisted of a thermopile in a hemispherical trap, designed by Blevin and Brown of CSIRO, for use in a Stefan–Boltzman constant determination [50]. The development of large area pyroelectric detectors enabled substantial improvements in the hemispherical trap concept [51] (Fig. 8.16). And still further improvements, especially in signal-to-noise performance, have been achieved more recently using a wedge configuration (Fig. 8.17) [52]. The Eisenman cone,

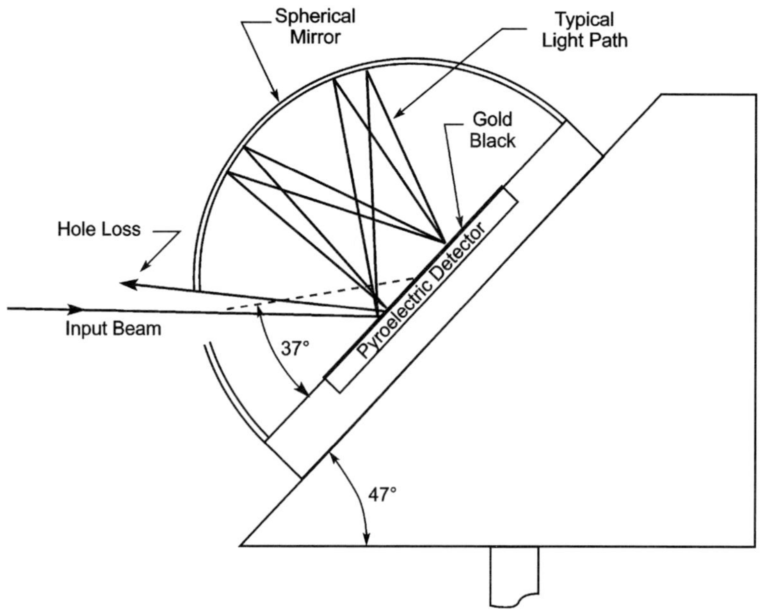


FIG. 8.16. Hemispherical trap.

the hemispherical trap with pyroelectric detector, and the wedge trap have been compared, directly or indirectly, with a conclusion that they are independent of wavelength to about $\pm 1\%$ from the visible to at least 10 μm.

Reflective trap configurations can also be used effectively with semiconductor diode detectors. One of the most successful of these has been a design developed by Ed Zalewski and Richard Duda [53] of NBS (Fig. 8.18), that uses four silicon photodiodes of a type that had previously been shown to have essentially 100% *internal* quantum efficiency across most of the visible spectrum [54, 55]. The four detectors of the Zalewski–Duda trap capture essentially all the incident light and result in a device with virtually 100% external quantum efficiency.

Reflective traps can also be designed to accommodate diverging beams, for example the direct output of a diode laser or optical fiber [56–58]. These typically involve the use of concave mirrors, and have been designed with collection efficiencies up to 99.9% for numerical apertures as great as 0.24 (13.9° half-angle cone).

The multiple reflections that occur in most reflective traps often lead to a small polarization dependence in the response, and much of the light not

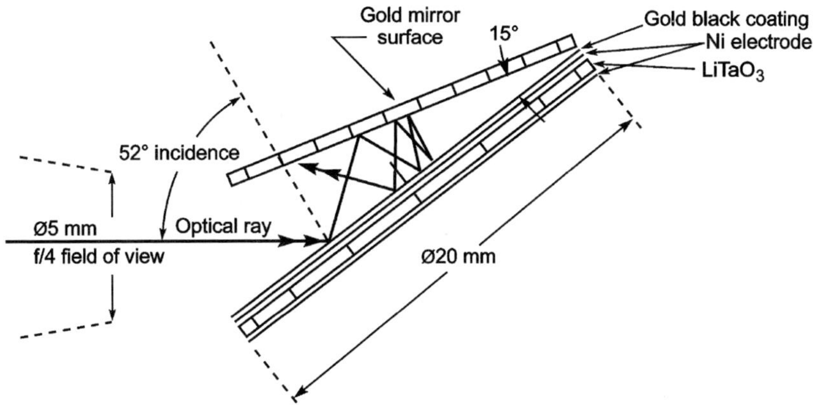


FIG. 8.17. Wedge pyroelectric trap.

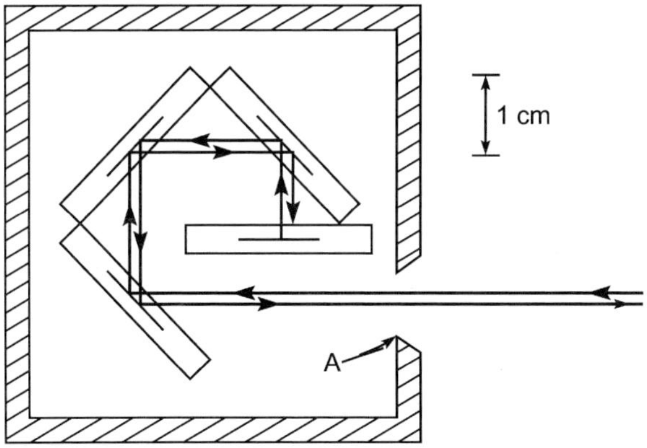


FIG. 8.18. 100% QE detector.

absorbed in the detectors is typically reflected back toward the source and is therefore difficult to measure accurately. Both of these issues in evaluating uncertainty can be addressed by transmission trap designs in which detector orientation minimizes polarization effects, and residual light is transmitted through the structure and can be directly measured. This concept was apparently conceived by Christopher Cromer, at NIST, and has been explored in several laboratories [59–62].

## 8.4 Comparison Methods and Linearity Issues

When comparing standards or other instruments at the same power or energy level, and when a stable, well characterized, laser is available, the most appropriate method of comparison is often direct substitution. The instruments are used alternately to determine the output of the laser until enough data are accumulated that measurement uncertainties can be evaluated. For cw lasers that are not sufficiently stable for this purpose, it is often possible to incorporate a power-stabilizing device into the measurement system. These devices function by sampling the laser power and adjusting to a constant level through a feedback-controlled variable attenuator. Stabilizers can often reduce low-frequency power fluctuations to the point that they are an insignificant part of the uncertainty in comparison.

It is much more difficult to stabilize the output of a low-repetition-rate pulsed laser, so a more typical approach is to use a calibrated beam splitter or one sort or another. And when the instruments must be compared at different levels—perhaps dictated by their respective ranges of linearity—either a calibrated attenuator or a calibrated beam splitter is required.

High quality conventional beamsplitters, used carefully, can meet both of these needs, but for highly coherent, collimated laser radiation, it is often better to use a wedged beamsplitter (Fig. 8.19) [63–66]. The wedged beamsplitter generates a family of beams with power ratios that can be calculated relative to the incident power. A wedge angle of $\sim 2^\circ$ with a thickness of $\sim 5\,\mathrm{mm}$ or more makes interference effects negligible in most cases. The angle of incidence ($8.76^\circ$ in the case shown) is typically chosen for experimental convenience as the angle at which the $+1$ beam (first external reflection from front surface) and the $+2$ beam (beam reflected once from each of the first and back surface, exiting in the forward direction) are collinear.

The ratios between various beams can also be determined experimentally by placing appropriate instruments in the two beams, determining the apparent ratio, switching the instruments, and measuring the ratio again. The best estimate of the true ratio is the average of the two measured ratios. This process is also a valuable quality assurance test for calibration systems based on the beam splitter. The experimentally determined ratios at a wavelength of 633 nm for fused silica, which is the most commonly used beamsplitter in the near ultra-violet, visible, and near infrared, are given in Table 8.1. Additional orders are often observable, but scattered light commonly limits the accuracy that can be obtained with them.

Instruments used to measure the output of lasers, especially those used in optical communications, are often designed to have a very large dynamic range, perhaps nine decades (90 dB) or more. Determining and describing

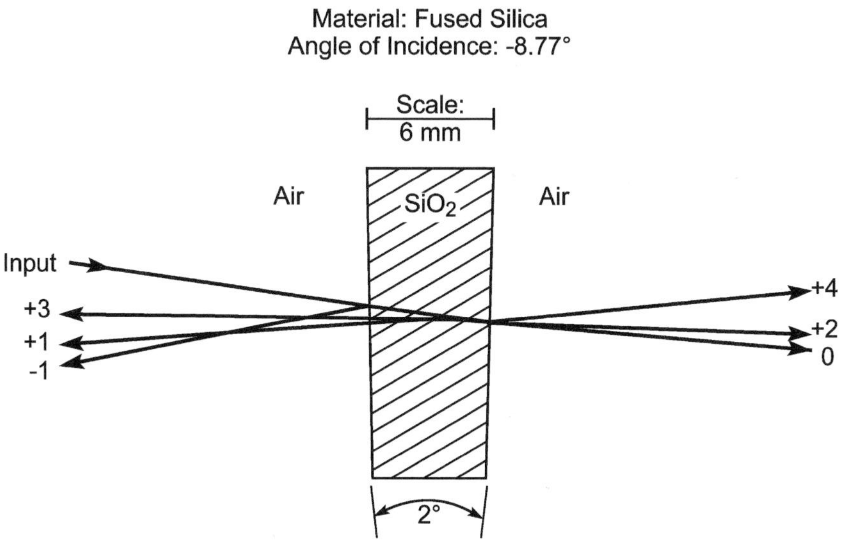


FIG. 8.19. Wedged beamsplitter attenuator.

TABLE 8.1. Attenuation versus Order for a Wedged Beamsplitters

| Order | Attenuation |
|---|---|
| 0 | 1.075 |
| −1 | 26.08 |
| +1 | 28.52 |
| +2 | $8.222 \times 10^2$ |
| +3 | $2.379 \times 10^4$ |
| +4 | $6.864 \times 10^5$ |

*Note*: $n = 1.5$; wedge angle, 2°; angle of incidence, 8.76°.

the linearity of such instruments thus becomes an important part of the calibration process.

Detector linearity has been widely studied [67–70]. For instruments measuring laser power, it is usual [71, 72] to quantify the nonlinearity as the difference between the response at an arbitrary power ($P$) and the response at a reference power ($P_r$), divided by the response at the reference power,

$$\Delta_{NL}(P; P_r) \equiv \frac{R(P) - R(P_r)}{R(P_r)} = \frac{V(P)P_r}{V(P_r)P} - 1 \qquad (8.5)$$

where the quantity $R(P)$ is the output of a instrument at a specific power, $V(P)$, divided by that power, $R(P) = V(P)/P$ which corresponds to the

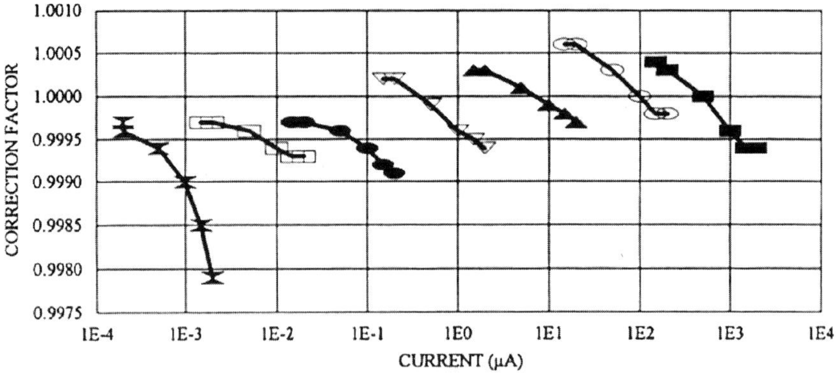


FIG. 8.20. Calibration factor data for a multi-range optical power meter.

traditional definition of responsivity, in the case of a simple detector. One can, equivalently and often more conveniently, state the definition in the inverse, and model the behavior as a polynomial [72]. Several measurement methods can be used [72]. These approaches are useful at the component level, where the focus is on the calibration of the device.

In the calibration of multi-range power meters, range-to-range discontinuities complicate the description of the results, and the most useful presentation of results may be a series of calibrations for each range of the instrument. An example [73] of such data is shown in Figure 8.20.

## 8.5 Choices in Traceability

Providing traceable calibrations of instruments at levels ranging from the lowest detectable to those used to process or destroy materials, at wavelengths from the deep ultraviolet to the far infrared and with widely varying spatial and temporal characteristics, is a challenging requirement. But with the array of primary standards and transfer standards available, there are, in almost every instance, a variety of appropriate methods of relating the calibration of a laser radiometer to electrical parameters. Most of them fall into one of two general approaches, which each have established and potential advantages but which, with current technology, typically yield similar results.

One approach is the use of a range of primary standards, such as those described in Section 9.2, tailored to a specific set of laser characteristics. In most cases, calibrations can be made directly against the primary standards, or perhaps with a single transfer standard. This is the general approach followed at NIST. The advantage is that the resulting calibration systems are relatively simple and reliable; they require relatively little maintenance and quality assurance procedures are comparatively simple. There is a limitation to this approach, however, in that primary standards suitable for direct calibration at the parameter levels typically required must be operated at room temperature, and are therefore fundamentally less accurate than cryogenic radiometers. One may hope that cryogenic radiometers with broader operating ranges may be developed, but it is unlikely that it will ever be possible to use them directly to provide a significant fraction of laser radiometer calibrations.

The other approach is to use a cryogenic radiometer as the primary standard in all calibration traceability chains, with as many transfer standards as necessary to perform calibrations at the levels and with the parameters required. This approach is presently followed by several NMIs and is well illustrated by the traceability chain used for certain laser radiometer calibrations at PTB (Fig. 8.21). Often, as in the illustration, several transfer standards are required to transfer from the cryogenic radiometer to the radiometer to be calibrated. While this approach requires much more complex measurement systems and greater efforts in quality assurance, it offers the potential that, with the development of transfer standards that have greater accuracy and broader ranges of operating parameters, the traceability chain can be shortened and calibration accuracy can be improved.

## 8.6 Optical Fiber Power Meters

A substantial portion of the growing demand for traceable laser power calibrations is associated with radiometers in which the radiation enters the instrument through an optical fiber, typically terminated with a connector. This complication, along with the fact that typical instruments are designed to operate over as many as 10 decades of power and very broad wavelength ranges, leads to additional challenges in their design and calibration [74].

The input geometry of an optical fiber power meter is typically similar to that shown conceptually in Figure 8.22, with the fiber positioned close to the window of a large area photodiode, usually based on silicon, germanium, or indium gallium arsenide, depending on the spectral region of interest. This design requires close attention to several of the issues described earlier in this chapter, especially the spatial and angular dependence of the detector, and the potential for interference effects.

The area of the detector must be large enough to fully collect the diverging radiation from the fiber. The fundamental mode of a single mode fiber is approximately Gaussian in shape and may, depending on the core size and refractive index profile, diverge with a half-cone angle ranging from 3° to 5°.

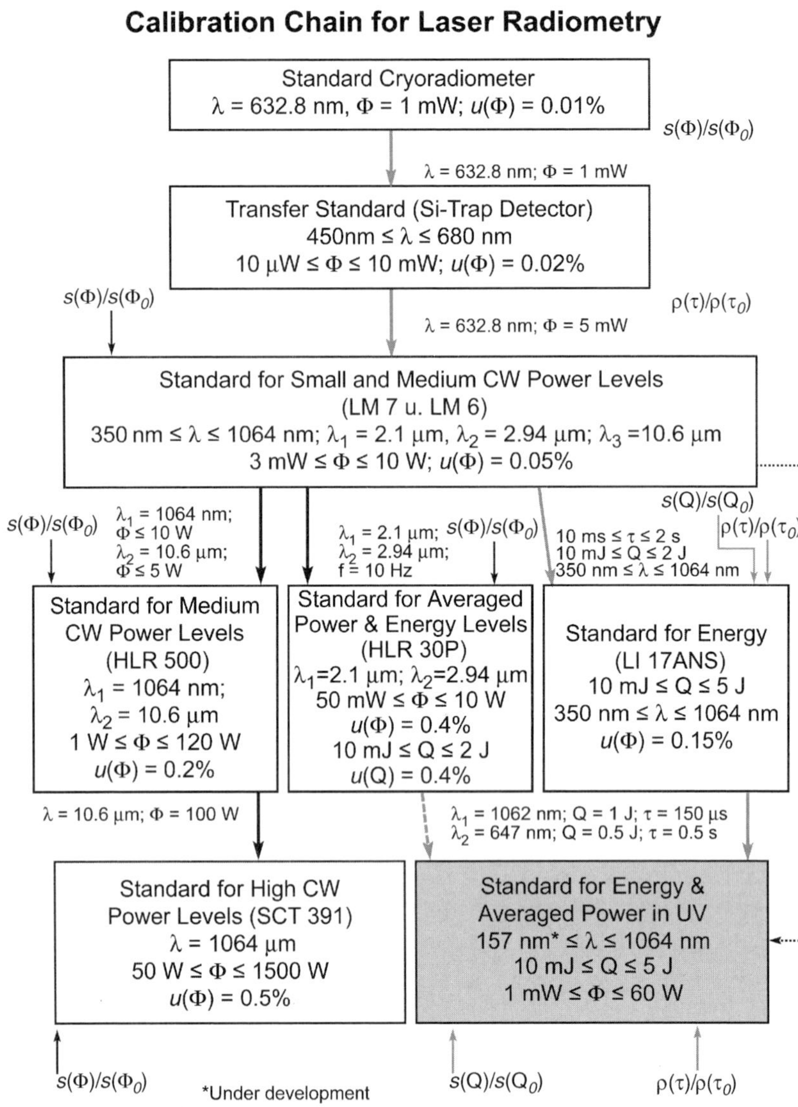


FIG. 8.21. PTB calibration traceability chains for high power cw lasers and pulsed UV lasers. (Source: Stefan Kück, PTB.)

Multimode fibers, despite their larger core diameters, typically exhibit larger divergence angles, often around 12°. In an imprecise analogy to microscope systems, the angular divergence (or acceptance) angle of a multimode fiber is often specified using the concept of "numerical aperture," which is the sine

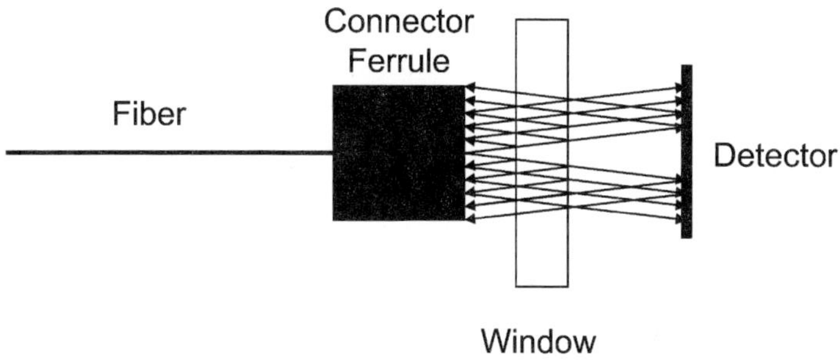


FIG. 8.22. Conceptual drawing of the input to an optical fiber power meter.

of the divergence angle. When used with sufficiently coherent light, multimode fibers also display a highly structured and dynamic intensity pattern resulting from the interference of hundreds of modes typically excited.

Avoiding interference from reflections between and among the surfaces of the detector, the window, and the fiber is a particularly important design issue. It is complicated by the fact that coatings that can be used to suppress reflections must be designed to perform over a broad range of angles and wavelengths.

With a power meter designed to address these issues, calibration is typically similar to that of other laser power meters, and at most NMIs is carried out by direct substitution against a transfer standard calibrated against a cryogenic radiometer. The transfer standard must meet the same requirements as the power meter, with its linearity known over the full dynamic range of the instrument to be calibrated. For this purpose, NIST uses an ECPR, and typically provides calibrations with uncertainties of around 0.3%, depending on the properties of the instrument being calibrated, or a trap detector, providing similar uncertainty [57, 75].

## 8.7 Laser Beam Characteristics

Although a full description of the propagation characteristics of laser radiation [76] is well beyond the scope of this chapter, a working knowledge of laser propagation can be valuable in performing radiometric measurements with lasers. Some useful information can be found in an international standard for laser beam characteristics [77].

The fundamental mode of many lasers has a Gaussian irradiance profile given by the expression

$$I(r) = e^{-2r^2/w^2(z)} \tag{8.6}$$

where $r$ is the radius measured from the axis of propagation and $w$ a parameter commonly known as the "spot size" that varies along the direction of propagation.

If such a beam is centered on a circular aperture of diameter $2a$, the fraction of light transmitted is given by

$$\frac{I(a)}{I(\infty)} = 1 - e^{-2a^2/w^2} \tag{8.7}$$

which is plotted in Figure 8.23.

An important property of a Gaussian beam is that its profile remains Gaussian as it propagates, including propagation through planar surfaces

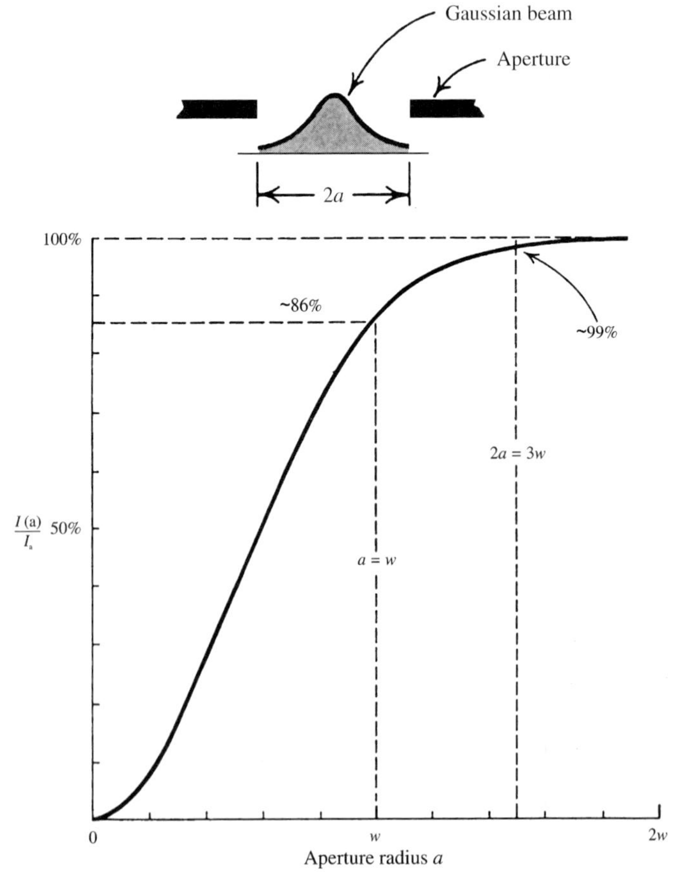


FIG. 8.23. Fractional transmission of a Gaussian beam through a circular aperture.

and ideal lenses. In free space, the spot size of a collimated Gaussian beam (a point along the direction of propagation at which the wavefronts are planar), will increase from $w_0$ according to the expression (Fig. 8.24)

$$w(z) = w_0\sqrt{1 + \left(\frac{\lambda z}{\pi w_0^2}\right)^2}$$

$$\approx \frac{\lambda z}{\pi w_0} \quad \text{for } z = \frac{\pi w_0^2}{\lambda} \tag{8.8}$$

Alternatively, using an ideal lens with focal length $f$, a collimated Gaussian beam with spot size $W_0$ can be focused to a smaller spot given by (Fig. 8.25):

$$w_0 = \frac{\lambda f}{\pi W_0} \tag{8.9}$$

Some lasers, particularly higher power cw lasers and many pulsed lasers oscillate in one or more higher order Hermite Gaussian modes. Each of these could be analyzed individually, but a more typical approach for radially symmetric beams is to specify a parameter known as the "beam propagation

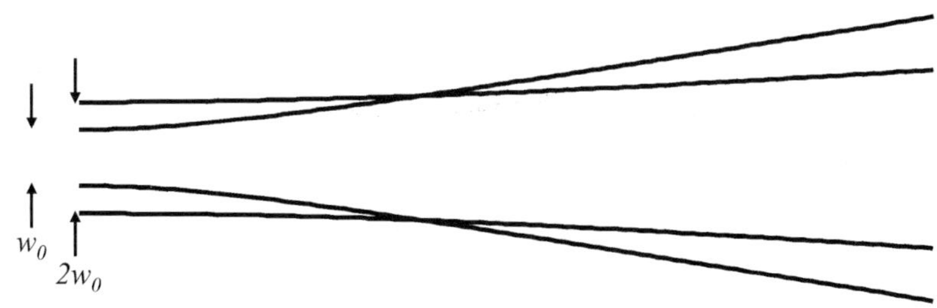


FIG. 8.24. Divergence of a Gaussian beam.

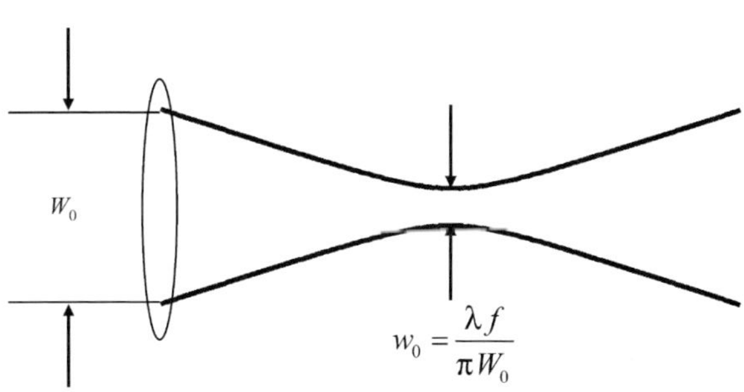


FIG. 8.25. Focusing of a Gaussian beam.

factor," commonly designated $M^2$, which is determined experimentally for a given beam [77]. Using the $M^2$ concept, the spot size of an initially collimated beam will grow as

$$w_0(z) \approx \frac{M^2 \lambda z}{\pi w_0} \quad \text{for } z = \frac{\pi w_0^2}{\lambda} \tag{8.10}$$

and a collimated beam can be focused to a spot size given by

$$w_0 = \frac{M^2 \lambda f}{\pi W_0} \tag{8.11}$$

$M^2$ is determined by a minimum of three measurements on a focused beam [77].

Some laser beams can be extraordinarily complex, lacking radial symmetry and exhibiting astigmatic focusing characteristics. These require more complicated descriptions, though similar approaches can be used in some cases.

## 8.8 Waveform Measurements

Although, as suggested above, most measurements in laser radiometry are concerned with pulse energy or average power, there are times when it is necessary to determine temporal or waveform characteristics of laser radiation. One example is to determine the peak pulse power or shape of a laser pulse and another is to the need to determine waveforms in digital systems.

In many cases, these needs can be addressed using commonly available photodiodes, provided appropriate steps are taken to maintain linearity, address any dynamic optical characteristics that might be present (e.g. polarization or wavelength), and avoid damage. Since size, responsivity, and speed are generally competing design parameters, it is important to choose a photodiode for a particular application.

One of the most widely used detectors for waveform measurements is the p–i–n photodiode [78]. The diode is reverse-biased and the radiation is absorbed in the depletion region, creating electron–hole pairs that are swept into the external circuit. Generally, p–i–n photodiodes can be modeled as current sources in parallel with a large resistance and a small capacitance (Fig. 8.26), with perhaps a small inductance in series. For high speed measurements, where the photodiode is typically coupled into a 50 Ω transmission line, the resistance can generally be ignored.

The time necessary to sweep the carriers from the depletion region and the capacitance associated with both the depletion region and packaging are the principal factors that inherently determine the speed of the detector.

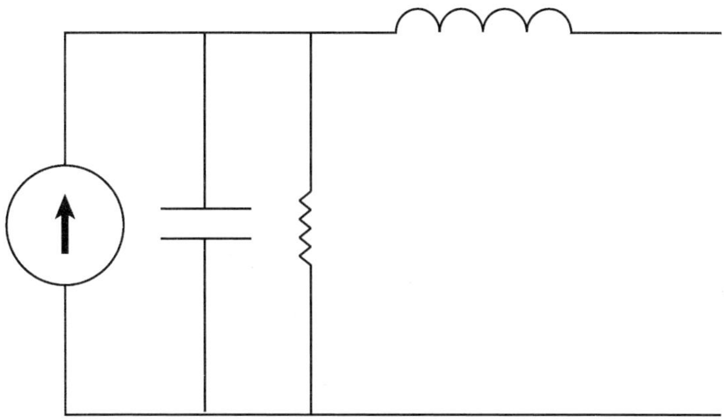

FIG. 8.26. Equivalent circuit of a photodiode.

Increasing the reverse bias voltage generally decreases the transit time and also lowers the capacitance of the depletion region. Packaging is critical, in that the bias must be applied in a way that does not substantially increase capacitance and inductance, and the structure must be compatible with impedance matching into a transmission line.

An optical (power) waveform, $s(t)$, striking the detector will produce a current, $i(t)$ that is the product of the detector responsivity $R(\lambda)$ (in amps per watt) and the convolution of the optical signal with the impulse response of the detector, $h(t)$. The output voltage is thus

$$v(t) = i(t)R_L = R(\lambda)\{s(t) * h(t)\}R_L \tag{8.12}$$

where $R_L$ is the load resistance. One common detector specification is the impulse response, $h(t)$ or, reducing the specification to one number, the FWHM duration of the impulse response. This is illustrated in Figure 8.27a [79].

The Fourier transforms of $s(t)$ and $v(t)$, $S(f)$ and $V(f)$, are related through the Fourier transform of the impulse response, $H(f)$,

$$V(f) = R(\lambda)S(f)H(f)R_L \tag{8.13}$$

where $S(f)$ and $H(f)$ are complex functions. With an electrical spectrum analyzer, the parameter measured is electrical power, which is proportional to $V^2(f)$, and from this measurement, the magnitude of either $S(f)$ or $H(f)$ can be determined, if the other is known.

The magnitude of $H(f)$ is another common detector specification. And often, it too is reduced to a single number, the bandwidth, but there is a common ambiguity in this specification. Some specifications quote the frequency at which $H(f)$ drops 3 dB from its low frequency value; this is

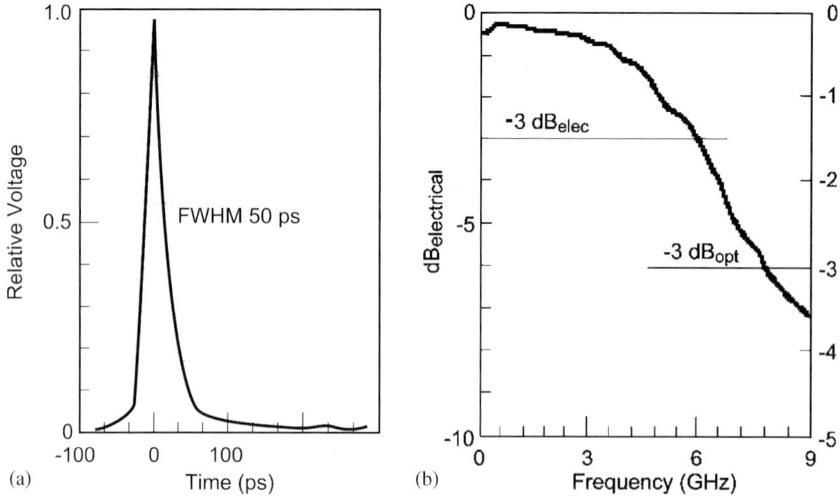


FIG. 8.27. (a) Example detector impulse response. (b) Example detector frequency response. © Agilent Technologies

commonly known as the "optical bandwidth." Other specifications quote the frequency at which $H^2(f)$ drops 3 dB from its low frequency value; this is commonly known as the "electrical bandwidth." The relationship between these parameters is illustrated in Figure 8.27b [79].

While the relationship between impulse response and the magnitude of the frequency response will depend on the details of the shapes of both functions, simple estimates are often made. Typically, the product of the impulse response width at half maximum and the−3 dB optical power bandwidth will fall in the range 0.3–0.7. For determining pulse shape, the duration of the impulse response must be substantially shorter than the duration of the pulse to be measured—the ratio will depend on the accuracy desired and the shape of both the impulse response and the waveform to be measured. High-speed photodiodes with optical bandwidths specified in the range of 50–70 GHz and impulse responses as short as about 7 ps are commercially available.

When measuring single pulses, one is generally limited to the use of real-time oscilloscopes, some of which have storage features capable of digitizing the waveform. The fastest commercially available instruments typically have bandwidths in the range 6–8 GHz.

For measuring single events with shorter temporal resolution, one alternative is to use a streak camera [80]. Streak cameras function by producing an image in which the brightness of the image in one direction is proportional to the intensity of the pulse versus time. Several technologies have

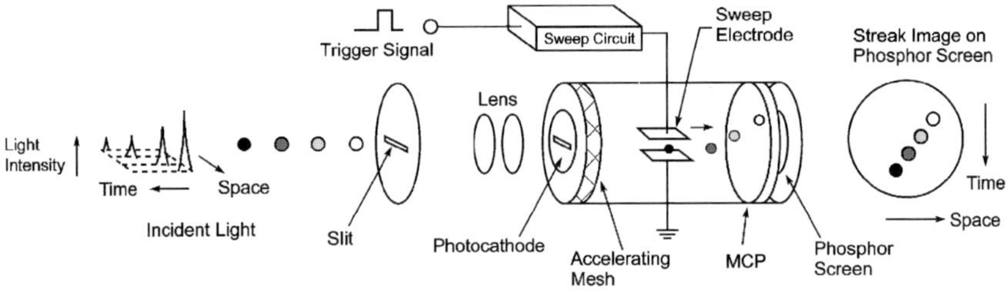


FIG. 8.28. Schematic of a streak camera. © Hamamatsu Photonics K.K.

been used to produce the image, including simple deflection of the collimated laser radiation across a recording device (e.g. film). The fastest streak cameras are similar to early real-time oscilloscopes except that the electrons are produced by the laser pulse striking a photocathode. They are then accelerated and swept across a phosphor, producing a streak in which the intensity of the phosphor emission along its length is proportional to the intensity of the pulse versus time (Fig. 8.28). Commercially available streak cameras may have a temporal resolution of less than 1 ps.

When measuring stable repetitive pulses, electrical sampling oscilloscopes can provide, about an order of magnitude, shorter temporal resolution than real-time oscilloscopes. These are commercially available with bandwidths of up to 70 GHz, approximately matching the speeds of the fastest available photodiodes.

Optical sampling of repetitive waveforms using nonlinear optical processes can provide measurements with a frequency response approaching 1 THz, or a temporal resolution of 1 ps or less [81, 82]. This is typically achieved by sampling the test waveform in a nonlinear optical material with a yet-briefer pulse from another laser; the resulting waveform can then be detected with a photodiode that need not be fast enough to resolve the sampled waveform. A recent implementation [82] of optical sampling for monitoring waveforms used in optical communications systems is shown in Figure 8.29; it uses 100 fs pulses from an erbium-doped optical fiber laser as the sampling pulses and sum-frequency generation in a $MgO–LiNbO_3$ crystal as the sampling mechanism. Sampled data from this system is shown in Figure 8.30. Optical sampling oscilloscopes with bandwidths between 500 GHz and 1 THz are available from several instrument manufactures.

The sampling techniques described above all rely on the availability of an event (pulse) that is substantially briefer than the event to be measured. When measuring pulses briefer than the picosecond range, this becomes very difficult and, when approaching the femtosecond range, is virtually impossible.

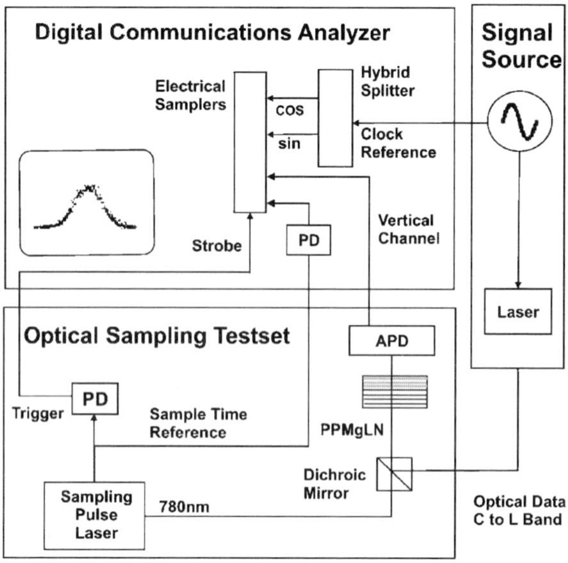


FIG. 8.29. Schematic of an optical sampling system [82]. © IEEE

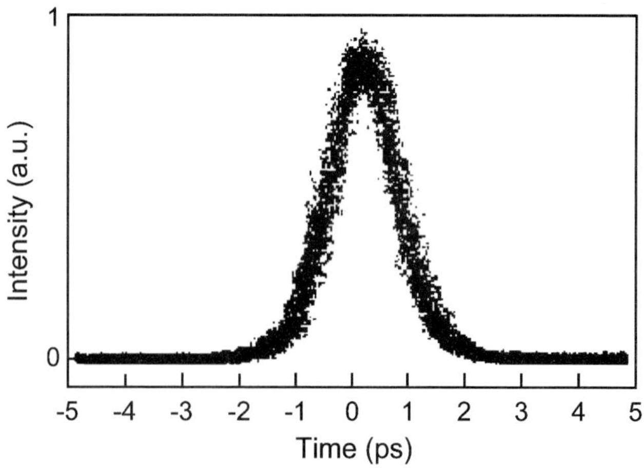


FIG. 8.30. Sample data from the system shown in Figure 8.29, showing the ability to measure pulses with durations of about 1 ps and to characterize pulse jitter in the range of 100 fs. © IEEE

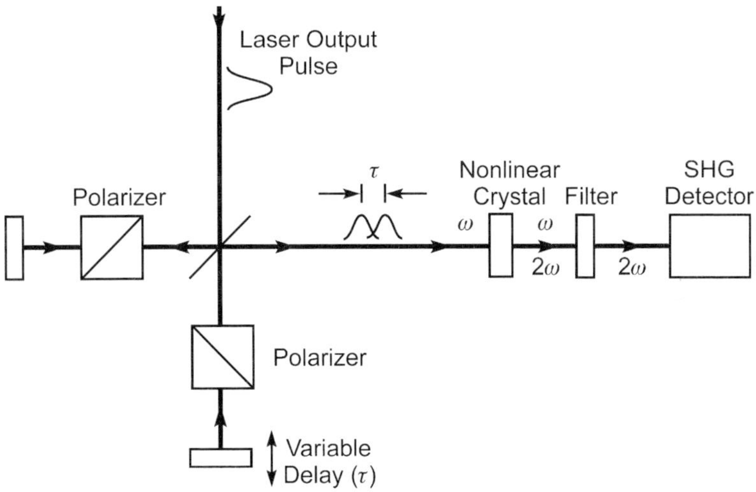


FIG. 8.31. Schematic of system for determining the autocorrelation of a laser pulse.

For these cases, it is common to use techniques in which the waveform to be measured is compared to itself. One method, first developed in the 1970s and still widely used today, is to determine the autocorrelation of the pulse using second harmonic generation [83] (Fig. 8.31).

In this measurement, a laser pulse is divided into two pulses, one of which can be delayed relative to the other, and the two pulses are used together to generate, in a crystal, a pulse at the second harmonic of the laser frequency. The efficiency of the second harmonic generation depends on the degree of overlap of the two parts of the pulse, and by scanning the delay, a measure of the pulse duration is obtained.

Autocorrelation is relatively easy to use and provides a reasonable estimate of the pulse duration. Autocorrelators capable of providing information on pulses with durations of 20 fs or less are commercially available. If the shape of the pulse is known, a priori, they can provide very good measurements of the pulse duration, but this is rarely the case. They typically do not provide accurate information about the details of the pulse shape. Figure 8.32 shows the autocorrelation function for several laser pulses, illustrating its insensitivity to pulse shape [84].

For the accurate characterization of pulscs substantially briefer than 1 ps, additional information is required. The range of possible methods is broad and beyond the scope of this chapter [85, 86].

One group of techniques is known as Frequency Resolved Optical Gating (FROG) [87]. In FROG, the output from an autocorrelator is analyzed spectrally, thus yielding the spectrum of the optical pulse versus delay.

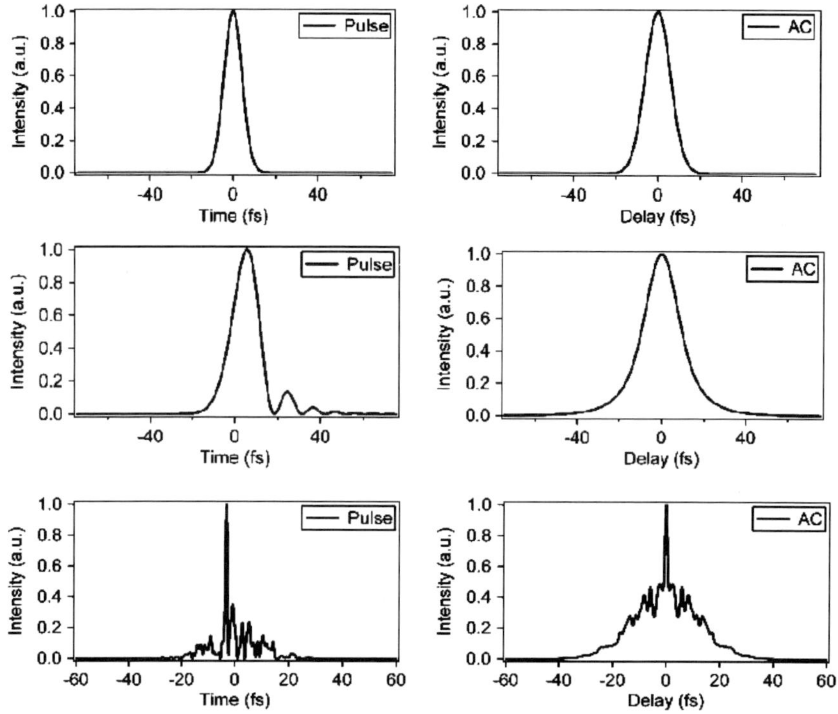


FIG. 8.32. Pulse waveforms (left) and their corresponding autocorrelation functions (right). © R. Trebino

From this data, the pulse characteristics are computed through mathematical methods known as two-dimensional-phase retrieval.

Another group of techniques involves reconstructing the pulse waveform directly from the magnitude and phase spectrum of the pulses [88]. The magnitude, which consists of a comb of discrete frequencies (modes), is obtained easily with an optical spectrum analyzer. The phase can be obtained by any of the several techniques, most of which involve measuring the relative phase of pairs of modes.

## 8.9 Summary

Although laser radiometry involves, and is complicated by, many issues not encountered in other branches of radiometry, the field has matured substantially over the last 40 years. Well characterized primary standards—comparators that relate absorbed optical power or energy to dissipated electrical power or energy—are now available for a wide variety of laser characteristics and their functionality and accuracy continues to be improved. A broad range of transfer standards is available to extend their range and enable additional measurements. NMIs worldwide use these technologies to provide traceable measurements that have been shown to be consistent with those of other NMIs. Measurements of parameters uniquely associated with lasers, for example, beam and waveform characteristics, are also well developed, and sometimes supported by internationally agreed standard procedures. Nonetheless, advances in laser technology continue and, with them, the need for advances in laser radiometry.

## References

*1.* (a) T. H. Maiman, Stimulated optical radiation in ruby, *Nature* 187, 493–494 (1960);
(b) T. H. Maiman, R. H. Hiskins, I. J. D'Haenens, C. K. Asawa, and V. Evtuhov, Stimulated optical emission in fluorescent solids II: Spectroscopy and stimulated emission in ruby, *Phys. Rev.* 123, 1151–1157 (1961).

*2.* E. Snitzer, Glass lasers, *Proc. IEEE* 54, 1249–1261 (1966).

*3.* T. S. Clement, S. A. Diddams, and D. J. Jones, Lasers, ultrafast pulse technology, in "Encyclopedia of Physical Science and Technology," 3rd edition, 499–510. Academic Press, San Diego, 2002.

*4.* See as examples: S. R. Gunn, Calorimetric measurements of laser energy and power, *J. Phys. E, Sci. Instrum.* 6, 105–114 (1973);
P. J. Batemen, The measurement of laser power and energy, IEE Conf. on Lasers and their applications (London), 40-1–40-7 (1964);
G. Birnbaum and M. Birnbaum, Measurement of laser power and energy, *Proc. IEEE* 55, 1026–1031 (1967);
D. E. Killick, D. A. Batemen, D. R. Brown, T. S. Moss, and E. T. de la Perrelle, Power and energy measuring techniques for solid state lasers, *Infrared Phys.* 6, 85–109 (1966).

*5.* F. Hengstberger, "Absolute Radiometry: Electrically Calibrated Thermal Detectors of Optical Radiation." Academic Press, Boston, 1989.

*6.* H. L. Callendar, The radio balance. A thermoelectric balance for the absolute measurement of radiation, with applications to radium and its emanation, *Proc. Phys. Soc. (London)* 23, 1–34 (1911).

*7.* J. Guild, Investigations of absolute radiometry, *Proc. Roy. Soc. A.* 161, 1–38 (1937).

*8.* E. J. Gillham, Recent investigations in absolute radiometry, *Proc. Roy. Soc. (London) Ser. A.* 269, 249–276 (1962).

*9.* Tingye Li and S. D. Sims, A calorimeter for energy measurements of optical masers, *Appl. Opt.* 1, 325–328 (1962).
*10.* J. A. Calbiello, An optical calorimeter for laser energy measurements, *Proc. IEEE* 51, 611–612 (1963).
*11.* J. G. Edwards, An accurate carbon cone calorimeter for pulsed lasers, *J. Sci. Instrum.* 44, 835–838 (1967).
*12.* E. K. Damon and J. T. Flynn, A liquid calorimeter for high-energy lasers, *Appl. Opt.* 2, 163–164 (1963).
*13.* D. A. Jennings, Calorimetric measurement of pulsed laser output energy, *IEEE Trans. Instrum. Meas.* IM-15, 161–164 (1966).
*14.* D. A. McSparron, C. A. Douglas, and H. L. Badger, "Radiometric Methods for Measuring Laser Output." NBS Technical Note 418, 1967.
*15.* J. G. Edwards, An accurate carbon cone calorimeter for pulsed lasers, *J. Sci. Instrum.* 44, 835–838 (1967);
J. G. Edwards, A glass disk calorimeter for pulsed lasers, *J. Phys. E: Sci. Instrum.* 3, 452–454 (1970);
J. G. Edwards and R. Jefferies, Power and energy monitor for pulsed lasers, *J. Phys. E: Sci. Instrum.* 4, 580–584 (1971);
J. G. Edwards, A standard calorimeter for pulsed lasers, *J. Phys. E: Sci. Instrum.* 8, 663–665 (1975);
J. G. Edwards, Recent developments and problems in detection for measurement of laser outputs, *Proc. SPIE* 234, 12–16 (1980).
*16.* K. Sakurai, Y. Mitsuhashi, and T. Honda, A laser microcalorimeter, *IEEE Trans. Instrum. Meas.* IM-16, 212–219 (1967).
*17.* K. Möstl, PTB (retired), personal communication, 2004; also K. Möstl, Standards for measuring the output poser of cw lasers, Proc. 7th Int. Symp Tech. Comm. Photon-Detectors, Braunschweig, FRG, May 1976.
*18.* J. G. Edwards, A standard calorimeter for pulsed lasers, *J. Phys. E: Sci. Instrum.* 8, 663–665 (1975);
T. Honda and M. Endo, International comparison of laser power at 633 nm, *IEEE J. Quantum Electron.* QE-14, 213–215 (1978).
*19.* E. D. West and K. L. Churney, Theory of isoperibol calorimetry for laser power and energy measurements, *J. Appl. Phys.* 41, 2705–2712 (1970).
*20.* E. D. West, W. E. Case, A. L. Rasmussen, and L. B. Schmidt, A reference calorimeter for laser energy measurements, *J. Res. Nat. Bur. Stan.* 76 A, 13–26 (1972);
E. D. West and W. E. Case, Current status of NBS low-power laser energy measurement, *IEEE Trans. Instrum. Meas.* IM-23, 422–425 (1974).
*21.* E. D. West, "Data Analysis for Isoperibol Laser Calorimetry." NBS Technical Note 396, February 1971.

*22.* C. L. Cromer, NIST Optoelectronics Division, unpublished.
*23.* R. W. Faaland and M. L. Naiman, Laser power standards: a comparison of two scales, *IEEE Trans. Instrum. Meas.* IM-36, 455–457 (1987); R. L. Gallawa, J. L. Gardner, D. H. Nettleton, K. D. Stock, T. H. Ward, and Xiaoyu Li, Pilot study for the international intercomparison of responsivity scales at fibre optic wavelengths, *Metrologia* 28, 151–154 (1991).
*24.* "NIST Calibration Services Users Guide, Section on Lasers and Components used with Lasers." NIST Special Publication, 250, issued annually.
*25.* Chapters 2 and 3 (this volume).
*26.* See, for example, J. E. Martin, N. P. Fox, and P. J. Key, A cryogenic radiometer for absolute radiometric measurements, *Metrologia* 21, 147–155 (1985);
T. R. Gentile, J. M. Houston, J. E. Hardis, C. L. Cromer, and A. C. Parr, National Institute of Standards and Technology high-accuracy cryogenic radiometer, *Appl. Opt.* 35, 1056–1068 (1996).
*27.* D. Livigni, "High-Accuracy Laser Power and Energy Meter Calibration Service." NIST Special Publication, 250–262 (2003).
*28.* D. J. Livigni, C. L. Cromer, T. R. Scott, B. C. Johnson, and Z. M. Zhang, Thermal characterization of a cryogenic radiometer and comparison with a laser calorimeter, *Metrologia* 35, 819–827 (1998).
*29.* J. Lehman, I. Vayshenker, D. J. Livigni, and J. Hadler, Intamural comparison of NIST laser and optical fiber power calibrations, *J. Res. Natl. Inst. Stand. Tech.* 109, 291–298 (2004).
*30.* E. D. West and L. B. Schmidt, "A System for Calibrating Laser Power Meters for the Range 5–1000 W." NBS Techical Note 685, 1977.
*31.* X. Li, NIST, 325 Broadway, Boulder, CO 80305, personal communication and recent calibration results.
*32.* X. Li, T. R. Scott, C. L. Cromer, D. Keenan, F. Brandt, and K. Mostl, Power measurement standards for high-power lasers: comparison between the NIST and the PTB, *Metrologia* 37, 445–447 (2000).
*33.* R. L. Smith, T. W. Russell, W. E. Case, and A. L. Rasmussen, A calorimetch for high-power CW lasers, *IEEE Trans. Instrum. Meas.* IM-21, 434–438 (1972);
G. E. Chamberlain, P. A. Simpson, and R. L. Smith, Improvements in a calorimeter for high-power cw lasers, *IEEE Trans. Instrum. Meas.* IM-27, 81–86 (1978).
*34.* S. R. Gunn, Volume-absorbing calorimeters for high-power laser pulses, *Rev. Sci. Instrum.* 45, 936–943 (1974).
*35.* D. L. Franzen and L. B. Schmidt, Absolute reference calorimeter for measuring high power laser pulses, *Appl. Opt.* 15, 3115–3122 (1976).

*36.* J. G. Edwards, An accurate carbon cone calorimeter for pulsed lasers, *J. Sci. Instrum.* 44, 835–838 (1967).

*37.* J. G. Edwards, A standard calorimeter for pulsed lasers, *J. Phys. E: Scientific Instruments* 8, 663–665 (1975).

*38.* R. W. Leonhardt and T. R. Scott, Deep-UV excimer laser measurements at NIST, *Proc. SPIE* 2439, 448–459 (1995);
R. W. Leonhardt, "Calibration Service for Laser Power and Energy at 248 nm." NIST Technical Note 1395 (1998).

*39.* Shao Yang, D. Keenan, H. Laabs, and M. Dowell, A 193 nm detector nonlinearity measurement system at NIST, *Proc. SPIE* 5040, 1651–1656 (2003).

*40.* M. L. Dowell, R. D. Jones, H. Laabs, C. L. Cromer, and R. D. Morton, New developments in excimer laser metrology at 157 nm, *Proc. SPIE* 4689, 63–69 (2002).

*41.* C. L. Cromer, NIST, Optoelectronics Division, unpublished.

*42.* R. W. Whatmore, Pyroelectric devices and materials, *Rep. Prog. Phys.* 49, 1335–1386 (1986).

*43.* G. W. Day, C. A. Hamilton, R. L. Peterson, R. J. Phelan, Jr., and L. O. Mullen, Effects of poling conditions on responsivity and uniformity of polarization of $PVF_2$ pyroelectric detectors, *Appl. Phys. Lett.* 23, 456 (1974);
G. W. Day, C. A. Hamilton, P. M. Gruzensky, and R. J. Phelan, Jr., Performance and characteristics of polyvinylidene fluoride pyroelectric detectors, *Ferroelectrics* 10, 99–102 (1976).

*44.* R. J. Phelan, Jr., and A. R. Cook, Electrically calibrated pyroelectric optical radiation detector, *Appl. Opt.* 10, 2492–2500 (1973).

*45.* C. A. Hamilton, G. W. Day, and R. J. Phelan, Jr., "An Electrically Calibrated Pyroelectric Radiometer System." NBS Technical Note 678 (1976).

*46.* W. R. Blevin and W. J. Brown, Black coatings for absolute radiometers, *Metrologia* 2, 139–143 (1966).

*47.* L. Harris, "The Optical Properties of Metal Blacks and Carbon Blacks." Massachusetts Institute of Standards and Technology, Cambridge, 1967.

*48.* R. Stain, W. E. Schneider, W. R. Waters, and J. K. Jackson, Some factors affecting the sensitivity and spectral response of thermoelectric (radiometric) detectors, *Appl. Opt.* 4, 703–710 (1965).

*49.* W. L. Eisenman, R. L. Bates, and J. D. Merriam, Black radiation detector, *J. Opt. Soc. Am.* 53, 729–734 (1963);
W. L. Eisenman and R. L. Bates, Improved black radiation detector, *J. Opt. Soc. Am.* 54, 1280–1281 (1964).

*50.* W. R. Blevin and W. J. Brown, A precise measurement of the Stefan–Boltzman constant, *Metrologia* 7, 15–29 (1971).
*51.* G. W. Day, C. A. Hamilton, and K. W. Pyatt, A convenient, spectrally flat reference detector for the visible to 12 μm region, *Appl. Opt.* 15, 1865–1868 (1976).
*52.* J. H. Lehman, Pyroelectric trap detector for spectral responsivity measurements, *Appl. Opt.* 36, 9117–9118 (1997).
*53.* E. F. Zalewski and C. R. Duda, Silicon photodiode device with 100% external quantum efficiency, *Appl. Opt.* 22, 2867–2873 (1983).
*54.* E. F. Zalewski and G. Geist, Silicon photodiode absolute spectral response self-calibration, *Appl. Opt.* 19, 1214–1216 (1980).
*55.* J. Geist, E. F. Zalewski, and A. R. Schaefer, Spectral response self-calibration and interpolation of silicon photodiodes, *Appl. Opt.* 19, 3795–3799 (1980).
*56.* J. Lehman, J. Sauvageau, I. Vayshenker, C. Cromer, and K. Foley, *Meas. Sci. Tech.* 9, 1694–1698 (1988).
*57.* J. H. Lehman and X. Li, A transfer standard for optical fiber power metrology, *Appl. Opt.* 38, 7164–7166 (1999).
*58.* J. H. Lehman and C. L. Cromer, Optical trap detector for calibration of optical fiber power meters: Coupling efficiency, *Appl. Opt.* 41, 6531–6536 (2002).
*59.* J. L Gardner, Transmission trap detectors, *Appl. Opt.* 33, 5914–5918 (1994).
*60.* J. L. Gardner, A four-element transmission trap detector, *Metrologia* 32, 469–472 (1995/1996).
*61.* T. Kübarsepp, P. Kährä, and E. Ikonen, Characterization of a polarization-independent transmission trap detector, *Appl. Opt.* 36, 2807–2812 (1997).
*62.* J. H. Lehman and C. L. Cromer, Optical tunnel-trap detector for radiometric measurements, *Metrologia* 37, 477–480 (2000).
*63.* Y. Beers, "The Theory of the Optical Wedge Beam Splitter." NBS Monograph 146, 1974.
*64.* D. L. Franzen, Precision beam splitters for $CO_2$ lasers, *Appl. Opt.* 14, 647–652 (1975).
*65.* B. L. Danielson and Y. Beers, "Laser Attenuators for the Production of Low Power Beams in the Visible and 1.06 μm Regions." NBS Technical Note 677, 1976.
*66.* B. L. Danielson, "Measurement Procedures for the Optical Beam Splitter Attenuation Device BA-1." NBS Internal/Interagency Report, NBSIR 77-858, 1977.
*67.* L. Doslovi and F. Righini, Fast determination of the nonlinearity of photodetectors, *Appl. Opt.* 19, 3200–3203 (1980).

68. A. R. Schaefer, E. F. Zalewski, and J. Geist, Silicon detector nonlinearity and related effects, *Appl. Opt.* 22, 1232–1236 (1983).
69. R. D. Saunders and J. B. Shumaker, Automated radiometric linearity tester, *Appl. Opt.* 23, 3504–3506 (1984).
70. J. Fischer and Lei Fu, Photodiode nonlinearity measurement with an intensity stabilized laser as a radiation source, *Appl. Opt.* 32, 4187–4190 (1993).
71. International Electrotechnical Commission Standard IEC 61315: "Calibration of Fibre-Optic Power Meters."
72. Shao Yang, I. Vayshenker, Xiaoyu Li, T. R. Scott, and M. Zander, "Optical Detector Nonlinearity: Simulation." NIST Technical Note, 1376 (1995).
73. I. Vayshenker, NIST, 325 Broadway, Boulder, CO 80305, Personal Communication, Example data from a NIST calibration.
74. Calibration of optical fiber power meters, IEC Standard 61315, International Electrotechnical Commission, Geneva, Switzerland.
75. I. Vayshenker, Xiaoyu Li, D. J. Livigni, T. R. Scott, C. L. Cromer, "Optical Fiber Power Meter Calibrations at NIST." NIST Special Publication 250–254 (2000).
76. See for example, A. E. Siegman, "Lasers." University Science Books, 1986.
77. "Terminology and Test Methods for Lasers," ISO Standard 11146, International Organization for Standardization (ISO), Geneva, Switzerland, 1999.
78. J. E. Bowers and C. A. Bussus, Jr., Ultrawide-band long-wavelength p-i-n photodetectors, *J. Lightwave Tech.* LT-3, 1339–1350 (1987).
79. Technical Specifications, 83440B/CD High-Speed Lightwave Converters, Agilent Technologies, 2002.
80. See for example, C. Belzile, J. C. Kieffer, C. Y. Cote, T. Oksenhendler, and D. Kaplan, Jitter-free subpicosecond streak cameras, *Rev. Sci. Instrum.* 73, 1617–1620 (2002).
81. For an early example of optical sampling, demonstrating its use with a single-shot waveform, see G. C. Vogel, A. Savage, and M. A. Duguay, Picosecond optical sampling, *IEEE J. Quantum Electron.* QE-10, 642–646 (1974).
82. For more recent work on optical sampling, see R. L. Jungerman, G. Lee, O. Buccafusca, Y. Kaneko, N. Itagaki, R. Shioda, A. Harada, Y. Nihei, and G. Sucha, 1-THz bandwidth C- and L-band optical sampling with a bit rate agile timebase, *IEEE Photonics Tech. Lett.* 14, 1148–1150 (2002), and references therein.
83. E. P. Ippen and C. V. Shank, Techniques for measurement, Chapter 3 in "Ultrashort Light Pulses" (S. L. Shapiro, Ed.), 2nd edition. Springer-Verlag, New York, 1984.

*84.* R. Trebino, "Frequency-Resolved Optical Gating: The Measurement of Ultrashort Laser Pulses." Springer, New York, 2002.

*85.* R. P. Trebino and I. A. Walmsley, Eds., Generation, amplification, and measurement of ultrashort laser pulses, *Proc. SPIE* 1 2116, (1994).

*86.* C. Dorrer and I. A. Walmsley, Concepts for the temporal characterization of short optical pulses, forthcoming in *J. Appl. Signal Processing*.

*87.* R. Trebino, K. W. DeLong, D. N. Fittinghoff, J. N. Sweetser, M. A. Krumbügel, and B. A. Richman, Measuring ultrashort laser pulses in the time-frequency domain using frequency-resolved optical gating, *Rev. Sci Instrum.* 68, 3277–3295 (1997).

*88.* S. A. Diddams, X. M. Zhao, and J.-C. M. Diels, Pulse measurements without optical nonlinearities, *Proc. SPIE* 2116, 238–244 (1994).

# 9. DIFFRACTION EFFECTS IN RADIOMETRY*

Eric L. Shirley
National Institute of Standards and Technology, Gaithersburg, Maryland, USA

## 9.1 Introduction and Definitions

### 9.1.1 The Context of Traditional Radiometry

Radiometry is the measurement of energy in the form of electromagnetic radiation (i.e., light). We can specify such a measurement further by geometrical definitions that delineate the radiation that is measured. Two "end-point" delineations are self-evident in the definitions of radiance and irradiance. Radiance (denoted by $L$) is defined as the power that is emitted per projected unit area of source per steradian. Irradiance (denoted by $E$) is defined as the power incident per unit area of a surface. Radiance, irradiance and other quantities are further discussed in Chapter 1. The definitions of radiance and irradiance illustrate how traditional radiometry frequently involves transfer of electromagnetic energy from points on one surface to points on a different surface.

### 9.1.2 Throughput of an Optical Setup

For a given optical setup, it is standard to relate source radiance ($L$) to power reaching the detector ($\Phi$) by a measurement equation [1]. In this chapter, the measurement equation can have the form

$$\Phi_\lambda(\lambda) = T(\lambda) L_\lambda(\lambda) \tag{9.1}$$

A spectral quantity such as $\Phi_\lambda(\lambda)$ is the power per unit wavelength at wavelength $\lambda$, according to the pattern,

$$\Phi = \int_0^\infty \mathrm{d}\lambda \; \Phi_\lambda(\lambda) \tag{9.2}$$

The quantity $T(\lambda)$ is the "throughput" of the optical setup under consideration. As a starting approximation, we can compute $T(\lambda)$ using ray-tracing according to geometrical optics. We then have $T(\lambda) = T_0 M(\lambda)$, where $T_0$ is the "geometrical throughput" of an optical setup, a purely geometrical entity, with no spectral dependence, and $M(\lambda)$, which has an

*Contribution of the National Institute of Standards and Technology.



EXPERIMENTAL METHODS IN THE PHYSICAL SCIENCES, vol. 41
ISSN 1079-4042 DOI: 10.1016/S1079-4042(05)41009-7

ideal value of unity, can be introduced to account for optical filters, lens transmittance, mirror reflectance, and so forth. For simple setups, it is often possible to express $T_0$ in closed form (see Chapter 1 for an example).

#### 9.1.2.1 Diffraction effects

Because electromagnetic radiation actually propagates as a wave entity instead of according to geometrical (ray) optics, we generally have $T(\lambda) \neq T_0 M(\lambda)$. The relationship between $T(\lambda)$ and $T_0$ can instead be given in the form

$$T(\lambda) = F(\lambda) T_0 M(\lambda) = [1 + \varepsilon_{\text{diff}}(\lambda)] T_0 M(\lambda) \tag{9.3}$$

The quantity $F(\lambda)$ is a ratio that would be unity in the ideal case of geometrical optics. The quantity $\varepsilon_{\text{diff}}(\lambda)$ is the difference between $F(\lambda)$ and unity. This difference and its impact on radiometric measurements shall be referred to as "diffraction effects" on spectral throughput or spectral power. We can also consider diffraction effects on spectral and total irradiance, point-to-point propagation of radiation, and other quantities.

Accounting for the difference between $F(\lambda)$ and unity, and its impact on measurement results shall be referred to as including "diffraction corrections." From Eq. (9.3), we may deduce that many quantities, including detector responsivity, source radiance, and geometrical throughput (as influenced, say, by the area of an aperture under test), can be inferred from a radiometric measurement. This requires sufficient knowledge of all other quantities affecting the measurement. In general, diffraction is one factor that affects the outcome. Hence, diffraction effects may need to be taken into account.

In the case of complex radiation, such as is encountered when measuring the radiance of a laboratory blackbody source, diffraction effects on the spectrally integrated signal are more relevant than diffraction effects at just one wavelength. Suppose that the output signal $S$ of a sensor (for instance, this signal may be an output voltage) is given by $S = \int_0^\infty d\lambda\, R(\lambda)\Phi_\lambda(\lambda)$, which involves a linear combination of the spectral power $\Phi_\lambda(\lambda)$ reaching the detector at each wavelength weighted by the sensor's relative spectral response, $R(\lambda)$. The diffraction effects on the signal become

$$\langle F \rangle = 1 + \langle \varepsilon_{\text{diff}} \rangle = 1 + \frac{\int_0^\infty d\lambda [F(\lambda) - 1] R(\lambda)\Phi_\lambda(\lambda)}{\int_0^\infty d\lambda\, R(\lambda)\Phi_\lambda(\lambda)} \tag{9.4}$$

### 9.1.3 Chapter Organization

In what follows, aspects of diffraction effects are further discussed within the context of radiometry. Diffraction theory is first discussed, with concepts and conceptual pictures of diffraction as it affects radiometry being laid out. Only the very core essentials are discussed, with much detail left to the references.

A simplified model for diffraction effects is derived that is appropriate for describing Fresnel and Fraunhofer diffraction in systems that can be treated using Gaussian optics [2, 3]. The model is applicable to diffraction by single apertures and lenses as well as by several optical elements in series, which can often be more relevant to practical radiometry.

Nonetheless, the main diffraction effects often result from that by one optical element, which is frequently a circular aperture or lens that is placed between a coaxial circular "effective" source and "effective" detector. For example, a defining aperture that is in front of the opening of a blackbody cavity might be treatable as a Lambertian source for purposes of diffraction effects that occur downstream from the defining aperture. Therefore, diffraction effects on the cylindrically symmetrical source–aperture–detector (SAD) problem are discussed in more detail.

The radiometry literature attests to the considerable attention paid to diffraction effects. Some of this work, much of it recent, is mentioned. The role of diffraction in radiometry is still a dynamic and rapidly evolving field. Because diffraction effects are especially important at longer wavelengths, the extension of radiometry into the infrared has driven much of the need for improvement, coupled with the need for ever higher accuracy, thanks to the innovations in cryogenic radiometry. The reader should therefore also search the literature written after this volume.

Finally, brief mention is made of novel and/or unconventional radiation sources, such as synchrotrons. For these sources, novel coherence that is not present in incoherent sources such as blackbodies may require reassessing the applicability of conventional ways of thinking about diffraction effects.

## 9.2 Theories of Diffraction

Here, a "theory of diffraction" could operationally involve electromagnetic wave propagation in the complex geometry of an optical setup. However, we do not really want a complete description of such propagation. For purposes of diffraction studies, we want to estimate how the actual propagation differs from that predicted by geometrical optics, and how these differences affect measurements. Distilling just this information is the goal of diffraction analysis.

A complete description of the propagation of electromagnetic radiation would not only require solving Maxwell's equations that govern the radiation within free space and other media [4]. We would also need to describe the coupled motion of induced dielectric polarization and electrical currents within optical elements. Even now, this task has been carried out, for instance, in the fields of photonic band structure and optical nanostructures, only for simple or highly simplified systems [5]. Some simplifications are usually required to describe electromagnetic wave propagation in conventional optical systems.

### 9.2.1 Motivation of the Scalar Helmholtz Equation

Within a region of free space or some other uniform medium, the electric field $\mathbf{E}(\mathbf{x}, t)$ has the general form,

$$\mathbf{E}(\mathbf{x}, t) = \int_{-\infty}^{\infty} \mathrm{d}\omega \int_S \mathrm{d}^2\hat{\mathbf{q}} \sum_{\mu} X_{\mu}(\hat{\mathbf{q}}, \omega)\hat{\mathbf{e}}_{\mu}(\hat{\mathbf{q}}, \omega) \exp\{\mathrm{i}\omega[n_{\mu}(\hat{\mathbf{q}}, \omega)\hat{\mathbf{q}} \cdot \mathbf{x}/c - t]\} \quad (9.5)$$

Here $\mathbf{x}$ and $t$ are space and time coordinates, $\omega$ the angular frequency, $S$ the unit sphere, $\hat{\mathbf{q}}$ a directional unit vector, $\hat{\mathbf{e}}_{\mu}(\hat{\mathbf{q}}, \omega)$ a polarization vector for polarization $\mu$, $n_{\mu}(\hat{\mathbf{q}}, \omega)$ an index of refraction, and $c$ the speed of light in vacuum. Furthermore, i denotes the square root of $-1$, and we use the convention of time evolution having the form, $\exp(-\mathrm{i}\omega t)$. A parameter $X_{\mu}(\hat{\mathbf{q}}, \omega)$ is the complex amplitude for a particular combination of frequency, polarization and direction of propagation. The magnetic induction $\mathbf{B}(\mathbf{x}, t)$ may be found from $\mathbf{E}(\mathbf{x}, t)$ using Maxwell's equations, which formally completes one aspect of describing electromagnetic wave propagation.

We now consider monochromatic radiation. We can sum any calculated result for the electromagnetic field over frequency components to describe the behavior of complex radiation, once we describe the behavior of monochromatic radiation. We also now restrict ourselves for simplicity to isotropic media, so that, for a given value of $\omega$, the magnitude of the angular wavevector is simply $q = n(\omega)\omega/c$. From Maxwell's equations, we can derive the Helmholtz wave equation that is obeyed by the individual vector components of the electric field $\mathbf{E}(\mathbf{x}, t)$ and magnetic induction $\mathbf{B}(\mathbf{x}, t)$:

$$\left.\begin{aligned} [\nabla^2 + q^2]\mathbf{E}(\mathbf{x}, t) &= 0 \\ [\nabla^2 + q^2]\mathbf{B}(\mathbf{x}, t) &= 0 \end{aligned}\right\} \quad (9.6)$$

In optical setups, electromagnetic radiation frequently flows past certain regions in one general direction. In other instances, such as when radiation is reflected by a mirror, quantities such as $\mathbf{E}(\mathbf{x}, t)$ can be decomposed into incident and reflected parts, each of which obeys the wave equation and can be considered separately except at the mirror surface. We therefore consider a field $\mathbf{E}(\mathbf{x}, t)$ that is associated with radiation flowing along one general direction.

We next re-express this field as a superposition of spherical waves originating from sources outside the region under consideration. From that, we can show that the electromagnetic energy current density, related to the Poynting vector, is approximately described as a superposition of the squares of the transverse parts of $\mathbf{E}(\mathbf{x}, t)$ associated with both polarizations.

For each polarization state, the energy density $W$ and energy current density $\mathbf{J}$ can be reasonably modeled in terms of a fictitious scalar wave field, $U(\mathbf{x}, t)$. For the case of monochromatic scalar radiation, this field also obeys the Helmholtz wave equation,

$$[\nabla^2 + q^2]U(\mathbf{x}, t) = 0 \tag{9.7}$$

In a given region, if the energy density is $W = C|U(\mathbf{x}, t)|^2$, where $C$ is a scaling constant, the energy current density would be $\mathbf{J} = [c/(n(\omega)q)]\,\mathrm{Re}\{CU^*(\mathbf{x}, t)\nabla U(\mathbf{x}, t)\}$. An asterisk denotes complex conjugation, and "Re" indicates the real part of an expression. For radiation that is nominally propagating along the direction $\hat{\mathbf{e}}$, we have $\mathbf{J} \cong (cW/n(\omega))\hat{\mathbf{e}}$. This means that the irradiance at a surface can be approximated as being proportional to the time-averaged part of $W$ associated with an incident wave.

### 9.2.2 Kirchhoff's Integral Formula

Instead of solving the Helmholtz wave equation completely for the scalar radiation field, Kirchhoff (see [4], pp. 427–429, [6]) solved this equation approximately, as a boundary-value problem using Green's function techniques. One Green's function for the Helmholtz equation, once the value of $q$ is assumed, is

$$G(\mathbf{x}, \mathbf{x}') = \frac{\exp(\mathrm{i}q|\mathbf{x} - \mathbf{x}'|)}{|\mathbf{x} - \mathbf{x}'|} \tag{9.8}$$

This Green's function obeys the inhomogeneous equation,

$$(\nabla^2 + q^2)G(\mathbf{x}, \mathbf{x}') = -4\pi\delta^3(\mathbf{x} - \mathbf{x}') \tag{9.9}$$

Here the Laplacian is taken with respect to the first argument, and the Dirac delta function is used, with $\delta^3(\mathbf{x} - \mathbf{x}') = \delta(x - x')\delta(y - y')\delta(z - z')$. We henceforth suppress time dependence for monochromatic radiation, so that $U(\mathbf{x}, t) = U(\mathbf{x})e^{-\mathrm{i}\omega t}$ is abbreviated as $U(\mathbf{x})$. The problem to be solved to find $U(\mathbf{x})$ is illustrated with the aid of Figure 9.1. The figure shows a cut-away view of $\Omega$, a region of three-dimensional space enclosed by a two-dimensional boundary surface, $Q$. Radiation is not emitted anywhere within $\Omega$, but it can enter $\Omega$ through an aperture whose area $Ap$ lies on $Q$. $Ap$ can be the surface of a mirror, aperture or lens, and can be flat or curved.

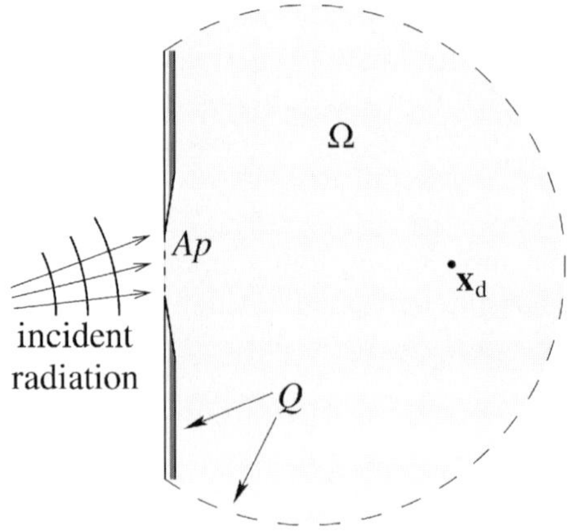


FIG. 9.1. Generic context of Kirchhoff boundary-value problem.

The value $U(\mathbf{x})$ and inwardly directed normal derivative of $U(\mathbf{x})$, $\partial U(\mathbf{x})/\partial n$, are assumed to be equal to the value and normal derivative of an incident wave on $Ap$, respectively denoted by $U_i(\mathbf{x})$ and $\partial U_i(\mathbf{x})/\partial n$, and to be equal to zero everywhere else on $Q$. If we set $\mathbf{x}' = \mathbf{x}_d$, where $\mathbf{x}_d$ is a point of interest in $\Omega$, multiplication of Eq. (9.7) on the left by $G(\mathbf{x}, \mathbf{x}_d)$ and integration with respect to $\mathbf{x}$ throughout $\Omega$ gives

$$\int_\Omega d^3\mathbf{x} G(\mathbf{x}, \mathbf{x}_d)(\nabla^2 + k^2)U(\mathbf{x}) = 0 \tag{9.10}$$

Similarly, multiplication of Eq. (9.8) on the left by $U(\mathbf{x})$ and the same integration gives

$$\int_\Omega d^3\mathbf{x} U(\mathbf{x})(\nabla^2 + q^2)G(\mathbf{x}, \mathbf{x}_d) = -4\pi U(\mathbf{x}_d) \tag{9.11}$$

Subtracting Eq. (9.11) from Eq. (9.10) and application of Green's theorem with the assumed boundary conditions gives

$$\begin{aligned} U(\mathbf{x}_d) &= \frac{1}{4\pi}\int_\Omega d^3\mathbf{x}[G(\mathbf{x}, \mathbf{x}_d)(\nabla^2 + q^2)U(\mathbf{x}) - U(\mathbf{x})(\nabla^2 + q^2)G(\mathbf{x}, \mathbf{x}_d)] \\ &= \frac{1}{4\pi}\int_\Omega d^3\mathbf{x}\{\nabla \cdot [G(\mathbf{x}, \mathbf{x}_d)\nabla U(\mathbf{x}) - U(\mathbf{x})\nabla G(\mathbf{x}, \mathbf{x}_d)]\} \\ &= \frac{1}{4\pi}\int_Q d^2\mathbf{x}[U(\mathbf{x})\partial G(\mathbf{x}, \mathbf{x}_d)/\partial n - G(\mathbf{x}, \mathbf{x}_d)\partial U(\mathbf{x})/\partial n] \\ &\cong \frac{1}{4\pi}\int_{Ap} d^2\mathbf{x}[U_i(\mathbf{x})\partial G(\mathbf{x}, \mathbf{x}_d)/\partial n - G(\mathbf{x}, \mathbf{x}_d)\partial U_i(\mathbf{x})/\partial n] \end{aligned} \tag{9.12}$$

The fact that contributions from every part of $Q$ other than $Ap$ can be neglected requires more discussion than we provide here. This detail is available, for instance, in Born and Wolf [7].

From Eq. (9.12), we see that, if values are known or at least assumed for $U_i(\mathbf{x})$ and $\partial U_i(\mathbf{x})/\partial n$ on a given optical element, we can find $U(\mathbf{x}_d)$ and its derivatives anywhere within $\Omega$. We can use Eq. (9.12) to find $U(\mathbf{x}_d)$ and its derivatives on the *next* optical surface (or aperture area) downstream from $Ap$, by temporarily considering the next element as being within $\Omega$. Repeated, iterative application of Eq. (9.12) in this fashion can help trace the flow of radiation through an optical setup. In many cases, we can assume that the radiation field originates as the radiation from infinitely many mutually incoherent point sources located on the area of a source. Radiation originating at $\mathbf{x}_s$ initially has the form $U(\mathbf{x}) = U_0 \exp(iq\,|\mathbf{x} - \mathbf{x}_s|)/|\mathbf{x} - \mathbf{x}_s|$.

It is probably safe to say that all or nearly all analyses of diffraction effects in radiometry are equivalent to some application of Eq. (9.12). Diffraction effects on the irradiance on a detector area can be inferred from how the value of $W$ or $\mathbf{J}$ computed according to Eq. (9.12) differs from the "ideal" value. The ideal value can be related to the distribution of rays traced from the source through the optical system according to geometrical optics that are incident on a given area. Perfect focusing as found in geometrical optics can lead to infinite ray densities at various places. It is for this reason that the distribution of incident rays can be most safely described in fashions such as the quantity of rays falling on a finite area.

#### *9.2.2.1 Some outstanding issues*

The last step in Eq. (9.12) is not rigorous, because the assumed boundary conditions are arbitrary. In fact, it can even be shown that if any solution of the Helmholtz equation $U(\mathbf{x})$ and its normal derivative $\partial U(\mathbf{x})/\partial n$ are both zero everywhere on a finite surface such as the portion of $Q$ other than $Ap$, then $U(\mathbf{x})$ must be zero everywhere in $\Omega$ (see [4], pp. 429–432).

One class of remedies to this problem is to change the boundary conditions, such as by assuming that only a certain linear combination $\alpha U(\mathbf{x}) + \beta \partial U(\mathbf{x})/\partial n$ is known on $Q$. This leads to the Rayleigh–Sommerfeld variants of the Kirchhoff diffraction theory [8]. This class of remedies requires different Green's functions involving image sources within $\Omega$ that help satisfy the assumed boundary conditions. Such Green's functions cannot be found in all instances, but have been formulated for planar $Ap$. The boundary-diffraction wave (BDW) formulation that is discussed in Section 9.2.3 has been generalized to several boundary conditions and several different types of incident wave fronts on a planar $Ap$. (Introducing curvature of *Ap* can be similar to warping wave fronts by means of adjusting their phase on a fixed, planar *Ap*.)

More rigorous theoretical treatments of diffraction have also been carried out. Sommerfeld considered diffraction of s- and p-polarized light at the straight edge of an infinitely conducting half-plane [9]. Bethe [10] and Bouwkamp [11] pioneered diffraction by very small, circular holes in perfectly conducting planes, and Levine and Schwinger [12] have carried out related investigations. Braunbek [13] has also attempted to refine Kirchhoff's theory, and Mielenz [14] has recently pursued a rigorous investigation of diffraction effects in close proximity to an aperture and transmission functions of small apertures reminiscent of the work by Levine and Schwinger. There have been other investigations as well, many of which have been compiled by Oughstun [15].

The above investigations lead to two main conclusions that summarize the state of affairs for most diffraction effects in radiometry. First, despite any formal mathematical inconsistencies, the Kirchhoff, Rayleigh–Sommerfeld and more rigorous treatments often predict similar diffraction effects, especially if one is considering a radiation field far from *Ap* and along a direction nearly normal to *Ap* and not too different from the natural path of geometrical rays. Fortunately, this qualification is met in most instances in practical radiometry.

Second, the Kirchhoff theory is bound to break down in the case of very small apertures, if the aperture diameter is at most a few times larger than the wavelength. Exactly how this breakdown occurs remains a problem of theoretical and experimental interest, especially when very small apertures are used in far-infrared measurements. At the time of this writing, measurement uncertainties remain too large to definitively assess the breakdown of Kirchhoff's theory. These uncertainties can have causes ranging from mundane problems such as measurement noise and limited repeatability to issues as fundamental as the ambiguity of what defines an aperture's edge and corresponding area, finite aperture thickness, and imprecise knowledge of appropriate boundary conditions for real apertures.

### 9.2.3 Boundary-Diffraction-Wave Formulation

Maggi [16], Rubinowicz [17] and Miyamoto and Wolf [18] pioneered a reformulation of Eq. (9.12) that re-expresses the radiation field in the following fashion:

$$U(\mathbf{x}) = U_G(\mathbf{x}) + U_B(\mathbf{x}) \tag{9.13}$$

The first term is called the "geometrical wave." It is the continuation of the wave incident on *Ap* in the illuminated region of $\Omega$ and zero in the shadow region. Therefore, $U_G(\mathbf{x})$ and $U_B(\mathbf{x})$ have canceling discontinuities at the boundary between the illuminated and shadow regions. These regions are indicated in Figure 9.2 for several incident waves. The second term is called the BDW, because it re-expresses diffraction effects on $U(\mathbf{x})$ as a line integral around $\Gamma$, the perimeter of $Ap$.

Formulas for $U_B(\mathbf{x})$ vary depending on the incident wavefront [19]. As an example, we can consider the case illustrated in Figure 9.3. If the incident

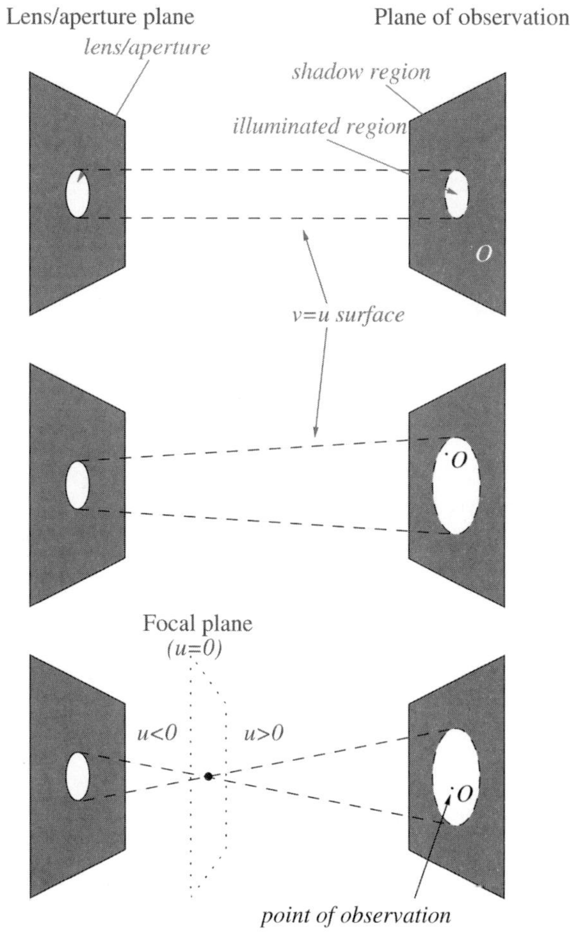


FIG. 9.2. Illuminated and shadow regions for radiation that is a plane wave, diverging spherical wave and converging spherical wave downstream from a lens or aperture. Dashed lines indicate boundary between regions. Cross sections of regions are illustrated in the plane of observation.

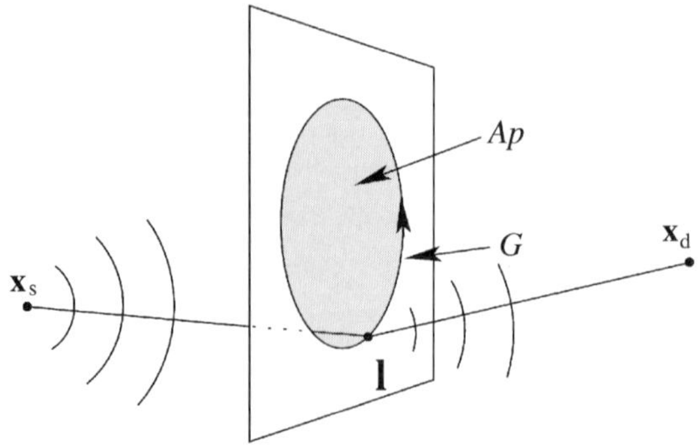


FIG. 9.3. Geometrical construction for BDW formulation.

wave is a spherical wave originating from point $\mathbf{x}_s$ as the spherical wave

$$U_i(\mathbf{x}) = U_0\left(\frac{\exp(iq|\mathbf{x} - \mathbf{x}_s|)}{|\mathbf{x} - \mathbf{x}_s|}\right) \tag{9.14}$$

one has $U_G(\mathbf{x}_d) = U_i(\mathbf{x}_d)$ in the illuminated region, $U_G(\mathbf{x}_d) = 0$ in the shadow region, and

$$U_B(\mathbf{x}_d) = \frac{U_0}{4\pi}\oint_\Gamma \frac{d\mathbf{l}\cdot(\mathbf{d}\times\mathbf{s})}{ds + \mathbf{d}\cdot\mathbf{s}}\left(\frac{\exp(iqs)}{s}\right)\left(\frac{\exp(iqd)}{d}\right) \tag{9.15}$$

Here $\mathbf{l}$ is a point on $\Gamma$, we have $\mathbf{d} = \mathbf{l} - \mathbf{x}_d$ and $\mathbf{s} = \mathbf{l} - \mathbf{x}_s$, and the line integral is performed in the right-hand sense about the forward direction of propagation. Besides being easier to evaluate numerically than a double integral, the single integral in Eq. (9.15) also provides better insight into the behavior and asymptotic properties of the BDW. The integrand in Eq. (9.15) has singularities when $\mathbf{x}_d$ is near a geometrical shadow boundary. This requires special care by the practitioner, and points to certain difficulties with the BDW formulation.

### 9.2.4 Geometrical Theory of Diffraction

Keller and co-workers [20] largely spawned the field of the geometrical theory of diffraction (GTD) in a classic series of papers in 1960s. Two classic volumes on this are the volumes by James [21] and by Borovikov [22]. In a sense, the GTD picture gives results that bear strong similarity to what would result from evaluating the BDW using asymptotic methods such as the stationary-phase approximation to evaluate the line integral. The GTD models the radiation that reaches $\mathbf{r}_d$ as corresponding to a geometrical ray and/or very few rays that are bent at certain points on aperture edges and other places. In geometrical optics, Fermat's least-time principle dictates the path taken by a ray through an optical system. Likewise, the stationary nature of the total optical path length at points on edges in the BDW integrand can be deduced from a generalization of Fermat's principle to the diffracted part of a radiation field.

If we go one step beyond the GTD, continuing to work backwards from solving the full Maxwell's equations with all boundary conditions, one logical finishing point would be to arrive at geometrical optics. Luneburg [23] has presented a systematic development of the successive steps and approximations that can bridge Maxwell's equations and geometrical optics. Luneburg's contribution provides a nearly seamless path from fundamental physical equations to geometrical optics. A similar critique of logical steps and approximations made at each stage might provide insight into future practical approximations describing diffraction effects in radiometry.

## 9.3 Practical Diffraction Calculations

The Kirchhoff diffraction theory is usually adequate for estimating diffraction effects in radiometry. We can make several additional, simplifying, mathematical approximations to streamline analytic and numerical calculations without significant loss of accuracy. In this section, we discuss some of these approximations, including those related to Gaussian optics and developments that lead to the Fraunhofer and Fresnel approximations.

### 9.3.1 Unfolding and Neglect of Obliquity Factors

Note how the optical setup in Figure 9.4a can be "unfolded" into a nearly equivalent optical system shown in Figure 9.4b, with the following characteristics: a length that is much larger than its width, and radiation that is incident on or proceeds downstream from surfaces of optical elements (including aperture areas) at angles close to normal. This justifies using the following approximations in Eq. (9.12):

$$\frac{\partial G(\mathbf{x}, \mathbf{x}_\mathrm{d})}{\partial n} \cong -\mathrm{i}qG(\mathbf{x}, \mathbf{x}_\mathrm{d}) \tag{9.16}$$

and

$$\frac{\partial U_\mathrm{i}(\mathbf{x})}{\partial n} \cong +\mathrm{i}qU_\mathrm{i}(\mathbf{x}) \tag{9.17}$$

Hence, Eq. (9.12) can be rewritten as

$$U(\mathbf{x}_\mathrm{d}) \cong \frac{1}{\mathrm{i}\lambda}\int_{Ap} \mathrm{d}^2\mathbf{x}\, U_\mathrm{i}(\mathbf{x})G(\mathbf{x}, \mathbf{x}_\mathrm{d}) \tag{9.18}$$

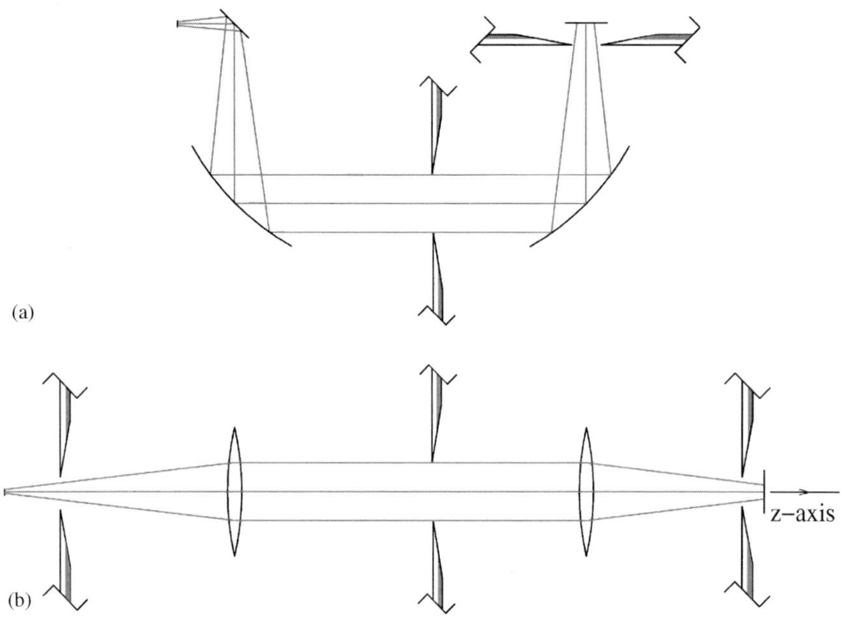


FIG. 9.4. Unfolding of an optical setup: (a) actual setup, (b) unfolded setup.

### 9.3.2 Gaussian-Optics Approximation of a Spherical Wave

The Green's function given in Eq. (9.8) has the form of a spherical wave that involves the distance between $\mathbf{x}$ and $\mathbf{x}'$ in the exponent and in the denominator. This distance can be expanded in a Taylor expansion,

$$\begin{aligned}\left|\mathbf{x}-\mathbf{x}'\right| &= [(x-x')^2+(y-y')^2+(z-z')^2]^{1/2}\\ &= \left|z-z'\right| + \frac{(x-x')^2+(y-y')^2}{2|z-z'|} + \cdots\\ &\cong \left|z-z'\right| + \left(\frac{x^2+y^2}{2|z-z'|}\right) + \left(\frac{x'^2+y'^2}{2|z-z'|}\right) - \left(\frac{xx'+yy'}{|z-z'|}\right)\end{aligned} \tag{9.19}$$

As shown in Figure 9.4b, the $z$-axis is the optical axis of the unfolded setup, with propagation generally along the positive-$z$ direction. The Taylor expansion in Eq. (9.19) assumes that distances along the $z$-axis between points on successive optical elements are much larger than the corresponding distances along transverse directions.

In Gaussian or paraxial optics, a distance in the exponent is approximated by the four terms shown in Eq. (9.19), but a distance in the denominator is approximated by the first term only:

$$G(\mathbf{x}, \mathbf{x}') \cong \frac{\exp\left\{ iq\left[ |z - z'| + \left( \frac{x^2 + y^2}{2|z - z'|} \right) + \left( \frac{x'^2 + y'^2}{2|z - z'|} \right) - \left( \frac{xx' + yy'}{|z - z'|} \right) \right] \right\}}{|z - z'|} \tag{9.20}$$

This combination of approximations has several advantages. One regarding its efficacy is "self-consistency," which is discussed later. A key practical advantage is that the approximate Green's function is proportional to a generalized Gaussian-type exponential function. We can use this fact to great advantage, because of the insight that is available regarding integration of Gaussian functions. This becomes especially clear when we realize that such integration is done over two-dimensional surfaces. The right-hand side of Eq. (9.20) is the product of four factors: an overall prefactor, a Gaussian that depends on the transverse coordinates $x$ and $y$ on one optical element, a corresponding Gaussian that depends on $x'$ and $y'$, and an exponential function involving the inner product of both sets of transverse coordinates, $xx' + yy'$.

One can also easily include curvature of optical surfaces in Gaussian-optics in an analogous approximate fashion. A curved element is approximated as a thin lens. A ray passing through such an element is treated as a ray passing through a constant-$z$ surface, but with a modified effective optical path length depending on the transverse coordinates $x$ and $y$ where it crosses the surface. This modification is done by the replacement

$$U_i(x, y, z) \to U_i(x, y, z) \exp\left[ -\frac{iq(x^2 + y^2)}{2f} \right] \tag{9.21}$$

after $U_i$ is computed and before we iterate Eq. (9.18) to the next element. The parameter $f$ is the signed focal length of the optical element. A non-curved element has $f = \pm\infty$.

### 9.3.3 Fraunhofer and Fresnel Diffraction

In the case of diffraction of a spherical wave originating at $\mathbf{x}$ by one optical element in free space shown in Figure 9.5, we have

$$\begin{aligned} U(x', y', z') &\\ \cong \frac{U_0 e^{iq(d+d')}}{i\lambda d d'} \iint_{Ap} d^2\mathbf{x}'' \exp\Bigg\{ iq\Bigg[ &\frac{(x - x'')^2 + (y - y'')^2}{2d} \\ &+ \frac{(x' - x'')^2 + (y' - y'')^2}{2d'} - \frac{x''^2 + y''^2}{2f} \Bigg] \Bigg\} \end{aligned} \tag{9.22}$$

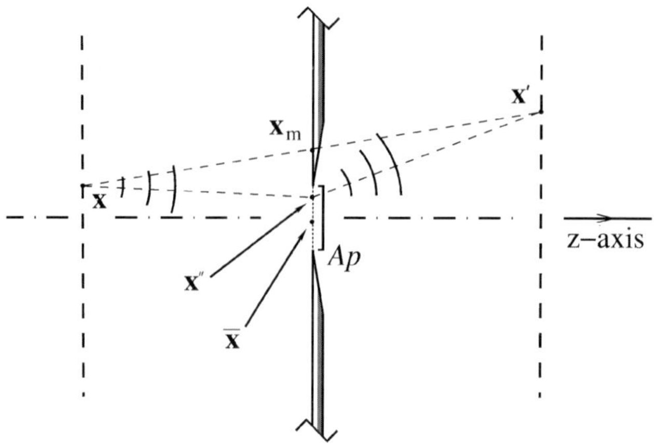


FIG. 9.5. Parameters pertinent to diffraction by a single optical element.

This approximates the initial spherical wave originating at $\mathbf{x}$ and $G(\mathbf{x}'', \mathbf{x}')$ according to Gaussian-optics and allows for focusing effects. It can be rearranged by introducing

$$x_{\mathrm{m}} = \frac{xd' + x'd}{d + d'}, \qquad y_{\mathrm{m}} = \frac{yd' + y'd}{d + d'} \tag{9.23}$$

the coefficient

$$C = \frac{1}{2d} + \frac{1}{2d'} - \frac{1}{2f} \tag{9.24}$$

and

$$x'_{\mathrm{m}} = x_{\mathrm{m}}\left(1 - \frac{1}{2Cf}\right), \quad y'_{\mathrm{m}} = y_{\mathrm{m}}\left(1 - \frac{1}{2Cf}\right) \tag{9.25}$$

If we indicate the distance along the geometrical optical path from $\mathbf{x}$ to $\mathbf{x}'$ as

$$L_0 = d + d' + \frac{(x - x')^2 + (y - y')^2}{2(d + d')} + \frac{x_{\mathrm{m}}x'_{\mathrm{m}} + y_{\mathrm{m}}y'_{\mathrm{m}}}{2f} \tag{9.26}$$

we have the convenient result,

$$U(x', y', z') \cong \frac{U_0 \mathrm{e}^{\mathrm{i}qL_0}}{\mathrm{i}\lambda dd'} \iint_{Ap} \mathrm{d}x''\mathrm{d}y'' \exp\{\mathrm{i}qC[(x'' - x'_{\mathrm{m}})^2 + (y'' - y'_{\mathrm{m}})^2]\} \tag{9.27}$$

Here $(x'' - x'_{\mathrm{m}}, y'' - y'_{\mathrm{m}}, 0)$ is the position of a point on the element area $Ap$ relative to $(x'_{\mathrm{m}}, y'_{\mathrm{m}}, z'')$. By Fermat's principle, the geometrical optical path from $\mathbf{x}$ to $\mathbf{x}'$ intersects the $z = z''$ plane at $(x'_{\mathrm{m}}, y'_{\mathrm{m}}, z'')$.

Consider a point $(\bar{x}, \bar{y}, z'')$ near the center of $Ap$. Then

$$(x'' - x'_{\text{m}})^2 = [(x'' - \bar{x}) + (\bar{x} - x'_{\text{m}})]^2$$
$$= (x'' - \bar{x})^2 + 2(\bar{x} - x'_{\text{m}})(x'' - \bar{x}) + (\bar{x} - x'_{\text{m}})^2$$

and

$$(y'' - y'_{\text{m}})^2 = [(y'' - \bar{y}) + (\bar{y} - y'_{\text{m}})]^2$$
$$= (y'' - \bar{y})^2 + 2(\bar{y} - y'_{\text{m}})(y'' - \bar{y}) + (\bar{y} - y'_{\text{m}})^2$$

imply that we have

$$U(x', y', z') \cong \frac{U_0 e^{iq\{L_0 + C[(\bar{x} - x'_{\text{m}})^2 + (\bar{y} - y'_{\text{m}})^2]\}}}{i\lambda d d'} \times \iint_{Ap} dx'' dy'' e^{iqC[2(\bar{x} - x'_{\text{m}})(x'' - \bar{x}) + 2(\bar{y} - y'_{\text{m}})(y'' - \bar{y}) + (x'' - \bar{x})^2 + (y'' - \bar{y})^2]} \tag{9.28}$$

This involves a prefactor and an integral of an exponential function. The argument in the exponent has terms that are linear and quadratic in $(x'' - \bar{x})$ and $(y'' - \bar{y})$. For a sufficiently small aperture, only the linear terms matter, and $U(x', y', z')$ is proportional to the Fourier transform of a function that is unity on $Ap$ and zero elsewhere. This function is sometimes called the "aperture function." In the Fraunhofer approximation, we keep only the linear terms. In the Fresnel approximation, we keep the linear and quadratic terms. As $C$ approaches zero, which occurs when $\mathbf{x}$ and $\mathbf{x}'$ are in conjugate planes, $x'_{\text{m}}$ and $y'_{\text{m}}$ diverge proportionally to $C^{-1}$, and the quadratic terms inside the exponent vanish. Hence, only the linear terms matter. The divergent terms in the exponent in the prefactor all cancel one another when $C$ approaches zero, so that the exponent reaches a limit.

### 9.3.4 Diffraction Effects for Multiple Elements

If there are multiple optical elements in series between the point $\mathbf{x}_{\text{s}}$ at which radiation is emitted and the point $\mathbf{x}_{\text{d}}$, generalization of the foregoing analysis gives

$$U(\mathbf{x}_{\text{d}}) \cong \frac{U_0 e^{iq(d_0 + d_1 + \cdots + d_N)}}{(i\lambda)^N d_0 d_1 \ldots d_N} \iint_{Ap1} dx_1 dy_1 \ldots \iint_{ApN} dx_N dy_N \times \exp\left[iq \sum_{\mu,\nu=0}^{N+1} B_{\mu\nu}(x_\mu x_\nu + y_\mu y_\nu)\right] \tag{9.29}$$

Various parameters are defined as indicated in Figure 9.6. The convention of $(x_{\text{s}}, y_{\text{s}}, z_{\text{s}}) = (x_0, y_0, z_{\text{s}})$ and $(x_{\text{d}}, y_{\text{d}}, z_{\text{d}}) = (x_{N+1}, y_{N+1}, z_{\text{d}})$ is used in the sum in

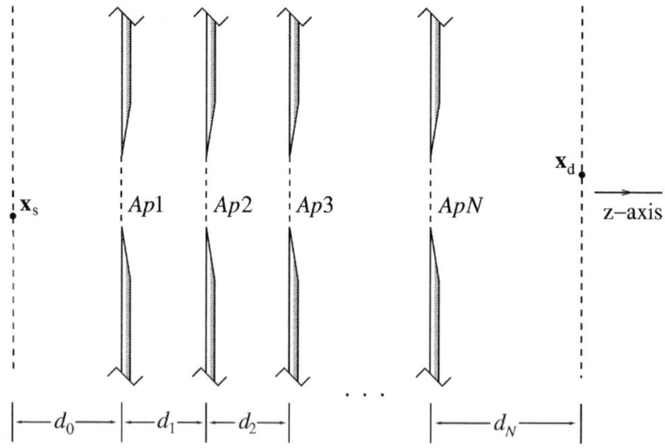


FIG. 9.6. Optical setup with multiple apertures that can diffract radiation in series.

Eq. (9.29), with

$$B_{\mu\nu} = [(d_{\mu-1}^{-1} + d_{\mu}^{-1} - f_{\mu}^{-1})\delta_{\mu\nu} + d_{\mu-1}^{-1}\delta_{\mu-1,\nu} + d_{\mu}^{-1}\delta_{\mu+1,\nu}]/2 \tag{9.30}$$

The Gaussian form of Eq. (9.29) permits progress along analytical and numerical lines regarding diffraction effects in optical systems featuring such a series of elements. Examples of this are provided later.

### 9.3.5 Self-Consistency of Gaussian-Optics

If we integrate Eq. (9.27) over the entire $z = z''$ plane,

$$\tilde{U}(x', y', z') \cong \frac{U_0 e^{iqL_0}}{i\lambda d d'} \int_{-\infty}^{+\infty} dx'' \int_{-\infty}^{+\infty} dy'' \exp\{iqC[(x'' - x'_m)^2 + (y'' - y'_m)^2]\} \tag{9.31}$$

we obtain the same result as if $Ap$ covered the entire plane. We can evaluate Eq. (9.31) analytically. Explicitly writing the value of $C$ for $f = \pm\infty$ gives

$$\tilde{U}(x', y', z') = \frac{U_0 e^{iqL_0}}{dd'} \left(\frac{1}{d} + \frac{1}{d'}\right)^{-1} = \frac{U_0 e^{iqL_0}}{d + d'} \tag{9.32}$$

and

$$\left|\tilde{U}(x', y', z')\right|^2 = \frac{|U_0|^2}{(d + d')^2} \tag{9.33}$$

Eqs. (9.32) and (9.33) give the same result as would be expected from geometrical optics, so that Fresnel diffraction is "self-consistent" (see [3], pp. 148–152). Changing any interval over which a spherical wave propagates freely into two subintervals on opposite sides of an infinite clear aperture does not change results. It may not be immediately obvious that this "check" works, because Eqs. (9.18) and (9.20) involve three different approximations. However, the combination of all three of these approximations leads to the above self-consistency. There is a further self-consistency of Gaussian-optics in the form of a conservation law. Suppose we take Eq. (9.18) and integrate $|U(\mathbf{x}_d)|^2$ over the entire $z = z_d$ plane. The result is

$$\begin{aligned}
&\int_{-\infty}^{+\infty} dx_d \int_{-\infty}^{+\infty} dy_d |U(\mathbf{x}_d)|^2 \\
&= \frac{1}{\lambda^2}\int_{-\infty}^{+\infty} dx_d \int_{-\infty}^{+\infty} dy_d \int_{Ap} d^2\mathbf{x} \int_{Ap} d^2\mathbf{x}'[U_i(\mathbf{x})G(\mathbf{x}, \mathbf{x}_d)]^* U_i(\mathbf{x}')G(\mathbf{x}', \mathbf{x}_d) \\
&= \frac{1}{\lambda^2 d^2}\int_{Ap} d^2\mathbf{x} \int_{Ap} d^2\mathbf{x}'[U_i(\mathbf{x})]^* U_i(\mathbf{x}') \\
&\quad\times \int_{-\infty}^{+\infty} dx_d \int_{-\infty}^{+\infty} dy_d \exp\left\{\frac{iq[(x' - x_d)^2 - (x - x_d)^2 + (y' - y_d)^2 - (y - y_d)^2]}{2d}\right\} \\
&= \frac{1}{\lambda^2 d^2}\int_{Ap} d^2\mathbf{x} \int_{Ap} d^2\mathbf{x}'[U_i(\mathbf{x})]^* U_i(\mathbf{x}') \exp\left[\frac{iq(x'^2 - x^2 + y'^2 - y^2)}{2d}\right] \\
&\quad\times \int_{-\infty}^{+\infty} dx_d \exp\left\{\left[\frac{iq(x - x')}{d}\right]x_d\right\} \int_{-\infty}^{+\infty} dy_d \exp\left\{\left[\frac{iq(y - y')}{d}\right]y_d\right\} \\
&= \int_{Ap} d^2\mathbf{x} \int_{Ap} d^2\mathbf{x}'[U_i(\mathbf{x})]^* U_i(\mathbf{x}')\delta(x - x')\delta(y - y')
\end{aligned} \tag{9.34}$$

or

$$\int_{-\infty}^{+\infty} dx_d \int_{-\infty}^{+\infty} dy_d |U(\mathbf{x}_d)|^2 = \int_{Ap} d^2\mathbf{x} |U_i(\mathbf{x})|^2 \tag{9.35}$$

Thus, the integrated spectral power falling on the area of one optical element is conserved when it reaches the plane of the next optical element. This implies a transmission function of unity even for very small apertures, which indicates a breakdown of the Kirchhoff diffraction theory in the small-aperture limit. However, this also ensures that the theoretical diffraction effects on total transmitted spectral flux approach zero correctly in the opposite, small-$\lambda$ limit.

### 9.3.6 Limitations of Approximations

The Gaussian-optics version of the Kirchhoff diffraction theory is widely found and used in the radiometric literature with good success. This version often predicts diffraction effects very similar to results obtained without making the approximations found in Eqs. (9.18), (9.20) and (9.21).

One should also be mindful of when a Gaussian-optics picture should fail. It breaks down when describing diffraction effects for very small apertures, mainly because the Kirchhoff theory itself breaks down. It does not describe aberrations correctly, so their effects must be otherwise considered. The present Gaussian-optics implementation of the Kirchhoff theory should also break down when obliquity factors become appreciably different from unity or when distances between optical elements do not greatly exceed the wavelength. However, because the correct results are obtained in the limit of optical elements whose areas extend over entire constant-$z$ planes, and because the total integrated flux is conserved in the manner demonstrated above, a Gaussian-optics picture can be useful in many practical situations (for which a cursory assessment of its validity might suggest otherwise).

## 9.4 The SAD Problem

Even in complicated optical setups, diffraction effects can arise mainly from one optical element. In other setups, the effects might be approximately described as the sum of effects of several different elements considered individually. In either type of situation, diffraction effects on the throughput of an optical system can be related to the diffraction effects that would occur in fictitious three-element setups that consist of an extended Lambertian *source*, *aperture* and *detector*. Such setups and the related diffraction problem may be denoted by the acronym, SAD. The SAD problem is one of the most studied diffraction problems in optics and radiometry. Furthermore, its study has been very profitable.

Consider the setup illustrated in Figure 9.7(a). A blackbody cavity placed behind a defining aperture illuminates a detector. To reduce the stray light reaching the detector, two non-limiting apertures are also placed strategically between the defining aperture and detector. Three SAD combinations are shown in Figure 9.7(b)–(d). Appropriate combination of diffraction effects arising in cases of the fictitious combinations of three indicated elements can account reasonably well for overall diffraction effects in the real setup with relatively little effort. In Figure 9.7(b), the defining aperture limits the throughput of the model optical system. In Figure 9.7(c) and (d), the non-limiting apertures would tend to increase the throughput of the model optical systems. For the non-limiting apertures, the effective source is the defining aperture. Depending on the geometry and wavelength, the sum of all diffraction effects can lead to a throughput that is smaller or larger than that expected geometrically.

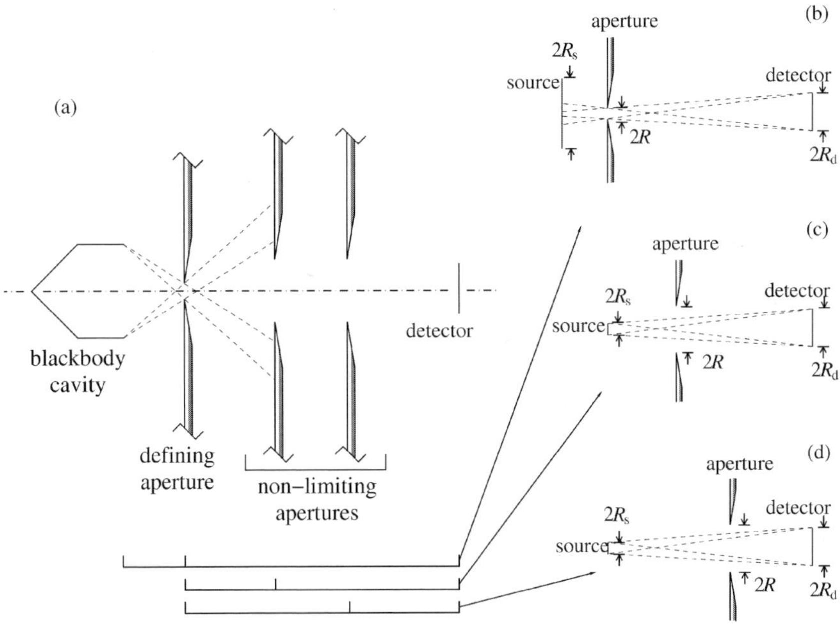


FIG. 9.7. Optical setup conceptually treated as three SAD setups for purposes of diffraction effects.

## 9.4.1 Lommel's Treatment for a Point Source

The ratio $U(x',y',z')/U_0$ implied by Eq. (9.28) is a function of $x'_{\rm m}$ and $y'_{\rm m}$ symmetric with respect to simultaneous exchange $\mathbf{x}$ with $\mathbf{x}'$ and $d$ with $d'$:

$$\frac{U(x',y',z')}{U_0} = S(x'_{\rm m}, y'_{\rm m}) = \frac{e^{iqL_0}}{i\lambda dd'} \iint_{Ap} dx''dy'' e^{iqC[(x''-x'_{\rm m})^2+(y''-y'_{\rm m})^2]} \tag{9.36}$$

We now assume that $Ap$ is circular with radius $R$ and centered on the $z$-axis. Introducing polar coordinates, $r'' = (x''^2 + y''^2)^{1/2}$, $r' = (x'^2_{\rm m} + y'^2_{\rm m})^{1/2}$, and $\theta$, the relative polar angle between $(x'', y'')$ and $(x'_{\rm m}, y'_{\rm m})$, we obtain

$$S(x'_{\rm m}, y'_{\rm m}) = S(r') = \frac{e^{iqL_0}}{i\lambda dd'} \int_0^R dr''r'' \int_0^{2\pi} d\theta e^{iqC(r''^2+r'^2-2r''r'\cos\theta)} \tag{9.37}$$

In the $q \to \infty$ limit, we have $S(r') \sim S_G(r')$, with

$$S_G(r') = \frac{e^{iqL_0}\Theta(R/r'-1)}{2dd'C} \tag{9.38}$$

Here $\Theta(x)$ is the Heaviside step function.

Introducing the unit-less parameters $\rho = r''/R$ (which varies from 0 to 1 on *Ap*), $u = 2qCR^2$, and $v = |2qCRr'|$, we have

$$S(r') = \frac{R^2 \exp[\mathrm{i}q(L_0 + Cr'^2)]}{\mathrm{i}\lambda dd'} \int_0^1 \mathrm{d}\rho\rho \int_0^{2\pi} \mathrm{d}\theta \mathrm{e}^{\mathrm{i}u\rho^2/2 - \mathrm{i}v\rho\cos\theta} \tag{9.39}$$

Lommel showed that the above integrals can be expressed using Neumann series' of Bessel functions [24], and the resulting functions are traditionally called Lommel functions of two arguments. Lommel's treatment is also given by Born and Wolf (see [7], pp. 435–442). Here, the quantity of greatest interest is the squared ratio, $|S(r')/S_G(0)|^2$. The quantity $|S(r')/S_G(0)|^2 - 1$ gives the diffraction effects on point-to-point propagation of radiation, normalized so that the geometrically allowed propagation is unity.

The Lommel functions are

$$U_n(u, v) = \sum_{s=0}^{\infty} (-1)^s (u/v)^{n+2s} J_{n+2s}(v) \tag{9.40}$$

and

$$V_n(u, v) = \sum_{s=0}^{\infty} (-1)^s (v/u)^{n+2s} J_{n+2s}(v) \tag{9.41}$$

where $n$ is an integer. $U_n(u, v)$ and $V_n(u, v)$ can be related to each other by $U_n(u, v) = V_n(v^2/u, v)$ and $V_n(u, v) = U_n(v^2/u, v)$. These functions also have other useful symmetries, including $U_n(u, v) = U_n(u, -v)$, $V_n(u, v) = V_n(u, -v)$, $U_n(u, v) = (-1)^n U_n(-u, v)$, and $V_n(u, v) = (-1)^n V_n(-u, v)$. Recent algorithms to evaluate Lommel functions of two arguments efficiently and to varying degrees of accuracy are described by Mielenz [25], Shirley and Terraciano [26], Shirley and Chang [27], and Edwards and McCall [28].

It is traditional to express the value of $|S(r')/S_G(0)|^2$ using either of two equivalent formulas:

$$\left|\frac{S(r')}{S_G(0)}\right|^2 = U_1^2(u, v) + U_2^2(u, v) \tag{9.42}$$

or

$$\left|\frac{S(r')}{S_G(0)}\right|^2 = 1 + V_0^2(u, v) + V_1^2(u, v) - 2V_0(u, v)\cos\left[\frac{1}{2}\left(u + \frac{v^2}{u}\right)\right] - 2V_1(u, v)\sin\left[\frac{1}{2}\left(u + \frac{v^2}{u}\right)\right] \tag{9.43}$$

Equation (9.42) tends to be more convenient if the geometrical optical path between $\mathbf{x}$ and $\mathbf{x}'$ is obstructed (which implies $r' > R$). Equation (9.43) tends to be more convenient if that path is not blocked (which implies $r' < R$).

In the BDW formulation, when we have $r' > R$, the combination $U_1^2(u,v) + U_2^2(u,v)$ on the right-hand side of Eq. (9.42) results from the square of the BDW $U_B$. Likewise, when we have $r' < R$, the leading "1" on the right-hand side of Eq. (9.43) results from the square of the geometrical wave $U_G$, $V_0^2(u,v) + V_1^2(u,v)$ results from the square of $U_B$, and the remaining terms result from interference of $U_G$ with $U_B$.

#### *9.4.1.1 Asymptotic properties of Lommel's result*

In the small-$\lambda$ regime, when the geometrical optical path between $\mathbf{x}$ and $\mathbf{x}'$ is not blocked, the asymptotic behavior of Bessel functions of a large positive argument,

$$J_m(v) \sim \left(\frac{2}{\pi v}\right)^{1/2} \cos(v - m\pi/2 - \pi/4) \tag{9.44}$$

gives

$$V_0(u,v) \cong \left(\frac{2}{\pi v}\right)^{1/2} \frac{\cos(v - \pi/4)}{1 - v^2/u^2} \tag{9.45}$$

and

$$V_1(u,v) \cong \left(\frac{2}{\pi v}\right)^{1/2} \left(\frac{v}{u}\right) \frac{\sin(v - \pi/4)}{1 - v^2/u^2} \tag{9.46}$$

Substitution of these approximations into Eq. (9.43) gives

$$\left|\frac{S(r')}{S_G(0)}\right|^2 \cong 1 + \frac{1}{\pi v}\left[\frac{1 + v^2/u^2}{(1 - v^2/u^2)^2}\right] + \frac{1}{\pi v}\left[\frac{\sin(2v)}{1 - v^2/u^2}\right] - \left(\frac{2}{\pi v}\right)^{1/2} \times \left\{\frac{\cos[(u+v)^2/(2u) - \pi/4]}{1 + v/u} + \frac{\cos[(u-v)^2/(2u) + \pi/4]}{1 - v/u}\right\} \tag{9.47}$$

When the geometrical optical path is blocked, we have

$$U_1(u,v) \cong \left(\frac{2}{\pi v}\right)^{1/2} \left(\frac{u}{v}\right) \frac{\sin(v - \pi/4)}{1 - u^2/v^2} \tag{9.48}$$

$$U_2(u,v) \cong -\left(\frac{2}{\pi v}\right)^{1/2} \left(\frac{u}{v}\right)^2 \frac{\cos(v - \pi/4)}{1 - u^2/v^2} \tag{9.49}$$

and

$$\left|\frac{S(r')}{S_G(0)}\right|^2 \cong \frac{u^2}{\pi v^3}\left[\frac{1 + u^2/v^2}{(1 - u^2/v^2)^2}\right] - \frac{u^2}{\pi v^3}\left[\frac{\sin(2v)}{1 - u^2/v^2}\right] \tag{9.50}$$

While Eqs. (9.42) and (9.43) are formally equivalent, the asymptotic approximation being used here requires that Eq. (9.47) or Eq. (9.50) is used only in the respective case of non-blockage or blockage of the geometrical optical path from $\mathbf{x}$ to $\mathbf{x}'$. In each case, Eq. (9.47) or (9.50) breaks down as $r'$ approaches $R$.

Equations (9.47) and (9.50) illustrate the main behavior of diffraction effects on point-to-point propagation of radiation using simple functions. These equations contain both non-oscillatory and oscillatory terms. The non-oscillatory terms often matter the most. Cumulative effects of oscillatory terms can be largely self-canceling upon $\mathbf{x}$ sampling the source area, such as in the case of an extended source, $\mathbf{x}'$ sampling the finite detector area, and/or $\lambda$ sampling the relevant spectrum in the case of complex radiation.

### 9.4.2 Wolf's Formula for Integrated Flux

Wolf [29] introduced the function, $L(u, v_0)$, which is the fraction of the flux that is incident on $Ap$ that subsequently falls on a circular area $A'$ of the $z = z'$ plane. For all points $\{\mathbf{x}'\}$ on $A'$, we have $v < v_0$, and we have $v = v_0$ for points on the perimeter of $A'$. In analogy with Lommel's result, $L(u, v_0)$ can be given by one expression that is more convenient when the geometrical optical paths between points on the perimeter of $A'$ and $\mathbf{x}$ are blocked, implying $v_0 > |u|$, and by a different expression that is more convenient when these paths are not blocked, implying $v_0 < |u|$. First, we introduce the functions,

$$Q_{2s}(v_0) = \sum_{p=0}^{2s} (-1)^p [J_p(v_0) J_{2s-p}(v_0) + J_{p+1}(v_0) J_{2s+1-p}(v_0)] \tag{9.51}$$

and

$$Y_n(u, v_0) = \sum_{s=0}^{\infty} (-1)^s (n + 2s)(v_0/u)^{n+2s} J_{n+2s}(v_0) \tag{9.52}$$

For $v_0 > |u|$, we have

$$L(u, v_0) = 1 - \sum_{s=0}^{\infty} \frac{(-1)^s}{2s+1} \left(\frac{u}{v_0}\right)^{2s} Q_{2s}(v_0) \tag{9.53}$$

For $v_0 < |u|$, we have

$$\begin{aligned} L(u, v_0) = {} & \left(\frac{v_0}{u}\right)^2 \left[1 + \sum_{s=0}^{\infty} \frac{(-1)^s}{2s+1} \left(\frac{v_0}{u}\right)^{2s} Q_{2s}(v_0)\right] \\ & - \frac{4}{u}\left\{ Y_1(u, v_0) \cos\left[\frac{1}{2}\left(u + \frac{v_0^2}{u}\right)\right] \right. \\ & \left. + Y_2(u, v_0) \sin\left[\frac{1}{2}\left(u + \frac{v_0^2}{u}\right)\right]\right\} \end{aligned} \tag{9.54}$$

This formula is typeset incorrectly in several editions of Born and Wolf's classic text.

#### 9.4.2.1 *Asymptotic properties of Wolf's result*

Focke [30] considered the asymptotic properties of the above integrated flux, based on work by Schwarzschild [31], van Kampen [32] and Wolf's treatment, and obtained

$$L(u, v_0) \sim 1 - \frac{2}{\pi}\left(\frac{v_0}{v_0^2 - u^2}\right) \tag{9.55}$$

which is valid for large $v_0$ subject to having $v_0 > |u|$.

One may also consider the more general asymptotic expansion for a Bessel function of large non-negative argument [33],

$$J_m(v) \sim \left(\frac{2}{\pi v}\right)^{1/2}\left[\cos\zeta \sum_{s=0}^{\infty} \frac{(-1)^s A_{2s}(m)}{v^{2s}} - \sin\zeta \sum_{s=0}^{\infty} \frac{(-1)^s A_{2s+1}(m)}{v^{2s+1}}\right] \tag{9.56}$$

with $\zeta = v - m\pi/2 - \pi/4$, $A_0(m) = 1$, and $A_s(m) = (4m^2 - 1^2)(4m^2 - 3^2)\cdots[4m^2 - (2s-1)^2]/[8^s(s!)]$ for all other $s$, and apply this approximation to Wolf's result. For $|w| < 1$, this gives

$$\begin{aligned}\sum_{s=0}^{\infty} \frac{(-1)^s}{2s+1} w^{2s} Q_{2s}(v) = {} & \frac{2\sigma_0}{\pi v} - \frac{\sigma_0 \cos(2v)}{\pi v^2} \\ & - \frac{16\sigma_4 + 32\sigma_3 + 8\sigma_2 - 8\sigma_1 - 3\sigma_0}{12\pi v^3} \\ & + \left(\frac{8\sigma_2 + 8\sigma_1 - \sigma_0}{4\pi v^3}\right)\sin(2v) \\ & + \left(\frac{64\sigma_4 + 128\sigma_3 - 16\sigma_2 - 80\sigma_1 + 9\sigma_0}{32\pi v^4}\right) \\ & \times \cos(2v) + O(v^{-5})\end{aligned} \tag{9.57}$$

The $\sigma_k$ parameters are defined by

$$\sigma_k = \sum_{s=0}^{\infty} s^k w^{2s} = \left(w^2 \frac{\mathrm{d}}{\mathrm{d}(w^2)}\right)^k \frac{1}{1 - w^2} = \frac{W_k(w^2)}{(1 - w^2)^{k+1}} \tag{9.58}$$

The first six $W$-polynomials are

$$\begin{gathered} W_0(x) = 1,\ W_3(x) = x^3 + 4x^2 + x, \\ W_1(x) = x,\ W_4(x) = x^4 + 11x^3 + 11x^2 + x, \\ W_2(x) = x^2 + x,\ W_5(x) = x^5 + 26x^4 + 66x^3 + 26x^2 + x \end{gathered} \tag{9.59}$$

Application of Eq. (9.56) also gives

$$
\begin{aligned}
Y_1(u,v) = \frac{v^2}{2u}\left(\frac{2}{\pi v}\right)^{1/2}\Bigg[&\left(\frac{2\sigma_0+4\sigma_1}{v}\right)\sin(v-\pi/4) \\
&+\left(\frac{3\sigma_0+22\sigma_1+48\sigma_2+32\sigma_3}{4v^2}\right)\cos(v-\pi/4) \\
&+\left(\frac{15\sigma_0+62\sigma_1-160\sigma_2-960\sigma_3-1280\sigma_4-512\sigma_5}{64v^3}\right) \\
&\times\sin(v-\pi/4)\Bigg]+O(v^{-7/2})
\end{aligned}
\tag{9.60}
$$

and

$$
\begin{aligned}
Y_2(u,v) = -\frac{v}{2}\left(\frac{2}{\pi v}\right)^{1/2}\Bigg[&\left(\frac{4\sigma_1}{v}\right)\sin(v+\pi/4)+\left(\frac{-\sigma_1+16\sigma_3}{2v^2}\right)\cos(v+\pi/4) \\
&+\left(\frac{-9\sigma_1+160\sigma_3-256\sigma_5}{32v^3}\right)\times\sin(v+\pi/4)\Bigg]+O(v^{-7/2})
\end{aligned}
\tag{9.61}
$$

As mentioned earlier and as noted by Focke, oscillatory terms can be less relevant and even undesirable to include when considering diffraction effects, because they can be self-canceling upon spatial or spectral integration, and when sampled at discrete wavelengths they can lead to randomly biased results.

### 9.4.3 Diffraction Effects on Spectral Throughput

The above analysis lays the groundwork for estimating diffraction effects on spectral throughput of optical systems. In this Section, we show how this is done for the SAD problem and its application to treat systems with varying degrees of complexity.

#### *9.4.3.1 SAD case*

The preceding analysis considered flux arising from a point source, but the finite extent of the source may also need to be taken into account. It is helpful to introduce the parameters, $v_s = qR_sR/d$, $v_d = qR_dR/d'$, $v_M = \max(v_s, v_d)$, and $\sigma = \min(v_s, v_d)/\max(v_s, v_d)$. $R_s$ and $R_d$ are the source and detector radii, respectively. Overall cylindrical symmetry is assumed. Edwards and McCall [28] have noted the connection between Wolf's result and the spectral power reaching the detector. As long as we have $|u|/v_M < 1-\sigma$ or $|u|/v_M > 1+\sigma$, the ratio of the spectral power incident on the detector to the source spectral radiance is given by the source-detector-symmetric expression:

$$\frac{\Phi_\lambda(\lambda)}{L_\lambda(\lambda)} = D\int_{-1}^{1}\frac{\mathrm{d}x\{(1-x^2)[(2+\sigma x)^2-\sigma^2]\}^{1/2}L(u, v_\mathrm{M}(1+\sigma x))}{1+\sigma x} \tag{9.62}$$

The leading factor is $D = 4\pi^3 R^4 R_\mathrm{s}^2 R_\mathrm{d}^2/[d_\mathrm{s}^2 d_\mathrm{d}^2(\lambda v_\mathrm{M})^2]$, which is source-detector-symmetric and independent of $\lambda$. Integration over $x$ can be done using Gauss–Chebyshev quadrature. In a recent calculation that is discussed in the next section, Edwards and McCall have also considered $1-\sigma < |u|/v_\mathrm{M} < 1+\sigma$.

#### *9.4.3.2 Case of other optical systems*

The preceding SAD problem was relatively self-contained, and the diffraction effects could be quantified in considerable detail. In more complicated optical systems, it may not be possible to break down diffraction effects, in a simple way, into those arising in SAD problems and related analyses without making gross simplifications and incurring significant inaccuracies. However, with care, SAD and related analyses can be used in many—though not all—radiometric situations in multistage optical systems, including one that arose in an actual blackbody calibration [34].

Two examples of the complexities that can arise are illustrated here. Figure 9.8 shows an optical setup that was considered when modeling diffraction effects in an actual blackbody calibration. Diffraction at the edges of the non-limiting apertures Ap 1 and 3 enhanced the irradiance at the detector. The simulated relative excess irradiance at points on the detector is illustrated as a percentage in the bottom panel of Figure 9.9, for the cases of only Ap 3 being present and both apertures being present.

In the top panel, the difference between the two results is shown, both as a dashed curve and as a solid curve based on a combination of the BDW formulation and asymptotic analysis in the spirit of the GTD. The distance of a point on the detector from the optical axis is indicated in the figure by $r_\mathrm{d}$. Near $r_\mathrm{d} = 8.2$ mm, a step is visible in the top panel of Figure 9.9. This can be explained in terms of one of the GTD-type rays bent by Ap 1 being geometrically blocked by the boundary of Ap 3. The model in Reference [34] attempts to address this type of vignetting effect in diffraction calculations and to address some higher-order diffraction effects in terms of SAD and related analyses.

The second example of possible complexities is similar. Figure 9.10 depicts an experimental setup used by Boivin [35] to study diffraction effects of four non-limiting apertures, $A_1$, $A_2$, $A_3$ and $A_4$, placed in series between a source and detector. The source featured a tungsten lamp that was located behind a 1 mm diameter aperture. In a simulation of the diffraction effects,

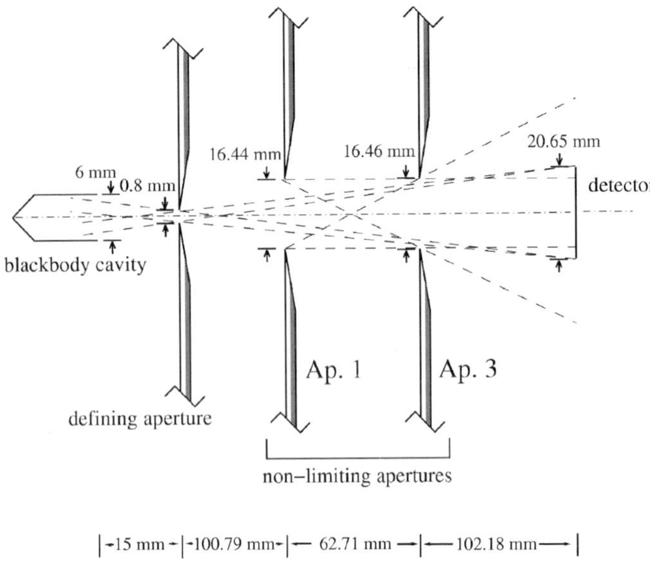


FIG. 9.8. Optical setup used to treat diffraction in a blackbody calibration.

this source was modeled as if the aperture was illuminated by plane waves with angles of incidence not exceeding 0.020 rad from normal. Agreement between measured and simulated diffraction effects indicated that the simulated model of the source was reasonable [26].

The flux reaching the detector for a given plane wave is depicted as a function of the angle of incidence in Figure 9.11. It is normalized so that the total flux incident on the source defining aperture is that aperture's area ($\pi/4$ mm$^2$). Six results are shown, featuring only $A_1$, only $A_2$, only $A_3$, only $A_4$, none, or all of the four non-limiting apertures being included in the simulation. The effect of each non-limiting aperture is to deflect excess radiation onto the detector, especially when the perimeter of that aperture is geometrically illuminated through the source defining aperture. Therefore, the effects of a non-limiting aperture can depend on the angle subtended by the simulated lamp filament at the source defining aperture. This subtlety might not be addressed correctly in a casual application of the SAD analysis, but could be treated appropriately using the method found in Reference [34].

## 9.5 Impacts of Diffraction Effects on Radiometry

We begin with a cursory survey of the radiometry literature dealing with diffraction effects. This survey highlights the insight gained into the role of

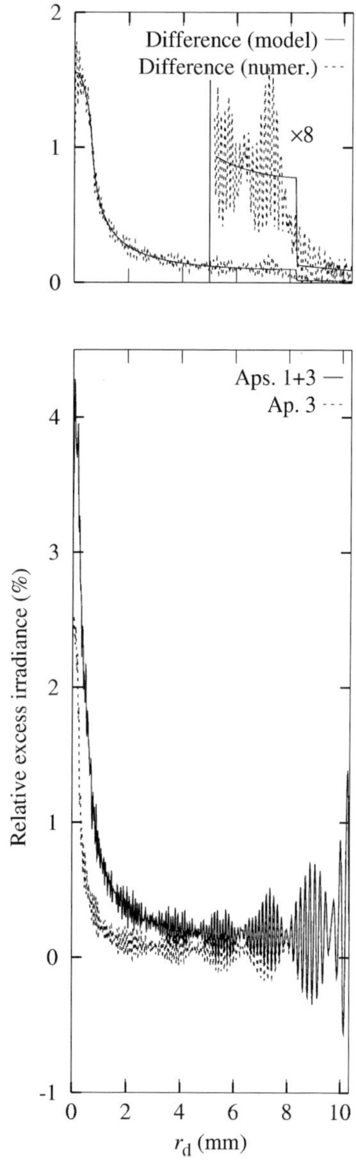


FIG. 9.9. Diffraction-induced irradiance because of non-limiting apertures in Figure 9.8.

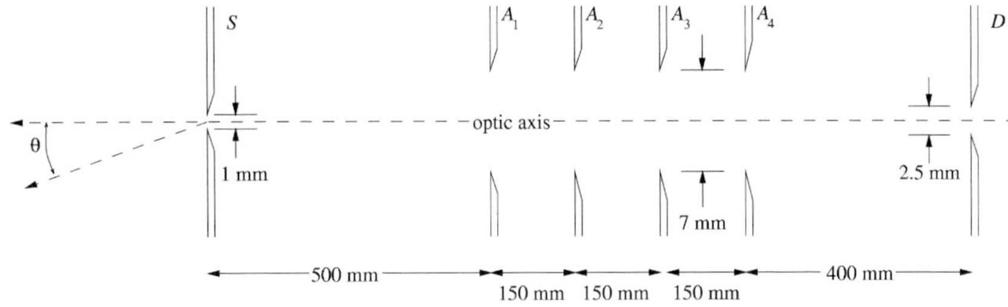


FIG. 9.10. Optical setup featuring four apertures in series studied by Boivin.

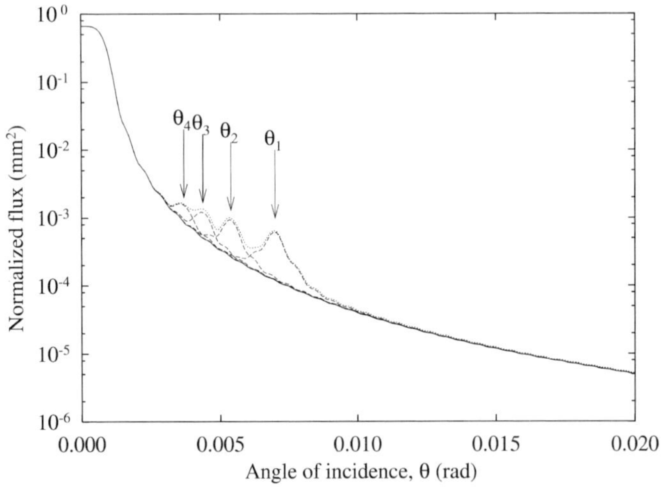


FIG. 9.11. Normalized flux versus plane-wave angle of incidence $\theta$ in setup in Figure 9.10. At $\theta = \theta_1$, the perimeter of $A_1$ is geometrically illuminated, etc.

diffraction effects and the cumulative developments regarding how to estimate or mitigate them. Because this chapter attempts to report the current state of affairs, it should be noted that some of the earlier developments or aspects thereof that are cited in this Section are no longer current.

Much of diffraction theory is formulated for monochromatic radiation, yet radiometric applications frequently involve complex radiation. This mismatch is not fully resolved, but some methods to overcome this mismatch are discussed in Section 9.5.2. These include the *effective-wavelength approximation* and other recent methods for more directly finding total irradiance and power.

Section 9.5.3 mentions three recent examples of diffraction modeling in support of actual radiometric measurements. In order of increasing complexity, these situations involve solar radiometry, spectral calibration of a radiometric telescope, and the already mentioned calibration of a reference blackbody source.

### 9.5.1 Radiometry Literature Survey

In 1881, Rayleigh [36] derived the fraction of flux in the Airy diffraction pattern enclosed within a circular area in the case of a circular aperture. Rayleigh obtained the well-known result,

$$L(0, v_0) = 1 - J_0^2(v_0) - J_1^2(v_0) \tag{9.63}$$

This was one of the earliest "modern" analyses of diffraction effects in optics. The Airy pattern is a consequence of Fraunhofer diffraction. A strong motivation for analyzing the Airy pattern was its impact on the resolving powers of optics. However, an internally consistent measurement of, say, the spectral irradiance of celestial objects, also requires accounting for diffraction losses, because $v_0$ depends on wavelength.

The 1885 work by Lommel and 1950s work by Wolf and Focke generalize the Rayleigh result for integrated flux to the case of Fresnel diffraction (or of not being in a focal plane). Some other analyses are included in the volume compiled by Oughstun and cited in the above references. These developments provided a good working model of diffraction effects to study the impact on radiometry. Much of the analysis in the context of radiometry considers the SAD problem, but other topics have also been addressed.

#### *9.5.1.1 Study of the SAD problem*

A survey of the radiometric literature shows that a remarkable fraction of the work on diffraction effects was done at or in association with, various national measurement institutes. In 1962, Sanders and Jones [37] discussed the role of diffraction effects and the need to account for them in the context of the then-current "problem of realizing the primary standard of light." Shortly thereafter, Ooba [38] studied the related diffraction effects experimentally.

In 1970, Blevin [39] considered the diffraction loss in the context of the SAD problem for the case of an axial point source, implying $\sigma = 0$, and under-filled detector, implying $|u| < v_M$. (When discussing analysis of the SAD problem, mathematical formulas from the radiometric literature are presented using the SAD notation of this chapter.) Blevin derived a diffraction loss consistent with $L(u, v_M) \cong 1 - 2v_M/[\pi(v_M^2 - u^2)]$, in agreement with earlier work, and confirmed this experimentally for at least two experimental geometries. Blevin used a broadband source and detector, and the diffraction analysis was based on using an *effective-wavelength approximation*, an idea that has sometimes been credited to Blevin and is discussed in Section 9.5.3. Note that a more up-to-date evaluation of $L(u, v_M)$ is discussed in Section 9.4.2.

In 1972, Steel et al. [40] provided an approximate extension of Blevin's result to finite $\sigma$ and analyzed the two cases, $|u| < (1-\sigma)v_M$ and $|u| > (1+\sigma)v_M$. Their results were also extended to allow for the effective-wavelength approximation. Their analysis of the case of a non-limiting aperture ($|u| > (1+\sigma)v_M$) is flawed, as noted by Boivin [41].

Steel et al. also referred to the case of having a limiting aperture as "Case 1" and the case of having a non-limiting aperture as "Case 2." This terminology has become quite standard, so much so that the factor $F$ that describes diffraction effects is often denoted by $F_1$ or $F_2$. Recently, Edwards and McCall [28] have also considered the intermediate case of $(1-\sigma)v_M < |u| < (1+\sigma)v_M$, which can be referred to as "Case 3" and which leads to an associated $F_3$. Edwards and McCall made an insightful geometric construction shown in Figure 9.12 that differentiates the three cases if there is a non-focusing aperture (i.e., $f = \pm\infty$).

In a series of several papers in the 1970s [35, 41–44], Boivin studied diffraction effects along the same lines. Boivin primarily analyzed effects of non-limiting apertures, which are used to reduce stray light. As long as a

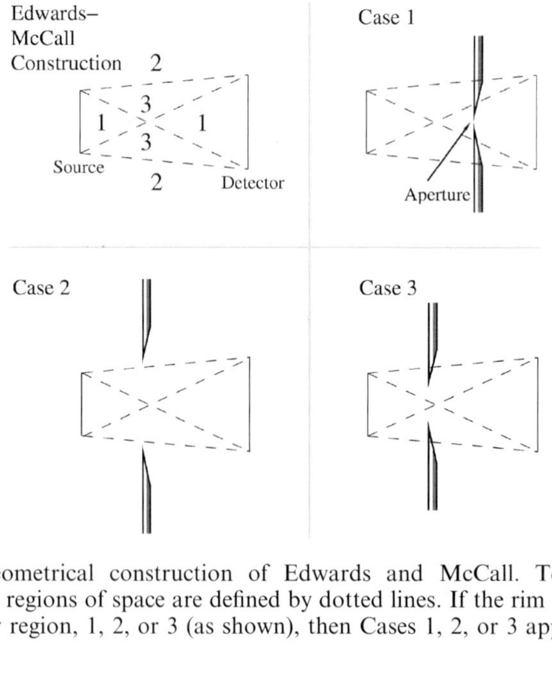


FIG. 9.12. Geometrical construction of Edwards and McCall. Top left: three (disconnected) regions of space are defined by dotted lines. If the rim of an aperture lies within any region, 1, 2, or 3 (as shown), then Cases 1, 2, or 3 applies.

detector is significantly overfilled, Boivin found that a non-limiting aperture gives $F_2(u, v_M) \cong 1 + 2/(\pi v_M)$, both experimentally and theoretically. A more precise result is $F_2(u, v_M) \cong 1 + 2/[\pi v_M(1 - v_M^2/u^2)] + \cdots$, as described in Section 9.4.2. To help establish his reported approximation for $F_2$, Boivin varied $v_s/v_d$, the ratio of angles subtended by the source and detector at the center of a non-limiting aperture in an experimental SAD setup. The ratio $v_s/v_d$ was sometimes smaller and sometimes larger than unity.

Boivin recommended that, if possible, a practitioner should use as few apertures as possible in an optical setup, with at most one aperture giving rise to substantial diffraction effects. This is consistent with trying to ensure that the extensive analysis applied to the SAD problem can be almost directly brought to bear in a real optical setup. Boivin also recommended against letting parts of an over-filled detector's area be anywhere near a non-limiting aperture's geometrical shadow boundary. This is at variance with the recent, novel idea of Edwards and McCall, discussed below. (Using the same reasoning by Boivin, a practitioner should also not let an under-filled detector's perimeter be anywhere near the penumbra or fully illuminated region in the case of a limiting aperture.)

Based on an idea attributed to Purcell and Koomen [45], Boivin [35] also experimentally demonstrated that using toothed instead of circular non-limiting apertures can reduce the associated diffraction effects. This was motivated by observing that toothing of an aperture perimeter introduces dephasing and concomitant self-cancellation of $U_B$ at a detector. This idea was subsequently explored and extended theoretically [46], but the author knows of no subsequent applications using toothed non-limiting apertures. It can be difficult to manufacture such apertures, the largest diffraction effects in radiometry are often losses because of limiting apertures, and the effects of non-limiting apertures can often be substantially reduced by modifying the design of an optical setup.

The author has reinvestigated the SAD problem for point and extended sources [47], as well as monochromatic and complex radiation, including radiation from a Planckian source [48]. This has led to more detailed expressions for $F(\lambda)$ and analogous expressions for $\langle F \rangle$ in the case of a Planckian source in the Fraunhofer and Fresnel regimes. A 1998 work introduced the formula found in Eq. (9.62), but the accompanying formulas for $L(u, v_M)$ have been superceded by subsequent work.

Edwards and McCall [28] also recently considered the SAD problem, introducing an algorithm for calculating diffraction effects that is applicable to Case 1–3. These workers consider diffraction effects as a function of intermediate aperture radius for a fixed extended source and detector. Noting that diffraction losses occur for $|u| < (1 - \sigma)v_M$, whereas diffraction gains tend to occur for $|u| > (1 + \sigma)v_M$, Edwards and McCall deduced that diffraction effects on throughput must cross zero at some point when Case 3 applies.

Such a zero crossing can be advantageous and feasible to arrange in some instances in radiometry. A zero crossing should occur nearly simultaneously for a wide range of wavelengths, so that diffraction effects and concomitant uncertainties in measurement results with complex radiation could be greatly reduced. However, a practitioner must also take into account the greater complexity involved in calculating geometrical throughput of the optical setup in Case 3. Furthermore, there are optical setups (for instance, a collimator involving a blackbody cavity, small defining aperture, and collimating mirror to simulate a remote object) for which Case 1 appears unavoidable.

#### *9.5.1.2 Other studies*

People have also considered more elaborate optical setups than those that can be mapped onto the SAD problem. In 2001, Suárez-Romero et al. [49] and Shirley and Terraciano [26] independently treated diffraction by a series of optical elements. The former authors made use of the cross-spectral density [50] and analyzed the problem of two apertures in series. The latter authors showed how a practitioner can analyze diffraction in multi-staged cylindrically symmetrical systems using Gaussian-optics and full Kirchhoff and Rayleigh–Sommerfeld treatments, with an arbitrary number of apertures in series. Two examples of the application of the latter development were already mentioned in Section 9.4.3.2, and two additional examples are mentioned in Sections 9.5.3.1 and 9.5.3.2.

Much less work regarding diffraction effects in radiometry has been done for cases without cylindrical symmetry. It is well known that the multiple integrations in Eq. (9.29), which does not assume cylindrical symmetry, can be done one aperture at a time by use of fast Fourier-transform (FFT) techniques [51]. Care must be taken during such a procedure to appropriately account for curved perimeters of optical surfaces, in order to faithfully reproduce the part of a wave front corresponding to $U_{\mathrm{B}}$. In the case of cylindrical symmetry, the double loops over radial coordinates $r$ and $s$ implicit in Eq. (9.25) of Reference [26] can also be accelerated by FFT methods. This can be realized by using logarithmic radial meshes that are spaced at regular intervals of the logarithm of $r$ or $s$ within a Gaussian-optics analysis.

### 9.5.2 Complex Radiation

Virtually all of the analysis presented thus far in this chapter has involved monochromatic radiation, but many radiometric applications involve complex radiation. In the diffraction analysis for complex radiation, the effective-wavelength approximation has been of longstanding use, and methods recently developed for directly evaluating diffraction effects on total irradiance and total power have also been recently introduced [48, 52].

### 9.5.2.1 Effective-wavelength approximation

Formulas for diffraction effects in Section 9.4 and the lowest-order expressions cited above indicate that $\varepsilon_{\text{diff}}(\lambda)$ can scale approximately linearly with $\lambda$. Assuming exact proportionality, Blevin noted that Eq. (9.4) becomes

$$\langle F \rangle = F(\lambda_e) \tag{9.64}$$

with the *effective wavelength* $\lambda_e$ being

$$\lambda_e = \frac{\int_0^\infty d\lambda \lambda R(\lambda) M(\lambda) L_\lambda(\lambda)}{\int_0^\infty d\lambda R(\lambda) M(\lambda) L_\lambda(\lambda)} \tag{9.65}$$

Note that the product $R(\lambda)M(\lambda)$ can include a factor that realizes effects of the spectral luminous efficiency function $V(\lambda)$. Hence, as shown by Blevin, the effective-wavelength approximation can be used in subfields of radiometry other than absolute radiometry, such as photometry. Because $F(\lambda)$ contains oscillatory terms that tend to average to nearly zero upon integration over wavelength, better approximations to Eq. (9.4) are often obtained in the effective-wavelength approximation if $F(\lambda_e)$ is computed with oscillatory terms omitted. Otherwise, their inclusion at a discrete wavelength ($\lambda_e$) is not representative of the result found after averaging over a spectrum.

### 9.5.2.2 Diffraction effects on total irradiance

If we introduce

$$f(l) = \iint_{Ap1} dx_1 dy_1 \ldots \iint_{ApN} dx_N dy_N \times \delta\left(d_0 + d_1 + \cdots + d_N + \sum_{\mu,\nu=0}^{N+1} B_{\mu\nu}(x_\mu x_\nu + y_\mu y_\nu) - l\right) \tag{9.66}$$

Eq. (9.29) can be rewritten as

$$U(\mathbf{x}_d) \cong \frac{U_0}{(i\lambda)^N d_0 d_1 \ldots d_N} \int_{-\infty}^{+\infty} dl e^{iql} f(l) \tag{9.67}$$

Rearranging and squaring Eq. (9.67) gives

$$X(\lambda;\mathbf{x}_\mathrm{d},\mathbf{x}_\mathrm{s}) = \left|\frac{U(\mathbf{x}_\mathrm{d})}{U_0}\right|^2 \cong (\lambda^N d_0 d_1 \ldots d_N)^{-2} \int_{-\infty}^{+\infty} \mathrm{d}l \int_{-\infty}^{+\infty} \mathrm{d}l' \mathrm{e}^{\mathrm{i}q(l-l')} f(l) f(l') \tag{9.68}$$

Here $X(\lambda;\mathbf{x}_\mathrm{d},\mathbf{x}_\mathrm{s})$ is a type of transfer function.

The detector-responsivity-weighted irradiance at $\mathbf{x}_\mathrm{d}$,

$$E'(\mathbf{x}_\mathrm{d}) = \int_0^\infty \mathrm{d}\lambda\, R(\lambda) E_\lambda(\lambda;\mathbf{x}_\mathrm{d}) \tag{9.69}$$

may be considered as arising from the sum of contributions from area elements of the source, according to

$$E'(\mathbf{x}_\mathrm{d}) = \int_{A_\mathrm{s}} \mathrm{d}^2\mathbf{x}_\mathrm{s} \int_0^\infty \mathrm{d}\lambda\, R(\lambda) X(\lambda;\mathbf{x}_\mathrm{d},\mathbf{x}_\mathrm{s}) L_\lambda(\lambda;\mathbf{x}_\mathrm{s}) \tag{9.70}$$

We may substitute the definition of $X(\lambda;\mathbf{x}_\mathrm{s},\mathbf{x}_\mathrm{s})$ given in Eq. (9.68) into the integral Eq. (9.70). If we change the variable of integration over the spectrum from $\lambda$ to $q$, using $q = 2\pi/\lambda$ and $\mathrm{d}\lambda = -2\pi\, \mathrm{d}q/q^2$, after a little rearrangement we obtain

$$\begin{aligned} E'(\mathbf{x}_\mathrm{d}) = {} & \frac{(2\pi)^{1-2N}}{(d_0 d_1 \ldots d_N)^2} \int_{A_\mathrm{s}} \mathrm{d}^2\mathbf{x}_\mathrm{s} \int_{-\infty}^{+\infty} \mathrm{d}l \int_{-\infty}^{+\infty} \mathrm{d}l' f(l) f(l') \int_0^\infty \mathrm{d}q \\ & \times q^{2N-2} R(\lambda) L_\lambda(\lambda;\mathbf{x}_\mathrm{s}) \mathrm{e}^{\mathrm{i}q(l-l')} \end{aligned} \tag{9.71}$$

Dependence of $f(l)$ and $f(l')$ on $\mathbf{x}_\mathrm{s}$ is implicit.

If $L_\lambda(\lambda;\mathbf{x}_\mathrm{s})$ is separable regarding its spectral and spatial dependences, according to $L_\lambda(\lambda;\mathbf{x}_\mathrm{s}) = a(\lambda) b(\mathbf{x}_\mathrm{s})$, the integration in Eq. (9.71) is analogously separable. In this case, evaluation of Eq. (9.71) is greatly simplified. For a Planckian source with emissivity $\varepsilon$, giving

$$L_\lambda(\lambda;\mathbf{x}_\mathrm{s}) = \frac{\varepsilon c_1}{\pi\lambda^5}\left[\exp\left(\frac{c_2}{\lambda T}\right) - 1\right]^{-1} \tag{9.72}$$

and a spectrally flat detector with $R(\lambda) = R_0$, we have

$$\begin{aligned} E'(\mathbf{x}_\mathrm{d}) &= \frac{(2\pi)^{-4-2N} \varepsilon c_1 R_0}{\pi (d_0 d_1 \ldots d_N)^2} \int_{A_\mathrm{s}} \mathrm{d}A_\mathrm{s} \int_{-\infty}^{+\infty} \mathrm{d}l \int_{-\infty}^{+\infty} \mathrm{d}l' f(l) f(l') \int_0^\infty \frac{\mathrm{d}q\, q^{2N+3} \mathrm{e}^{\mathrm{i}q(l-l')}}{\mathrm{e}^{\beta q} - 1} \\ &= \frac{(2\pi)^{-5-2N} (2N+3)! \varepsilon c_1 R_0}{(d_0 d_1 \ldots d_N)^2} \int_{A_\mathrm{s}} \mathrm{d}A_\mathrm{s} \int_{-\infty}^{+\infty} \mathrm{d}l \\ &\quad \times \int_{-\infty}^{+\infty} \mathrm{d}l' f(l) f(l') S_{2N+4}(\beta, l - l') \end{aligned} \tag{9.73}$$

with $\beta = c_2/(2\pi T)$ and $S_v(x,y) = \sum_{n=1}^{\infty}[(nx + \mathrm{i}y)^{-v} + (nx - \mathrm{i}y)^{-v}]$.

The remaining integrals have smoother integrands, permitting coarser numerical sampling, than their monochromatic counterparts, which have intrinsically oscillatory integrands. The utility of Eq. (9.73) has been demonstrated in Reference [52], which also shows how $S_4(x, y)$ in the present notation can be efficiently evaluated. It is also in principle possible to generalize Eq. (9.73) to optical setups with other forms of $R(\lambda)L_\lambda(\lambda; \mathbf{x}_s)$ while continuing to exploit separability of $L_\lambda(\lambda; \mathbf{x}_s)$.

Expressions for integrated total flux in the case of a Planckian source have also been derived. These are analogous to Wolf's expressions for integrated spectral flux. As an important example, in the case of Fraunhofer diffraction by a limiting circular aperture, the fraction of total power incident on the aperture that reaches a circular detector is [48]

$$F(\sigma) = 1 - \frac{4\zeta(3)A}{6\pi\zeta(4)} + \frac{A^3 \log_e A}{24\pi\zeta(4)} + \frac{(3 - 2\gamma - 6\log_e 2)A^3}{48\pi\zeta(4)} + O(A^5 \log_e A) \quad (9.74)$$

with $A = c_2/[(1+\sigma)v_M \lambda T]$, $\zeta(z)$ being the Riemann zeta function, and $\gamma \cong 0.577216$ the Euler's constant. The parameter $A$ depends on $\sigma$, $T$, and the geometry. However, $A$ does not depend on $v_M$ or $\lambda$ separately, because we have $v_M \propto \lambda^{-1}$, so that the product $\lambda v_M$ is a wavelength-independent, geometrical entity. In the spirit of Eq. (9.62), this gives

$$\frac{\Phi}{L} = D \int_{-1}^{1} \frac{dx\{(1 - x^2)[(2 + \sigma x)^2 - \sigma^2]\}^{1/2} F(\sigma x)}{1 + \sigma x} \quad (9.75)$$

which relates total power to total source radiance including diffraction effects.

### 9.5.3 Practical Examples

Three practical examples of diffraction effects in radiometry are presented below. These are solar radiometry, blackbody calibration of a telescope used to measure spectral irradiance of celestial objects, and calibration of a standard reference blackbody.

#### *9.5.3.1 Solar radiometry*

The layouts of two different absolute solar radiometers are shown in Figure 9.13. The PMO6 radiometer [53] features a view-limiting aperture that acts as a non-limiting aperture between the extended source (sun) and the defining aperture placed in front of an absolute radiometer. The SAD geometry is defined by $R = 4.25$ mm, $R_d = 2.5$ mm, $d' = 95.4$ mm, $d \cong 1.5 \times 10^{14}$ mm, and $R_s \cong 6.75 \times 10^{11}$ mm. If we model the sun as a 5900 K Planckian source, the corresponding version of Eq. (9.75) for the Fresnel regime and a non-limiting aperture gives $\langle \varepsilon_{\text{diff}} \rangle \cong 0.0012798$,

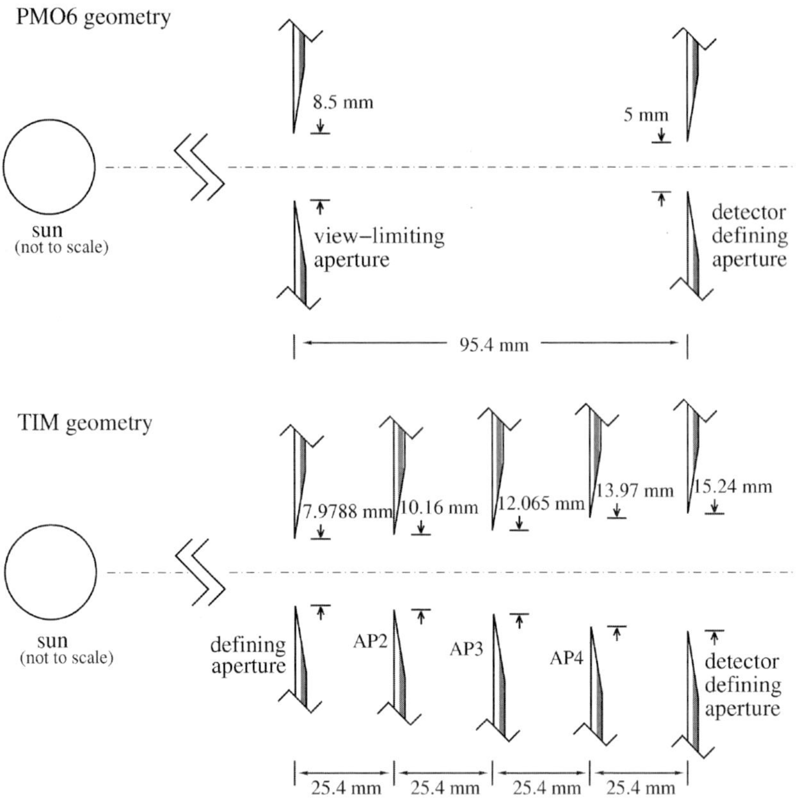


FIG. 9.13. Layouts of two solar radiometers.

implying that about 0.13% excess power reaches the detector compared to that expected from the geometrical throughput. This result automatically includes integration with respect to wavelength over the entire spectrum. In comparison, the effective-wavelength approximation gives $\langle \varepsilon_{\text{diff}} \rangle \cong 0.0012720$ if oscillatory terms are neglected and $\langle \varepsilon_{\text{diff}} \rangle \cong 0.0011656$ if oscillatory terms are (unadvisedly) included.

The total-irradiance monitor (TIM) instrument [54], which is also used in absolute solar radiometry, features a $R = 3.9894$ mm defining aperture that is $d' = 101.6$ mm from a $R_{\text{d}} = 7.62$ mm radiometer entrance. Three intervening, non-limiting apertures (labeled AP2, AP3, and AP4) are also noted. These non-limiting apertures are barely large enough to permit passage of all marginal rays from the perimeter of the defining aperture to the perimeter of the radiometer entrance. This makes it difficult to treat their diffraction effects using asymptotic methods. If the non-limiting apertures are ignored, the SAD problem leads to a diffraction loss of total power given by $\langle \varepsilon_{\rm diff} \rangle \cong -0.000432$. If the non-limiting apertures are taken into account, this loss is barely unchanged, based on calculations using the method of Reference [26].

#### *9.5.3.2 Calibration of a radiometric telescope*

Figure 9.14 shows an unfolded, model optical setup used to simulate diffraction effects in a blackbody calibration of a radiometric telescope [55]. A calibrated blackbody with a 10 mm radius diameter cavity opening (I) is 100 mm behind a $R_s = 1.778$ mm radius aperture (II), which is 150 mm from a $R_d = 6.35$ mm radius detector (IV). To help simulate a remote object, a pinhole aperture (III) with diameter $2R = 202$ μm or $2R = 262$ μm is located 23 mm downstream from the 1.778 mm aperture and 127 mm upstream from the detector. In the real optical setup, a fold mirror was located at the position corresponding to the model's detector, and a focusing mirror had one focus on the pinhole aperture and concentrated radiation onto a spectrometer entrance. In a field deployment, the telescope optics would have one focus at infinity and concentrate radiation onto the same spectrometer entrance in similar fashion.

If we treat the last three optical elements (II–IV) as a SAD setup, the diffraction effects indicated by the solid curves in Figure 9.15 are obtained. If instead the 100 mm separation between I and II is taken into account, and the system is treated according to Reference [26], the diffraction effects indicated by the dashed curves in Figure 9.15 are obtained. This further illustrates both the significance of diffraction effects in a measurement and the potential need to go beyond the SAD model.

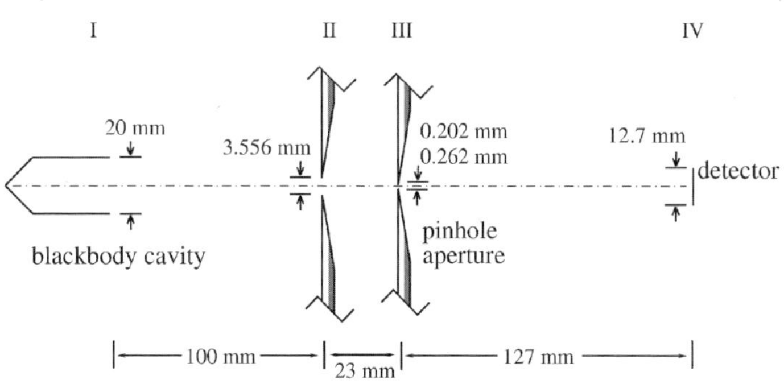


FIG. 9.14. Setup used to calibrate a radiometric telescope.

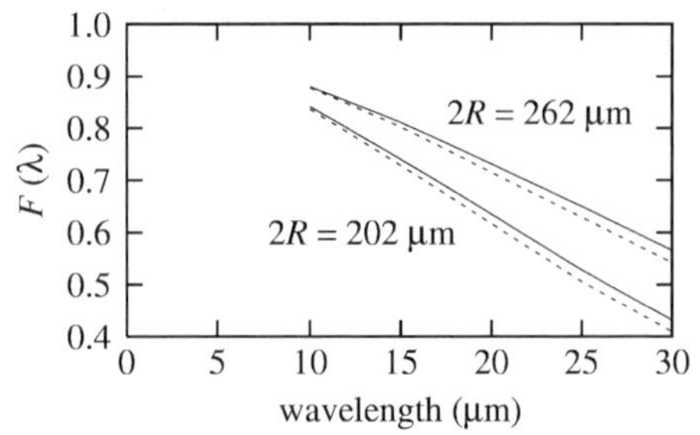


FIG. 9.15. Diffraction effects on spectral throughput of setup shown in Figure 9.14. The solid lines indicate a SAD treatment, whereas the dashed lines indicate a more complete treatment.

#### *9.5.3.3 Blackbody calibration*

Diffraction effects that arose in an actual blackbody calibration were already discussed in Section 9.4.3.2 and in Reference [34]. These included significant losses at the defining aperture, partially compensating gains because of non-limiting apertures, and the need to go beyond the SAD model as discussed further in Reference [34]. Other blackbody calibrations have repeatedly raised similar issues.

### 9.5.4 Diffraction Effects and Measurement Uncertainties

To illustrate the role of diffraction effects in measurement uncertainties, note that Eq. (9.3) gives

$$U_{\mathrm{R}}^2(T(\lambda)) = U_{\mathrm{R}}^2(F(\lambda)) + U_{\mathrm{R}}^2(T_0) + U_{\mathrm{R}}^2(M(\lambda)) + \cdots \qquad (9.76)$$

Here $U_{\mathrm{R}}(X)$ denotes the relative uncertainty of the quantity $X$, and terms involving correlations of the factors $F(\lambda)$, $T_0$ and $M(\lambda)$ are not explicitly noted.

Note that, for numerical reasons and for systematic reasons related to use of an approximate theory, the correct value of $F(\lambda)$ is not precisely known, and its uncertainty contributes to the overall uncertainty of a measurement. Obviously, ignoring diffraction effects does not reduce overall measurement uncertainties. Rather, theoretical and/or experimental diffraction analysis can reduce and better define the actual uncertainties by better determining the value of $F(\lambda)$.

It has been difficult to determine the uncertainty of the theoretical $F(\lambda)$, in part because measurements of $F(\lambda)$ intended to compare theory and the "correct" result have appreciable uncertainties. All indications are that the

Kirchhoff theory can be accurate to within a few percent of $\varepsilon_{\mathrm{diff}}(\lambda)$, say about 5%. However, this estimate remains to be confirmed.

Statistical correlations that are suppressed in Eq. (9.76) can also be important, and it is desirable to ensure that correlations and other systematic effects are kept as similar as possible between characterization of optical instrumentation and its application. In this way, diffraction effects, statistical correlations in the measurement equation and other aspects of performance can be experimentally and/or theoretically determined and compensated. The transmittance of such information is an indispensable part of reports on measurements. To accomplish this, Wyatt et al. note [1]: "A cardinal rule of calibration is that one should calibrate...under the same conditions [as usage]."

## 9.6 Radiometry of Novel Sources

Thus far, the discussion in this chapter has been phrased in terms of tracing the energy flow from points $\{\mathbf{r}_\mathrm{s}\}$ on a "source" surface to points $\{\mathbf{r}_\mathrm{d}\}$ on a "detector" surface. Between the source and detector, the wave nature of the radiation was respected and approximately described using the Kirchhoff theory. By summing irradiance at the detector over source points $\{\mathbf{r}_\mathrm{s}\}$, mutual incoherence of waves originating at different points was implicitly assumed.

Walther [56] called attention to this and related issues in 1968. To help generalize the classical, local definition of radiance, Walther introduced a "generalized radiance function." For radiation emitted from around a point $\mathbf{r}_\mathrm{s}$ into the direction $\hat{\mathbf{s}}$, Walther's function can be given by

$$B(\mathbf{r}_\mathrm{s}, \hat{\mathbf{s}}, \omega) = \frac{\cos\theta}{\lambda^2} \int \mathrm{d}^2\mathbf{r}' \, W^{(0)}(\mathbf{r}_\mathrm{s} + \mathbf{r}'/2, \mathbf{r}_\mathrm{s} - \mathbf{r}'/2, \omega) \exp(-ik\hat{\mathbf{s}} \cdot \mathbf{r}') \qquad (9.77)$$

where

$$W^{(0)}(\mathbf{x}, \mathbf{x}', \omega) = \int_{-\infty}^{+\infty} \mathrm{d}\tau \exp(+\mathrm{i}\omega\tau) \langle V(\mathbf{x}, t) V^*(\mathbf{x}', t + \tau) \rangle \qquad (9.78)$$

is the cross-spectral density for two points $\mathbf{x}$ and $\mathbf{x}'$ at frequency $\omega$, angle brackets indicate temporal averaging, and $V$ denotes a radiation field amplitude function in the same spirit as does $U$. There are pathologies associated with Walther's generalized radiance function, as discussed by others [57]. However, it does help to bridge the gap between real sources and the traditional, "local" definition of radiance used in classical radiometry.

An important result of further analysis made possible by Eqs. (9.77) and (9.78) is that traditional radiometric concepts such as radiance require no substantial modification for many incoherent sources, including blackbodies. This is discussed by Nugent and Gardner [58], Gardner [59], and Mielenz [60], whose work was stimulated in part by issues raised by Wolf [61].

Currently, there is also interest in using a synchrotron as a standard radiation source [62]. The spectral and spatial characteristics of synchrotron radiation can be computed if the electron energy, electron beam current and geometrical parameters of the electrons' orbit are known (see [4], pp. 672–679). In this case, the radiation in a plane near an electron is certainly not incoherent, and the concept of a generalized radiance function must be reconsidered. However, once the radiation field is known, for instance, at the first aperture through which the radiation passes, the Kirchhoff theory should still be useful for understanding subsequent propagation of this and other novel forms of radiation.

## References

*1.* C. L. Wyatt, V. Privalsky, and R. Datla, in "Recommended Practice; Symbols, Terms, Units and Uncertainty Analysis for Radiometric Sensor Calibration. NIST Handbook 152." U.S. Govt. Printing Office, Washington, DC, USA, 1998.
*2.* W. T. Welford, "Geometrical Optics." North-Holland Publishing Company, Amsterdam, 1962.
*3.* A. Walther, "The Ray and Wave Theory of Lenses." Cambridge University Press, Cambridge, UK, 1995.
*4.* J. D. Jackson, "Classical Electrodynamics," 2nd edition. Wiley, New York, USA, 1975.
*5.* S. G. Johnson and J. D. Joannopoulos, "Photonic Crystals: The Road from Theory to Practice." Kluwer Academic Publishers, Boston, USA, 2002.
*6.* G. Kirchhoff, Zur Theorie der Lichtstrahlen, *Ann. Physik* 254, 663–695 (1883).
*7.* M. Born and E. Wolf, "Principles of Optics." Pergamon Press Ltd., Oxford, UK, 1965.
*8.* A. Sommerfeld, "Optics." Academic Press, New York, USA, 1954.
*9.* A. Sommerfeld, Mathematische theorie der diffraction, *Math. Ann.* 47, 317–374 (1896).
*10.* H. A. Bethe, Theory of diffraction by small holes, *Phys. Rev.* 66, 163–182 (1944).
*11.* C. J. Bouwkamp, On Bethe's theory of diffraction by small holes, *Philips Res. Rep.* 5, 321–332 (1950).

*12.* H. Levine and J. Schwinger, On the theory of diffraction by an aperture in an infinite plane screen I, *Phys. Rev.* 74, 958–974 (1948); On the theory of diffraction by an aperture in an infinite plane screen II, *Phys. Rev.* 75, 1423–1432 (1949).

*13.* W. Braunbek, Neue Näherungsmethode für die Beugung am ebenen Schirm, *Z. Phys.* 127, 381–390, 1950); Zur Beugung an der Kreisscheibe, *Z. Phys.* 127, 405–415 (1950); Zur Beugung an der kreisförmigen Öffnung, *Z. Phys.* 138, 80–88 (1954).

*14.* K. D. Mielenz, Optical diffraction in close proximity to plane apertures. I. Boundary-value solutions for circular apertures and slits, *J. Res. Natl. Inst. Stand. Technol.* 107, 355–362 (2002); Optical diffraction in close proximity to plane apertures. II. Comparison of half-plane diffraction theories, *J. Res. Natl. Inst. Stand. Technol.* 108, 57–68 (2003); Optical diffraction in close proximity to plane apertures. III. Modified, self-consistent theory, *J. Res. Natl. Inst. Stand. Technol.* 109, 457–464 (2004).

*15.* K. E. Oughstun, "Selected Papers on Scalar Wave Diffraction." SPIE Optical Engineering Press, Bellingham, USA, 1992.

*16.* G. A. Maggi, *Ann. Mat.* IIa(16), 21ff (1888).

*17.* A. Rubinowicz, Die Beugungswelle in der Kirchhoffschen Theorie der Beugungserscheinungen, *Ann. Physik.* 53, 257–278 (1917).

*18.* K. Miyamoto, E. Wolf, Generalization of the Maggi–Rubinowicz theory of the boundary diffraction wave—Part I, *J. Opt. Soc. Am.* 52, 615–625 (1962); Generalization of the Maggi–Rubinowicz theory of the boundary diffraction wave—Part II, *J. Opt. Soc. Am.* 52, 626–637 (1962).

*19.* E. Wolf and E. Marchand, Comparison of the Kirchhoff and the Rayleigh–Sommerfeld theories of diffraction at an aperture, *J. Opt. Soc. Am.* 54, 587–594 (1964).

*20.* J. B. Keller, Geometrical theory of diffraction, *J. Opt. Soc. Am.* 52, 116–130 (1962).

*21.* G. L. James, "Geometrical Theory of Diffraction." Peter Peregrinus Ltd., London, UK, 1986.

*22.* V. A. Borovikov and B. Ye. Kinber, "Geometrical Theory of Diffraction." Inst. Elect. Engineers, London, UK, 1994.

*23.* R. K. Luneburg, "Mathematical Theory of Optics." Cambridge University Press, London, UK, 1966.

*24.* E. Lommel, Die Beugungserscheinungen einer kreisrunden Oeffnung und eines kreisrunden Schirmschens theoretisch und experimentell Bearbeitet, *Abh. Bayer. Akad.* 15, 233–328 (1885).

*25.* K. D. Mielenz, Algorithms for Fresnel diffraction at rectangular and circular apertures, *J. Res. Natl. Inst. Stand. Technol.* 103, 497–509 (1998).

26. E. L. Shirley and M. L. Terraciano, Two innovations in diffraction calculations for cylindrically symmetrical systems, *Appl. Opt.* 40, 4463–4472 (2001).
27. E. L. Shirley and E. K. Chang, Accurate efficient evaluation of Lommel functions for arbitrarily large arguments, *Metrologia* 40, S5–S8 (2003).
28. P. Edwards and M. McCall, Diffraction loss in radiometry, *Appl. Opt.* 42, 5024–5032 (2003).
29. E. Wolf, Light distribution near focus in an error-free diffraction image, *Proc. Roy. Soc. A* 204, 533–548 (1951).
30. J. Focke, Total illumination in an aberration-free diffraction image, *Optica Acta* 3, 161–163 (1956).
31. K. Schwarzschild, *Sitz. Ber. Bayer. Akad. Wiss.* 28, 271 (1898).
32. N. G. van Kampen, An asymptotic treatment of diffraction problems, *Physica* 14, 575–589 (1949).
33. F. W. J. Olver, "Asymptotics and special functions." A.K. Peters, Ltd., Wellesley, USA, 1997.
34. E. L. Shirley, Intuitive diffraction model for multistaged optical systems, *Appl. Opt.* 43, 735–743 (2004).
35. L. P. Boivin, Reduction of diffraction errors in radiometry by means of toothed apertures, *Appl. Opt.* 17, 3323–3328 (1978).
36. Lord Rayleigh, On images formed without reflection or refraction, *Phil. Mag.* 11, 214–218 (1881).
37. C. L. Sanders and O. C. Jones, Problem of realizing the primary standard of light, *J. Opt. Soc. Am.* 52, 731–742 (1962).
38. N. Ooba, Experimental study on diffraction in the measurement of the primary standard of light, *J. Opt. Soc. Am.* 54, 357–361 (1964).
39. W. R. Blevin, Diffraction losses in radiometry and photometry, *Metrologia* 6, 39–44 (1970).
40. W. H. Steel, M. De, and J. A. Bell, Diffraction corrections in radiometry, *J. Opt. Soc. Am.* 62, 1099–1103 (1972).
41. L. P. Boivin, Diffraction corrections in radiometry: Comparison of two different methods of calculation, *Appl. Opt.* 14, 2002–2009 (1975).
42. L. P. Boivin, Diffraction losses associated with Tungsten lamps in absolute radiometry, *Appl. Opt.* 14, 197–200 (1975).
43. L. P. Boivin, Diffraction corrections in the radiometry of extended sources, *Appl. Opt.* 15, 1204–1209 (1976).
44. L. P. Boivin, Radiometric errors caused by diffraction from circular apertures: Edge effects, *Appl. Opt.* 16, 377–384 (1977).
45. J. D. Purcell and M. J. Koomen, "Coronagraph with Improved Scattered-Light Properties. Report of NRL Progress." Govt. Printing Office, Washington, DC, USA, 1962.

*46.* E. L. Shirley and R. U. Datla, Optimally toothed apertures for reduced diffraction, *J. Res. Natl. Inst. Stand. Technol.* 101, 745–753 (1996).

*47.* E. L. Shirley, Revised formulas for diffraction effects with point and extended sources, *Appl. Opt.* 37, 6581–6590 (1998).

*48.* E. L. Shirley, Fraunhofer diffraction effects on total power for a Planckian source, *J. Res. Natl. Inst. Stand. Technol.* 106, 775–779 (2001); Diffraction corrections in radiometry: Spectral and total power, and asymptotic properties, *J. Opt. Soc. Am. A* 21, 1895–1906 (2004).

*49.* J. G. Suárez-Romero, E. Tepíchin-Rodríguez, and K. D. Mielenz, Cross-spectral density propagated through a circular aperture, *Metrologia* 38, 379–384 (2001).

*50.* L. Mandel and E. Wolf, "Optical Coherence and Quantum Optics." Cambridge University Press, Cambridge, UK, 1975.

*51.* J. W. Goodman, "Introduction to Fourier Optics," 2nd edition. McGraw-Hill, New York, USA, 1996.

*52.* E. L. Shirley, Diffraction effects on broadband radiation: Formulation for computing total irradiance, *Appl. Opt.* 43, 2609–2620 (2004).

*53.* R. W. Brusa and C. Fröhlich, Absolute radiometers (PMO6) and their experimental characterization, *Appl. Opt.* 25, 4173–4180 (1986).

*54.* G. M. Lawrence, G. Kopp, G. Rottman, J. Harder, T. Woods, and H. Loui, Calibration of the total irradiance monitor, *Metrologia* 40, S78–S80 (2003).

*55.* F. C. Witteborn, M. Cohen, J. D. Bregman, D. H. Wooden, K. Heere, and E. L. Shirley, Spectral irradiance calibration in the infrared. XI. Comparison of $\alpha$ Bootis and 1 Ceres with a laboratory standard, *Astron. J.* 117, 2552–2560 (1999).

*56.* A. Walther, Radiometry and coherence, *J. Opt. Soc. Am.* 58, 1256–1259 (1968).

*57.* E. Wolf, Coherence and radiometry, *J. Opt. Soc. Am.* 68, 6–17 (1978).

*58.* K. A. Nugent and J. L. Gardner, Radiometric measurements and correlation-induced spectral changes, *Metrologia* 29, 319–324 (1992).

*59.* J. L. Gardner, Partial coherence and practical radiometry, *Metrologia* 30, 419–423 (1993).

*60.* K. D. Mielenz, "Wolf shifts" and their physical interpretation under laboratory conditions, *J. Res. Natl. Inst. Stand. Technol.* 98, 231–240 (1993).

*61.* E. Wolf, Invariance of the spectrum of light on propagation, *Phys. Rev. Lett.* 56, 1370–1372 (1986).

*62.* G. Ulm, Radiometry with synchrotron radiation, *Metrologia* 40, S101–S106 (2003); P.-S. Shaw, U. Arp, H. W. Yoon, R. D. Saunders, A. C. Parr, and K. R. Lykke, A SURF beamline for synchrotron source-based absolute radiometry, *Metrologia* 40, S124–S127 (2003).

# 10. THE CALIBRATION AND CHARACTERIZATION OF EARTH REMOTE SENSING AND ENVIRONMENTAL MONITORING INSTRUMENTS*

James J. Butler
NASA's Goddard Space Flight Center, Greenbelt, Maryland, USA

B. Carol Johnson
National Institute of Standards and Technology, Gaithersburg, Maryland, USA

Robert A. Barnes
Science Applications International Corporation, Beltsville, Maryland, USA

## 10.1 Introduction

### 10.1.1 The Use of Satellite Instruments in Earth Remote Sensing and Environmental Monitoring

The use of remote-sensing instruments on orbiting satellite platforms in the study of Earth science and environmental monitoring was officially inaugurated with the April 1, 1960 launch of the Television Infrared Observation Satellite (TIROS) [1]. The first TIROS accommodated two television cameras and operated for only 78 days. However, the TIROS program, in providing in excess of 22,000 pictures of the Earth, achieved its primary goal of providing Earth images from a satellite platform to aid in identifying and monitoring meteorological processes. This marked the beginning of what is now over four decades of Earth observations from satellite platforms.

Satellite-based Earth remote-sensing can be defined as the measurement of reflected and emitted radiation from the Earth using instruments on satellite platforms. These measurements are input to climate models, and the model results are analyzed in an effort to detect short- and long-term changes and trends in the Earth's climate and environment, to identify the cause of those changes, and to predict or influence future changes. Examples of short-term climate change events include the periodic appearance of the El Nino-Southern Oscillation (ENSO) in the tropical Pacific Ocean [2] and the spectacular eruption of Mount Pinatubo on the Philippine island of Luzon in 1991. Examples of long-term climate change events, which are

*Parts of this chapter are a contribution of the National Institute of Standards and Technology.



EXPERIMENTAL METHODS IN THE PHYSICAL SCIENCES, vol. 41
ISSN 1079-4042 DOI: 10.1016/S1079-4042(05)41010-3

more subtle to detect, include the destruction of coral reefs, the disappearance of glaciers, and global warming. Climatic variability can be both large and small scale and can be caused by natural or anthropogenic processes. The periodic El Nino event is an example of a natural process, which induces significant climatic variability over a wide range of the Earth. A classic example of a large-scale anthropogenic influence on climate is the well-documented rapid increase of atmospheric carbon dioxide occurring since the beginning of the Industrial Revolution [3]. An example of the study of a small-scale anthropogenic influence in climate variability is the Atlanta Land-use Analysis Temperature and Air-quality (ATLANTA) project [4]. This project has found that the replacement of trees and vegetation with concrete and asphalt in Atlanta, Georgia, and its environs has created a microclimate capable of producing wind and thunderstorms. A key objective of climate research is to be able to distinguish the natural versus human roles in climate change and to clearly communicate those findings to those who shape and direct environmental policy.

The advantages of using satellite platforms in Earth remote-sensing and environmental monitoring are made clear when one considers the challenging nature of the measurement problem. The Earth is a complex, integrated, changing physical system driven by a principle energy source, the Sun, which interacts with the components of the Earth system, namely the land, oceans, atmosphere, and cryosphere. From a remote-sensing perspective, the Earth appears as a temporally variable, spatially and spectrally variegated object. Earth climate and environmental studies, therefore, require the acquisition of large numbers of observations to be properly understood. Instruments placed on satellites in Earth orbit satisfy temporal Earth remote-sensing requirements in that they are capable of performing frequent observations. They also satisfy spatial remote-sensing requirements in that they monitor areas of the Earth larger than those capable of being monitored by ground-based, airborne, or balloon-borne instruments.

On-orbit measurements made by Earth remote-sensing instruments are exo-atmospheric radiances, irradiances, or reflectances measured at the remote-sensing instruments' apertures. The Earth's atmosphere, through the processes of absorption, re-emission, reflectance, and scattering, plays a significant role in the amount of solar energy impinging on the Earth surface. Figure 10.1 shows the solar irradiance at the top of the atmosphere and through the atmosphere at an altitude 2 km above mean sea level. Figure 10.1 also shows the irradiance produced by a blackbody at 5870 K, the approximate equivalent radiance temperature of the Sun. Absorption features seen in the irradiance data obtained by looking through the atmosphere are attributed primarily to the following gaseous molecules: $O_3$ below 360 nm, $O_2$ centered at 762.1 nm, $H_2O$ centered at 720, 810, 940, 1100,

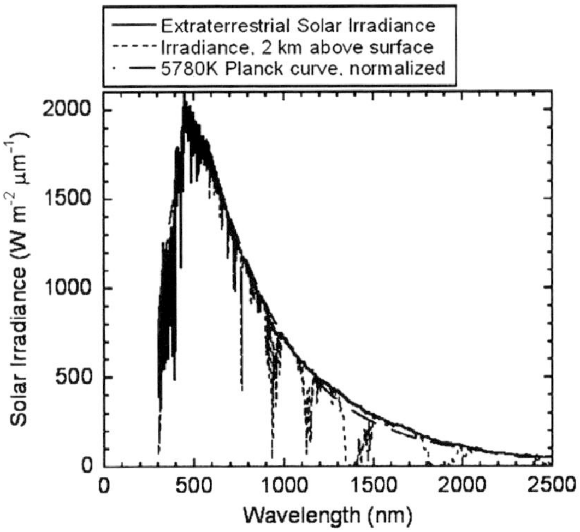


FIG. 10.1. Spectral solar irradiance as a function of wavelength at the top of the Earth's atmosphere (i.e. extraterrestrial) and at an altitude of 2 km looking through the Earth's atmosphere. A Planck curve for a 5870 K blackbody, the approximate color temperature of the Sun, is also shown for comparison. The Planck curve is normalized to the irradiance data at 650 nm.

1400, and 1900 nm, and $CO_2$ centered at 1400 and 2000 nm. In Figure 10.1, the absorption feature beginning near 2500 nm extends to 3300 nm and is attributed primarily to $H_2O$ and $CO_2$. Also visible in this figure are the atmospheric window bands centered at 1050, 1250, 1650, and 2300 nm. An example of the complexity of the interaction of incident solar flux with the Earth's atmosphere and surface is shown in Figure 10.2, which shows the typical radiation budget of the Earth [5]. Since Earth scientists are typically interested in processes occurring at the Earth's surface, at-satellite measurements must be transferred to the Earth's surface. This involves correcting the at-satellite measurements for the presence of the Earth's atmosphere using an atmospheric model, inputs to which include temperature, pressure, concentrations of absorbing and scattering atmospheric molecular species, aerosol optical depth, and water vapor content for the region of interest. Inputs to atmospheric correction models are obtained from a variety of ground-based, air-borne, balloon-borne, and satellite-based instruments.

### 10.1.2 Launch and Space Environmental Effects on Remote-Sensing Instruments

In addition to the spatial, spectral, and temporal complexities of remotely sensing the Earth system, the on-orbit deployment and operation of satellite

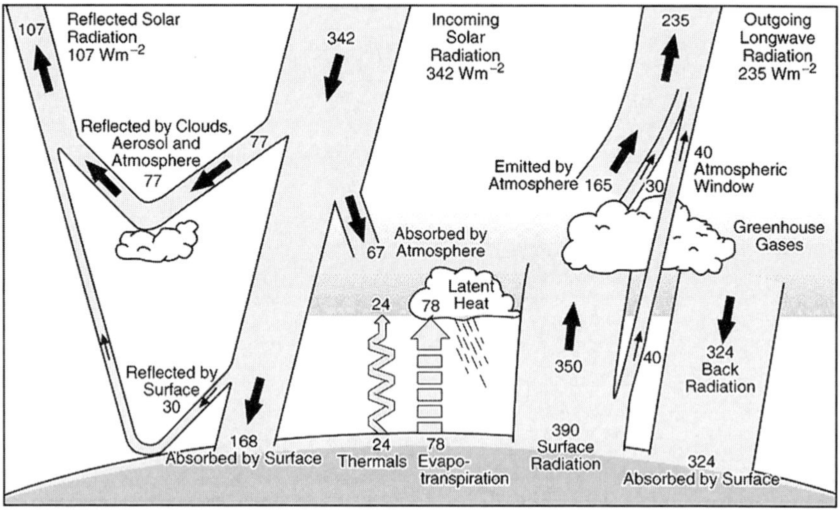


FIG. 10.2. The Earth's energy balance [5]. The left side of the figure depicts the shortwave (i.e. ultraviolet through shortwave infrared) albedo of the Earth, or that fraction of the incident TSI reflected by the Earth, while the right side depicts long-wave emitted processes. The incoming solar radiation of 342 W/m$^2$ is equal to a solar constant of 1368 W/m$^2$ divided by a factor of 4, taking into account the total surface area of an assumed spherical Earth (i.e. $4\pi r^2$ versus $\pi r^2$). Figure reproduced with the permission of the American Meteorological Society and the authors.

instruments present a unique set of additional challenges. Remote-sensing instruments and their spacecraft must first survive the vibrational and acoustic stresses of the launch process and entry into the vacuum of space. Once on-orbit, the instruments are exposed to ionizing radiation and coronal mass ejection (CME) protons produced by the Sun. On-orbit instruments are also exposed to impacts by orbital debris, oxidizing chemical species, such as atomic oxygen, and high-energy electrons and protons, heavy ions, and cosmic ray flux, which can cause instrument on-orbit Single Event Upsets (SEUs) and Single Event Latchups (SELs). Instruments operating in the vacuum environment of space are subject to outgassing of various molecular species, which, in the presence of solar radiation, can be baked or "solarized" onto exposed instrument optical components and surfaces. The on-orbit operation of remote-sensing instruments occurs in a zero gravitational, continuously changing thermal environment. While the effects of the latter are extensively modeled and tested pre-launch, the effects of the former cannot be tested. The net effect on the orbiting instrument is usually a degradation of the overall system radiometric responsivity. A key aspect in the use of satellite remote-sensing data in the study of the Earth is the ability to distinguish between those on-orbit instrument changes and actual Earth geophysical changes.

### 10.1.3 Target Calibration Uncertainties, Measurement Stability and Continuity in Climatic Remote Sensing and Environmental Monitoring

The ability to monitor, understand, and subsequently predict short- and long-term climate and environmental processes is related to the quality of the data produced by the remote-sensing instrument or instruments used to measure those processes. While it is possible to study short-term processes using a single generation of a remote-sensing satellite instrument, long-term, decadal processes require measurements from successive generations of instruments, which are often non-identical and potentially operate in different on-orbit environments. The global nature of the study of climate and environmental processes requires consistent calibration of successive generations of remote-sensing instruments against a common scale and the careful characterization of those instruments. Calibration is defined as the process of quantitatively defining an instrument's system response to known, controlled signal inputs. Characterization is defined as the set of operations or processes used to quantitatively understand the operation of an instrument. For a satellite instrument, characterization involves determination of the instrument's response as a function of the gamut of operating and viewing conditions experienced by that instrument on-orbit. Instrument measurement accuracy, precision, and stability requirements are directly derived from Earth science and environmental monitoring requirements and differ depending on the particular geophysical process being studied. The study of geophysical processes using instruments on satellite platforms often requires the measurement of small-scale signals and signal changes superimposed upon larger signals. For this reason, the accuracy, precision, and stability requirements imposed upon satellite remote-sensing instruments and, logically, upon the instrumentation used in their calibration and characterization are, from purely a metrological standpoint, often state of the art. Table 10.1 presents the accuracies and stabilities of remote-sensing satellite instrument measurements required in the study of the listed geophysical parameters [6].

### 10.1.4 The Importance of Traceability in Earth Remote Sensing

The study of climate and environmental change is global in nature and involves measurements made by remote-sensing instruments from a number of countries. The ability to relate and compare instrument measurements depends on the pre-launch and on-orbit calibration of those instruments against a common, internationally recognized physical scale.

TABLE 10.1. Accuracies and Stabilities of Satellite Instrument Measurements Required for the Determination of Listed Earth Geophysical Parameters

| Geophysical Parameters | Satellite Instrument Type | Accuracy | Stability Per Decade |
|---|---|---|---|
| Solar Irradiance | Broadband Radiometer | 1.5 $W/m^2$ | 0.3 $W/m^2$ |
| Surface Albedo | Visible Radiometer | 5% | 1% |
| Net Solar Radiation: Top of Atmosphere | Broadband Radiometer | 1 $W/m^2$ | 0.3 $W/m^2$ |
| Spectral Thermal Radiance | Infrared Spectroradiometer | 0.1 K | 0.04 K |
| Cloud Base Height | Visible/Infrared Radiometer | 1 K | 0.2 K |
| Cloud Effective Particle Size Distribution | Visible/Infrared radiometer | 3.7 μm: Water, 5%; Ice, 10% | 3.7 μm: Water, 1%; Ice, 2% |
| | | 1.6 μm: Water, 2.5%; Ice, 5% | 1.6 μm: Water, 0.5%; Ice, 1% |
| Cloud Optical Thickness | Visible Radiometer | 5% | 1% |
| Cloud Top Height, Pressure and Temperature | Infrared Radiometer | 1 K | 0.2 K |
| Tropospheric Temperature | Infrared Radiometer | 0.5 K | 0.04 K |
| Stratospheric Temperature | Infrared Radiometer | 1 K | 0.08 K |
| Water Vapor | Infrared Radiometer | 1 K | 0.03 K |
| Total Column Ozone | Ultraviolet/Visible Spectrometer | 2% (spectrally dependent) 1% (spectrally independent) | 0.2% |
| Stratospheric Ozone | Ultraviolet/Visible Spectrometer | 3% | 0.6% |
| Tropospheric Ozone | Ultraviolet/Visible Spectrometer | 3% | 0.1% |
| Aerosols | Visible Radiometer/Polarimeter | Radiometric: 3% Polarimetric: 0.5% | Radiometric: 1% Polarimetric: 0.25% |
| Carbon Dioxide | Infrared Radiometer | 3% | Forcing: 1%; Sources/sinks: 0.25% |
| Ocean Color | Visible Radiometer | 5% | 1% |
| Sea Surface Temperature | Infrared Radiometer | 0.1 K | 0.01 K |
| Sea Ice Area | Visible Radiometer | 12% | 10% |
| Snow Cover | Visible Radiometer | 12% | 10% |
| Vegetation | Visible Radiometer | 2% | 0.8% |

The remote-sensing goals of acquiring high-quality, global remote-sensing data over a time period of decades require that the calibration of these instruments be traceable to metrological physical scales maintained by national measurement institutes (NMIs) such as the National Institute of Standards and Technology (NIST) and the National Physical Laboratory (NPL). NMIs realize the physical quantities that define the International System of Units (SI) and, as such, represent the expertise in fundamental metrology. Interactions between NMIs through the Treaty of the Meter and related key measurement intercomparisons ensure a global consistency in measurement science. The level of interaction of NMIs with remote-sensing calibration facilities occurs on several levels. Traceability of remote-sensing instrument calibration to the SI is accomplished at the first level through the use of reference information, special publications, databases, calibration services, and standard reference materials readily available from NMIs. The state-of-the-art metrological requirements on remotely sensed data have prompted some remote-sensing programs and instrument calibration facilities to invite greater involvement by NMIs in their remote-sensing instrument calibration efforts. This level of involvement by NMIs includes coordinating training courses, workshops, and conferences and direct participation in and coordination of radiometric measurement campaigns and artifact measurement round-robins [7]. The advantages of these programs include direct validation of instrument calibration methods through NMI measurements (i.e. shorter traceability chains and lower calibration uncertainties) and access to NMI metrology programs, which are permanent across multiple remote-sensing mission lifetimes. The NMIs also benefit from close interaction with remote-sensing calibration programs in that the state-of-the-art calibration requirements of the instruments often lead to the incorporation of improved measurement technologies by the NMIs which are then made available to the larger international remote-sensing measurement community.

### 10.1.5 The Measurement Equation Approach in Earth Remote Sensing

Characterization procedures are not limited to the radiometric sources used to calibrate satellite instruments. For accurate results and proper assessment of measurement uncertainties, the flight instrument must also be characterized. A systematic, proven approach to the characterization of remote-sensing instruments is to first determine the instrument's measurement equation. The measurement equation of an instrument is a mathematical expression that describes the roles and effects of all influencing parameters. Table 10.2 lists a number of influencing parameters, which must be measured in the process of instrument characterization. In addition to the parameters listed in Table 10.2, time is an essential variable, as it relates to

TABLE 10.2. Satellite Instrument Characterization Parameters

| Radiometric | Spectral | Spatial |
|---|---|---|
| Dynamic range | Spectral responsivity | Pointing |
| | Within band | Accuracy |
| | Out-of-band | Knowledge |
| Linearity | Wavelength stability | Spectral band registration |
| | Within band | Within band |
| | Band-to-band | Band-to-band |
| Signal-to-noise | Wavelength accuracy and precision | Spatial responsivity |
| | | Within field |
| | | Out-of-field |
| Stability | | Spatial response uniformity |
| Short-term | | Within field |
| Long-term | | Out-of-field |
| Crosstalk | | Modulation transfer function |
| Optical | | |
| Electrical | | |
| Polarization responsivity | | |

both the measurement precision (i.e. short-term stability) and drift or degradation (i.e. long-term stability). Temperature of the surrounding environment, intervening medium, or instrument components is also a significant influencing parameter. Certain parameters influence the measurement process more than others.

As a first approximation in the measurement equation approach, influencing parameters can be treated independently, allowing for separate characterizations to be made for each. This assumption must be confirmed in an effort to identify and test cross-parameter dependencies, and experiments must be designed to minimize possible systematic effects. Examples of these tests include measurements that determine the degree to which a satellite sensor's output is independent of integration time or temperature. An example of reducing sensitivity to systematic effects would be to fully characterize the response of an instrument over its entire field of view, which means to approximately $\pm 55^\circ$ for nadir observations from a typical low Earth orbit to the Earth's limb. The measurement equation ultimately forms the basis of a satellite instrument's radiometric mathematical model. The instrument mathematical model is used pre-launch to validate that an instrument design will meet specifications and on-orbit to assist in understanding instrument operation.

In the case of remote-sensing instruments, the response of an instrument to incoming photon flux is usually expressed in digital numbers (DNs). For the measurement of radiance, this response in general can be described mathematically by the measurement equation

$$\mathrm{DN}_{i,j} = G\iiiint L(x, y, \lambda, t)R(x, y, \lambda, t)\mathrm{d}x\,\mathrm{d}y\,\mathrm{d}\lambda\,\mathrm{d}t \tag{10.1}$$

where $\mathrm{DN}_{i,j}$ is the DN output by instrument detector $i$ in band $j$, $G$ the instrument detector and digitization gain, $L(x, y, \lambda, t)$ the spectral radiance at the instrument entrance aperture, and $R(x, y, \lambda, t)$ the instrument spectral responsivity. For simplicity and completeness, the limits of integration for each variable in Eq. (10.1) and for variables in the equations which follow are assumed to be $+\infty$ to $-\infty$. For an instrument with narrow bandwidth channels, the assumption can be made that the variables in Eq. (10.1) do not have a strong wavelength dependence. In addition, the time and spatial variables in Eq. (10.1) can be ignored if the scene radiance and detector uniformity are spectrally uniform and temporally invariant. These assumptions lead to the following simplified equation describing an instrument's digital output

$$\mathrm{DN}_{i,j} = GA_{i,j}\Omega L(\lambda)\Delta\lambda\eta t\tau\rho \tag{10.2}$$

where $A_{i,j}$ is the area of detector $i$ in band $j$, $\Omega$ the instrument acceptance solid angle, $\Delta\lambda$ the bandwidth, $\eta$ the detector quantum efficiency in electrons per incident photon, $t$ the integration time, $\tau$ the instrument optical transmission, and $\rho$ the instrument optical reflectance. For optical systems employing refractive and reflective optics, the instrument optical transmission and reflection can be represented by

$$\tau = \prod_{1}^{N} \tau_i \tag{10.3}$$

and

$$\rho = \prod_{1}^{M} \rho_i \tag{10.4}$$

where $\tau_i$ and $\rho_i$ are the transmission and reflection of the $i$th of $N$ transmissive and $M$ reflective optical elements, respectively. Instrument response non-linearity, zero radiance response (i.e. background), focal plane temperature effects, or response versus scan angle effects are not shown in Eq. (10.2). These quantities are determined in pre-launch instrument characterization tests and are incorporated in instrument radiometric math models and in the production of measured radiances.

Equation (10.2) can be rewritten as

$$L(\lambda) = \mathrm{DN}_{i,j}m \tag{10.5}$$

where

$$m = \frac{1}{GA_{i,j}\Omega\Delta\lambda\eta\tau\rho} \tag{10.6}$$

is the inverse of the product of the instrument responsivity and gain. It is determined pre-launch for an end-to-end remote-sensing instrument by viewing uniform sources of known radiance, such as well-characterized and calibrated integrating sphere sources and blackbodies. It is also determined pre-launch through remote-sensing subsystem characterization measurements of quantities such as mirror reflectance, polarization responsivity, and spectral radiance responsivity. These subsystem level characterization measurements are used as input to instrument radiometric mathematical models used to validate the system level pre-launch calibration and in the calculation of instrument measurement uncertainty. The quantity, $m$, in Eq. (10.5) is monitored on-orbit using stable, uniform on-board sources of known radiance.

Equation (10.1) is written for the detection of radiance. For remote-sensing measurements in the reflective solar wavelength region (i.e. 200–2500 nm), Eq. (10.1) can also be written in terms of the relative quantity of reflectance,

$$\mathrm{DN}_{i,j} = G\iiiint E_{\mathrm{sun}}(x, y, \lambda)\mathrm{BRDF}(\theta_{\mathrm{i}}, \varphi_{\mathrm{i}}, \theta_{\mathrm{r}}, \varphi_{\mathrm{r}}, \lambda, t) \times R(x, y, \lambda, t)\mathrm{d}x\,\mathrm{d}y\,\mathrm{d}\lambda\,\mathrm{d}t \tag{10.7}$$

where $E_{\mathrm{sun}}$ is the solar spectral irradiance, BRDF the bidirectional reflectance distribution function of either the Earth scene or a reflectance standard, and $\theta_{\mathrm{i}}$, $\varphi_{\mathrm{i}}$ and $\theta_{\mathrm{r}}$, $\varphi_{\mathrm{r}}$ the elevation and azimuthal angles for incident and reflected solar flux, respectively. In reflectance-based Earth remote-sensing measurements, the instrument is used as a transfer radiometer between two diffuse surfaces both illuminated by the Sun, namely, the Earth scene and an optically diffuse reflectance standard. Assuming narrow bandwidth channels and spectrally and spatially uniform and temporally invariant detectors, instrument measurements of an Earth scene and a diffuser can be described by

$$\mathrm{DN}_{i,j,\mathrm{ES}} = G\,\mathrm{BRDF}_{\mathrm{ES}}(\theta_{\mathrm{i}}, \varphi_{\mathrm{i}}, \theta_{\mathrm{r}}, \varphi_{\mathrm{r}}, \lambda, t)E_{\mathrm{sun}}(\lambda)\,A_{i,j}\,\Omega\,\Delta\lambda\eta\tau\rho \tag{10.8}$$

and

$$\mathrm{DN}_{i,j,\mathrm{SD}} = G\,\mathrm{BRDF}_{\mathrm{SD}}(\theta_{\mathrm{i}}, \varphi_{\mathrm{i}}, \theta_{\mathrm{r}}, \varphi_{\mathrm{r}}, \lambda, t)E_{\mathrm{sun}}(\lambda)\,A_{i,j}\,\Omega\,\Delta\lambda\eta\tau\rho\Gamma \tag{10.9}$$

where $\mathrm{DN}_{i,j,\mathrm{ES}}$ and $\mathrm{DN}_{i,j,\mathrm{SD}}$ are the digital numbers output by instrument detector $i$ in band $j$ for the Earth scene, ES, and solar diffuser, SD, respectively; $\mathrm{BRDF}_{\mathrm{ES}}$ and $\mathrm{BRDF}_{\mathrm{SD}}$ the bidirectional reflectance distribution functions for Earth scene and solar diffuser views, and $\Gamma$ accounts for on-orbit degradation of the solar diffuser BRDF. Taking the ratio of Eqs. (10.8) and (10.9),

$$\mathrm{BRDF_{ES}} = \frac{\mathrm{DN_{ES}}}{\mathrm{DN_{SD}}}\,\mathrm{BRDF_{SD}}\Gamma \tag{10.10}$$

On-orbit degradation of the solar diffuser, Γ, is determined using dedicated on-board detector-based monitor hardware, multiple diffuser surfaces exposed to the sun for variable duty cycles, or repeated views of stable or well-characterized celestial and ground-based targets, such as the Moon or desert playas.

For remote-sensing in the thermal infrared region beyond 2500 nm, the brightness temperature of an Earth scene is determined relative to the brightness temperature of an on-board, well-characterized blackbody source. Assuming a simple infrared remote-sensing instrument comprised of a scan mirror, transmissive optics, an on-board blackbody located in the instrument scan cavity, and a filtered infrared detector, the infrared detection of Earth scenes can be mathematically described by

$$\begin{aligned}\mathrm{DN}_{i,j,\mathrm{ES}} &= G\{[\rho_{\mathrm{M}}(\theta_{\mathrm{i}})L(T_{\mathrm{ES}}) + \rho_{\mathrm{M}}(\theta_{\mathrm{i}})\varepsilon_{\mathrm{CAV}}(\varphi_{\mathrm{i}})L(T_{\mathrm{CAV}}) \\ &\quad + \varepsilon_{\mathrm{M}}(\theta_{\mathrm{i}})L(T_{\mathrm{M}})]\tau_{\mathrm{OPT}} + \varepsilon_{\mathrm{OPT}}L(T_{\mathrm{OPT}})\} + \mathrm{DN_{offset}}\end{aligned} \tag{10.11}$$

where $\mathrm{DN}_{i,j,\mathrm{ES}}$ is the digital signal from instrument detector $i$ in band $j$ while viewing an Earth scene, $G$ the instrument gain, $\rho_{\mathrm{M}}(\theta_{\mathrm{i}})$ the reflectance of the instrument scan mirror in the direction $\theta_{\mathrm{i}}$, $L(T_{\mathrm{ES}})$ the radiance of the Earth scene at temperature $T_{\mathrm{ES}}$, $\varepsilon_{\mathrm{CAV}}(\varphi_{\mathrm{i}})$ the emissivity of the scan cavity in the direction of the scan mirror $\varphi_{\mathrm{i}}$, $L(T_{\mathrm{CAV}})$ the radiance of the scan cavity at temperature $T_{\mathrm{CAV}}$, $\varepsilon_{\mathrm{M}}(\theta_{\mathrm{i}})$ the emissivity of the scan mirror in the direction $\theta_{\mathrm{i}}$, $L(T_{\mathrm{M}})$ the radiance of the scan mirror at temperature $T_{\mathrm{M}}$, $\tau_{\mathrm{OPT}}$ the transmission of the instrument optics, $\varepsilon_{\mathrm{OPT}}$ the emissivity of the instrument optics; $L(T_{\mathrm{OPT}})$ the radiance of the instrument optics at temperature $T_{\mathrm{OPT}}$, and $\mathrm{DN_{offset}}$ the electronic noise offset in the instrument. On the right-hand side of Eq. (10.11), the first term represents the scene radiance reflected by the scan mirror. The second term represents the radiance emitted by the scan cavity and reflected by the scan mirror into the instrument. The third term represents the radiance emitted by the mirror, and the fourth term represents the radiance emitted by the instrument optics.

For instrument views of the on-board blackbody, the detected infrared radiance can be represented by

$$\begin{aligned}\mathrm{DN}_{i,j,\mathrm{BB}} &= G\{[\rho_{\mathrm{M}}(\theta_{\mathrm{BB}})\varepsilon_{\mathrm{BB}}(\theta_{\mathrm{BB}})L(T_{\mathrm{BB}}) + \rho_{\mathrm{M}}(\theta_{\mathrm{BB}})\varepsilon_{\mathrm{CAV}}(\varphi_{\mathrm{BB}})L(T_{\mathrm{CAV}}) \\ &\quad + \varepsilon_{\mathrm{M}}(\theta_{\mathrm{BB}})L(T_{\mathrm{M}})]\tau_{\mathrm{OPT}} + \varepsilon_{\mathrm{OPT}}L(T_{\mathrm{OPT}})\} + \mathrm{DN_{offset}}\end{aligned} \tag{10.12}$$

where $DN_{i,j,BB}$ is the digital signal from instrument detector $i$ in band $j$ while viewing the on-board blackbody, $\rho_M(\theta_{BB})$ the reflectance of the instrument scan mirror in the direction $\theta_{BB}$, $\varepsilon_{BB}(\theta_{BB})$ the emissivity of the blackbody, $L(T_{BB})$ the radiance of the blackbody at temperature $T_{BB}$, $\varepsilon_{CAV}(\varphi_{BB})$ the emissivity of the scan cavity in the direction of the blackbody $\varphi_{BB}$, and $\varepsilon_M(\theta_{BB})$ the emissivity of the scan mirror in the direction $\theta_{BB}$. On the right-hand side of Eq. (10.12), the first term represents the blackbody radiance reflected by the scan mirror, the second represents the radiance emitted by the scan cavity and reflected by the scan mirror, the third represents the radiance emitted by the mirror, and the fourth term represents the radiance emitted by the instrument optics.

For instrument views of deep space, the detected infrared radiance can be represented by

$$\begin{aligned} DN_{i,j,SV} &= G\{[\rho_M(\theta_{SV})\varepsilon_{SV}(\theta_{SV})L(T_{SV}) + \rho_M(\theta_{SV})\varepsilon_{CAV}(\varphi_{SV})L(T_{CAV}) \\ &\quad + \varepsilon_M(\theta_{SV})L(T_M)]\tau_{OPT} + \varepsilon_{OPT}L(T_{OPT})\} + DN_{offset} \end{aligned} \tag{10.13}$$

where $DN_{i,j,SV}$ is the digital signal from detector $i$ in band $j$ when looking at deep space, $\rho_M(\theta_{SV})$ the reflectance of the instrument scan mirror in the direction $\theta_{SV}$, $\varepsilon_{SV}(\theta_{SV})$ the emissivity of the blackbody, $L(T_{SV})$ the radiance of deep space at its temperature $T_{SV}$, $\varepsilon_{CAV}(\varphi_{SV})$ the emissivity of the scan cavity in the direction $\varphi_{SV}$; and $\varepsilon_M(\theta_{SV})$ the emissivity of the scan mirror in the direction $\theta_{SV}$.

Assuming that the infrared radiance from deep space is essentially zero, the infrared radiance from the scene can be calculated by subtracting Eq. (10.13) from Eq. (10.11) and grouping similar terms:

$$\begin{aligned} DN_{i,j,ES} - DN_{i,j,SV} &= G\tau\{\rho_M(\theta_i)L(T_{ES}) - [\rho_M(\theta_{SV})\varepsilon_{CAV}(\varphi_{SV}) - \rho_M(\theta_{ES})\varepsilon_{CAV}(\varphi_{ES})]L(T_{CAV}) \\ &\quad - [\varepsilon_M(\theta_{SV}) - \varepsilon_M(\theta_i)]L(T_M)\} \end{aligned} \tag{10.14}$$

Likewise, the infrared radiance from the on-board blackbody can be calculated by subtracting Eq. (10.13) from Eq. (10.12) and grouping similar terms:

$$\begin{aligned} DN_{i,j,BB} - DN_{i,j,SV} &= G\tau\{\rho_M(\theta_{BB})\varepsilon(\theta_{BB})L(T_{BB}) - [\rho_M(\theta_{SV})\varepsilon_{CAV}(\varphi_{SV}) \\ &\quad - \rho_M(\theta_{BB})\varepsilon_{CAV}(\varphi_{BB})]L(T_{CAV}) - [\varepsilon_M(\theta_{SV}) - \varepsilon_M(\theta_{BB})]L(T_M)\} \end{aligned} \tag{10.15}$$

The second and third terms on the right-hand side of Eqs. (10.14) and (10.15) are essentially zero. This leads to

$$DN_{i,j,ES} - DN_{i,j,SV} = G\tau\rho_M(\theta_i)L(T_{ES}) \tag{10.16}$$

$$DN_{i,j,BB} - DN_{i,j,SV} = G\tau\rho_M(\theta_{BB})\varepsilon_{BB}(\theta_{BB})L(T_{BB}) \tag{10.17}$$

The changing on-orbit thermal environment of the instrument and drift in the detector responsivity requires frequent views of the on-board blackbody for instrumental calibration. The ratio of the measured infrared Earth scene radiance and the measured blackbody radiance comprises the basic measurement approach. From this ratio, the infrared Earth scene radiance is calculated according to

$$L(T_{\mathrm{ES}}) = \frac{\rho_{\mathrm{M}}(\theta_{\mathrm{BB}})\varepsilon_{\mathrm{BB}}(\theta_{\mathrm{BB}})L(T_{\mathrm{BB}})}{\rho_{\mathrm{M}}(\theta_{\mathrm{i}})[\mathrm{DN}_{i,j,\mathrm{BB}} - \mathrm{DN}_{i,j,\mathrm{SV}}]}[\mathrm{DN}_{i,j,\mathrm{ES}} - \mathrm{DN}_{i,j,\mathrm{SV}}] \tag{10.18}$$

### 10.1.6 The Importance of Multiple Measurement Methodologies in Earth Remote Sensing

Confidence in the correct interpretation of Earth science remote-sensing data in the study of Earth science and environmental processes requires confidence in the quality of the on-orbit data used to study those processes. Confidence in the accuracy of on-orbit remote-sensing data is increased if those data are validated through the use of multiple, independent measurement methodologies and approaches. Multiple measurement methodologies include intercomparisons of measurements from instruments on the same or different satellites and from ground-based, balloon-based, and airborne validation campaigns. Validation can therefore be defined as the process of assessing, by independent means, the quality of data products derived from satellite instrument measurements.

This chapter provides examples of current, established approaches in the pre-launch and on-orbit radiometric calibration, spectral characterization and calibration, and validation of fundamental metrological measurements made by Earth remote-sensing instruments. Included are brief descriptions of innovative and state-of-the-art approaches in the field including references, where appropriate. The examples are confined to optical measurements made in the air ultraviolet through thermal infrared wavelength regions from 190 nm to 100 μm. The chapter concludes by identifying a number of challenging areas in the calibration and characterization of remote-sensing instruments, several of which are currently the objects of significant and promising research in remote-sensing metrology.

## 10.2 The Role of Pre-Launch Calibration and Characterization

### 10.2.1 The Air Ultraviolet and Solar Reflective Range (190–2500 nm)

#### *10.2.1.1 Introduction*

Remote-sensing of the Earth at wavelengths between 190 and 2500 nm involves detecting that portion of incident solar radiation, which is either backscattered or reflected in the direction of the satellite instrument. The wavelength region between 190 and 400 nm, a region in this chapter referred to as the air ultraviolet, is important in studies of upper atmospheric chemistry and solar physics. The region between 190 and 300 nm is strongly dominated by atmospheric ozone absorption. At visible wavelengths between 400 and 700 nm, near-infrared wavelengths between 700 and 1000 nm, and shortwave infrared wavelengths between 1000 and 2500 nm, the remotely sensed processes of transmission, absorption, and reflection of incident solar radiation can be used to qualitatively and quantitatively identify a wide range of water, land, vegetative, and atmospheric conditions and properties. It is the reflectance spectrum, or the fractional amount of reflected incident solar radiation as a function of wavelength, of geophysical features that provides the basis of characterizing and identifying these conditions and properties on the Earth.

Ultraviolet measurements from satellites, the Space Shuttle, and the ground have been used to measure stratospheric profiles and total columns of ozone [8]. The importance of ozone monitoring, its biological impacts and anthropogenic causes began in 1978 with the launch of the Total Ozone Mapping Spectrometer (TOMS)/Solar Backscatter UltraViolet (SBUV) experiment on Nimbus-7 [9]. The Nimbus-7 TOMS instrument viewed the Earth at six ultraviolet wavelengths between 312.3 and 379.9 nm at nadir and measured total column ozone. The Nimbus-7 SBUV instrument viewed the Earth in 12 UV wavelengths between 255 and 340 nm at nadir and measured total column ozone and ozone profiles. The success of these instruments led to the 1989 National Plan for Stratospheric Monitoring, which mandated continuing ozone measurements on the National Oceanic and Atmospheric Administration (NOAA) Polar Orbiting Environmental Satellites (POES) using SBUV and SBUV/2 instruments [9, 10]. International recognition of the importance of ozone measurements led to flights of a number of overlapping instruments. This included the TOMS instruments on the Earth Probes (EP), Meteor, and the Advanced Earth Observing Satellite (ADEOS) platforms, SBUV/2 instruments on NOAA-9, -11, -16, and -17, the Global Ozone Monitoring Experiment (GOME) and Scanning Imaging Absorption Spectrometer for Atmospheric Chartography (SCIAMACHY) instruments on the European Remote-Sensing Satellite-2 (ERS-2) and the Environmental Satellite (ENVISAT) [11–13], respectively, the Ozone Monitoring Instrument (OMI) on Earth Observing System (EOS) Aura [14], the SAGE instruments on the Earth Radiation Budget Satellite (ERBS) and Meteor [15], and the upcoming flight of the Ozone Mapping and Profiling Suite (OMPS) instrument on the National Polar Orbiting Environmental Satellite System (NPOESS) and NPOESS Preliminary Project (NPP) satellites [16]. Continuous monitoring of ozone concentrations by these instruments enabled the detection of depletion in global ozone of 2–3% [17, 18] and 5–8% in upper stratospheric ozone [19]. Moreover, data from these satellite instruments complemented by both ground-based and airborne atmospheric chemistry measurements indicated that these depletions are anthropogenic in origin [20] due primarily to the release and slow diffusion of man-made chlorofluorocarbons into the stratosphere. Data from these satellites and instruments input to atmospheric chemistry models predicted that a phase-out of chlorofluorocarbon emissions would restore ozone concentrations to pre-1980 levels by 2030 [21]. Early evidence of this turnaround and recovery is of extreme interest to a number of atmospheric scientists [22, 23].

Visible, near-infrared, and shortwave-infrared measurements from satellite platforms have been used to identify, examine, and monitor a number of features on and above the Earth. Ocean color, aerosols, clouds, vegetation, water vapor, snow, and ice are examples of Earth geophysical properties that have been studied through the detection of transmitted, absorbed, or reflected solar radiation by instruments on satellite platforms. The ability of Earth remote sensing in the visible through shortwave infrared to examine these and other geophysical properties and processes has resulted in the on-orbit deployment of a large number of satellite instruments. The need to monitor these processes for purposes of climate research or environmental study has produced a correspondingly large number of temporally overlapping heritage sensors. The large number of satellite remote-sensing instruments operating in this wavelength region preclude embarking on an extensive review as part of this chapter; however, examples of two heritage missions and their geophysical relevance are provided. Marine phytoplankton use carbon dioxide that has settled into the ocean for photosynthesis, making the oceans Earth's primary storage sinks for that greenhouse gas. Marine phytoplankton also respond rapidly and often dramatically to environmental change. Remote-sensing of the chlorophyll pigment of marine phytoplankton provides a sensitive measure of that change. The Coastal Zone Color Scanner (CZCS) on Nimbus-7, operating from 1978 to 1986, produced the first, high-quality global distribution map of ocean chlorophyll. The measurement of chlorophyll by CZCS was continued by the Japanese Ocean Color Temperature Scanner (OCTS) on ADEOS from 1996 to 1997, the Sea Viewing Wide Field of View Sensor (SeaWiFS) on OrbView2 from 1997 to present, the Moderate Resolution Imaging Spectroradiometer (MODIS) on Terra and Aqua from 1999 and 2002 to present, the Medium Resolution Imaging Spectrometer (MERIS) on ENVISAT from 2002 to present, and the Global Imager (GLI) on ADEOS-II from 2002 to 2003. A second, excellent example of a heritage system of remote-sensing satellite instruments operating in the reflective solar wavelength region is

Landsat. Since 1972, Landsat has provided the remote-sensing community with a continuous stream of data and images of the Earth's land surface and coastal regions. Landsat data have been used to monitor a variety of environmental processes including, but by no means limited to, deforestation, population growth, and a variety of natural disasters. The first instrument, originally called the Earth Resources Technology Satellite (ERTS) and later renamed Landsat-1, included a 4 band MultiSpectral Scanner (MSS) and a return beam vidicon (RBV). The MSS was included on Landsats-2 and -3, launched in 1975 and 1978, respectively. Landsats-4 and -5, launched in 1982 and 1984, respectively, were equipped with an improved MSS called the Thematic Mapper (TM) [24]. The TM added three new bands and had improved spatial resolution over the MSS design. Landsat-5 is currently still in operation. Landsats-6 and -7 were equipped with TM instruments containing 7 channels and a high-resolution panchromatic sensor. Landsat-6 was launched in 1993 and never achieved orbit; Landsat-7 was launched in 1999 and is currently in operation.

#### *10.2.1.2 The radiometric calibration of UV earth remote-sensing instruments*

A variety of laboratory or facility-based sources can be used in the radiance and irradiance calibration of ultraviolet remote-sensing instrumentation between 190 and 400 nm. The spectral power distribution of synchrotron emission from relativistic electrons in a storage ring can be expressed in terms of the electron current, the orbital radius, and the magnetic flux density. The spectral radiance of a blackbody, where thermal emission is in equilibrium with the solid-cavity material, is known in terms of its thermodynamic temperature and the measurement wavelength. However, temperatures are limited to about 3000 K using graphite sources with inert gas purge at ambient pressure, and solid blackbody sources are practical for the spectral region longer than 250 nm. An exception is emission from optically thick resonance lines of trace elements in high-temperature plasma sources (~12,000 K), which are blackbody line sources for calibration at discrete wavelengths in the ultraviolet [25]. A third class of ultraviolet source is the simple plasma for which the necessary atomic or molecular data are known. For example, a high-temperature (~20,000 K), wall-stabilized, hydrogen arc is a source of known spectral radiance in the air ultraviolet from 190 to 360 nm, given the continuum emission coefficient and the plasma length [25, 26]. It is also possible to generate sources with known relative spectral distributions. An example of this type of source is realized through the electron impact excitation of atoms and molecules combined with knowledge of cross-sections, branching ratios, and transition probabilities [27].

The most common sources used in the calibration of ultraviolet remote-sensing instruments are transfer standard sources. Transfer standard sources are portable and easier to use than the calculable sources described above; once calibrated they can be used to disseminate spectral radiance, irradiance, or intensity values. Emission lines from metal-rare gas hollow cathode sources are used as radiant intensity standards, and operation in the vacuum ultraviolet is possible [28, 29]. The argon mini-arc is a secondary standard that has a line-free continuum between 194 and 330 nm [30]. Originally developed with windows as a spectral radiance standard, later modification included windowless operation, useful for the air ultraviolet region, and for use as an irradiance standard [31, 32].

Deuterium arc lamps, which utilize a low-pressure plasma, are common transfer standards in the ultraviolet [33]. NIST issues modified commercial 30 W $D_2$ lamps that are calibrated for spectral irradiance from 200 to 400 nm [34]. It is also possible to assign spectral radiance values to $D_2$ lamps, and they have been used as transfer standards in international intercomparisons of spectral radiance [35].

From 250 through 400 nm, NIST issues a 1000 W tungsten quartz halogen (TQH) lamp, type FEL [36] as a standard of spectral irradiance. The double-coiled tungsten filament, enclosed by a quartz envelope and surrounded by a halogen-doped inert gas, results in a stable source with a relative spectral distribution equivalent to a blackbody at about 3000 K. The FEL lamp is the most common source used in radiance and irradiance calibrations of remote-sensing instruments operating from 250 to 2500 nm; further discussion of this source can be found later in this chapter. For spectral radiance standards, gas-filled tungsten strip lamps are possible [36].

Ultraviolet satellite instruments used to determine total column ozone and ozone profiles measure the Earth's albedo, or the ratio of the Earth's backscattered radiance to incoming solar irradiance, at a number of wavelengths between 230 and 425 nm. From 230 to 340 nm, the absorption cross-section of ozone increases rapidly with increasing wavelength. The accuracy of ozone measurements made by these satellite instruments depends on the accuracy of their pre-launch albedo calibrations, linearity characterizations, and wavelength calibrations. Consistency in the pre-launch calibration and characterization of ozone measuring instruments on the same and different spacecraft is critical in continuing the decadal ozone data record.

In the pre-launch time frame, ultraviolet albedo measuring satellite instruments are calibrated for radiance and irradiance in the laboratory. The radiance calibration determines the instrument response in its on-orbit radiance measurement mode, that is, while looking at the Earth. The irradiance calibration determines the instrument response in its on-orbit irradiance measurement mode, that is, while measuring incident solar irradiance reflected off or transmitted through its on-board solar diffuser. In the radiance calibration, the satellite instrument views a laboratory diffuse target with a known spectral BRDF illuminated by an irradiance standard FEL lamp. Confidence in this calibration is established by employing a series of standard lamps and diffuse targets and comparing the results [37]. More recently, an integrating sphere has been employed in this calibration [38, 39]. The integrating sphere was a 50.8 cm diameter aluminum shell with a 20.3 cm diameter output aperture. The interior of the sphere was coated with barium sulfate paint and was internally illuminated by eight 200 W tungsten halogen bulbs. In this approach, the satellite instrument was used to transfer the irradiance calibration from an irradiance standard FEL lamp to the aperture of the integrating sphere. The instrument viewed a diffuse target initially illuminated by a spectral irradiance standard FEL lamp and then by the integrating sphere. The spectral radiance of the sphere was calculated using the technique described by Walker et al. [40] employing the radii of the integrating sphere and instrument apertures and their separation distance. The integrating sphere, calibrated for spectral radiance, was then viewed directly by the satellite instrument. This technique effectively eliminated the uncertainty in the knowledge of the BRDF of the diffuse target. The agreement between the integrating sphere-based and established diffuser-based techniques was on the order of 1%.

In the pre-launch irradiance calibration, an irradiance standard FEL lamp is used to illuminate the satellite instrument's on-board diffuser in an optical geometry similar to that of the Sun on-orbit. A goniometric calibration of the instrument and its on-board diffuser is also performed to determine the instrument's angular response. While the ratio measurement methodology of the albedo measurement causes many variables to cancel, uncertainty in the BRDF of the on-board diffuser remains. Therefore, determination of the BRDF, and any changes in the BRDF, of the on-board diffuser is critical in the on-orbit measurement of incident solar irradiance and the calculation of atmospheric ozone concentration.

#### *10.2.1.3 Spectral characterization of UV earth remote-sensing instruments*

Many of the ultraviolet albedo measuring satellite instruments are dispersive instruments employing either fixed or scanning gratings or prisms to spectrally resolve the backscattered and incident solar ultraviolet light. For these instruments, spectral calibration involves establishing their absolute wavelength scales and determining their spectral responsivities. Key to this process is the determination of an instrument's slit function, or the monochromatic image of its entrance and exit slits. The slit function determines the spectral resolution of the instrument. For a scanning instrument, the instrument grating or prism is scanned while viewing a monochromatic source calibrated for wavelength. For a fixed instrument, incident monochromatic light, calibrated for wavelength, is scanned over the instrument's operating channels. Most commonly, this monochromatic light is produced using a broadband light source and a monochromator, with the spectral bandpass of this light being smaller than the bandpass of the satellite instrument channel being measured. For this reason, this monochromator-based technique often suffers from low signal to noise. By monitoring the output of the monochromatic calibration source during the calibration using a second reference detector calibrated for spectral responsivity, the spectral responsivity of the remote-sensing instrument can be determined.

Several innovative approaches in the spectral calibration of ultraviolet remote-sensing instruments have appeared in the literature. System level responsivity measurements, particularly with monochromators equipped with broadband sources, often require a long time to acquire sufficient in-band and out-of-band data. It is also difficult for system level spectral characterizations to fill the instrument's entrance pupil. In the spectral calibration of OMI on EOS Aura, a xenon lamp illuminated echelle grating was used to produce multiple monochromatic lines at high grating orders [41]. The multiple spectral lines produced by this system enabled the complete operational wavelength range of OMI to be scanned using a step size of 0.01 nm.Using this source, the slit function of OMI was determined through the instrument's nadir, Sun, and calibration ports. Tunable ultraviolet lasers are high photon flux, monochromatic sources capable of providing a high signal-to-noise measurement of spectral responsivity. Slit functions for the six spectral bands of the QuikTOMS instrument were measured using tunable ultraviolet light obtained by frequency doubling the output of a ring dye laser pumped by a $Nd:VO_4$ laser [42]. In addition, this technique successfully established the wavelength centers for the six QuikTOMS bands to better than 0.1 nm.

#### *10.2.1.4 The radiometric calibration of visible/near infrared/shortwave infrared earth remote-sensing instruments*

In the visible through shortwave infrared, remote-sensing instruments are calibrated for irradiance and radiance in the pre-launch timeframe using a variety of filament lamps deployed in stand-alone configurations, with reflectance targets, or inside integrating spheres. Similar to work in the ultraviolet, the 1000 W quartz tungsten halogen FEL lamp is the most common irradiance standard source. Historically, the spectral irradiance values of working standard FEL lamps at NIST were assigned using a chain of comparisons based on the spectral radiance of a blackbody at the freezing temperature of gold [36]. In 2000, the procedure was revised; now the spectral irradiance values of the working standard lamps are assigned by direct comparison to the spectral irradiance produced from a high-temperature blackbody. The blackbody temperature is determined using a set of absolute filter radiometers [43]. The great advantage to the new method is a reduction in the uncertainties in spectral irradiance by a factor of between 2 and 10.

For decades, diffuse targets illuminated by irradiance standard lamps and internally illuminated integrating spheres have been used as radiance standards for the calibration of satellite instruments from the visible through the shortwave infrared [44, 45]. The use of lamps and diffuse targets in the calibration of remote-sensing instruments was thoroughly reviewed as part of the Fourth SeaWiFS Intercalibration Round-robin Experiment (SIRREX-4) in May 1995 [46]. Briefly, accurate radiance calibration using a lamp-illuminated diffuse target requires measurement of the BRDF of the target over the range of incident and view angles, spatial locations, and wavelengths corresponding to the on-orbit operational instrument configuration. Following BRDF characterization, the target must be stored, handled, and deployed in such a manner as not to contaminate or change its reflectance. At the typical distances that irradiance lamps are used with diffuser targets (i.e. 50–100 cm), the distribution of the lamp irradiance across the panel is not uniform. This non-uniformity coupled with the angular or goniometric response of the instrument under calibration must be characterized and understood. Lastly, the use of a 1000 W FEL lamp in a laboratory requires meticulous attention to baffling and stray light control.

With integrating spheres, internal diffuse sphere coatings, commonly barium sulfate paint or polytetrafluoroethylene (PTFE), scatter the light from filament lamps that provide the source of their illumination. The spheres are designed to provide a source of uniform flux for instruments with large entrance apertures. Care is taken to provide spatial and angular uniformity across the exit aperture of the calibration spheres. Over the past decades, incremental improvements have been made to these radiance sources [47]; but the basic design of the spheres has remained the same. In general, the spheres are calibrated with transfer radiometers that view, in an alternating fashion, the output aperture of the sphere and a source of known spectral radiance. The calibration references in the visible through shortwave infrared for the spheres use standard irradiance FEL lamps. As discussed previously, the irradiance from these lamps can be scattered from diffuse reflecting plaques to provide the radiance reference for the transfer radiometer [48]. Alternatively, it is possible for the reference lamp to be viewed directly by the transfer radiometer [40]. Over the past decade, the development of stable, portable transfer radiometers has allowed independent verifications of the spectral radiance outputs from these calibration spheres. In a recent experiment [49], the measured radiances by these radiometers confirmed the calibration of such a sphere at the 3% level in the visible and near infrared and at the 4% level in the shortwave infrared.

A principal shortcoming of calibration spheres and diffuse targets illuminated using filament lamps is the mismatch of the distribution of their spectral radiance from that of Earth scenes viewed by the satellite instruments that they calibrate. The lamps produce a maximum flux output in the near infrared, around 1000 nm, and relatively little output in the blue and green portions of the spectrum, near 400 and 500 nm. For Earth observing satellites, blue and green Earth scenes are particularly important. A new type of calibration sphere using a spectrally tunable light source has been developed to resolve the problem with spectral mismatch [50]. The sphere uses solid-state light emitting diodes (LEDs) for illumination. Individual diodes produce light with different colors throughout the visible spectrum. Combinations of these diodes allow the production of radiance spectra corresponding to the range of Earth scenes viewed from the orbit. In addition, the diodes of each color are sufficiently bright, that when used in combination, they are able to match the light levels from the brightest Earth scenes. An externally controlled, multiple channel power supply provides the current to the LEDs in the set, which numbers 144 diodes in the current application. A fiber-coupled spectroradiometer monitors the sphere output, controlling the power supply and the spectrum from the sphere by determining the differences of its measured values from an input target spectrum.

#### *10.2.1.5 The spectral characterization of visible/near infrared/shortwave infrared earth remote-sensing instruments*

Measurement methodologies for the determination of the spectral responsivity of remote-sensing instruments operating in the visible through shortwave infrared are largely identical to those previously described for the ultraviolet, differing only in the sources employed. Briefly, the choice of measurement methodology depends on whether the instrument employs a fixed or scanning dispersive device, such as a grating or prism, or a series of fixed wavelength filters. For fixed grating or prism and filter instruments, a monochromator equipped with a quartz tungsten halogen lamp with an output bandwidth narrower than the remote-sensing instrument to be calibrated is scanned over each instrument channel. By monitoring the output of this system with a calibrated reference detector, the spectral responsivity of the remote-sensing instrument is determined. The signal to noise obtained with this approach in the visible through shortwave infrared is typically higher than that obtained in the ultraviolet due to higher source photon flux and optical throughput at those wavelengths. The absolute wavelength scale and spectral responsivity of a scanning instrument can be determined using a monochromator equipped with a tungsten lamp, gas discharge lamps containing mercury and rare gases such as xenon, argon, krypton, and neon.

An alternative to using monochromator-based sources or atomic discharge lamps for the measurement of spectral responsivity uses tunable lasers [51]. At NIST, the facility providing this measurement service is referred to by the acronym, SIRCUS, for Spectral Irradiance and Radiance Responsivity Calibrations with Uniform Sources. The laser sources in the SIRCUS facility provide a high signal-to-noise determination of the wavelength scale and absolute spectral responsivity of an instrument. In SIRCUS, the laser-based light source is monochromatic, and the wavelength of the source is tunable from 200 to 1800 nm. The response of an instrument under test is calibrated against a reference detector with known spectral responsivity. These reference detectors are calibrated on a wavelength-by-wavelength basis over the full range of their response, including wavelength regions where their response is small. As the requirements for Earth-observing satellite instruments continue to develop, the isolation of the measurements of individual instrument channels to narrow ranges of wavelengths becomes increasingly important. The laser-based calibration method provides the basis for techniques to separate a channel's response in a desired operating wavelength region from its response to wavelengths outside of that region (i.e. the channel's spectral out-of-band responsivity). A bright, spectrally tunable light source covering a broad wavelength range, as discussed above, provides reference spectra that can test those techniques.

### 10.2.2 The Thermal Emissive Range (2500 nm to 100 μm)

#### *10.2.2.1 Introduction*

Observations of the Earth in the thermal infrared spectral region are important for quantitative analysis of the Sun–Earth system, because the temperatures of terrestrial systems result in thermal emission for wavelengths greater than about 2.5 μm. In addition, the total fraction of solar radiation, which can be approximated by a blackbody distribution at 5870 K, is less than about 3.4% for wavelengths longer than 2.5 μm; reflected solar flux is therefore not a major source of contamination.

The list of existing and historical sensors or programs with infrared channels is extensive. In the United States, NOAA specifies and operates sensors for near-real time storm warnings and other evaluations primarily using satellites in geostationary orbits (e.g. the Geostationary Operational Environmental Satellite (GOES) series), and for longer-term weather forecasting and climate studies primarily using satellites in low Earth orbits (e.g. the POES series). Currently, key instruments with Thermal Infrared (TIR) channels are the Advanced Very High Resolution Radiometer (AVHRR) and the High-Resolution Infrared Radiation Sounder (HIRS), which are in the POES series, and the GOES Imager and Sounder, which replaced the GOES Visible Infrared Spin Scan Radiometer (VISSR) in 1994. Operational observations from polar orbits for civilian (e.g. POES) and military (e.g. the Defense Meteorological Satellite Program (DMSP)) programs are combined for future missions as the NPOESS program. The first of satellite platform in the NPOESS series, termed the NPOESS Preparatory Project or NPP, has four sensors, two of which have thermal infrared channels: the Visible/Infrared Imager Radiometer Suite (VIIRS) and the Cross-track Infrared Sounder (CrIS). The next major series in the GOES program, termed GOES-R, is scheduled for enhanced capabilities such as increased spectral, temporal, and spatial resolution in the thermal infrared. In NASA's EOS program, the MODIS, which is on the Terra and Aqua platforms, has heritage with AVHRR and HIRS. The Clouds and the Earth's Radiant Energy System (CERES) instrument, on the Tropical Rainfall Measuring Mission (TRMM), Terra, and Aqua platforms, and the Atmospheric Infrared Sounder (AIRS), on Aqua, are other examples. The TM and Enhanced Thematic Mapper Plus (ETM+) instruments on the Landsat series of satellites have one channel in the thermal infrared. A similar listing of sensors could be assembled for missions of the European Space Agency (ESA) and other countries.

A major application of absolute spectral radiance measurements in the thermal infrared is the determination of temperature. The simplest example of deriving the temperature of an object from absolute radiometric measurements is the observation of a blackbody with no atmospheric absorption—the spectral radiance and total exitance are given by Planck's law or the Stefan–Boltzmann law, respectively. Measurements in the thermal infrared of Earth scenes must be designed to account for, or provide corrections for, physical effects such as atmospheric absorption and thermal emission, diurnal variation in the solar illumination, the thermal radiative properties of the target object, and sources of non-thermal emission.

Regions of low atmospheric absorption, or "windows", are used to determine surface temperatures. The primary windows for this in the thermal infrared are from 3.8 to 4.0 μm and from 10.2 to 12.5 μm. Correction for atmospheric absorption by water vapor is required; typically, measurements of the water vapor feature at 6.3 μm are used for this purpose. The emittance of the surface must be known, since it is not an ideal blackbody radiator. For sea surface temperature (SST) measurements, which are typically performed at 11 and 12 μm, emittance values for seawater are known, but there is a significant "skin effect" that must be accounted for accurate determinations of the bulk temperature. The temperatures of land, snow, or ice can be determined in similar ways if the target is classified properly so its emittance can be assigned.

Values for the temperature of the atmosphere as a function of altitude are determined by radiance measurements at multiple discrete wavelengths that overlap a broad absorption line of a well-mixed atmospheric gas such as $CO_2$. For these spectrally resolved measurements of atmospheric absorption features from space, the optical depth at the measurement wavelength is a strong function of wavelength—near the center of the absorption feature, the optical depth is small with the observed thermal emission originating high in the atmosphere; measurements at wavelengths closer to the edge of the absorption feature sample the temperature of the atmosphere at lower altitudes. This technique is termed atmospheric sounding [52]. Two features in $CO_2$ are typically used for temperature sounding, one near 4.47 μm and the other near 14.95 μm. These studies began with filter radiometry. The finite number of measurement wavelengths that can be placed in filter radiometers such as HIRS limits the vertical resolution of the temperature retrievals. High-resolution grating or interferometric systems do not suffer this limitation. AIRS is a grating spectrometer with coverage from 3.74 to 15.4 μm [53]. Use of Fourier transform spectrometers for sounding of the Earth's atmosphere began in 1969 with the Infrared Interferometer Spectrometer (IRIS) on Nimbus [54]. Currently, the limb-scanner named as the Michelson Interferometer for Passive Atmospheric Sounding (MIPAS) is on the ESA's ENVISAT [55] and the Tropospheric Emission Spectrometer (TES) is on EOS Aura. In the near future, the Infrared Atmospheric Sounding Interferometer (IASI) will fly on ESA's Meteorological Operational satellite (MetOP) and the CrIS on NPP/NPOESS. The increased spectral resolution using CrIS is expected to improve the vertical resolution in the temperature profiles by a factor of 2 or 3 [56].

Another broad application of thermal infrared measurements involves cloud studies, see Menzel [57] for example. On-orbit observations at 10.7 μm give the brightness temperature of the top of the cloud. Since the cloud-top temperature is close to the atmospheric temperature, and the temperature profile is known, the height of the cloud can be determined. Cloud detection algorithms incorporate visible imagery with the thermal infrared imagery. In a typical image, pixels that are bright in the visible channels and dim (i.e. cold) in the thermal infrared channels are probably clouds, since, compared to the surface below, clouds reflect more incident solar flux but are cooler because of the cloud top's position in the atmosphere.

Many other products are derived using thermal infrared measurements, such as the vertical distribution of water vapor, total ozone levels, thermal inertial studies, and the Earth's radiation budget. As shown in Figure 10.2, the emitted thermal radiation and the incoming and reflected solar flux are the three components of the Earth's radiation budget. Values for and variations in these parameters are critical for accurate understanding of global climate change. The CERES instrument is designed to measure these parameters [58].

#### *10.2.2.2 The radiometric calibration of thermal infrared remote-sensing instruments*

The most common calculable sources for the thermal emissive spectral range are based on blackbody physics. Several blackbody sources are described in Chapter 5. For ideal blackbodies, Planck's law specifies the spectral radiance from an ideal blackbody in terms of its thermodynamic temperature and the measurement wavelength. In "fixed-point" blackbodies, the temperature is set by surrounding the blackbody cavity by a pure, molten metal that is undergoing a phase transition. With the temperature at a constant value during the transition, the spectral radiance is calculable. Values for the relevant temperatures of the reference materials are given in The International Temperature Scale of 1990 (ITS-90) [59]. Reference points in the temperature range of interest are the triple points of Hg, $H_2O$, and the melting point of Ga.

Fixed-point blackbodies are expensive, tedious to operate, and exhibit substantial temperature gaps in their coverage. The diameter of the cavity exit aperture is generally small compared to the entrance apertures of Earth-observing sensors. Instead, variable temperature blackbody standards are constructed with various materials and instrumented with calibrated contact thermometers, such as thermistors or platinum resistance thermometers (PRTs). Thus the traceability of spectral radiance values is established using temperature standards that are calibrated according to ITS-90.

If a cavity is black with unity emittance and its temperature is uniform, the uncertainty of the spectral radiance is determined only by the uncertainty of the temperature and associated fundamental constants. Real blackbodies are only an approximation of this ideal, and therefore they must be characterized to determine the emittance, spatial and angular uniformity, stability, and so forth. For large-area blackbodies, it is difficult to achieve unity emittance. If the background environment and the flight instrument are not cooled to low temperatures ($\sim$80 K), the reflection of the background radiance from the blackbody must be considered.

In general, a blackbody source is designed to function either as a laboratory reference standard or as a component of the flight sensor. Pre-flight measurements by the sensor of the laboratory reference blackbody source are typically performed in a thermal-vacuum chamber over a range of blackbody temperatures. During flight, measurements by the sensor of the internal "on-board" blackbody source are used to account for changes in the sensor responsivity. In addition, measurements of the internal blackbody by the sensor during the pre-flight calibration establish traceability to the reference blackbody, provided the uncertainties in each step, which may include corrections for bias and other effects that are difficult to quantify (e.g. the stability of the on-board source, the effects of launch, the variation in reflectance of the scan mirror with angle of incidence, or stray sources of illumination), are described in complete detail. Many sensors are also designed to have a view of space during part of the measurement procedure so that the sensor output to a cold source (i.e. approximately zero radiance) can be determined.

The laboratory reference blackbody is designed to achieve an emittance as close to unity as possible and to provide full-aperture illumination for the sensor over its dynamic range by operation at different temperatures. Isothermal cavities with small exit apertures compared to their overall length are preferred, since this geometry allows for multiple reflections, resulting in higher emittance compared to flat-plate sources [60]. Cavity-type blackbody sources were used, for example, to calibrate CERES [58] and MODIS [61]. However, long structures increase the size and cost of the thermal-vacuum chamber, and a common compromise is to use a flat-plate structure with grooves in combination with a temperature-controlled, cylindrical baffle. This design concept was used to calibrate, for example, the Improved Stratospheric and Mesospheric Sounder (ISAMS) [62] and the Advanced Spaceborne Thermal Emission and Reflection Radiometer (ASTER) that is on Terra [63]. The on-board blackbody sources are generally of the flat-plate design, in order to comply with the sensor's size and mass restrictions.

The result of detailed characterization and modeling are spectral radiance values, traceable to temperature standards, with the uncertainty components arising from all possible sources: the measurement of the bulk temperature, the emittance of the cavity based on its geometric design and optical properties of the coating, the effect of temperature gradients, the effect of incident irradiance, and so forth. The temperature sensors sometimes can be calibrated once they are mounted to the blackbody by use of an isothermal bath [64]. The cavity emittance can be calculated for a range of parameters, including non-isothermal conditions, using Monte Carlo methods [65]. The specular and diffuse reflectance of the coating or assembled cavity can be measured using established techniques in IR spectrophotometry; however, the low reflectance values may result in uncertainties that are larger than desired. A complete, system level thermal and radiative analysis of the blackbody source is often performed to identify and quantify sources of bias. Finally, the spectral radiance of the blackbody source can be determined within the exit aperture as a function of position, view angle, and operating temperature. Comparison of the output as a function of size of the exit aperture is also informative.

Advances in detectors have made it possible to verify the output of blackbody sources by direct measurement of the spectral radiance or exitance. These measurements are critical for validating the characterization and calibration of the blackbody source. The NIST Low Background Infrared Calibration Facility (LBIR) uses an absolute cryogenic radiometer to determine the radiance temperature of a user's blackbody from exitance measurements [66, 67]. In some cases the radiometrically determined temperatures were in poor agreement with the blackbody contact thermometry, resulting in improvements to the blackbody design.

It is not always possible to send a source to an external facility for calibration, and in the thermal infrared, differences in thermal environment between the host chamber and the NIST facility may result in misleading results. The portable Thermal Infrared Transfer Radiometer (TXR) was developed by NIST with support from NASA/EOS and the Department of Defense (DOD) so that the spectral radiance of a blackbody source could be verified at two wavelengths—5 μm using a photovoltaic indium antimonide (InSb) detector and 10 μm using a photovoltaic mercury cadmium telluride (MCT or HgCdTe) detector [68]. The detectors, filters, and reflective optics are built into a liquid-nitrogen cryostat with a ZnSe window. A variable-temperature, vacuum compatible blackbody that can rotate in front of the TXR window under computer control is used to monitor the stability of the TXR during deployments. A convenient feature of the TXR is that it can be operated either in a vacuum chamber or on a laboratory bench under typical laboratory conditions (i.e. an "ambient" environment).

Calibration of the TXR has been performed using multiple approaches. Initially, the NIST water-bath blackbody [69] was used. More recently, the TXR was calibrated in the NIST Medium Background Infrared (MBIR) facility [70] by using its cryogenic blackbody. Note both of these methods establish traceability to temperature standards. At the present time, the TXR uncertainty is $<0.1\,\mathrm{K}$ at $300\,\mathrm{K}$ ($k = 2$). In the future, the spectral characterization will be improved using tunable laser measurements on the NIST Infrared SIRCUS facility (see Chapter 4), and the MBIR cryogenic blackbody source will be validated using an electrical substitution radiometer (ESR) based on high-$T_c$ superconducting temperature sensors (see Chapter 2).

The TXR has been used on multiple occasions to verify thermal infrared spectral radiance values. In July 1999 and August 2001, it was deployed to the remote-sensing radiometric calibration facility at Los Alamos National Laboratory (LANL) in support of DOE programs [71, 72]. In July 2001, the TXR was deployed to the GOES-Imager radiometric calibration chamber at

ITT in Ft. Wayne, Indiana [73]. The ITT chamber is not equipped with cooled shrouds. The reference blackbody for the GOES Imager is the Earth Calibration Target (ECT), and a cold blackbody source is used for offset determinations. The ITT procedure involves correcting for temperature gradients in the ECT, which were believed to be driven by the thermal background of the chamber [74]. During the TXR deployment, measurements were made of the cold blackbody, and the ECT and the TXR check standard blackbody at a range of temperatures. An analysis procedure was developed that enabled parameterization of the results in terms of a non-unity emittance and a temperature gradient in the ECT. The results are in qualitative agreement with the existing GOES model, and more analysis is planned. More recently, in the Fall of 2003, the TXR was used to measure a blackbody source at Santa Barbara Remote Sensing (SBRS) in Goleta, California in support of DOD research. Then, early in 2004, the TXR measured an additional blackbody source at SBRS [75]. This source was used to calibrate MODIS [53], and it will be used to calibrate the VIIRS on NPP/NPOESS. The experiment was interesting in that a variable temperature scene plate was used and this information was exploited to determine the blackbody emissivity [75].

#### *10.2.2.3 The spectral characterization of thermal infared remote-sensing instruments*

The measurement facilities and procedures necessary for sensor characterization in the thermal infrared are similar to those in the visible/near infrared and the shortwave infrared, with a couple of major exceptions. First, at ambient ($\sim$300 K) temperatures, all objects are sources of radiant flux, so care must be taken to separate signal from background, even when chopping shutters are used. Second, radiometric artifacts or systems that play a critical role in the visible/near infrared and the shortwave infrared, such as reliable and relatively inexpensive transfer standard radiometers employing silicon (Si) photodiode trap detectors and commercially available, broadly tunable lasers, are not available.

General approaches for thermal infrared spectral characterization methods are given in texts such as in Wyatt [76]. For example, the determination of the sensor's response to quasi-monochromatic flux that fills the entrance pupil and is varied spectrally to cover the full range of possible detector response defines the spectral responsivity function. The wavelength and bandwidth of the input flux is determined by the source, which is usually a continuous source filtered by a monochromator or interferometer. A broadband, spectrally flat, calibrated detector is used to correct for the variation of output flux with wavelength. Typically, the sensor is operated in the thermal-vacuum chamber, and best results will be obtained if the relevant temperatures (e.g. the instrument housing temperature, electronic components, scan mirror, etc.) are varied over the range of values expected during flight. There are numerous examples in the literature [61, 63, 77].

The determination of an instrument's spectral responsivity function is often calculated using measurement results of the individual components, but this may not account for the effects of component temperature, inter-reflections, diffraction, scatter, and differences in the *f*/# of the incident flux. System level measurements that fill the entrance pupil, provided sufficient flux can be produced, are more accurate [78]. Accuracy is essential: in atmospheric sounding sensors, such as HIRS, small discrepancies in the wavelength scale of the channel can induce large errors in temperature retrievals (e.g. a $3\,\mathrm{cm}^{-1}$ shift in center frequency at 14.95 μm results in a 10 K error for stratospheric temperatures [79]). Because HIRS is an operational system, results from sensors on different platforms are often intercompared. However, real but unquantified differences in the sensors' spectral responsivities make this very difficult [80]. As part of this study, NIST measured filter witness samples for NOAA-N′ HIRS at four temperatures between 15 and 30°C with a geometry matched to the HIRS (*f*/8) geometry [81]. The observed differences with the results provided by the vendor are a likely explanation for some of the large observed inter-satellite radiances [80].

Recent advances in tunable lasers and transfer detectors with responsivity values traceable to absolute cryogenic radiometry have resulted in a Spectral Irradiance and Radiance responsivity Calibrations using Uniform Sources (SIRCUS) facility at NIST. Because of the high flux levels, the narrow bandpass, low uncertainty in wavelength, unpolarized nature of the flux, and the ability to chop at the source using a SIRCUS-type facility, the uncertainties in the sensor's radiometric calibration and spectral characterization can be greatly reduced. There are many examples of applications for remote sensing in the spectral region from about 380–1000 nm [82]. Measurements with tunable lasers are also possible out to about 5 μm, and several filter radiometers have been characterized [83]. Plans call for extension of the spectral range past 5 μm, and this would greatly benefit a wide range of remote-sensing instruments.

## 10.3 The Measurement of Total Solar Irradiance (TSI)

### 10.3.1 The Importance of TSI Measurements in Climate Studies

The driving energy source for the Earth system is the Sun. The interaction of the Sun's radiant flux with the Earth's land, oceans, atmosphere, and cryosphere influence a number of processes on the Earth including climate, weather, photosynthesis, temperature, and ocean dynamics. The radiant output of the Sun varies with wavelength and time [84]. One example of temporal solar variation with time is the 11-year solar cycle discovered in 1843 by Heinrich Schwabe by counting sunspot activity over a 17-year period [85]. Long-term variations in the radiant output of the Sun are potential causes of natural climate change [86], while short-term variations are important in gaining a better understanding of the fundamental physics of the Sun. Understanding the long-term and often subtle variations and trends in the Sun's radiant output, therefore, is critically important in being able to differentiate overall between natural and anthropogenic climate change processes on Earth. The importance of measurements of long-term change in the Sun's radiant output was underscored in a 1994 publication by the National Research Council on research priorities for solar influences on global change [87]. In that publication, the NRC emphasized the importance of solar measurements by recommending that highest priority be given to "monitoring total and spectral solar irradiance from an uninterrupted series of spacecraft radiometers employing in-flight sensitivity tracking".

The key parameter in the measurement and monitoring of the radiant output of the sun is TSI. In general, TSI can be defined as the power of all optical wavelengths reaching the Earth from the Sun. TSI can be more precisely defined as the radiant energy emitted by the Sun over all spectral regions falling each second on 1 $m^2$ at the mean Earth–Sun distance. Measurements of TSI have historically been made using ground-based instruments, balloon-borne instruments, aircraft instruments, shuttle-based instruments, and satellite instruments [88–98]. Comparisons of the long history of TSI measurements have led to the realization that high-accuracy TSI measurements need to be made above the Earth's atmosphere from satellite platforms. Long-duration, high-accuracy measurements of TSI from space began with the launch of the Earth Radiation Budget Experiment (ERBE) on the Nimbus-7 satellite in November 1978. High-accuracy TSI measurements by ERBE were realized through the use of electrical substitution radiometry. Table 10.3 lists the long-term TSI satellite instruments, all of which employed electrical substitution radiometry, and their on-orbit measurement data records beginning with the Nimbus-7 ERBE. Figure 10.3 presents the historical TSI database produced by several of these instruments. From the data in Figure 10.3, the instruments have shown good internal consistency or precision based on their ability to reproduce the relative magnitudes of the 11-year solar cycles.

However, the differences in the absolute values of the TSI reported by these instruments are several times the 1.3 $W/m^2$ amplitude of a typical solar cycle and in several cases are larger than the instruments' reported

TABLE 10.3. Total Solar Irradiance (TSI) Satellite Instruments and Associated Data Records: 1978 to Present

| TSI Instrument | Satellite/Platform | TSI Data Record (mm/dd/yy or mm/yy) |
|---|---|---|
| Earth Radiation Budget (ERB) Experiment | Nimbus-7 | 11/16/78 to 12/13/93 |
| Active Cavity Radiometer Irradiance Monitor I (ACRIM I) | Solar Maximum Mission (SMM) | 2/16/80 to 61/89 |
| Earth Radiation Budget Experiment (ERBE) | Earth Radiation Budget Satellite (ERBS) | 10/25/84 to present |
| Earth Radiation Budget Experiment (ERBE) | National Oceanic and Atmospheric Administration-9 (NOAA-9) | 1/23/85 to 12/20/89 |
| Earth Radiation Budget Experiment (ERBE) | National Oceanic and Atmospheric Administration-10 (NOAA-10) | 10/22/86 to 12/1/87 |
| Active Cavity Radiometer Irradiance Monitor II (ACRIM II) | Upper Atmospheric Research Satellite (UARS) | 10/5/91 to 8/01 |
| Solar Variability experiment 2 (SOVA) | European Retrievable Carrier (EURECA) | 8/7/92 to 6/93 |
| Solar Constant (SOLCON) | ATLAS and Hitchhiker | 3/24/92 to present |
| Variability of solar Irradiance and Gravity Oscillations (VIRGO) | Solar and Heliospheric Observatory (SOHO) | 12/2/95 to present |
| Active Cavity Radiometer Irradiance Monitor III (ACRIM III) | Active Cavity Radiometer Irradiance Monitor Satellite (ACRIMSAT) | 4/5/00 to present |
| Total Irradiance Monitor (TIM) | Solar Radiation and Climate Experiment (SORCE) | 2/25/03 to present |

measurement uncertainties. These differences coupled with temporal gaps in the historical database have led to ambiguities and controversial assumptions in the overall determination of the absolute TSI. For example, both Willson and Mordvinov [99] and Frohlich and Lean [100] have attempted to span a 28-month gap in the measurements of TSI by the ACRIM1 and ACRIM2 instruments between solar cycles 21 and 23 using TSI measurements from other on-orbit instruments. In order to span the data gap, Willson and Mordvinov used Nimbus7/ERB results [97] while Frohlich and Lean used ERBS results [101]. The results of Willson and Mordvinov show an upward trend of 0.05% per decade in TSI between consecutive solar minima while the results of Frohlich and Lean do not. Willson has claimed

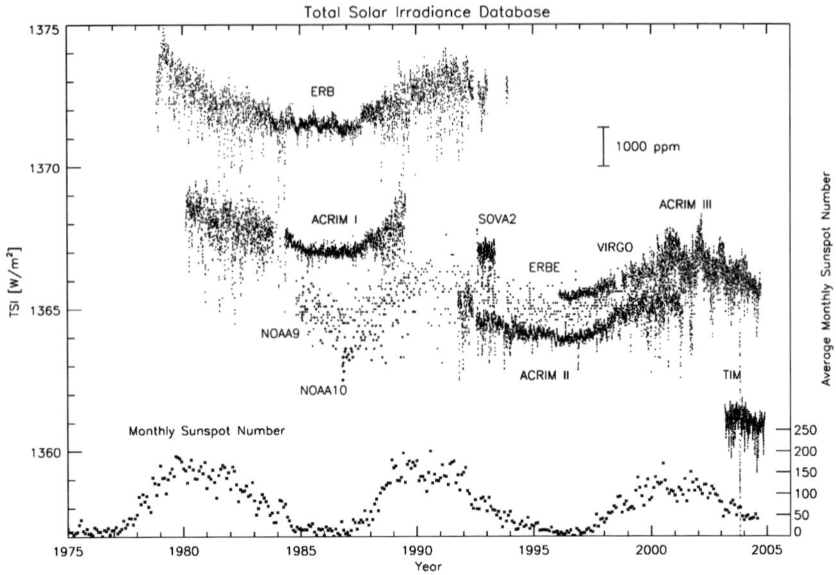


FIG. 10.3. Satellite instrument measurements of Total Solar Irradiance (TSI) since 1978. The frequency of sunspots is also plotted in the figure, showing the strong correlation between solar activity and TSI. Courtesy of the Laboratory for Atmospheric and Space Physics (LASP), University of Colorado.

the upward decade minimum-to-minimum trend to be significant and to have important implications for long-term climate studies. On a positive note, the differences in on-orbit TSI measurements have led to an increased realization by the TSI community of the importance of meticulous instrument pre-launch calibration and characterization, the accurate assessment of instrument on-orbit degradation, and the need for stronger cooperation between the scientific and national metrological laboratories in an effort to understand and reduce TSI measurement uncertainties.

## 10.3.2 The Operation of TSI Instruments

Since 1978, the majority of TSI measuring satellite instruments have been ESRs [90, 102]. These satellite instruments employ at least two identical cavity sensors. During operation, one cavity is "active" and the other serves as a reference. These cavities are highly emissive and thermally connected so that they initially experience the same thermal environment and hence the same temperature. The cavities are electrically heated to a common temperature. Upon exposure of the active cavity to the Sun, the incident solar flux is converted into heat by the highly absorptive cavity. This causes the temperature of the active cavity to rise relative to the reference, introducing an imbalance between the temperatures of the cavities. An amount of electrical heat is subtracted from the active cavity in order to induce temperature equivalence with the reference cavity. This amount of electric heat is equivalent to the amount of incident solar flux and proportional to the TSI.

From 1978 to 2002, TSI was measured by satellite-borne ESRs using a time domain analysis [91, 93, 95, 97]. In this method of operation, a shutter is opened at some time, $t$, exposing the active cavity to solar flux for a time equal to $\Delta t$, during which a signal offset in DNs, $\Delta$DN, is recorded proportional to the TSI through the standard watt. Measurements are made only after the cavity temperatures are completely stabilized following shutter openings or closings. Since February 2003, the Total Irradiance Monitor (TIM) instrument on the EOS Solar Radiation and Climate Experiment (SORCE) [96, 103, 104] has acquired TSI measurements using phase sensitive or lock-in detection. In this method of operation, all digital data are used during the shutter cycles and the irradiance is quantified using analysis in the frequency domain. The advantages of this approach include increased signal-to-noise and a simpler description of temporal and thermal system behavior.

### 10.3.3 The Calibration and Characterization of TSI Instruments

Given a typical on-orbit satellite instrument lifetime of 5 years, the approach toward monitoring long-term, decadal changes in TSI has been to employ an overlapping time series of satellite instruments. The required absolute accuracy and precision of the on-orbit measurement of TSI are 0.01 and 0.001%, respectively. From the historical TSI data shown in Figure 10.3, the agreement between instruments is of the order of 0.3%, a factor of 30 higher than the required absolute accuracy. Moreover, the 0.01% accuracy and 0.001% precision goals for the measurement of TSI are extremely ambitious when one considers that the best radiometric accuracy claimed by national measurement laboratories in the measurement of radiant flux in a laser beam using an absolute cryogenic radiometer is 0.01% [105] (see also Chapter 2). The TIM instrument on the EOS SORCE mission was designed to make TSI measurements with an absolute accuracy of 0.01% and a precision of 0.001% [86, 103]. The determination of the measurement uncertainty of the TIM instrument was based on the propagation of subsystem level uncertainties in the parameters, which comprise the TIM measurement equation and not on a system level measurement. Parameters included in the TIM analysis included uncertainties in the spacecraft/Sun distance, the thermal equivalence between solar and electrical heating of the detector cavities, on-board electrical standards, shutter waveform, instrument gain, cavity absorption and determination of its on-orbit degradation, diffraction effects, and the area of the limiting aperture.

Understanding the differences between the measurements of TSI by satellite instruments, as shown in Figure 10.3, depends on extensive pre-launch instrument characterization and an accurate assessment of on-orbit instrument degradation. A significant contributor in TSI measurement uncertainty and an important characterization parameter of TSI instruments is the accurate determination of the area of the instrument limiting aperture. Aperture area measurements for TSI instruments have been performed using both contact and non-contact methods [106]. Preliminary results have recently been published on an international comparison of aperture area measurements made by solar irradiance research groups [106]. The results of this study indicate that for the two participating laboratories, the aperture area measurements of the laboratories are consistently higher than those made by NIST, and differences between the laboratory measurements are greater than their combined measurement uncertainties. The average difference of the two laboratories from NIST are 0.013 and 0.065%. On-orbit degradation of the response of TSI instrumentation to solar flux is monitored using multiple cavity detectors exposed at different duty cycles over the mission lifetime. The rate of degradation of the emissivity of these cavities is dependent on the surface treatment of the interior of the cones. Before 2002, TSI instruments employed cavities with interiors painted with organic specular or diffuse black paints. The TIM instrument on EOS SORCE is the first TSI instrument to use a metallic nickel phosphide (NiP) black surface treatment on the absorbing interior of their cavities. Sudden changes in the emissivity/reflectance of the TIM cavity is monitored using photodiodes positioned to view the cavity interiors.

### 10.3.4 Future Developments and Directions in the Measurement of TSI

The influence of changes in TSI on climate will only be understood if there is a continued commitment to fly an overlapping long-term series of well-characterized and calibrated new and redundant instruments. The state-of-the-art requirements for the characterization and calibration of TSI instruments will require greater involvement by NMIs in the instruments' design, testing, and review processes. Pioneering work in comparing TSI instruments to SI has already begun [285, 286].

## 10.4 Spectral Solar Irradiance (SSI)

### 10.4.1 The Importance of Measurements of SSI in Earth Remote Sensing

Over the past several decades, there has been a concerted effort to understand the spectral irradiance from the Sun. Earth remote-sensing instruments do not measure this irradiance, a central component in interpreting Earth remote-sensing results. It is necessary in the conversion between radiance and reflectance data products [107], and, in some cases, in the intercomparison of results from remote-sensing instruments [108]. As a result, the solar spectral irradiance is incorporated into remote-sensing analysis procedures as an ancillary data set. In general, the quality of solar spectral irradiance measurements has improved over time. And in general, individual instruments and measurement campaigns do not provide the solar spectral irradiance over the full wavelength range required for the various types of Earth remote-sensing measurements. As a result, there are several sets of irradiance compilations in the literature. Four representative compilations are listed in Table 10.4.

Thekaekara [109] produced an early compilation, which at the time was given the informal titles of the "Standard Irradiance Table" and the "NASA Standard". The compilation was based on aircraft-based solar measurements and additional values from literature references in the ultraviolet and infrared [110]. A subsequent compilation by the World Radiation Center and the World Climate Research Program by Wehrli [111, 112] used ground-based measurements over wavelengths from 0.2 to 0.31 μm by Brasseur and Simon [113], from 0.31 to 0.33 μm by Arvesen et al. [114], from 0.33 to 0.87 μm by Neckel and Labs [115], from 0.87 to 2.5 μm by Arvesen et al. [114], and from 2.5 to 20 μm by Smith and Gottlieb [116]. As a reference spectrum for the Hubble Space Telescope, Colina et al. [117] compiled a set of measurements based on measurements by the Upper Atmosphere Research Satellite from 0.12 to 0.41 μm by Woods et al. [118], from 0.41 to 0.87 μm by Neckel and Labs [115], and from 0.87 to 0.96 μm by Arvesen et al. [114]. For wavelengths from 0.96 to 2.5 μm, Colina et al. [117] used a model spectrum citing deficiencies in the available measured solar irradiance spectra.

The Moderate Resolution Atmospheric Transmittance and Computer Model (MODTRAN) [119] includes several compiled solar irradiance spectra. With the need for calculations at a spectral resolution unobtainable by

TABLE 10.4. Four Representative Solar Irradiance Spectra

| Solar Irradiance Spectrum, Year and Reference | Wavelength Range (μm) | Spectrum Type |
|---|---|---|
| Thekaekara, 1974 [109] | 0.115–400 | Compilation (measurements) |
| Wehrli, 1985 [111] | 0.2–20 | Compilation (measurements) |
| Colina et al., 1996 [117] | 0.12–2.5 | Compilation (measurements and computations) |
| Thuillier et al., 2003 [124] | 0.2–2.4 | Measurements |

current solar-monitoring instruments, these irradiance spectra are derived, in part, by computation [120, 121]. Computation is a standard procedure for the production of very high spectral resolution solar irradiance spectra.

### 10.4.2 Recent Measurements of SSI

Recently, Thuillier has published sets of solar spectral irradiance measurements made from the Space Shuttle. The measurements were made using three spectrometers, with the ultraviolet measurements published in 1997 [122], the visible and near infrared results published in 1998 [123], and the full set of measurements published as a single solar spectrum from 0.2 to 2.4 μm published in 2003 [124]. These results have been designated as the reference solar spectrum for the Committee on Earth Observing Satellites (CEOS). However, there is no single spectrum mandated for universal use.

The Spectral Irradiance Monitor (SIM) [125, 126] was launched in January 2003 onboard the EOS SORCE satellite. The measurements from SIM should provide the next step in the determination of the solar irradiance spectrum from the ultraviolet to the shortwave infrared. However, for instruments that measure in the mid-wave infrared, 3.5 to 4 μm (e.g. MODIS [127]), the best available measurements of the solar irradiance are three decades old (see Table 10.4). This marks a fundamental deficiency in our understanding of the solar spectral irradiance.

## 10.5 Transferring Pre-Launch Calibration and Characterization to On-Orbit Operation

The possibility of change in the calibration of satellite instruments from the time of their laboratory characterization to the start of on-orbit operation is a principal source of uncertainty in their measurements. The determination of this change generally involves the measurement of an artifact within the instrument, both before and after the transfer to orbit. Several attempts have been made using onboard lamps as the artifact. The results of the Advanced Land Imager (ALI) [128] on the EO-1 platform are typical of those attempts. For ALI there was a noticeable increase in the output of the lamp in the visible and near infrared at the start of on-orbit operations. This change was attributed to the loss of convective cooling within the lamp and the zero gravity environment on-orbit. The reduced cooling increased the temperature of the lamp's filament and, thus, the lamp's output. For the MODIS instrument onboard the Terra spacecraft, an effort was made to account for the change in filament temperature of on-board lamps in the Spectroradiometric Calibration Assembly (SRCA) from pre-launch to on-orbit. This was done in a series of pre-launch experiments by tracking the radiance from the reference lamp as a function of the resistance of the lamp's filament [129]. However, the results of the on-orbit measurements with the reference lamp have not been presented. This is due to the decision to use the pre-launch characterization of the instrument's solar diffuser as the reference for the sensor's measurements in the visible, near infrared, and shortwave infrared wavelength regions [130, 131]. This decision negated the plan to use the SRCA to transfer the MODIS pre-launch radiance calibration to orbit. For Terra MODIS, there was no attempt to determine changes in the characteristics of the onboard diffuser during the transfer to orbit.

The Sea-viewing Wide Field-of-view Sensor (SeaWiFS) used measurements of the Sun outdoors at the instrument manufacturer's facility to predict the instrument outputs during solar measurements immediately after launch [132]. Because an on-board diffuser plate is required for these measurements, the experiment measured changes in the instrument-diffuser system. The largest uncertainty in the experiment comes from the determination of the atmospheric transmission in the pre-launch measurements and the overall uncertainty in the experiment is 3%. For the eight SeaWiFS bands, the initial instrument outputs averaged 0.8% higher than expected with a standard deviation of 0.9%. Within the uncertainty of the experiment, there were no changes in the responses of the SeaWiFS bands from the completion of the instrument's manufacture to its insertion into orbit.

Remote-sensing instruments operating at thermal infrared wavelengths typically use their on-board blackbody sources to transfer their pre-launch laboratory radiance calibrations to on-orbit operation. As outlined in Section 10.2.2.2, the radiance scale from a high-quality, laboratory blackbody is transferred pre-launch to the instrument's on-board blackbody. With proper handling and storage of the instrument before, during, and after integration onto the spacecraft, the instrument's on-board blackbody is assumed to maintain its scale through launch. For high-quality on-board blackbodies, such as those employing cavity-based designs, the pre-launch radiance calibration of the on-board blackbody can be carried directly into orbit.

## 10.6 The Role of Post-Launch Calibration and Characterization

### 10.6.1 The Ultraviolet and Solar Reflective Range

#### *10.6.1.1 Air ultraviolet and solar reflective on-board radiometric calibration*

The pre-launch radiance and irradiance calibrations of remote-sensing instruments are transferred to on-orbit operation and monitored over their complete mission lifetimes. Techniques for performing on-orbit calibration and monitoring of remote-sensing instruments operating in the ultraviolet and solar reflective wavelength range between 190 and 2500 nm employ dedicated hardware systems such as reflective and transmissive solar diffuse targets, lamp sources, and views of extensive, natural targets such as playas, snow fields, deserts, thick clouds, and celestial targets such as the Moon and stars.

*10.6.1.1.1 Solar diffusers.* On-board solar diffuse targets illuminated by the sun are commonly used to perform and monitor on-orbit radiance and irradiance calibrations of remote-sensing instruments. With knowledge of the SSI, the solar irradiance incident elevation and azimuthal angles, the instrument view angle on the diffuser, and the BRDF in the case of a reflective diffuser or the bidirectional transmission distribution function (BTDF) in the case of a transmissive diffuser, the radiance reflected by the diffuser into the satellite instrument is given by

$$L(\theta_\mathrm{i}, \varphi_\mathrm{i}, \theta_\mathrm{s}, \varphi_\mathrm{s}, \lambda) = \mathrm{BRDF}(\theta_\mathrm{i}, \varphi_\mathrm{i}, \theta_\mathrm{s}, \varphi_\mathrm{s}, \lambda) E_\mathrm{sun}(\lambda) \cos(\theta_i)/d^2_{\mathrm{earth-sun}} \quad (10.19)$$

where $L$ is the radiance off or through the solar illuminated diffuser at wavelength, $\lambda$, $\theta_\mathrm{i}$ the incident elevation angle of solar illumination onto the diffuser, $\varphi_\mathrm{i}$ the incident azimuthal angle of solar illumination onto the diffuser, $\theta_\mathrm{s}$ the elevation angle at which the instrument views the diffuser, $\varphi_\mathrm{s}$ the azimuthal angle at which the instrument views the diffuser, $E_\mathrm{sun}$ the solar irradiance at wavelength, $\lambda$, at 1 astronomical unit (AU), and $d_{\mathrm{earth-sun}}$ the distance from the Earth to the Sun (in AU) at the time when the instrument views the diffuser. Incident and viewing elevation and azimuthal angles can be specified relative to the diffuser normal. Carefully sized and appropriately positioned in a satellite instrument, the solar-illuminated diffuser can provide a full-system, full-aperture, on-orbit radiometric calibration using the Sun—the same source which illuminates the remotely sensed Earth. In addition to being a full-aperture illuminator, solar diffusers have other desirable properties. These include being chemically and optically stable and a Lambertian, spectrally featureless, spatially uniform reflector or transmitter.

While the basic measurement application of solar diffusers in the on-orbit calibration of remote-sensing instruments has largely remained unchanged for roughly four decades, advances in the optical, chemical, and mechanical testing and properties of materials and coatings has led to the evolution and use of a variety of types of diffusers. Table 10.5 lists a number of reflective and transmissive diffuser materials and coatings flown on the instruments listed.

The diffuser materials and coatings listed in Table 10.5 can be categorized according to whether they are bulk scatterers (i.e. permit incident photons to

TABLE 10.5. Solar Diffuser Materials and Coatings

| Diffuser | Type | Satellite Instrument | Reference |
|---|---|---|---|
| Space-grade Spectralon | Reflective | Moderate Resolution Imaging Spectroradiometer Terra and Aqua (MODIS Terra and Aqua) | [133–135] |
| | | Multi-angle Imaging SpectroRadiometer (MISR) | [136, 137] |
| | | Medium Resolution Imaging Spectrometer (MERIS) | [138] |
| | | Advanced Land Imager (ALI) | [139] |
| | | Modular Optoelectronic Scanner (MOS) | [140] |
| | | Global Imager (GLI) | [141] |
| Roughened aluminum | Reflective | Total Ozone Mapping Spectrometer (TOMS) | [142] |
| | | Global Ozone Monitoring Experiment 1 and 2 (GOME 1 and GOME 2) | [143, 144] |
| | | Solar Backscatter Ultraviolet and Solar Backscatter Ultraviolet/2 (SBUV and SBUV/2) | [9] |
| | | Ozone Monitoring Instrument (OMI) | [145, 146] |
| | | Scanning Imaging Absorption Spectrometer for Atmospheric Cartography (SCIAMACHY) | [147] |
| YB71 paint | Reflective | Landsat Enhanced Thematic Mapper Plus (ETM+) | [148, 149] |
| | | Sea viewing Wide Field of View Sensor (SeaWiFS) | [150] |
| | | Visible and Infrared Scanner (VIRS) | [151] |
| | | Hyperion | [152] |
| Quartz plate volume diffuser | Reflective | Ozone Monitoring Instrument (OMI) | [145, 146] |
| Quartz diffuser | Transmissive | Ozone Monitoring Instrument (OMI) | [147, 148] |
| Mirror Aperture Mosaic | Reflective | Clouds and the Earth's Radiant Energy System (CERES) | [153, 154] |
| Perforated Plate | Transmissive | | [155] |

penetrate the diffuse material and undergo multiple scattering events before exiting) or surface scatterers (i.e. largely reflect incident photons in a single scattering event). Of the materials listed, Spectralon, a form of pressed and sintered PTFE manufactured by Labsphere, is the best bulk scatterer and most closely approximates a Lambertian scatterer. However, significant research and development was necessary to produce Spectralon of sufficient optical stability to be used on-orbit. This was largely attributed to the undesirable property of Spectralon to readily absorb hydrocarbon contaminants. This research and development work, spearheaded by NASA's Jet Propulsion Laboratory and Labsphere Incorporated, included extensive analyses of the manufacturing, handling, and on-orbit degradation of Spectralon [156–158] and directly led to the development of space-grade Spectralon.

On-orbit use of space-grade Spectralon diffusers has been largely confined to those remote-sensing instruments operating in the visible, near infrared, and shortwave infrared wavelength regions between 400 and 2500 nm. Spectralon is not used in the on-orbit calibration of ultraviolet instruments operating between 200 and 400 nm due to its tendency to undergo significant reflectance degradation in that wavelength region. In the ultraviolet region, the diffuser material of choice historically has been roughened aluminum with aluminum or an aluminum alloy overcoat. The reflectance stability of aluminum diffusers, particularly for ultraviolet applications, is thought to arise from the formation over time of a protective, optically stable aluminum oxide overcoat on the diffuser. The OMI, launched in July 2004 on the EOS Aura platform, is equipped with two reflective aluminum diffusers, a quartz transmissive diffuser, and a reflective quartz plate volume diffuser. The two reflective aluminum diffusers are used in the on-orbit irradiance calibration of the instrument. The quartz transmissive diffuser is used to uniformly illuminate the OMI entrance slit during a measurement of an on-board white light source. The quartz plate volume diffuser is used daily in measurements of the solar irradiance spectrum. Interestingly, spectral interference effects first seen in reflective aluminum diffuser measurements during the pre-launch characterization of OMI are greatly reduced using the quartz plate volume diffuser [159].

Using clever instrument optical designs, some remote-sensing instruments are able to use their solar illuminated diffusers to realize multiple radiance levels. A simple design to do this is provided by the MODIS instruments on Terra and Aqua [127]. The MODIS instruments are able to view their solar diffusers under the conditions of full solar illumination and 8% solar illumination. The 8% illumination level is realized using an on-board, deployable transmissive screen mounted on the instrument solar diffuser door. The ALI instrument on EO-1 deploys a slotted mask in front of the diffuser to realize multiple illumination levels [139]. The mask is comprised of seven slots of varying area designed to provide signal levels off the diffuser from 0 to 90% equivalent diffuse Earth reflectance. During on-orbit solar reflective calibrations, a slide covering the mask is withdrawn, enabling the incident solar flux to pass through the exposed slots onto the solar diffuser. Illuminated in this manner, the solar diffuser effectively provides a series of uniform scenes of increasing radiance.

A key aspect of the use of solar diffusers in the on-orbit radiance calibration of remote-sensing instruments is the ability to monitor on-orbit, temporal changes in the diffusers' reflective or transmissive optical properties. In order to do this, some Earth remote-sensing instruments have flown dedicated detector-based hardware in addition to multiple diffusers. An example of a detector-based system used to monitor solar diffuser reflectance degradation is the Solar Diffuser Stability Monitor (SDSM) flown on the MODIS Terra and Aqua instruments [134, 160]. The SDSM is a 5 cm diameter integrating sphere with nine filtered silicon photodiode detectors embedded in the sphere wall [160]. The detectors are filtered at a number of visible/near infrared wavelengths. In order to track on-orbit reflectance changes in the solar diffuser, the SDSM operates as a ratioing radiometer in that it is used to view both the Sun and the illuminated solar diffuser. The SDSM aperture is equipped with a 2% transmissive screen to bring the solar diffuser view signal and the Sun view signal to similar levels.

The TOMS instrument launched in 1991 on a Russian Meteor-3 spacecraft first demonstrated the approach of using multiple diffusers to track and model on-orbit reflectance changes [142, 161]. To do this, three roughened aluminum diffusers, designated cover, working, and reference, were flown on the TOMS instruments and each were exposed to the Sun for varying amounts of time. The cover diffuser was exposed to the Sun constantly; the working diffuser was exposed for a few minutes per week; and the reference diffuser was exposed for a few minutes every 15 days. The data from TOMS viewing the three diffusers were used successfully to determine the reflectance degradation behavior of the working diffuser. This multiple diffuser approach was adopted by the Meteor-3 TOMS follow-on instruments flown on the EP and QuikTOMS missions and by the recently launched OMI instrument on EOS Aura.

*10.6.1.1.2 On-board lamp systems.* While solar diffusers represent a passive approach to the on-orbit calibration and monitoring of satellite instrument radiance responsivity, lamps represent an active and sometimes more challenging source-based approach. Transferring a lamp's absolute radiometric scale from pre-launch to on-orbit and maintaining that scale over the course of a mission is more complicated for several reasons. The thermal operational environment experienced by a lamp during on-orbit operation is vastly different than the thermal environment surrounding a lamp during its pre-launch calibration in a laboratory, resulting in a difference between the pre-launch and on-orbit brightness temperature. A change in brightness temperature alters the spectral irradiance, with the greatest change occurring in the ultraviolet. The filament temperature is affected by convective heat transfer of the internal and external gases as well as by conductive and radiative heat transfer. On-orbit, external gases are absent and there is no gravity to drive convection of the gas within the lamp envelope. Exhausting the heat from the lamp requires careful thermal design within the remote-sensing instrument, and consideration must be given to the duty cycles, so that the lamp is fully stabilized without compromising the sensor's thermal environment. A spectral disadvantage of using lamps as on-board calibrators is their inability to provide a solar-like spectrum due to low emission at blue and green wavelengths. An advantage of using lamps for on-orbit calibration is the availability of performing a calibration whenever desired or necessary, including immediately before launch and after attaining orbit.

The Advanced Spaceborne Thermal Emission and Reflection Radiometer (ASTER) currently flying on the EOS Terra satellite employs lamp-based on-board calibrators for the calibration of its visible/near infrared (VNIR) and shortwave infrared (SWIR) instruments [162]. Briefly, ASTER uses two doubly redundant sets of silicon photodiode-monitored, halogen lamps in its on-orbit calibration. In the pre-launch timeframe, the radiance scale from a copper point blackbody was transferred to the ASTER VNIR on-board lamps using ultrastable transfer radiometers and an integrating sphere source. For the SWIR instrument, the radiance scale from a zinc point blackbody was transferred to the ASTER SWIR lamps in similar fashion. Since its launch in December 1999, the lamp-based on-board calibration systems of ASTER have been used to trace the VNIR and SWIR instrument on-orbit radiance response history to pre-launch. Moreover, the systems have detected a decrease in the responsivity of the VNIR bands while the SWIR bands have remained stable.

Many of the on-orbit radiometric applications of lamps have been restricted to their use in stability monitoring capacities. For example, the ALI instrument on EO-1 uses an on-board lamp-based source to monitor on-orbit radiometric stability [163]. The source comprises three Welch Allyn gas-filled lamps mounted in an integrating sphere. The output of the illuminated sphere fills the ALI focal plane. By sequentially powering down the lamps, the source provides three radiance level inputs to the ALI. The ALI lamp source has shown good on-orbit stability [163]. Also on the EO-1 spacecraft, the Hyperion instrument's Internal Calibration Source (ICS) uses four Welch Allyn quartz tungsten halogen lamps to illuminate the YB 71-painted back of the instrument telescope cover [152]. The telescope cover then functions as a diffuse source.

Remote-sensing instruments in the Landsat, ERBE, and SCAnner for RAdiation Budget (SCARAB) projects have long histories of using on-board lamp sources in monitoring changes in instrument radiance responsivity at visible, near infrared, and shortwave infrared wavelengths. The Landsat 5 TM and Landsat 7 ETM+ instruments used miniature quartz tungsten halogen lamps, the outputs from which were optically coupled to the instruments' focal planes using fiber optics and sapphire light rods [164, 165]. The non-scanning active cavity radiometers (ACRs) flown on the ERBE missions between 1984 and 1986 were equipped with a ShortWave Internal Calibration Source (SWICS) comprised of a silicon photodiode-monitored tungsten lamp coupled to fiber optics [166]. The output of the SWICS was used to monitor the irradiance responsivity of the ERBE Shortwave Medium Field-of-View (SMFOV) and Shortwave Wide Field-of-View (SWFOV) instruments. The ERBE follow-on instrument, the CERES instruments flown on the EOS Terra and Aqua satellites successfully continued this application of lamps as on-board responsivity monitors [167]. Similar to the ERBE and CERES instruments, the Scanner for Radiation Budget (SCaRaB) radiometers flown on the Russian Meteor 3M and RESURS platforms in the mid to late 1990s were designed to monitor the Earth's radiation budget at the top of the atmosphere. These instruments were equipped with multiple on-board lamps operated over an overlapping, wide range of duty cycles in an effort to monitor lamp degradation and to maintain a reference with the pre-launch calibration [168].

*10.6.1.1.3 Use of natural targets.* A third approach in monitoring the on-orbit radiometric calibration and degradation of remote-sensing instruments at visible through near-infrared wavelengths is the use of repeated views of natural targets. Ideally, the targets chosen for this application should provide a range of reflectances corresponding to that range of reflectances or radiances detected by the satellite instrument. In addition, the targets should be spectrally flat and spatially uniform. The targets must be of sufficient size to eliminate or reduce any sensor size-of-source effects. Finally, the atmosphere above the targets must be stable and capable of being modeled. Natural targets have been used for both absolute and relative calibration of satellite sensors. Ocean, cloud, and desert scenes have been used to absolutely calibrate the VISSR cameras on the Geostationary Operating Environmental Satellite 5 (GOES 5) and 6 (GOES 6) [169], the AVHRR on the NOAA-7, -9, -11, and -14 satellites [170–174] and the Systeme Pour l'Observation de la Terre (SPOT) Haute Resolution Visible (HRV) radiometers [175]. For ocean views, the technique is based on the dominance of molecular scattering over the ocean relative to the radiance originating from aerosol scattering, ocean reflected skylight, underwater reflectance, and ozone absorption. High bright clouds over ocean scenes are used as spectrally invariant on-orbit diffuse targets.

Non-extended natural sources have also been used in the on-orbit monitoring of satellite instrument stability and degradation in the ultraviolet through shortwave infrared. For example, the Moon has been successfully used by the SeaWiFS, MODIS, and ALI projects to monitor long-term degradation in instrument radiance responsivity in the visible through shortwave infrared [163, 176]. Similarly, stars have been used to monitor on-orbit degradation of the GOES 8 and 9 imagers and sounders in the visible [177] and the Solar Stellar Irradiance Calibration Experiment (SOLSTICE) in the ultraviolet [178–180].

#### *10.6.1.2 Solar reflective on-board spectral characterization and calibration*

Changes in the spectral calibration of satellite instrument wavelength bands are typically monitored on orbit using dedicated on-board hardware systems and wavelength-specific atmospheric and solar absorption features. The ESA's MERIS instrument performed on-orbit determinations of the central wavelengths of its bands using a solar illuminated Erbium-doped "pink" diffuser, by observing the atmospheric $O_2A$ absorption feature at 760 nm, and by observing number of solar Fraunhofer lines [181]. The MODIS Terra and Aqua instruments are equipped with a SRCA [182]. The SRCA is essentially an "instrument within an instrument" capable of performing not only spectral on-orbit calibrations, but also radiometric and spatial calibrations. The SRCA comprises a light source, monochromator, and collimator. The light source is an integrating sphere containing 10 and 30 W embedded lamps. The monochromator is a single grating Czerny-Turner instrument capable of producing monochromatic light from 400 to 2200 nm. The collimator is an on-axis inverted telescope. The spectral output of the monochromatic light from the SRCA is monitored using a didymium filtered, calibrated silicon photodiode detector. The unfiltered monochromatic output of the SRCA, which is viewed by the MODIS instrument, is monitored using a second reference silicon photodiode detector embedded in the collimator secondary mirror. The ratio of the signals from the calibration and reference photodiodes produce the known didymium absorption profile, providing an on-orbit confirmation of the SRCA output spectrum. Another approach to the on-orbit spectral calibration of remote-sensing instruments is provided by the ESA's GOME 1 and 2 instruments. These instruments have employed on-board, neon-filled hollow platinum cathode lamps for purposes of spectral calibration [143, 144].

### 10.6.2 The Thermal Emissive Range

#### *10.6.2.1 Thermal infrared on-board radiometric calibration*

In remote-sensing, the thermal infrared wavelength region extends from 2.5 to 100 μm. From 2.5 to 3.5 μm, the Earth's atmosphere is opaque, prohibiting remote-sensing measurements. Above 3.5 μm, the amount of solar radiation reflected by the Earth and its atmosphere is exceeded by the amount of radiation thermally emitted. It is above 3.5 μm that atmospheric sounding and sea and land surface temperature are manifested as small infrared signatures superimposed on what is thermally designed to be a slowly varying, well-characterized and calibrated instrument infrared background. The ability to distinguish these subtle infrared geophysical signals from background requires regular, repeated calibration of infrared channels using stable, well-characterized on-board sources.

In filter-based, prism- or grating-based, and interferometer-based infrared remote-sensing, on-board blackbody sources in combination with views of cold space are commonly used to account for changes in instrument radiance responsivity or gain and linearity, and offset. The on-board blackbody is a full-aperture, system-level calibrator that provides a stable reference temperature against which scene temperatures can be determined. The use of a blackbody requires, at a minimum, a two-point temperature calibration. In its simplest application, a blackbody can be operated either at the instrument surrounding temperature or at the temperature of the scene to be remotely sensed. The second temperature is commonly obtained by viewing cold space. Operating the blackbody at the temperature of its surrounds eliminates thermal gradients across the blackbody. Operating the blackbody at or near the temperature of the scene to be remotely sensed, effectively reduces concerns about response non-linearity. Blackbodies are also operated at two temperatures corresponding to the extreme temperatures of remotely sensed scenes. These blackbodies often are capable of being ramped between these two temperatures providing an on-orbit check of instrument linearity. For all instruments employing blackbodies, the linearity and gain of the satellite instrument's thermal infrared bands must be carefully determined in pre-launch testing. Moreover, since infrared measurements are obtained during day and night orbital segments, this non-linearity must be measured over the complete range of on-orbit instrument operating temperatures through carefully designed thermal/vacuum tests.

The uncertainty of the radiance from the on-board blackbody directly determines the uncertainty of the radiance measured by the satellite instrument. Therefore, the emissivity, temperature uniformity, and temperature measurement accuracy of the blackbody must be established pre-launch and known on-orbit. As shown in Table 10.6, a number of blackbody designs have been used in an effort to stay within often tight instrument mass and volume constraints while at the same time providing an isothermal, high-emissivity target. The thermal mass and position of the blackbody within the instrument should minimize thermal excursions from orbit to orbit. The background or thermal offset signal of the satellite instrument due to

TABLE 10.6. On-Board Blackbody Designs for Several Remote Sensing Instruments

| Blackbody Type | Remote Sensing Instruments | References |
|---|---|---|
| Cavity | Across Track Scanning Radiometer (ATSR) | [183] |
| | Advanced Track Scanning Radiometer-2 (ATSR-2) | |
| | Advanced Across Track Scanning Radiometer (AATSR) | |
| | Measurement of Pollution in the Troposphere (MOPITT) | [184] |
| | High Resolution Dynamics Limb Sounder (HIRDLS) | |
| | Tropospheric Emission Spectrometer (TES) | |
| | Atmospheric Infrared Sounder (AIRS) | [185] |
| Grooved | Moderate Resolution Imaging Spectroradiometer (MODIS Terra and Aqua) | [186–188] |
| | Clouds and the Earth's Radiant Energy System (CERES) | [189] |
| Flat-plate honeycomb | Advanced Very High Resolution Radiometer (AVHRR) | [190–192] |
| | Advanced Spaceborne Thermal Emission and Reflection Radiometer (ASTER) | [193–195] |
| | Geostationary Operational Environmental Satellites (GOES) | [196, 197] |

thermal emission from optics and instrument and spacecraft structures is measured on-orbit using space views and monitored using strategically placed temperature sensors throughout the instrument.

#### *10.6.2.2 Thermal infrared on-board spectral calibration*

On-orbit spectral calibration of infrared satellite instruments is performed using on-board hardware and established, known spectra of gaseous atmospheric molecules. For example, the grating-based AIRS instrument employs an on-board parylene-coated mirror. This mirror is used in retro-reflection enabling the instrument to view itself as a cold target modified by the known absorption spectrum of parylene. In this configuration, parylene provides AIRS with broad features suitable for trending instrument spectral stability [198]. The AIRS detector spectral response functions were determined through pre-flight testing to be insensitive to the predicted on-orbit temperature environment. Given this result, AIRS developed a technique to monitor the detector wavelength centroids by correlating measured upwelling atmospheric radiance spectra with modeled spectra [199]. The high spectral resolution of satellite interferometer-based infrared instruments enables on-orbit wavelength calibrations using measurements of upwelling atmospheric radiance and models of well-known sharp spectral features from molecular species such as $CO_2$, CO, $H_2O$, $CH_4$, $O_3$, and $N_2O$ [200–203]. The interferometer spectral scale, spectral sampling interval, and instrument line shape (ILS) are determined pre-launch and monitored on-orbit. In a manner similar to that used by high-resolution grating-based instruments, spectral scale is established through a comparison of measured and modeled atmospheric spectra. The stability of the spectral sampling interval is monitored on-orbit using a wavelength-stabilized laser source, often referred to as a metrology laser or source. The metrology laser on the IASI is a particularly novel design in which the output of a diode laser is locked onto an absorption line of $^{13}C_2H_2$ [204].

### 10.6.3 Vicarious calibration techniques

Vicarious calibration is a key component in the validation of the on-orbit calibration of Earth remote-sensing satellite instruments and in the assessment of the quality of their fundamental radiance and reflectance measurements. Vicarious calibration employs both ground-based and airborne sensors making simultaneous radiometric measurements of optically and spatially well-characterized Earth targets at the times of satellite instrument overpasses. Vicarious calibration is used to monitor on-orbit instrument calibration over full mission lifetimes and can be used to cross-calibrate different instruments on different or the same spacecraft. Because vicarious calibration uses the same illumination source as the satellite instrument, that is, the Sun, calibration source color temperature concerns are largely eliminated. Lastly, vicarious calibration is used to complement and validate results obtained from on-board calibration systems, effectively providing a valuable multiple measurement methodology to establish confidence in those on-board systems. In contrast, for instruments such as the AVHRR, the Polarization and Directionality of the Earth's Reflectances (POLDER) instrument, and the Systeme Probatoire Pour l'Observation de la Terre (SPOT) instruments, the lack of on-board calibration systems dictates a total reliance on vicarious calibration for their on-orbit calibration. The main drawback concerning vicarious calibration remains the limited frequency at which ground-based and airborne campaigns can be undertaken and the efficiency at which vicarious calibration results are made available to satellite instrument teams. The former issue is being addressed through research in the establishment of un-manned, instrumented vicarious calibration target sites [205]. The latter is being addressed through projects such as the development of Quality Assurance and Stability Reference (QUASAR) test sites [206].

Vicarious calibration sites have a number of desirable characteristics. The optical properties of a vicarious calibration site should include spectral and spatial homogeneity and time invariance. Time invariance requires the site to have a reasonably robust surface with minimum human intervention. While minimal human intervention often translates to a remote location for the site, the site must not be so remote that site accessibility by personnel and their instrumentation is impossible. In thermal infrared vicarious calibration, water sites are particularly attractive because they exhibit small temperature excursions over the timeframes of typical vicarious measurements. The ideal vicarious site should be located at a high altitude with clear skies overhead for a significant portion of the year. In order to eliminate unwanted adjacency effects, the site should be at least $3 \times 3$ pixels larger than the ground instantaneous field-of-view (GIFOV) of the satellite to be calibrated [207]. The site should provide a range of reflectances or radiances covering the operating range of the satellite instrument. This dynamic range requirement is often met only by employing a number of sites of different types. Reflectance from the site should be reasonably Lambertian. Examples of sites used in the vicarious calibration of satellite sensors in the visible through thermal infrared are provided in Table 10.7.

#### *10.6.3.1 Visible/near infrared/shortwave infrared techniques*

The vicarious calibration of satellite sensors with bands between 400 and 2500 nm is accomplished using ground-based or airborne measurements of the reflectance or radiance from vicarious calibration sites at the times of satellite instrument overpasses. The University of Arizona Optical Sciences Center's Remote-Sensing Group has developed and refined three vicarious calibration techniques: the reflectance-based, radiance-based, and irradiance-based (or improved reflectance-based) techniques [207]. In the reflectance-based technique, the nadir reflectance of the vicarious calibration site is measured using ground-based or airborne radiometers over an area corresponding to GIFOV of the overflying satellite instrument to be calibrated. During the course of the ground-based measurements, measurements are also made of the reflectance of a field-deployed panel whose reflectance was carefully measured in the laboratory. The site reflectances are ratioed to the known panel reflectances. For satellite instruments with non-nadir fields-of-view, such as the Multiangle Imaging SpectroRadiometer (MISR) instrument on the EOS Terra platform, the bidirectional reflectance factor (BRF) or the BRDF of the site must be measured using ground-based or airborne (e.g. AirMISR) instruments. Since vicarious calibration involves the comparison of at-satellite radiances, careful characterization of the atmosphere above the calibration site is crucial and requires a suite of additional

TABLE 10.7. Sites Used in the Vicarious Calibration of Earth Remote Sensing Instruments

| Site | Site Type | Vicarious Calibration Wavelength Region (s) | Reference |
|---|---|---|---|
| Lunar Lake, Nevada | Land | Vis/NIR/SWIR | [208] |
| Railroad Valley, Nevada | Land, water | Vis/NIR/SWIR/TIR | [209–212] |
| White Sands, New Mexico | Land | Vis/NIR/SWIR/TIR | [213, 214] |
| Edwards AFB, California | Land | Vis/NIR/SWIR/TIR | [215] |
| Ivanpah Playa, California | Land | Vis/NIR/SWIR/TIR | [216, 217] |
| Maricopa Agricultural Center, Arizona | Land | Vis/NIR/SWIR | [218] |
| Lake Tahoe, California | Water | Vis/NIR/SWIR/TIR | [219, 220] |
| La Crau, France | Land | Vis/NIR/SWIR | [221, 222] |
| Marine Optical Buoy Network, Hawaii | Water | Vis/NIR | [223] |
| Newell County Rangeland, Canada | Land | Vis/NIR/SWIR | [224] |
| Barreal Blanco, Argentina | Land | Vis/NIR/SWIR | [217] |
| Salar de Arizaro, Argentina | Land | Vis/NIR/SWIR | [225] |
| Pima County Fairgrounds, Arizona | Land | Vis/NIR/SWIR | [217] |
| Lake Frome, Australia | Land | Vis/NIR/SWIR | [226] |
| Strzelecki Desert, Australia | Land | Vis/NIR/SWIR | [227] |
| North Slope ARM Site, Alaska | Land | Vis/NIR/SWIR/TIR | [228] |
| Southern Great Plains ARM Site, Oklahoma | Land | Vis/NIR/SWIR/TIR | [229] |
| Tropical Western Pacific ARM Site | Water/land | Vis/NIR/SWIR/TIR | [230] |
| Salar de Uyuni, Bolivia | Water/land | Vis/NIR/SWIR | [231] |
| Uardry, Australia | Land | Vis/NIR/SWIR/TIR | [232] |
| Amburla, Australia | Land | Vis/NIR/SWIR/TIR | [232, 233] |
| Thangoo, Australia | Water/land | Vis/NIR/SWIR/TIR | [234] |
| Niobrara, Nebraska | Land | Vis/NIR/SWIR/TIR | [235] |
| Jornada, New Mexico | Land | Vis/NIR/SWIR/TIR | [236] |
| Salton Sea, California | Water/land | Vis/NIR/SWIR/TIR | [237] |
| Townsville-Kelso Reef, Australia | Water | TIR | [238] |
| Perth-Rottnest Island, Australia | Water | TIR | [238] |
| Dunghuang, China | Land | Vis/NIR/SWIR/TIR | [239] |

ground-based instruments in addition to the nadir-measuring radiometer. These instruments include the following: an all-sky camera taking real-time photos of the sky, a pyranometer acquiring measurements of TSI, a line-of-site radiometer measuring sky radiance in the direction of the satellite instrument, and a solar radiometer measuring variation of solar irradiance, the data from which are used to determine the input parameters to an atmospheric radiative transfer model. The output of the radiative transfer model is a top-of-the-atmosphere radiance, which, when convolved with the system spectral responsivity of a particular satellite instrument band, can be compared to the satellite measured radiance.

In the radiance-based technique, a well-calibrated, fully characterized radiometer is flown in the airspace below the satellite instrument to be calibrated and above the vicarious calibration site. In this measurement configuration, the airborne radiometer measures the upwelling radiance from the test site above a significant fraction of the aerosols and water vapor in the atmosphere. Simultaneously, measurements can be made on the ground of the site reflectance, and the effect of the intervening atmosphere on the airborne measurement can be derived. The aircraft measurements are corrected for the fraction of the atmosphere above the aircraft and are compared to the satellite radiance measurements. In addition to a reduced atmospheric effect, the radiance-based technique can cover a significantly larger area than the ground-based reflectance technique.

In the irradiance-based (or improved reflectance-based) technique, the diffuse-to-global irradiance is measured in addition to the test site reflectance, panel reflectance [240], and atmospheric characterization measurements made in the reflectance-based technique. The diffuse-to-global irradiance is determined by measuring the reflectance panel with direct solar irradiance blocked and un-blocked and is extrapolated to those satellite and solar zenith angles at the time of vicarious calibration. In this technique, the calculated at-satellite radiance has a reduced dependence on the accuracy of the size distribution of atmospheric aerosols.

While the reflectance-, radiance-, and irradiance-based vicarious calibration techniques have been developed and refined over bright land targets, the Marine Optical BuoY (MOBY) system has been developed for the vicarious calibration of ocean color satellite instruments [223]. For ocean color measurements from space, the water-leaving radiance is calculated from the top-of-the-atmosphere radiance, that is, from the radiance leaving the ocean–atmosphere system [241, 242]. Since the ocean is dark and ~90% of the top-of-the-atmosphere radiance comes from the atmosphere, it is necessary to have low uncertainty in the at-satellite radiances. MOBY provides water-leaving radiances, which, when combined with the atmospheric correction, are used to provide a calibration for the on-orbit radiances from the satellite instrument [243]. In this use, the calibration coefficients for the satellite instrument are adjusted so that the radiances from the satellite instrument–atmosphere system agree with the water-leaving radiances from MOBY. In general, this type of vicarious calibration is not applied in the derivation of land and atmosphere products from the ocean color satellite instrument [244].

MOBY contains two spectrographs that measure flux from 340 to 950 nm [223]. The blue spectrograph separates light into 512 channels from 340 to 640 nm, with the red spectrographs covering the spectrum from 650 to 950 nm in 512 channels. The upwelling radiance is measured from three arms extending from the buoy's central column, each at its own depth from the surface. The radiances from the arms are transmitted to the spectrographs via fiber-optic cables. These upwelling radiance measurements are used to calculate the water-leaving radiance [223, 243]. The characterization and calibration of MOBY has been carried out in a long-term collaboration with the NIST [82, 245].

Since the mid 1990s, the United States Geologic Survey (USGS) in Flagstaff Arizona has developed a photometric model of the Moon for use as a common, vicarious calibration target for satellite instruments that make remote-sensing measurements of the Earth [246–248]. This lunar reflectance model has been produced from ground-based measurements. The model accounts for the effects of solar and lunar distance, and of lunar phase and libration in the ground-based observations. Using this model of reflectance, phase, and libration, the irradiance for a given spacecraft's lunar observation can be calculated. Kieffer has shown that the lunar surface is photometrically stable over a period of $10^6$ years [249]. However, the actual comparisons of the model with the instruments include the time-dependent, cyclical changes in the lunar irradiance as viewed by the satellite instruments. Comparisons with the USGS lunar model have shown the Moon to be an exceptionally stable reference for instruments making long-term climate change measurements [250]. This has been demonstrated for measurements by individual instruments covering limited ranges of lunar phase angles [176]. The applicability of the Moon as a cross-calibration reference for instruments that measure at different phase angles has yet to be confirmed. In addition, calibration measurements to determine the absolute reflectance of the Moon [251] have yet to be completed.

#### *10.6.3.2 Thermal infrared techniques*

Ideal sites for the vicarious calibration of satellite sensors operating in the thermal infrared should possess known emissivity and large spatial and long temporal temperature uniformity. For these reasons, water sites are particularly well suited as thermal infrared vicarious calibration sites. Ground sites such as playas, deserts, and snow fields are also used as thermal infrared calibration targets, which span the low- to high-temperature calibrations of satellite instruments. The vicarious calibration of the thermal instrument in the Advanced Spaceborne Thermal Emission and Reflection Radiometer (ASTER) is a joint effort between research groups in Japan and the

United States that has advanced both radiance-based and temperature-based thermal infrared vicarious calibration techniques [252]. In the radiance-based technique, a well-calibrated radiometer is flown above the test site. The radiometer measures the thermal infrared radiance upwelling from the test site. With careful co-registration of the satellite and airborne instrument fields-of-view and proper accounting of any spatial out-of-field radiance contributions for each instrument, this radiance is corrected for residual scattering between the airborne radiometer and the satellite instrument and compared to the satellite measurements. In the temperature-based approach, a ground- or water-based sensor measures the thermal infrared radiance of the site. This measured radiance is corrected for downwelling atmospheric and along path radiance contributions using a radiative transfer model with either measured (e.g. temperature and humidity radiosonde measurements) or climatological atmospheric profiles. The radiance is also corrected for the non-unity emissivity of the surface. The surface temperature is determined by inverting the Planck equation using this corrected measured radiance. These temperatures are propagated to at-satellite instrument radiance using the aforementioned radiative transfer model. The predicted satellite instrument radiance is calculated by convolving this at-satellite instrument radiance with the system level spectral responsivities for those satellite instrument bands to be vicariously calibrated. The radiance measured by the satellite instrument is compared to this predicted radiance.

#### *10.6.3.3 Comparison of radiometers and sources used in vicarious calibration*

Determinations of the measurement uncertainties, stabilities, and repeatabilities of the radiometers and sources used in vicarious calibration are essential in ascertaining whether vicarious calibration results are of sufficient quality to be used either in the validation or determination of satellite instrument calibration coefficients. A key, first step in this process includes validating, in the laboratory, measurements made by vicarious calibration radiometers viewing well-calibrated, stable uniform radiance sources and determining the stability, repeatability, and uniformity of those calibration sources used before, during, and after radiometer field deployments. In addition, the wavelength accuracy, spectral and spatial stray light rejection, and spectral responsivity of the vicarious calibration radiometers should be characterized. NASA's EOS and NIST have coordinated a number of these laboratory-based comparisons [253–256]. In 1997, an intercomparison of field instruments operating at visible, near infrared, and shortwave infrared wavelengths was performed in the laboratory of the University of Arizona Optical Sciences Center's Remote-Sensing Group in advance of a field deployment to Lunar Lake and Railroad Valley, Nevada [254]. This intercomparison focused on the stability and short-term repeatability of the field instruments. The intercomparison revealed large variations in the response of certain field instruments, due to insufficient warm-up and basic instrument design deficiencies such as the lack of detector temperature stabilization. In 1998 and 2001, an intercomparison of field radiometers and blackbody calibration sources operating in the thermal infrared was held at the University of Miami's Rosenstiel School of Marine and Atmospheric Science. These ship-borne infrared radiometers measure the sea surface skin temperature, which is used to validate satellite retrieved SSTs. In the 1998 comparison, commercial and custom infrared radiometers measured the NIST calibrated, water-based blackbody. In the 2001 comparison, the NIST Thermal Transfer Radiometer (TXR) measured four blackbodies used to calibrate ship-borne radiometers. Following the laboratory measurements, the radiometers were mounted aboard the RV Walton Smith and made SST measurements during a two-day roundtrip between Miami and the Bahamas. These laboratory and ship-based experiments produced SST measurement discrepancies between the radiometers of less than 0.1 K at ambient temperatures and confirmed the ability of the radiometers to validate satellite SST measurements [257].

## 10.7 Cross-Calibration of Earth Remote-Sensing Instruments

Cross-calibration of instruments on the same or different satellite platforms is important in establishing confidence in the quality of satellite remote-sensing data. Cross-calibration can be used to validate that the satellite instruments are making remote-sensing measurements calibrated to the same physical scale or to establish the relative offsets between satellite instruments calibrated to different scales. Cross-calibration can be used to normalize remote-sensing measurements from different instruments, producing continuous remote-sensing data sets necessary for the study of short- and long-term climate change. Repeated cross-calibration between satellite instruments provides a means of monitoring their individual on-orbit operations over time, validating the information provided by their on-board calibration subsystems.

The cross-calibration of successive generations of the same instrument or different instruments on the same platform or different platforms ideally requires that a spatially and spectrally uniform, large, cross-calibration target scene be viewed under stable atmospheric and identical illumination conditions by each instrument at the same time. However, in the real world, such optically ideal targets do not exist. Instruments have different spectral responsivities and ground instantaneous fields-of-view (GIFOVs).

In addition, instruments, particularly on different platforms, are often unable to view the same target at precisely the same time. For these reasons, the spectral reflectance or radiance of the target and the system level spectral responsivities of the instrument bands must be accurately known [258]. Instrument scenes must be carefully co-registered. For non-simultaneous views, changing atmospheric and illumination conditions, including polarization effects, must be considered.

Cross-calibration studies of satellite instruments on the same platform and on different platforms employ a variety of instrumented and non-instrumented sites as common targets. Cross-calibration of the visible/near infrared/shortwave infrared channels of the MODIS, the Multi-angle Imaging Spectroradiometer (MISR), and the Advanced Spaceborne Thermal Emission and Reflection Radiometer on the EOS Terra platform has been performed by Thome et al. using coincident ground measurements at Railroad Valley, Nevada [212]. Frequent near-nadir coincident views of snow, ice, and open ocean near the north and south poles by polar orbiting satellites have also been used in instrument cross-calibrations [259]. Examples of instruments cross-calibrated using these scenes include the 11 and 12 μm bands on MODIS and the AVHRRs on NOAA-16 [260] and the visible/near infrared channels of GLI, MODIS, SeaWiFS, MERIS, and the AVHRRs on NOAA-16 and 17 [261]. The stable, consistent albedos provided by deep convective clouds have also been used to cross-calibrate those same AVHRR instruments and MODIS on Terra and Aqua [262]. Desert scenes have been used in cross-calibration studies of visible, near infrared, and shortwave infrared channels on MODIS and the AVHRR on NOAA-9 [263]; SeaWiFS, VGT, AVHRR and POLDER [264]; and MODIS and ATSR-2 [265].

Minnis et al. [266, 267] performed an extensive cross-calibration study using the visible and shortwave infrared on-board calibration systems of VIRS and CERES, respectively, in an effort to monitor the on-orbit calibration of a number of research and operational instruments in geostationary and low Earth orbits. Both VIRS and CERES are currently flying on the TRMM. The ability of the VIRS solar diffuser to accurately monitor the on-orbit, solar diffuser-based calibration of its 0.65 and 1.6 μm channels was validated through comparisons with stable, broadband shortwave calibrated radiances from CERES using ocean and desert scenes. Using common Earth scenes, a series of comparisons were then made among the 0.65 μm channel of VIRS, the 0.62 μm channel of the GOES-8 imager, the 0.645 μm channel of MODIS on the Terra satellite, the broad channel 1 of the Geostationary Meteorological Satellite (GMS) VISSR, and the 0.67 μm channel of the ATSR-2 on ERS-2. This technique was successfully extended by Minnis et al. [268] in a comparison of thermal infrared radiances measured by these instruments in the vicinity of the VIRS 3.7, 10.8, and 12.0 μm channels.

Formation flying of spacecraft provides a large number of excellent opportunities for the on-orbit cross-calibration of remote-sensing instruments. One example of this is the cross-calibration of the ETM+ on Landsat-7 and the TM on Landsat-5 in 1999 [269]. Following its launch on April 15, 1999, Landsat-7 was placed into an orbit 6 km below Landsat-5. This enabled the ETM+ and TM instruments to view hundreds of near-coincident scenes for purposes of cross-calibrating 3 visible, 1 near infrared, and 2 shortwave infrared channels. On April 14, 2003, the close, on-orbit flying formation of the EO-1 and Terra spacecraft was used to cross-calibrate a number of visible, near infrared, and shortwave infrared instruments using the Moon as a common target. The EO-1 and Terra spacecraft orbit the Earth at 705 km with an orbital inclination of 98.2 and 98.1°, respectively. More importantly, the equatorial crossing times of EO-1 and Terra are 10:01 and 10:30 Local Solar Time (LST). On April 14, 2003, these spacecraft performed maneuvers to enable the ALI and Hyperion instruments on EO-1 and the MODIS, MISR and ASTER instruments on Terra to view the Moon at essentially the same lunar phase and libration. In addition, the SeaWiFS instrument on the SeaStar satellite, at 705 km altitude, 98° inclination, and 12:00 LST equatorial crossing time, was maneuvered to also view the Moon ~17 min after the last instrument on Terra. Using the USGS lunar irradiance model as a reference comparison spectrum and to correct for small differences in lunar phase, the cross-calibration of these EO-1 and Terra instruments using the Moon reproduced the trends seen in previous cross-calibrations using vicarious techniques.

## 10.8 Continuing Issues and New Developments in Earth Remote Sensing

### 10.8.1 Stability, Absolute Calibration, and Characterization

The study of climate change requires a temporally continuous stream of remote-sensing measurements of the Earth made by multiple instruments on different spacecraft platforms. In an effort to understand and predict climate trends, the United States Climate Change Research Program has identified the need to accurately quantify the short- and long-term uncertainties in the fundamental measurements made by those Earth remote-sensing instruments [270]. The application of Earth remote-sensing data in the study of climate change is strongly dependent on the ability of instrument scientists and engineers to separate satellite instrument change from geophysical change. Realization of this goal is fundamentally dependent on the determination and monitoring of on-orbit instrument stability over orbit-to-orbit and mission timeframes. Without on-orbit instrument stability, or the ability to accurately monitor it, the next steps of absolutely calibrating a remote-sensing instrument and cross-calibrating of successive or similar follow-on instruments are difficult if not impossible to accomplish. The production of long-term high-quality geophysical data from multiple satellite instruments strongly depends on confidence that the satellite instruments have been calibrated against common physical standards and the ability of instruments to view the same well-characterized target. A lack or incomplete knowledge of the on-orbit stability of remote-sensing instruments often leads to a significant and often unforeseen investment in interpreting or perhaps salvaging data and often leads to severe delays in the release of validated data to the science public.

The determination of instrument stability is one of several exercises in instrument characterization, which must be addressed not only on-orbit but also pre-launch. Climate requirements on Earth remote-sensing data have driven not only instrument calibration, but also characterization requirements to state-of-the-art levels. Satellite instruments are complex electro-optical systems often with complicated focal planes continuously operating in the changing thermal and radiation environments of space. Complex instruments require more extensive characterization. For example, the increasing use of charge coupled device (CCD) detectors in remote-sensing instruments brings a set of new, additional challenges in instrument characterization. Characterization issues such as light leaks, ghost reflections, optical scattering or blooming, and polarization sensitivity must be understood pre-launch in order to prevent on-orbit "surprises" resulting in delays in the release of science data to the remote-sensing community.

Satellite instrument calibration and characterization ideally should be performed under environmental conditions that mimic on-orbit operation. Certain pre-launch instrument characterization tasks, such as setting instrument gains, can be performed in ambient. However, most characterization tasks must be performed in thermal/vacuum over a range of temperatures encompassing those experienced by the instrument during its on-orbit mission lifetime. Instrument on-orbit stability, particularly long-term stability, is a challenging characterization measurement, which the on-orbit operational environment is approximated by having the instrument view a stable radiance source over a series of repeated thermal cycles. Because instrument thermal-vacuum characterization tests require a significant block of time and are chronologically situated close to the date for instrument delivery to the spacecraft integrator, these tests are often truncated, feebly replaced by a less time consuming ambient surrogate test, or eliminated altogether. The net result is incomplete knowledge of sensor performance with potentially serious on-orbit consequences.

### 10.8.2 New Developments and Future Needs

Recent developments and applications of electro-optical technologies in the field of remote-sensing instrument calibration and characterization have increased the volume and have improved the overall quality of data from space-borne, ground-based, and airborne instrumentation. Close, international-scale, collaborative efforts between a number of remote-sensing institutions and agencies and NMIs have produced many of these advances. For example, the Spectral Irradiance and Radiance Responsivity Calibrations with Uniform Sources (SIRCUS) facility at NIST [51, 82, 271] and the National Laser Radiometry Facility (NLRF) at NPL are detector-based facilities capable of accurately determining the subsystem or system-level absolute spectral irradiance or radiance responsivity of remote-sensing instruments. The facilities' tunable, laser-based light sources, with their high photon fluxes enabling higher signal-to-noise measurements, improve on previous monochromator-based and piece-part calculation approaches for the determination of remote-sensing instrument responsivity. In an effort to adapt to fixed instrument deployment and flight schedules, NIST designed a portable version of their SIRCUS facility called "Travelling SIRCUS" capable of being deployed at satellite and vicarious calibration instrument facilities. In 2003, "Travelling SIRCUS" successfully characterized the spectral-out-of-band response of two CCD-based spectrographs for the MOBY project in Hawaii [272] and measured the relative in-band spectral profiles of eight channels of the Robotic Lunar Observatory's (ROLO's) visible/near infrared telescope system located at the USGS in Flagstaff, Arizona [251]. NMIs have also been actively involved in the pre-launch and on-orbit application of absolute cryogenic radiometry in elucidating the current 0.36% difference between on-orbit measurements of TSI. In the pre-launch laboratory calibration of TSI satellite instruments, the comparison of radiative power measurements made by satellite instruments and absolute cryogenic radiometers is being extensively discussed. NPL's Absolute Radiometric Measurements in Space (ARMS) project proposes to construct the first cryogenic solar radiometer to fly in space, potentially producing a 10-fold improvement in the TSI measurement uncertainty [273]. Another technologically promising area for improvement in remote-sensing data quality is the application of LEDs in the pre-launch and on-orbit calibration of remote-sensing instruments. NIST has developed an LED-based integrating sphere source with spectrally tunable output approximating the color distributions of remotely sensed scenes [50]. Improvements in brightness, stability, and spectral output coupled with low power consumption and insensitivity to gravitational effects, led to the examination of LEDs as possible on-board calibration sources for Earth remote-sensing instruments [274] and, ultimately, to their deployment on instruments such as OMI on EOS Aura. The OMI instrument uses LEDs in the vicinity of their focal planes to perform on-orbit evaluations of electronic gains and pixel performance [145, 146].

There are several areas in the calibration and characterization of Earth remote-sensing instruments that, with additional commitment on the part of remote-sensing agencies and institutions, would significantly improve data quality. One area is the continuing development of celestial objects such as the Moon, stars, and planets as common on-orbit calibration and characterization sources. Since 1993, the ROLO project at USGS in Flagstaff has acquired images of the Moon in 23 visible/near infrared and 9 shortwave infrared bands for purposes of producing models of lunar irradiance and radiance suitable for use in calibrating and cross-calibrating satellite instruments. The ROLO lunar irradiance data has been successfully used to monitor the long-term response degradation of the eight bands of the SeaWiFS instrument to 0.1% [176, 250]. Current work on ROLO is focused on reducing the 5–10% absolute calibration uncertainty of the telescope systems through the determination of the system-level absolute and relative spectral response for the ROLO bands and concurrent examination of the ROLO data reduction, atmospheric correction, and modeling software [251]. The absolute calibration scale for ROLO is currently based on historical, published photometric observations of the star Vega. Therefore, understanding the differences between the Vega-based calibration of ROLO and that using ground-based sources with radiant output traceable to the SI would provide an important link between the astronomical stellar and SI scales.

The radiance calibration specifications for remote-sensing instruments are usually written for uniform, extended scenes such as the output of an integrating sphere, lamp illuminated diffuse target, or blackbody. Unfortunately, most remote-sensing instruments view spatially inhomogeneous scenes of varying contrast. An aspect of quantitative remote-sensing, which is often neglected or not properly understood in instrument characterization, is the radiometric effect of an instrument's encircled energy response function or size of source effect. The size of source effect is a resolution-dependent component in any instrument's measurement uncertainty budget, which due to optical degradation can change over the course of a mission. Instrumentation to quantify the effect pre-launch through careful measurement of instrument point spread and modulation transfer functions should be carefully designed and calibrated [275, 276]. Moreover, the effect should be monitored on-orbit using high-contrast scenes of varying extent and views of celestial objects such as the Moon.

The specification document for a satellite instrument contractually presents to the instrument vendor the performance, testing, assurance, and calibration/characterization requirements for that instrument. For instruments used in the validation or vicarious calibration of satellite instrument measurements, no such specification documents exist. The degree to which ground-based, airborne, or balloon-borne validation instruments are calibrated and characterized is a function of the degree of importance the instrument user or institution places on calibration and characterization. Given the importance of the role field instruments have and will continue to play in the remote-sensing field, the remote-sensing community must be proactive in a number of areas. One area is the development of measurement protocols for field instruments to ensure the production of consistent measurements of sufficiently high quality. A second area is the increased use of workshops to inform and illustrate to members of the remote-sensing community good metrological practices. A third area is participation in measurement comparisons both in the laboratory and in the field. Excellent examples of activities in all these areas were the SIRREXs held on eight occasions between July 1992 and December 201 [46, 277–283]. The purpose of the SIRREX experiments were to ensure that radiometric standards used by institutions involved in the validation of SeaWIFS measurements were consistently calibrated to the same radiance and irradiance scale traceable to the Systeme Internationale (SI). Dovetailed into the SIRREX program was the equally important, clear articulation of protocols for the measurement of ocean color and environmental optical properties [284] by these institutions. These measurement protocols were based on the requirement to validate the SeaWiFS 5% water leaving radiance and 35% chlorophyll a concentration uncertainty specifications. Only through activities such as these will remote sensing data be of sufficiently high quality to be used to validate satellite measurements or to determine satellite instrument calibration coefficients.

## References


*1.* L. J. Allison and E. A. Neal, "Final Report on the TIROS 1 Meteorological Satellite System." NASA Technical Report R-131, NASA's Goddard Space Flight Center, Greenbelt, MD, 1962.

*2.* W. Quinn, V. T. Neal, and S. E. A. Mayolo, El Nino occurrences over the past four and a half centuries, *J. Geophys. Res.* 92, 14449–14461 (1987).

*3.* P. M. Baede, E. Ahlonsou, Y. Ding, and D. Schimel, "The Climate System: An Overview." Climate Change 2001: The Scientific Basis, Third Assessment Report of the Intergovernmental Panel on Climate Change, 2001.

*4.* D. A. Quattrochi, J. C. Luvall, and M. G. Estes, Jr., "Project ATLANTA (Atlanta Land Use Analysis: Temperature and Air quality)—A Study of How the Urban Landscape Affects Meteorology and Air Quality Through Time." 2nd Urban Environments Meeting, American Meteorological Society, Albuquerque, NM, 1998.

*5.* J. T. Kiehl and K. E. Trenberth, Earth's annual global mean energy budget, *Bull. Amer. Meteor. Soc.* 78, 197–208 (1997).

*6.* G. Ohring, B. Weilicki, R. Spencer, W. Emery, and R. Datla, (Eds.), "Satellite Instrument Calibration for Measuring Global Climate Change," p. 101. National Institute of Standards and Technology, Publication NISTIR 7047, Gaithersburg, MD, 2002.

*7.* J. J. Butler and B. C. Johnson, Calibration in the EOS project part 2: Implementation, *The Earth Observer* 8, 26–31 (1996).

*8.* R. P. Cebula, E. Hilsenrath, and B. Guenther, Calibration of the shuttle borne solar backscatter ultraviolet spectrometer, optical radiation measurements II, *Proc. SPIE* 1109, 205–218 (1989).

*9.* D. F. Heath, A. J. Krueger, H. R. Roeder, and B. D. Henderson, The solar backscatter ultraviolet and total ozone mapping spectrometer (SBUV/TOMS) for Nimbus G, *Opt. Eng.* 14, 323–331 (1975).

*10.* J. E. Frederick, R. P. Cebula, and D. F. Heath, Instrument characterization for the detection of long-term changes in stratospheric ozone: An analysis of the SBUV/2 Radiometer, *J. Atmos. Oceanic Technol.* 3, 472–480 (1986).

*11.* J. P. Burrows, M. Weber, M. Buchwitz, V. Rozanov, A. Ladstatter-Weissenmayer, A. Richter, R. De Beek, R. Hoogen, K. Bramstedt, K. W. Eichmann, M. Eisinger, and D. Perner, The global ozone monitoring experiment (GOME): Mission concept and first scientific results, *J. Atmos. Sci.* 56, 151–175 (1999).

*12.* H. Bovensmann, M. Buchwitz, J. Frerick, R. W. M. Hoogeveen, Q. Kleipool, G. Lichtenberg, S. Noel, A. Richter, A. Rozanov, V. V. Rozanov, J. Skupin, C. von Savigny, M. W. Wuttke, and J. P. Burrows, SCIAMACHY on ENVISAT: In-flight optical performance and first results, *Proc. SPIE* 5235, 160–173 (2004).

*13.* S. Noel, H. Bovensmann, J. P. Burrows, J. Frerick, K. V. Chance, A. P. H. Goede, and C. Muller, SCIAMACHY on ENVISAT-1, *Proc. SPIE* 3498, 94–104 (1998).

*14.* J. de Vries, G. H. J. van den Oord, E. Hilsenrath, M. B. te Plate, P. F. Levelt, and R. Dirksen, Ozone monitoring instrument, *Proc. SPIE* 4480, 315–325 (2002).

*15.* W. P. Chu and L. E. Mauldin, Overview of the SAGE III experiment, *Proc. SPIE* 3756, 102–109 (1999).

*16.* L. Flynn, C. Seftor, J. Larsen, and P. Xu, "The Ozone Mapping and Profiler Suite (OMPS)." Book chapter to be published by Springer-Verlag, 2005.

*17.* R. Stolarski, R. Bojkov, L. Bishop, C. Zerofos, J. Stachelin, and J. Zawodny, Measured trends in stratospheric ozone, *Science* 256, 342–349 (1992).

*18.* R. D. Bojkov, L. Bishop, W. J. Hill, G. C. Reinsel, and G. C. Tsao, A statistical trend analysis of revised Dobson total ozone data over the northern hemisphere, *J. Geophys. Res.* 95, 9785–9807 (1990).

*19.* L. L. Hood, R. D. McPeters, J. P. McCormack, L. E. Flynn, S. M. Hollandsworth, and J. F. Gleason, Altitude dependence of stratospheric ozone trends based on Nimbus-7 SBUV data, *Geophys. Res. Lett.* 21, 2667–2670 (1993).

*20.* World Meteorological Organization/United Nations Environment Programme (WMO/UNEP), "Scientific Assessment of Ozone Depletion, Global Ozone Research and Monitoring Project." Report No. 25, 1991.

*21.* World Meteorological Organization/United Nations Environment Programme (WMO/UNEP), "Scientific Assessment of Ozone Depletion, Global Ozone Research and Monitoring Project." Report No. 37, 1994.

*22.* E. C. Weatherhead, G. C. Reinsel, G. C. Tiao, C. H. Jackman, L. Bishop, S. M. Hollandsworth Frith, J. DeLuisi, T. Keller, S. J. Oltmans, E. L. Fleming, D. J. Wuebbles, J. B. Kerr, A. J. Miller, J. Herman, R. McPeters, R. M. Nagatani, and J. E. Frederick, Detecting the recovery of total column ozone, *J. Geophys. Res.* 105, 22201–22210 (2000).

*23.* G. C. Reinsel, E. C. Weatherhead, G. C. Tiao, A. J. Miller, R. M. Nagatani, D. J. Wuebbles, and L. E. Flynn, On detection of turnaround and recovery in trend for ozone, *J. Geophys. Res.* 107, D10, 4078 (2002).

*24.* B. L. Markham, J. C. Storey, D. L. Williams, and J. R. Irons, Landsat sensor performance: History and current status, *IEEE Trans. Geosci. Remote Sens.* 42, 269–2694 (2004).

*25.* W. R. Ott, P. Fieffe-Prevost, and W. L. Wiese, VUV Radiometry with hydrogen arcs. 1: Principle of the method and comparisons with blackbody calibrations from 1650 Angstroms to 3600 Angstroms, *Appl. Opt.* 12, 1618–1629 (1973).

26. W. R. Ott, K. Behringer, and G. Gieres, Vacuum ultraviolet radiometry with hydrogen arcs. 2: The high power arc as an absolute standard of spectral radiance from 124 nm to 360 nm, *Appl. Opt.* 14, 2121–2128 (1975).
27. J. M. Ajello, D. E. Shemansky, B. Franklin, J. Watkins, S. Srivastava, G. K. James, W. T. Simms, C. W. Hord, W. Pryor, W. McClintock, V. Argabright, and D. Hall, Simple ultraviolet calibration source with reference spectra and its use with the Galileo orbiter ultraviolet spectrometer, *Appl. opt.* 27, 890–914 (1988).
28. J. Hollandt, M. C. E. Huber, and M. Kuhne, Hollow-cathode transfer standards for the radiometric calibration of VUV telescopes of the solar and heliospheric observatory (SOHO), *Metrologia* 30, 381–388 (1993).
29. J. Hollandt, M. Kuhne, M. C. E. Huber, and B. Wende, High-current hallow-cathode source as a radiant intensity standard in the 40–125 nm wavelength range, *Appl. Opt.* 33, 68–71 (1994).
30. J. M. Bridges and W. R. Ott, Vacuum ultraviolet radiometry. 3: The argon mini-arc as a new secondary standard of spectral radiance, *Appl. Opt.* 16, 367–376 (1977).
31. J. M. Bridges, Development and calibration of UV/VUV radiometric sources, *Proc. SPIE* 1764, 262–270 (1993).
32. J. M. Bridges, J. Z. Klose, and W. R. Ott, Argon mini-arc source-recent developments, *J. Opt. Soc. Am.* 72, 1804 (1982).
33. R. D. Saunders, W. R. Ott, and J. M. Bridges, Spectral irradiance standard for the ultraviolet: The deuterium lamp, *Appl. Opt.* 17, 593–600 (1978).
34. J. H. Walker, R. D. Saunders, and A. T. Hattenburg, "Spectral Radiance Calibrations," SP 250-1. U.S. Department of Commerce, Gaithersburg, MD.
35. R. P. Lambe, R. D. Saunders, C. Gibson, J. Hollandt, and E. Tegeler, A CCPR international comparison of spectral radiance measurements in the air-ultraviolet, *Metrologia* 37, 51–54 (2000).
36. J. H. Walker, R. D. Saunders, J. K. Jackson, and D. A. McSparron, "Spectral Radiance Calibrations," SP 250-20. U.S. Department of Commerce, Gaithersburg, MD.
37. H. Park, A. Kreuger, E. Hilsenrath, G. Jaross, and R. Haring, Radiometric calibration of second generation total ozone mapping spectrometer, *Proc. SPIE* 2820, 162–173 (1996).
38. D. F. Heath, Z. Wei, W. K. Fowler, and V. W. Nelson, Comparison of spectral radiance calibrations of SBUV-2 satellite ozone monitoring

instruments using integrating sphere and flat-plate diffuser techniques, *Metrologia* 30, 259–264 (1993).

*39.* S. E. Janz, E. Hilsenrath, J. Butler, and R. P. Cebula, Uncertainties in radiance calibrations of backscatter ultraviolet (BUV) instruments as determined from comparisons of BRDF measurements and integrating sphere calibrations, *Metrologia* 32, 637–641 (1996).

*40.* J. H. Walker, C. L. Cromer, and J. T. McLean, A technique for improving the calibration of large-area sphere sources, *Proc. SPIE* 1493, 224–230 (1991).

*41.* K. Smorenburg, M. Dobber, E. Schenkeveld, R. Vink, and H. Visser, Slitfunction measurement optical stimulus, *Proc. SPIE* 4881, 511–520 (2003).

*42.* R. E. Haring, F. L. Williams, U. G. Hartman, D. Becker, G. Vanstone, H. Park, P. K. Bhartia, R. D. McPeters, G. Jaross, and M. Kowalewski, Spectral band calibration of the total ozone mapping spectrometer (TOMS) using a tunable laser technique, *Proc. SPIE* 4135, 421–431 (2000).

*43.* H. W. Yoon, C. E. Gibson, and P. Y. Barnes, Realization of the national institute of standards and technology detector-based spectral irradiance scale, *Appl. Opt.* 41, 5879–5890 (2002).

*44.* R. D. Jackson, T. R. Clarke, and M. S. Moran, Bidirectional calibration results for 11 spectralon and 16 barium sulfate reference reflectance panels, *Remote Sens. Environ.* 40, 231–239 (1992).

*45.* W. A. Hovis and J. S. Knoll, Characteristics of an internally illuminated calibration sphere, *Appl. Opt.* 22, 4004–4007 (1983).

*46.* B. C. Johnson, S. S. Bruce, E. A. Early, J. M. Houston, T. R. O'Brian, A. Thompson, S. B. Hooker, and J. L. Mueller, in "The Fourth SeaWiFS Intercalibration Round-robin Experiment, SIRREX-4, May 1995" (S. B. Hooker and E. R. Firestone, Eds.), NASA Tech. Memo 104566, Vol. 37, p. 65. NASA Goddard Space Flight Center, Greenbelt, MD.

*47.* S. W. Brown and B. C. Johnson, Development of a portable integrating sphere source for the earth observing system's calibration validation programme, *Int. J. Remote Sensing* 24, 215–224 (2003).

*48.* E. A. Early, P. Y. Barnes, B. C. Johnson, J. J. Butler, C. J. Bruegge, S. F. Biggar, P. R. Spyak, and M. M. Pavlov, Bidirectional reflectance round-robin in support of the earth observing system program, *J. Atmos. Oceanic Technol.* 17, 1077–1091 (2000).

*49.* J. J. Butler, B. C. Johnson, S. W. Brown, R. D. Saunders, S. F. Biggar, E. F. Zalewski, B. L. Markham, P. N. Gracey, J. B. Young, and R. A. Barnes, Radiometric measurement comparison on the integrating sphere source used to calibrate the moderate imaging spectroradiometer

(MODIS) and the Landsat 7 enhanced thematic mapper (ETM+), *J. Res. Natl. Inst. Stand. Technol.* 108, 199–228 (2003).
*50.* S. W. Brown and B. C. Johnson, Advances in radiometry for ocean color, *Proc. SPIE* 5151, 441–453 (2003).
*51.* S. W. Brown, G. P. Eppeldauer, and K. R. Lykke, NIST Facility for spectral irradiance and radiance responsivity calibrations with uniform sources, *Metrologia* 37, 579–582 (2000).
*52.* W. G. Rees, "Physical Principles of Remote Sensing." Cambridge University Press, Cambridge, UK, 2003.
*53* P. G. Morse, J. C. Bates, C. R. Miller, M. T. Chahine, F. O'Callaghan, H. H. Aumann, and A. R. Karnik, Development and test of the atmospheric infrared sounder (AIRS), *Proc. SPIE* 3759, 236–253 (1999).
*54.* R. A. Hanel, B. Schlachman, D. Rogers, and D. Vanous, Nimbus 4 Michelson interferometer, *Appl. Opt.* 10, 1376–1382 (1972).
*55.* M. Ridolfi, L. Magnani, M. Carlotti, and B. M. Dinelli, MIPAS-ENVISAT limb-sounding measurements: Trade-off study for improvement of horizontal resolution, *Appl. Opt.* 43, 5814–5824 (2004).
*56.* C. Cao, Personal communication.
*57.* W. P. Menzel, Cloud tracking with satellite imagery: From the pioneering work of Ted Fujita to the present, *Bull. Amer. Meteor. Soc.* 82, 33–47 (2002).
*58.* R. B. Lee, B. R. Barkstrom, G. L. Smith, J. E. Cooper, L. P. Kopia, R. W. Lawrence, S. Thomas, D. K. Pandey, and D. A. Crommelynck, The clouds and the Earth's radiant energy system (CERES) sensors and preflight calibration plans, *J. Atmos. Oceanic Technol.* 13, 300–313 (1996).
*59.* H. Preston-Thomas, *Metrologia*, 27, 3–10, 107 (1990).
*60.* J. R. Chandos and R. E. Chandos, Radiometric properties of isothermal, diffuse wall cavity sources, *Appl. Opt.* 13, 2142–2152 (1974).
*61.* B. Guenther, W. Barnes, E. Knight, J. Barker, J. Harnden, R. Weber, M. Roberto, G. Godden, H. Montgomery, and P. Abel, MODIS calibration: A brief review of the strategy for the at-launch calibration approach, *J. Atmos. Oceanic Technol.* 13, 274–285 (1996).
*62.* T. J. Nightingale and J. Crawford, A radiometric calibration system for the ISAMS remote sounding instrument, *Metrologia* 28, 233–237 (1991).
*63.* A. Ono, F. Sakuma, K. Arai, Y. Yamaguchi, H. Fujisada, P. N. Slater, K. J. Thome, F. D. Palluconi, and H. H. Kieffer, Pre-flight and in-flight calibration plan for ASTER, *J. Atmos. Oceanic Technol.* 13, 321–335 (1996).
*64.* D. A. Byrd, F. D. Michaud, S. C. Bender, A. L. Luettgen, R. F. Holland, W. H. Atkins, T. R. O'Brian, and S. R. Lorentz, Design,

manufacture and calibration of infrared radiometric blackbody sources, *Proc. SPIE* 2743, 216–226 (1996).

65. A. V. Prokhorov, Monte Carlo method in optical radiometry, *Metrologia* 35, 465–471 (1998).

66 A. C. Carter, T. M. Jung, A. Smith, S. R. Lorentz, and R. Datla, Improved broadband blackbody calibrations at NIST for low-background infrared applications, *Metrologia* 40, S1–S4 (2003).

67. R. U. Datla, M. C. Croarkin, and A. C. Parr, Cryogenic blackbody calibrations at the national institute of standards and technology low-background infrared calibration facility, *J. Res. NIST* 99, 77–87 (1994).

68. J. P. Rice and B. C. Johnson, The NIST EOS thermal-infrared transfer radiometer, *Metrologia* 35, 505–509 (1998).

69. J. B. Fowler, A 3rd-generation water bath based blackbody source, *J. Res. NIST* 100, 591–599 (1995).

70. J. B. Fowler, B. C. Johnson, J. P. Rice, and S. R. Lorentz, The new cryogenic vacuum chamber and black-body source for infrared calibrations at NIST's FARCAL facility, *Metrologia* 35, 323–327 (1998).

71. J. P. Rice, S. C. Bender, and W. H. Atkins, Thermal infrared scale verifications at 10 μm using the NIST EOS TXR, *Proc. SPIE* 4135, 96–107 (2000).

72. J. P. Rice, S. C. Bender, W. H. Atkins, and F. J. Lovas, Deployment test of the NIST EOS thermal-infrared transfer radiometer, *Int. J. Remote Sens.* 24, 367–388 (2003).

73. J. P. Rice, "Verification of the GOES Imager Absolute Calibration Model Using the NIST Thermal-infrared Transfer Radiometer," NIST memo. Gaithersburg, MD, 2003.

74. W.H. Farthing, "Imager S/N 03 Channel 2 Non-linearity," Swales Aerospace Memorandum. Beltsville, MD, 1993

75. J. P. Rice, "Measurements of the Raytheon Santa Barbara Remote Sensing Blackbody Calibration Source using the NIST Thermal-infrared Transfer Radiometer," NIST memo. Gaithersburg, MD, 2004.

76. C. L. Wyatt, "Radiometric Calibration: Theory and Methods." Academic Press, New York, 1978.

77. M. H. Weiler, L. Strow, S. Hannon, S. Gaiser, R. Schindler, K. Overoye, and H. H. Aumann, Spectral test and calibration of the atmospheric infrared sounder, *Proc. SPIE* 4483, 44–52 (2002).

78. S. Hansen, J. Peterson, R. Esplin, and J. Tansock, Component level prediction versus system level measurement of SABER relative response, *Int. J. Remote Sens.* 24, 389–402 (2003).

79. C. Cao, M. Weinreb, and S. Kaplan, "Verification of the HIRS Spectral Response Functions for More Accurate Atmospheric Sounding." Proceedings of Calcon, Utah State University, Logan, UT, 2004.

80. C. Cao, H. Xu, J. Sullivan, L. McMillin, P. Ciren, and Y.-T. Hou, *J. Atmos. Oceanic Technol.* 22, 381–395 (2005).
81. S. Kaplan, "Transmittance Measurements of 19 Witness Infrared Bandpass Filters from the HIRS/4 Instrument," NIST memo. Gaithersburg, MD, 2003.
82. S. W. Brown, G. P. Eppeldauer, J. P. Rice, J. Zhang, and K. R. Lykke, Spectral irradiance and radiance responsivity calibrations using uniform sources (SIRCUS) facility at NIST, *Proc. SPIE* 5542, 363–374 (2004).
83. G. P. Eppeldauer, J. P. Rice, J. Zhang, and K. R. Lykke, Spectral irradiance responsivity measurements between 1 μm and 5 μm, *Proc. SPIE* 5543, 248–257 (2004).
84. J. Lean, The Sun's variable radiation and its relevance for earth, *Ann. Rev. Astron. Astrophys.* 35, 33–67 (1997).
85. H. Schwabe, Solar observations during 1843, *Astron. Nachr.* 20, 495 (1843).
86. J. Lean and D. Rind, Climate forcing by changing solar radiation, *J. Climate* 11, 3069–3094 (1998).
87. National Research Council, "Solar Influences on Global Change. National Academy Press." Washington, DC, 1994.
88. C. Frohlich, History of solar radiometry and the world radiometric reference, *Metrologia* 28, 111–115 (1991).
89. K. Angstrom, The absolute determination of the radiation of heat with the electrical compensation pyroheliometer, with examples of the application of this instrument, *Astrophysics* 9, 332 (1899).
90. F. Hengstberger, "Absolute Radiometry: Electrically Calibrated Thermal Detectors of Optical Radiation." Academic Press, Boston, MA, 1989.
91. R. Willson and R. S. Helizon, EOS/ACRIM III instrumentation, *Proc. SPIE* 3750, 233–242 (1999).
92. R. C. Willson, Irradiance observation of SMM spacelab 1, UARS and ATLAS experiment, in "The Sun as a Variable Star: Solar and Stellar Irradiance Variations" (J. M. Pap and C. Frohlich, Eds.). Cambridge University Press, Cambridge, UK, 1994.
93. C. Frohlich, B. N. Anderson, T. Appourchaux, B. Berthomieu, D. A. Crommelynck, V. Domingo, A. Fichot, M. F. Finsterle, M. F. Gomez, D. Gough, A. Jimenez, T. Leifsen, M. Lombaerts, J. M. Pap, J. Provost, T. Roca Cortes, J. Romero, H. Roth, T. Sekii, U. TellJohann, T. Toutain, and C. Wehrli, in "The First Results from SOHO" (B. Fleck and Z. Svestka, Eds.). Kluwer Academic Publishers, Dordrecht, 1997.
94. J. C. Romero, C. Wehrli, and C. Frohlich, Solar total irradiance variability from SOVA2 on board EURECA, *Sol. Phys.* 152, 23–29 (1994).

*95.* D. A. Crommelynck and S. Dewitte, Metrology of total solar irradiance monitoring, *Adv. Space Res.* 24, 195–204 (1999).

*96.* G. M. Lawrence, G. Rottman, G. Kopp, J. Harder, W. McClintock, and T. Woods, The total irradiance monitor (TIM) for the EOS source mission, *Proc. SPIE* 4135, 215–224 (2000).

*97.* D. V. Hoyt, H. L. Kyle, J. R. Hickey, and R. H. Maschoff, The Nimbus 7 solar total irradiance: A new algorithm for its derivation, *J. Geophys. Res.* 97, 148–227 (1992).

*98.* H. L. Kyle, D. V. Hoyt, J. R. Hickey, and G. J. Valette, Nimbus-7 Earth Radiation Budget Calibration History—Part 1: The Solar Channels, NASA Reference Publication 1316, 1993.

*99.* R. C. Willson and A. V. Mordvinov, Secular total solar irradiance trend during solar cycles 21023, *Geophys. Res. Lett.* 30, 1199–1203 (2003).

*100.* C. Frohlich and J. Lean, The Sun's total irradiance: Cycles and trends in the past two decades and associated climate change uncertainties, *Geophys. Res. Lett.* 25, 4377–4380 (1998).

*101.* R. B. Lee III, Long-term total solar irradiance variability during sunspot cycle 22, *J. Geophys. Res.* 100, 1667–1675 (1995).

*102.* A. C. Parr, "A National Measurement System for Radiometry, Photometry and Pyrometry Based Upon Absolute Detectors," Optical Radiation Measurement, NIST Technical Note 1421, 1996.

*103.* G. M. Lawrence, G. Kopp, G. Rottman, J. Harder, T. Woods, and H. Loui, Calibration of the total irradiance monitor, *Metrologia* 40, S78–S80 (2003).

*104.* G. M. Lawrence, G. Rottman, J. Harder, and T. Woods, Solar total irradiance monitor, *Metrologia* 37, 407–410 (2000).

*105.* R. Goebel, M. Stock, and R. Kohler, "Report on the International Comparison of Cryogenic Radiometers Based on Transfer Detectors," BIPM Report B IPM-2000/9, 1–68, 2000.

*106.* B. C. Johnson, M. Litorja, and J. J. Butler, Preliminary results of aperture area comparison for exo-atmospheric solar irradiance, *Proc. SPIE* 5151, 454–461 (2003).

*107.* R. A. Barnes and E. F. Zalewski, Reflectance-based calibration of SeaWiFS. II. Conversion to radiance, *Appl. Opt.* 42, 1648–1660 (2003).

*108.* K. S. Thome, S. Biggar, and P. Slater, Effects of assumed solar spectral irradiance in intercomparisons of Earth observing sensors, *Proc. SPIE* 4540, 260–269 (2001).

*109.* M. P. Thekaekara, Extraterrestrial solar spectrum, 3000–6100 Angstroms at 1-Angstrom intervals, *Appl. Opt.* 13, 518–522 (1974).

*110.* M. P. Thekaekara, Extraterrestrial solar spectral irradiance, in "The Extraterrestrial Solar Spectrum" (A. J. Drummond and

M. P. Thekaekara, Eds.), 71–133. Institute of Environmental Sciences, Mt. Prospect, Illinois, 1973.

*111.* C. Wehrli, "Extraterrestrial Solar Spectrum," Publ. 615. Physikalisch-Meterologisches Observatorium Davos and World Radiation Center, Davos-Dorf, Switzerland, 1985.

*112.* C. Wehrli, "Spectral Solar Irradiance Data," World Climate Research Program (WRCP) Publication Series No. 7, WMO ITD No. 149, 119–126. World Meteorological Organization, Geneva, Switzerland, 1986.

*113.* G. Brasseur and P. C. Simon, Stratospheric chemical and thermal response to long-term variability in solar UV irradiance, *J. Geophys. Res.* 86, 7343–7362 (1981).

*114.* J. C. Arvesen, R. N. Griffen, Jr., and B. D. Pearson, Jr., Determination of extraterrestrial solar spectral irradiance from a research aircraft, *Appl. Opt.* 8, 2215–2232 (1969).

*115.* H. Neckel and D. Labs, The solar radiation between 3300 and 12500 Å, *Sol. Phys.* 90, 205–258 (1984).

*116.* E. V. P. Smith and D. M. Gottlieb, Solar flux and its variations, *Space Sci. Rev.* 16, 771–802 (1974).

*117.* L. Colina, R. C. Bolin, and F. Castelli, The 0.12–2.5 um absolute flux distribution of the Sun for comparison with solar analog stars, *Astronom. J.* 112, 307–315 (1996).

*118.* T. N. Woods, D. K. Prinz, G. J. Rottman, J. London, P. C. Crane, R. P. Cebula, E. Hilsenrath, G. E. Brueckner, M. D. Andrews, O. R. White, M. E. VanHoosier, L. E. Floyd, L. C. Herring, B. G. Knapp, C. K. Pankratz, and P. A. Reiser, The validation of the UARS solar ultraviolet irradiances: Comparison with the ATLAS 1 and 2 measurements, *J. Geophys. Res.* 101, 9541–9569 (1996).

*119.* A. Berk, L. S. Bernstein, and D. C. Robertson, "MODTRAN: A Moderate Resolution Model for LOWTRAN7," Tech. Report GL-TR-90-0122. Geophysical Directorate Phillips Laboratory, Hanscom AFB, Massachusetts, 1990.

*120.* R. L. Kurucz, The solar spectrum: Atlases and line identifications, in "Laboratory and Astronomical High Resolution Spectra" (A. J. Saval, R. Blomme, and N. Grevesse, Eds.), 17–31. Astronomical Society of the Pacific, San Francisco, 1995.

*121.* R. L. Kurucz, The Solar Irradiance by Computation, in "Proceedings of the Conference on Atmospheric Transmission Models (17th)," Tech. Memo. PL-TR-95-2060. (G. P. Anderson, R. H. Pikard and J. H. Chetwynd, Eds.), 333–334. Phillips Laboratory, Hanscomb Air Force Base, MA, 1995.

*122.* G. M. Thuillier, M. Hersé, P. C. Simon, D. Labs, H. Mandel, and D. Gillotay, Observation of the UV solar spectral irradiance between

200 and 350 nm during the Atlas I mission by the SOLSPEC spectrometer, *Sol. Phys.* 171, 283–302 (1997).

*123.* G. M. Thuillier, M. Hersé, P. C. Simon, D. Labs, H. Mandel, D. Gillotay, and T. Foujols, The visible solar spectral irradiance from 350 to 850 nm as measured by the SOLSPEC spectrometer during the Atlas I mission, *Sol. Phys.* 177, 41–61 (1998).

*124.* G. M. Thuillier, M. Hersé, D. Labs, T. Foujols, W. Peetermans, D. Gillotay, P. C. Simon, and H. Mandel, The solar spectral irradiance from 200 to 2400 nm as measured by the solspec spectrometer from the Atlas and Eureca missions, *Sol. Phys.* 214, 1–22 (2003).

*125.* J. W. Harder, G. M. Lawrence, G. J. Rottman, and T. N. Woods, Spectral irradiance monitor (SIM) for the SORCE mission, *Proc. SPIE* 4135, 204–214 (2000).

*126.* J. Harder, G. M. Lawrence, G. Rottman, and T. Woods, Solar spectral irradiance monitor (SIM), *Metrologia* 37, 415–418 (2000).

*127.* W. L. Barnes, T. S. Pagano, and V. V. Salomonson, Prelaunch characteristics of the moderate resolution imaging spectroradiometer (MODIS) on EOS-AM1, *IEEE Trans. Geosci. Remote Sens.* 36, 1088–1100 (1998).

*128.* J. A. Mendenhall, D. E. Lencioni, D. R. Hearn, and C. J. Digenis, Flight tests of the Earth observing-1 advanced land imager, Earth observing systems VII, *Proc. SPIE* 4814, 296–305 (2002).

*129.* H. N. Montgomery, N. Che, and J. Bowzer, Methodology for determining band radiance changes under SRCA illumination, *Proc. SPIE* 3870, 369–376 (1999).

*130.* J. A. Walker, Jr., M. M. Pavlov, and M. L. Byers, Characterization of the solar diffuser as the primary standard for on-orbit calibration of the moderate resolution spectroradiometer, sensors, systems and next-generation satellites II, *Proc. SPIE* 3498, 496–505 (1998).

*131.* B. Guenther, X. Xiong, V. V. Salomonson, W. L. Barnes, and J. Young, On-orbit performance of the earth observing system moderate resolution imaging spectroradiometer: First year of data, *Remote Sens. Environ.* 83, 16–30 (2002).

*132.* R. A. Barnes, R. E. Eplee, Jr., S. F. Biggar, K. J. Thome, P. N. Slater, and A. W. Holmes, SeaWiFS transfer-to-orbit experiment, *Appl. Opt.* 39, 5620–5631 (2000).

*133.* B. Guenther, W. Barnes, E. Knight, J. Barker, J. Harnden, R. Weber, M. Roberto, G. Godden, H. Montgomery, and P. Abel, MODIS calibration: A brief review of the strategy for the at-launch calibration approach, *J. Atmos. Ocean. Technol.* 13, 274–285 (1996).

*134.* X. Xiong, K. Chiang, J. Esposito, B. Guenther, and W. Barnes, MODIS on-orbit calibration and characterization, *Metrologia* 40, S89–S92 (2003).

*135.* W. L. Barnes, X. Xiong, B. W. Guenther, and V. Salomonson, Development, characterization and performance of the EOS MODIS sensors, *Proc. SPIE* 5151, 337–345 (2003).

*136.* C. J. Bruegge, N. C. L. Chrien, B. G. Chafin, D. J. Diner, and R. R. Ando, In-flight calibration of the EOS/Multi-angle imaging spectroradiometer (MISR), *Proc. SPIE* 419, 36–46 (2001).

*137.* N. L. Chrien, C. J. Bruegge, and R. R. Ando, Multi-angle imaging spectroradiometer (MISR) on-board calibrator (OBC) in-flight performance studies, *IEEE Trans. Geosci. Remote Sens.* 40, 1493–1499 (2002).

*138.* M. Rast and J. L. Bezy, The ESA medium resolution imaging spectrometer MERIS: A review of the instrument and its mission, *Int. J. Remote Sens.* 20, 1681–1702 (1999).

*139.* J. A. Mendenhall, D. E. Lencioni, and A. C. Parker, Radiometric calibration of the EO-1 advanced land imager, *Proc. SPIE* 3750, 117–131 (1999).

*140.* K.-H. Sumnich and H. Schwarzer, In-flight calibration of the modular optoelectronic scanner (MOS), *Int. J. Remote Sens.* 19, 3237–3259 (1998).

*141.* S. Kurihara, H. Murakami, K. Tanaka, T. Hashimoto, I. I. Asanuma, and J. Inoue, Calibration and instrument status of ADEOS-II global imager, *Proc. SPIE* 5234, 11–19 (2004).

*142.* G. Jaross, A. J. Krueger, R. P. Cebula, C. J. Seftor, U. Hartmann, R. Haring, and D. Burchfield, Calibration and postlaunch performance of the meteor 3/TOMS instrument, *J. Geophys. Res.* 100, 2985–2995 (1995).

*143.* R. Hoekstra, C. Olij, A. E. Zoutman, and H. G. Werij, Pre- and inflight calibration of GOME, *Proc. SPIE* 2957, 312–321 (1997).

*144.* J. Callies, E. Corpaccioli, M. Eisinger, A. Lefebvre, R. Munro, A. Perez-Albinana, B. Ricciarelli, L. Calamai, G. Gironi, R. Veratti, G. Otter, M. Eschen, and L. van Riel, GOME-2 the ozone instrument on-board the European METOP satellites, polarization science and remote sensing, *Proc. SPIE* 1–11 (2003).

*145.* P. Stammes, P. Levelt, J. de Vries, H. Visser, B. Kruizinga, C. Smorenburg, G. Leppelmeier, and E. Hilsenrath, Scientific requirements and optical design of the ozone monitoring instrument on EOS-CHEM, *Proc. SPIE* 3750, 221–232 (1999).

*146.* P. F. Levelt and R. Noordheok, OMI algorithm theoretical basis document, volume 1: OMI instrument, level 0-1b processor, calibration and operations, available at http://eospso.gsfc.nasa.gov/eos_hompage/for_scientists/atbd/docs/OMI/ATBD-OMI-01.pdf

*147.* E. Zoutman and C. Olij, Calibration approach for SCIAMACHY, *Proc. SPIE* 3117, 306–317 (1997).

*148.* B. L. Markham, W. C. Boncyk, D. L. Helder, and J. L. Barker, Landsat-7 enhanced thematic mapper radiometric calibration, *Can. J. Remote Sens.* 23, 318–332 (1997).

*149.* B. L. Markham, J. L. Barker, E. Kaita, and I. Gorin, Landsat-7 enhanced thematic mapper plus: Radiometric calibration and prelaunch performance, *Proc. SPIE* 3221, 170–177 (1997).

*150.* R. A. Barnes and E. F. Zalewski, Reflectance-based calibration of SeaWiFS, I. Calibration coefficients, *Appl. Opt.*, 1629–1647 (2003).

*151.* C. H. Lyu, W. L. Barnes, and R. A. Barnes, First results from the on-orbit calibrations of the visible and infrared scanner for the tropical rainfall measuring mission, *J. Atmos. Ocean. Technol.* 17, 385–394 (2000).

*152.* P. Jarecke, K. Yokohama, and P. Barry, On-orbit solar radiometric calibration of the hyperion instrument, *Proc. SPIE* 4480, 225–230 (2002).

*153.* R. B. Lee III, Flight solar calibrations using the mirror attenuator mosaic (MAM) low scattering mirror, *Proc. SPIE* 1493, 267–280 (1991).

*154.* R. B. Lee III, B. R. Barkstrom, G. L. Smith, J. E. Cooper, L. P. Kopia, and R. W. Lawrence, The clouds and the Earth's radiant energy system (CERES) sensors and preflight calibration plans, *J. Atmos. Ocean. Technol.* 13, 300–313 (1996).

*155.* J. C. Bremer, H. J. Wood, and G. Si, Attenation of direct solar radiation by a perforated plate for on-orbit calibration, *Proc. SPIE* 4135, 374–383 (2000).

*156.* A. E. Stiegman, C. J. Bruegge, and A. W. Springsteen, Ultraviolet stability and contamination analysis of spectralon diffuse reflectance material, *Opt. Eng.* 32, 799–804 (1993).

*157.* C. J. Bruegge, A. E. Stiegman, R. A. Rainen, and A. W. Springsteen, Use of spectralon as a diffuse reflectance standard for in-flight calibration of Earth-orbiting sensors, *Opt. Eng.* 32, 805–814 (1993).

*158.* E. Hilsenrath, H. Herzig, D. E. Williams, C. J. Bruegge, and A. E. Stiegman, Effects of space shuttle flight on the reflectance characteristics of diffusers in the near-infrared, visible and ultraviolet regions, *Opt. Eng.* 33, 3675–3682 (1994).

*159.* J. de Vries, E. C. Laan, E. Schenkeveld, G. van den Oord, and D. de Winter, OMI flight model performance test results, *Proc. SPIE* 4881, 137–146 (2003).

*160.* J. Q. Sun, X. Xiong, and B. Guenther, MODIS solar diffuser stability monitor performance, *Proc. SPIE* 4483, 156–164 (2001).

*161.* J. R. Herman, P. A. Newman, R. McPeters, A. J. Krueger, P. K. Bhartia, C. J. Seftor, O. Torres, G. Jaross, R. P. Cebula, D. Larko, and

C. Wellmeyer, Meteor 3 total ozone mapping spectrometer observations of the 1993 ozone hole, *J. Geophys. Res.* 100, 2973–2983 (1995).

*162.* F. Sakuma, A. Ono, M. Kudoh, H. Inada, S. Akagi, and H. Ohmae, ASTER on-board calibration status, *Proc. SPIE* 4881, 407–418 (2003).

*163.* J. A. Mendenhall, D. R. Hearn, D. E. Lencioni, C. J. Digenis, and L. Ong, Summary of the EO-1 ALI performance during the first 2.5 years on-orbit, *Proc. SPIE* 5151, 574–585 (2003).

*164.* B. L. Markham, J. C. Seiferth, J. Smid, and J. L. Barker, Lifetime responsivity behavior of the Landsat-5 thematic mapper, *Proc. SPIE* 3427, 420–431 (1998).

*165.* R. R. Turtle, Modified internal calibration lamp source for ETM + on Landsat 7, *Proc. SPIE* 3439, 423–430 (1998).

*166.* S. Thomas, K. J. Priestley, R. B. Lee III, P. L. Spencc, R. S. Wilson, A. Al-Hajjah, J. Paden, and D. K. Pandey, Performance studies of CERES sensors on earth science enterprise (ESE) terra mission using on-borad calibrations and other validation methods, *Proc. SPIE* 4886, 172–181 (2003).

*167.* R. B. Lee III, J. Paden, D. K. Pandey, R. S. Wilson, K. A. Bush, and G. L. Smith, On-orbit radiometric calibrations of the earth radiation budget experiment (ERBE) active-cavity radiometers on the earth radiation budget satellite (ERBS), *Proc. SPIE* 4814, 369–379 (2002).

*168.* J. L. Monge, R. Kandel, L. A. Pakhomov, and V. I. Adasko, The SCaRaB Earth radiation budget scanning radiometer, *Metrologia* 28, 261–264 (1991).

*169.* R. S. Fraser and Y. J. Kaufman, Calibration of satellite sensors after launch, *Appl. Opt.* 25, 1177–1185 (1986).

*170.* E. Vermote, R. Santer, P. Y. Deschamps, and M. Herman, In-flight calibration of large field of view sensors at short wavelengths using rayleigh scattering, *Int. J. Remote Sens.* 13, 3409–3429 (1992).

*171.* Y. J. Kaufman and B. N. Holben, Calibration of the AVHRR visible and near-IR bands by atmospheric scattering, ocean glint and desert reflection, *Int. J. Remote Sens.* 14, 21–52 (1993).

*172.* E. Vermote and Y. J. Kaufman, Absolute calibration of AVHRR visible and near-infrared channels using ocean and cloud views, *Int. J. Remote Sens.* 16, 2317–2340 (1995).

*173.* C. R. N. Rao and J. Chen, Inter-satellite calibration linkages for the visible and near-infrared channels of the advance very high resolution radiometer on NOAA-7, -9 and -11 spacecraft, *Int. J. Remote Sens.* 16, 1931–1942 (1995).

*174.* C. R. N. Rao and J. Chen, Post-launch calibration of the visible and near-infrared channels of the advanced very high resolution radiometer on the NOAA-14 spacecraft, *Int. J. Remote Sens.* 17, 2743–2747 (1996).

175. A. Meygret, S. Briottet, P. Henry, and O. Hagolle, Calibration of SPOT 4 HRVIR and VEGETATION camera over the rayleigh scattering, *Proc. SPIE* 4135, 302–313 (2004).
176. J. J. Butler, G. Meister, F. S. Patt, and R. A. Barnes, Use of the moon as a reference for satellite-based climate change measurements, *Proc. SPIE* 5570, 328–341 (2004).
177. J. C. Bremer, J. G. Baucom, H. Vu, M. P. Weinreb, and N. Pinkine, Estimation of long-term throughput degradation of GOES 8 & 9 visible channels by statistical analysis of star measurements, *Proc. SPIE* 3439, 145–154 (1998).
178. G. J. Rottman and T. N. Woods, In-flight calibration of solar irradiance measurements by direct comparison with stellar observations, *Proc. SPIE* 924, 136–143 (1988).
179. G. J. Rottman, T. N. Woods, and T. P. Sparn, Solar stellar irradiance comparison experiment I: 1. Instrument design and operation, *J. Geophys. Res.* 98, 10667–10677 (1993).
180. T. N. Woods, G. J. Rottman, and G. Ucker, Solar stellar irradiance comparison experiment I: 2. Instrument calibration, *J. Geophys. Res.* 98, 10679–10694 (1993).
181. S. Delwart, L. Bourg, and J. P. Huot, MERIS 1st year: Early calibration results, *Proc. SPIE* 5234, 379–390 (2003).
182. N. Che, X. Xiong, and W. Barnes, On-orbit spectral characterization results for terra MODIS reflective solar bands, *Proc. SPIE* 5151, 367–373 (2003).
183. J. Delderfeld, D. T. Llewellyn-Jones, R. Bernard, Y. deJavel, E. J. Williamson, I. Mason, D. R. Pick, and I. J. Barton, The along track scanning radiometer (ATSR) for ERS1, *Proc. SPIE* 589, 114–120 (1985).
184. R. Bouchard and J. Giroux, Test and qualification results on the MOPITT flight calibration sources, *Opt. Eng.* 36, 2992–3000 (1997).
185. T. S. Pagano, H. H. Aumann, D. E. Kagan, and K. Overoye, Prelaunch and in-flight calibration of the atmospheric infrared sounder (AIRS), *IEEE Trans. Geosci. Remote Sens.* 41, 265–273 (2003).
186. E. M. Sparrow and S. H. Lin, Absorption of thermal radiation in a V-groove cavity, *Int. J. Heat Mass Transfer* 5, 1111–1115 (1962).
187. X. Xiong, K. Chiang, J. Esposito, B. Guenther, and W. L. Barnes, MODIS on-orbit calibration and characterization, *Metrologia* 40, 89–92 (2003).
188. X. Xiong, K. Chiang, B. Guenther, and W. L. Barnes, MODIS thermal emissive bands calibration algorithm and on-orbit performance, *Proc. SPIE* 4891, 392–401 (2002).

*189.* S. Thomas, K. J. Priestley, R. B. Lee III, P. L. Spence, R. S. Wilson, A. Al-Hajjah, and D. K. Pandey, Performance studies of CERES sensors on earth science enterprise (ESE) terra mission using on-board calibrations and other validation methods, *Proc. SPIE* 4886, 172–181 (2003).

*190.* "AVHRR/2 Advanced Very High Resolution Radiometer Technical Description," NASA Contract Number NAS5-26771. ITT Aerospace/ Optical Division, Fort Wayne, IN, 1982.

*191.* A. Schwalb, "The TIROS-N/NOAA-G Satellite Series," NOAA Technical Memorandum NESS 95. Washington, DC, 1978.

*192.* A. Schwalb, "Modified version of the TIROS-N/NOAA A-G satellite series (NOAA E-J)-advanced TIROS-N (ATN)," NOAA Technical Memorandum NESS 116. Washington, DC, 1982.

*193.* A. Ono and F. Sakuma, ASTER Instrument Calibration Plan, *Proc. SPIE* 1939, 198–209 (1993).

*194.* T. Maekawa, O. Nishihara, Y. Aoki, K. Tsubosaka, and S. Kitamura, Design challenges of ASTER in the thermal infrared spectral region, *Proc. SPIE* 1939, 176–186 (1993).

*195.* K. Thome, K. Arai, S. Hook, H. Kieffer, H. Lang, T. Matsunaga, A. Ono, F. Palluconi, F. Sakuma, T. Takashima, H. Tonooka, S. Tsuchida, R. M. Welch, and E. Zalewski, ASTER preflight and inflight calibration and the validation of level 2 products, *IEEE Trans. Geosci. Remote Sens.* 36, 1161–1172 (1998).

*196.* C. Bradley, The GOES I-M system functional description, *NOAA Tech. Rep. NESDIS* 40, 1–126 (1988).

*197.* M. P. Weinreb, "Imager/Sounder Inflight Infrared Calibration and Visible Normalization," GOES I-M Operational Satellite Conference, Arlington, VA, U.S. Department of Commerce, NOAA, 1989, 397–404.

*198.* P. G. Morse, J. C. Bates, C. R. Miller, M. T. Chahine, F. O'Callaghan, H. H. Aumann, and A. R. Karnik, Development and test of the atmospheric infrared sounder (AIRS), *Proc. SPIE* 3759, 236–253 (1999).

*199.* S. L. Gaiser, H. H. Aumann, L. L. Strow, S. E. Hanson, and M. Weiler, In-flight spectral calibration of the atmospheric infrared sounder, *IEEE Trans. Geosci. Remote Sens.* 41, 287–297 (2003).

*200.* R. Beer, T. A. Glavich, and D. M. Rider, Tropospheric emission spectrometer for the earth observing system's aura platform, *Appl. Opt.* 40, 2356–2367 (2001).

*201.* M. Endemann, P. Gare, D. J. Smith, K. Hoerning, B. Fladt, and R. Gessner, MIPAS design overview and current development status, *Proc. SPIE* 2956, 124–135 (1997).

202. R. Gessner, D. J. Smith, M. Kolm, M. Endermann, and P. Gare, MIPAS onboard ENVISAT: Flight model performance, calibration and characterization, *Proc. SPIE* 4169, 133–143 (2001).
203. F. Henault, C. Buil, B. Chidaine, and D. Scheidel, Spaceborne infrared interferometer of the IASI instrument, *Proc. SPIE* 3437, 192–202 (1998).
204. L. Pujol, J. Lizet, and O. Sosnicki, Reference laser source for IASI interferometer, *Proc. SPIE* 4169, 153–158 (2001).
205. P. M. Teillet, K. J. Thome, N. Fox, and J. T. Morisette, Earth observation sensor calibration using a global instrumented and automated network of test sites (GIANTS), *Proc. SPIE* 4540, 246–254 (2001).
206. P. M. Teillet, D. N. H. Horler, and N. T. O'Neill, Calibration, validation and quality assurance in remote sensing: A new paradigm, *Can. J. Remote Sens.* 23, 401–414 (1997).
207. P. N. Slater, S. F. Biggar, K. J. Thome, D. I. Gellman, and P. R. Spyak, Vicarious calibration of EOS sensors, *J. Atmos. Oceanic Technol.* 13, 349–359 (1996).
208. K. Thome, S. Schiller, J. Conel, K. Arai, and S. Tsuchida, Results of the 1996 earth observing system vicarious calibration joint campaign at Lunar Lake Playa, Nevada (USA), *Metrologia* 35, 631–638 (1998).
209. K. P. Scott, K. J. Thome, and M. R. Browlee, Evaluation of the railroad valley playa for use in vicarious calibration, *Proc. SPIE* 2818, paper number 30 (1996).
210. K. J. Thome, S. F. Biggar, and W. Wisniewski, Cross comparison of EO-1 sensors and other earth resources sensors to Landsat-7 ETM+ using railroad valley playa, *IEEE Trans. Geosci. Remote Sens.* 41, 1180–1188 (2003).
211. A. A. Bannari, K. Omari, P. M. Teillet, and G. Fedosejevs, Multisensor and multiscale survey and characterization for radiometric spatial uniformity and temporal stability of railroad valley playa (Nevada) test site used for optical sensor calibration, *Proc. SPIE* 5234, 590–604 (2004).
212. K. J. Thome, S. F. Biggar, and H. J. Choi, Vicarious calibration of terra ASTER, MISR and MODIS, *Proc. SPIE* 5542, 290–299 (2004).
213. K. Thome, B. Markham, J. Barker, P. Slater, and S. Biggar, Radiometric calibration of Landsat, *Photogramm. Eng. Remote Sens.* 63, 853–858 (1997).
214. Z. Wan, Y. Zhang, X. Ma, M. D. King, J. S. Meyers, and X. Li, Vicarious calibration of the moderate resolution imaging spectroradiometer airborne simulator thermal infrared channels, *Appl. Opt.* 38, 6294–6306 (1999).

*215.* P. N. Slater, P. M. Teillet, and Y. Ding, "The Absolute Calibration of the Advanced Very High Resolution Radiometer," Semiannual Report. University of Arizona Optical Sciences Center, Tucson, AZ, 1988.

*216.* E. Villa-Aleman, R. J. Kureja, and M. M. Pendergast, Assessment of Ivanpah playa as a site for thermal vicarious calibration for the MTI satellite, *Proc. SPIE* 5093, 331–342 (2003).

*217.* S. F. Biggar, K. J. Thome, and W. Wisniewski, Vicarious radiometric calibration of EO-1 sensors by reference to high-reflectance ground targets, *IEEE Trans. Geosci. Remote Sens.* 41, 1174–1179 (2003).

*218.* M. S. Moran, R. D. Jackson, T. R. Clarke, J. Qi, F. Cabot, K. J. Thome, and B. Markham, Reflectance factor retrieval for landsat TM and SPOT HRV data for bright and dark targets, *Remote Sens. Environ.* 52, 218–230 (1995).

*219.* R. J. Parada, K. J. Thome, and R. P. Santer, Results of dark target vicarious calibration using lake Tahoe, Europto, *Proc. SPIE* 2957, 332–343 (1996).

*220.* S. J. Hook, G. Chander, J. A. Barsi, R. E. Alley, A. Abtahi, F. D. Palluconi, B. L. Markham, R. C. Richards, S. G. Schladow, and D. L. Helder, In-flight validation and recovery of water surface temperature with Landsat-5 thermal infrared data using an automated high-altitude lake validation site at lake Tahoe, *IEEE Trans. Geosci. Remote Sens.* 42, 2767–2776 (2004).

*221.* R. Santer, X. F. Gu, G. Guyot, J. L. Deuze, C. Devaux, E. Vermote, and M. Verbrugghe, SPOT calibration at the LaCrau Test site, *Remote Sens. Environ.* 41, 227–237 (1992).

*222.* R. P. Santer, C. Six, and J. P. Buis, Vicarious calibration on land site using automatic ground-based optical measurements: Application to SPOT-HRV, *Proc. SPIE* 4891, 524–534 (2003).

*223.* D. K. Clark, M. E. Feinholz, M. A. Yarbrough, B. C. Johnson, S. W. Brown, Y. S. Kim, and R. A. Barnes, Overview of the radiometric calibration of MOBY, *Proc. SPIE* 4483, 64–76 (2001).

*224.* P. M. Teillet, G. Fedosejevs, and R. P. Gauthier, Operational radiometric calibration of broadscale satellite sensors using hyperspectral airborne remote sensing of Prairie Rangeland, *Metrologia* 35, 639–641 (1998).

*225.* R. O. Green, B. E. Pavri, and T. G. Chrien, On-orbit radiometric and spectral calibration characteristics of EO-1 hyperion derived with an underflight of AVIRIS in situ measurements at Salar de Arizaro, Argentina, *IEEE Trans. Geosci. Remote Sens.* 41, 1194–1203 (2003).

*226.* S. Campbell, J. L. Lovell, D. L. B. Jupp, R. D. Graetz, E. A. King, G. Byrne, P. S. Barry, P. Jarecke, and J. Pearlman, The Lake Frome field campaign in support of hyperion instrument calibration and validation, *Proc. IGARSS*, 2593–2595 (2001).

227. R. M. Mitchell, D. M. O'Brien, M. Edwards, C. C. Elsum, and R. D. Graetz, Selection and initial characterization of a bright calibration site in the Strzelecki Desert, South Australia, *Can. J. Remote Sens.* 23, 342–353 (1997).
228. B. Zak, K. Stamnes, and K. Widener, "Site Scientific Mission Plan for the DOE/ARM North Slope of Alaska/Adjacent Arctic Ocean (NSA/AAO) Cloud and Radiation Testbed (CART)," U.S. Department of Energy, Office of Science, Office of Biological Research, ARM-00-002, 2000.
229. R. A. Peppler, D. L. Sisterson, and P. Lamb, "Site Scientific Mission Plan for the Southern Great Plains CART Site, January-June 2000," U.S. Department of Energy, Office of Science, Office of Biological Research, ARM-00-006, 2000.
230. W. E. Clements, L. A. Jones, C. N. Long, and T. P. Ackerman, "Tropical Western Pacific Site Scientific Mission Plan, July-December 2000," U.S. Department of Energy, Office of Science, Office of Biological Research, ARM-00-005, 2000.
231. F. J. Ponzoni, J. Zullo, Jr., R. A. C. Lamparelli, G. Q. Pellegrino, and Y. Arnaud, In-flight absolute calibration of the Landsat-5 TM on the test site Salar de Uyuni, *IEEE Trans. Geosci. Remote Sens.* 42, 2761–2766 (2004).
232. A. J. Prata and R. P. Cechet, An assessment of the accuracy of land surface temperature determination from the GMS-5 VISSR, *Remote Sens. Environ.* 67, 1–14 (1999).
233. A. J. Prata, I. F. Grant, R. P. Cechet, and G. F. Rutter, Five years of shortwave radiation budget measurements at a continental land site in Southeastern Australia, *J. Geophys. Res. Atmos.* 103, 26093–26106 (1998).
234. A. J. Prata, "Precipitable Water Retrieval from Multi-filter Rotating Shadowband Radiometer Measurements," CSIRO Atmospheric Research Technical Paper No. 47, 2000.
235. S. E. Black, D. L. Helder, and S. J. Schiller, Irradiance-based cross-calibration of Landsat-5 and Landsat-7 thematic mapper sensors, *Int. J. Remote Sens.* 24, 287–304 (2003).
236. K. M. Havstad, W. P. Kustas, A. Rango, J. C. Ritchie, and T. J. Schmugge, Jornada experimental range: A unique arid land location for experiments to validate satellite systems, *Remote Sens. Environ.* 74, 13–25 (2000).
237. T. Matsunaga, T. Nonaka, Y. Sawabe, M. Moriyama, H. Tonooka, and H. Fukasawa, "Cross and Vicarious Calibration Experiments for

Terra ASTER and Landsat-7 ETM+ in the Thermal Infrared Region Using Hot Ground Targets," in Proceedings of the CEReS International Symposium on Remote Sensing of the Atmosphere and Validation of Satellite Data, Chiba, Japan, 2001, 143–147.

*238.* M. Edwards and D. Llewellen-Jones, "The AATSR Validation Programme: An Overview," in Proceedings of Envisat Validation Workshop, Frascati, Italy, December 9–13, 2003.

*239.* D. Wu, Y. Yin, Z. Wang, X. Gu, M. Verbrugghe, and G. Guyot, "Radiometric Characterization of Dunhuang Satellite Calibration Test Site (China)," in Physical Measurements and Signatures in Remote Sensing. Balkema, Rotterdam, 1997, 151–160.

*240.* S. F. Biggar, R. P. Santer, and P. Slater, Irradiance-based calibration of imaging sensors, *Proc. IGARSS*, 1283–1285 (1990).

*241.* R. H. Evans and H. R. Gordon, Coastal zone color scanner system calibration—A retrospective examination, *J. Geophys. Res.* 99, 7293–7303 (1994).

*242.* D. K. Clark, H. R. Gordon, K. K. Voss, Y. Ge, W. Brokenow, and C. Trees, Validation of atmospheric correction over oceans, *J. Geophys. Res.* 102, 209–217 (1997).

*243.* R. E. Eplee, W. D. Robinson, S. W. Bailey, D. K. Clark, P. J. Werdell, M. Wang, R. A. Barnes, and C. R. McClain, Calibration of SeaWiFS. II. Vicarious techniques, *Appl. Opt.* 40, 6701–6718 (2001).

*244.* R. A. Barnes, R. E. Eplee, Jr., G. M. Schmidt, F. S. Patt, and C. R. McClain, The calibration of SeaWiFS. I. Direct techniques, *Appl. Opt.* 40, 6682–6700 (2000).

*245.* S. W. Brown, B. C. Johnson, M. E. Feinholz, M. A. Yarbrough, S. J. Flora, K. R. Lykke, and D. C. Clark, Stray-light correction in spectrographs, *Metrologia* 40, S81–S84 (2003).

*246.* H. H. Kieffer and R. L. Wildey, Establishing the Moon as a spectral radiance standard, *J. Atmos. Oceanic Technol.* 13, 360–375 (1996).

*247.* H. H. Kieffer and J. M. Anderson, Use of the Moon for spacecraft calibration over 350–2500 nm, *Proc. SPIE* 3498, 325–336 (1998).

*248.* H. H. Kieffer, T. C. Stone, R. A. Barnes, S. Bender, R. E. Eplee, Jr., J. Mendenhall, and L. Ong, On-orbit radiometric calibration over time and between spacecraft using the Moon, *Proc. SPIE* 4881, 287–298 (2003).

*249.* H. H. Kieffer, Photometric stability of the lunar surface, *Icarus* 130, 323–327 (1997).

*250.* R. A. Barnes, R. E. Eplee, Jr., F. S. Patt, H. H. Kieffer, T. C. Stone, G. Meister, J. J. Butler, and C. R. McClain, Comparison of SeaWiFS measurements of the Moon with the U.S. geological survey lunar model, *Appl. Opt.* 43, 5838–5854 (2004).

*251.* T. C. Stone and H. H. Kieffer, Assessment of uncertainty in ROLO lunar irradiance for on-orbit calibration, *Proc. SPIE* 5542, 300–310 (2004).
*252.* F. Palluconi, H. Tonooka, S. Hook, A. Abtahi, R. Alley, T. Thompson, G. Hoover, and S. Zadourian, EOS ASTER thermal infrared band vicarious calibration, *Proc. SPIE* 4540, 255–259 (2001).
*253.* J. J. Butler and R. A. Barnes, The use of transfer radiometers in validating the visible to shortwave infrared calibrations of radiance sources used by instruments in NASA's earth observing system, *Metrologia* 40, S70–S77 (2003).
*254.* S. W. Brown, B. C. Johnson, H. W. Yoon, J. J. Butler, S. Biggar, P. Spyak, K. Thome, E. Zalewski, M. Helmlinger, C. Bruegge, S. Schiller, G. Fedosejevs, R. Gauthier, S. Tsuchida, and S. Machida, Radiometric comparison of field radiometers in support of the 1997 lunar Lake Nevada experiment to determine surface reflectance and top-of-atmosphere radiance, *Remote Sens. Environ.* 77, 257–376 (1997).
*255.* R. Kannenberg, IR instrument comparison workshop at the Rosenstiel School of marine and atmospheric science (RSMAS), *The Earth Observer* 10, 51–54 (1998).
*256.* J. P. Rice, J. J. Butler, B. C. Johnson, P. J. Minnett, T. J. Nightingale, S. J. Hook, A. Abtahi, C. J. Donlon, and I. J. Barton, The Miami 2001 infrared radiometer calibration and intercomparison: 1. Laboratory characterization of blackbody targets, *J. Atmos. Oceanic Technol.* 21, 268–283 (2004).
*257.* I. J. Barton, P. J. Minnett, K. A. Maillet, C. J. Donlon, S. J. Hook, A. T. Jessup, and T. J. Nightingale, The Miami 2001 infrared radiometer calibration and intercomparison: 2. Ship-board results, *J. Atmos. Oceanic Technol.* 21, 258–267 (2004).
*258.* P. Teillet, G. Fedosejevs, and K. J. Thome, Spectral band difference effects on radiometric cross-calibration between multiple satellite sensors in the Landsat solar-reflective spectral domain, *Proc. SPIE* 5570, 307–316 (2004).
*259.* C. Cao and M. Weinreb, Predicting simultaneous nadir overpasses among polar-orbiting satellites for the intersatellite calibration of radiometers, *J. Atmos. Oceanic Technol.* 21, 537–542 (2004).
*260.* C. Cao and A. Heidinger, Inter-comparison of the longwave infrared channels of MODIS and AVHRR/NOAA-16 using simultaneous nadir observations at orbit intersections, *Proc. SPIE* 4814, 306–316 (2002).
*261.* J. Nieke, T. Aoki, T. Tanikawa, H. Motoyoshi, M. Hori, and Y. Nakajima, Cross-calibration of satellite sensors over snow fields, *Proc. SPIE* 5151, 406–414 (2003).

*262.* D. R. Doelling, L. Nguyen, and P. Minnis, On the use of deep convective clouds to calibrate AVHRR data, *Proc. SPIE* 5542, 281–289 (2004).

*263.* A. K. Heidinger, C. Cao, and J. Sullivan, Using moderate resolution imaging spectroradiometer (MODIS) to calibrate advanced very high resolution radiometer reflectance channels, *J. Geophys. Res.* 107, 11-1-10 (2002).

*264.* F. Cabot, O. Hagolle, C. Ruffel, and P. Henry, Use of a remote sensing data repository for in-flight calibration of optical sensors over terrestrial targets, *Proc. SPIE* 3750, 514–523 (1999).

*265.* C. R. N. Rao, C. Cao, and N. Zhang, Inter-calibration of the moderate resolution imaging spectroradiometer and the along track scanning radiometer-2, *Int. J. Remote Sens.* 24, 1913–1924 (2003).

*266.* P. Minnis, L. Nguyen, D. R. Doelling, D. F. Young, W. F. Miller, and D. P. Kratz, Rapid calibration of operational and research meteorological satellite imagers. Part I: Evaluation of research satellite visible channels as references, *J. Atmos. Oceanic Technol.* 19, 1233–1249 (2002).

*267.* L. Nguyen, D. R. Doelling, P. Minnis, and J. K. Ayers, Rapid technique to cross-calibrate satellite imager visible channels, *Proc. SPIE* 5542, 227–235 (2004).

*268.* P. Minnis, L. Nguyen, D. R. Doelling, D. F. Young, W. F. Miller, and D. P. Kratz, Rapid calibration of operational and research meteorological satellite imagers. Part II: Comparison of infrared channels, *J. Atmos. Oceanic Technol.* 19, 1250–1266 (2002).

*269.* P. M. Teillet, J. L. Barker, B. L. Markham, R. R. Irish, G. Fedosejevs, and J. C. Storey, Radiometric cross-calibration of the Landsat-7 ETM + and Landsat-5 TM sensors based on tandem data sets, *Remote Sens. Environ.* 78, 39–54 (2001).

*270.* Strategic Plan for the U.S. Climate Change Science Program, Climate Change Science Program and Subcommittee on Global Change Research, Washington, DC, 2003.

*271.* G. P. Eppledauer, S. W. Brown, T. C. Larason, M. Racz, and K. R. Lykke, Realization of a spectral radiance responsivity scale with a laser-based source and Si radiance meters, *Metrologia* 37, 531–534 (2000).

*272.* S. W. Brown, B. C. Johnson, S. J. Flora, M. E. Feinholz, R. A. Barnes, Y. S. Kim, K. R. Lykke, and D. K. Clark, Stray light correction of the marine optical buoy, in "Ocean Optics Protocols for Satellite Ocean Color Validation, Rev. 4, Vol VI: Special Topics in Ocean Optics Protocols and Appendices," NASA TM-2003-211621/Rev. 4-Vol. VI, 87–124. Greenbelt, MD, 2003.

273. D. Lizius, Absolute radiometric measurements in space, *ORM News* 2, 5 (1996).
274. J. Nieke, M. Solbrig, K. H. Sumnich, G. Zimmermann, and H. P. Roser, Spaceborne spectrometer calibration with LEDs, *Proc. SPIE* 4135, 384–394 (2000).
275. E. C. Wack and J. E. Baum, Radiometric error in GOES 8 imager data due to sensor MTF, *Proc. SPIE* 3439, 155–164 (1998).
276. S. Qiu, G. Godden, X. Wang, and B. Guenther, Satellite-earth remote sensor scatter effects on earth scene radiometric accuracy, *Metrologia* 37, 411–414 (2000).
277. J. L. Mueller, "The First SeaWiFS Intercalibration Round-robin Experiment, SIRREX-1, July 1992," (S. B. Hooker and E. R. Firestone, Eds.). NASA Tech. Memo 104566, Vol. 14. NASA Goddard Space Flight Center, Greenbelt, MD.
278. J. L. Mueller, B. C. Johnson, C. L. Cromer, J. W. Cooper, J. T. McLean, S. B. Hooker, and T. L. Wesphal, "The Second SeaWiFS Intercalibration Round-robin Experiment, SIRREX-2, June 1993," (S. B. Hooker and E. R. Firestone, Eds.), NASA Tech. Memo 104566, Vol. 16. NASA Goddard Space Flight Center, Greenbelt, MD.
279. J. L. Mueller, B. C. Johnson, C. L. Cromer, S. B. Hooker, J. T. McLean, and S. F. Biggar, "The Third SeaWiFS Intercalibration Round-robin Experiment, SIRREX-3, September 1994," (S. B. Hooker, E. R. Firestone and J. G. Acker, Eds.), NASA Tech. Memo 104566, Vol. 34. NASA Goddard Space Flight Center, Greenbelt, MD.
280. B. C. Johnson, H. W. Yoon, S. S. Bruce, P. S. Shaw, A. E. Thompson, S. B. Hooker, R. A. Barnes, R. Eplee, S. Maritorena, and J. L. Muller, "The Fifth SeaWiFS Intercalibration Round-robin Experiment (SIRREX-5) July 1996," (S. B. Hooker and E. R. Firestone, Eds.), NASA Tech. Memo 1999-206892, Vol. 7. NASA Goddard Space Flight Center, Greenbelt, MD.
281. T. Riley and S. Bailey, "The Sixth SeaWiFS Intercalibration Round-robin Experiment (SIRREX-6) August-December 1997," (S. B. Hooker and E. R. Firestone, Eds.), NASA Tech. Memo 1998-206878, Vol. x. NASA Goddard Space Flight Center, Greenbelt, MD.
282. S. B. Hooker, S. McLean, J. Sherman, M. Small, G. Lazin, G. Zibordi, and J. W. Brown, "The Seventh SeaWiFS Intercalibration Round-robin Experiment (SIRREX-7) March 1999," (S. B. Hooker and E. R. Firestone, Eds.), NASA Tech. Memo 1998-206878, Vol. 17. NASA Goddard Space Flight Center, Greenbelt, MD.
283. G. Zibordi, D. D'Alimonte, D. W. van der Linde, J. F. Berthon, S. Hooker, J. L. Mueller, and G. Lazin, "The Eighth SeaWiFS Intercalibration Round-robin Experiment (SIRREX-8) September-December

2001," (S. B. Hooker and E. R. Firestone, Eds.), NASA Tech. Memo 2002-206892, Vol. 21. NASA Goddard Space Flight Center, Greenbelt, MD.

*284.* J. L. Mueller and R. W. Austin, "Ocean Optics Protocols for SeaWiFS Validation," (S. B. Hooker and E. R. Firestone, Eds.), NASA Tech. Memo 104566, Vol. 5. NASA Goddard Space Flight Center, Greenbelt, MD.

*285.* J. Romero, N. P. Fox, and C. Fröhlich, First comparison of the solar and an SI radiometric scale, *Metrologia*, 28, 125–128 (1991).

*286.* J. Romero, N. P. Fox, and C. Fröhlich, Improved comparison of the World Radiometric Reference and the SI radiometric scale, *Metrologia*, 32, 523–524 (1995).

# APPENDIX A. EXAMPLE: CALIBRATION OF A CRYOGENIC BLACKBODY*

Raju U. Datla, Eric L. Shirley, Albert C. Parr
National Institute of Standards and Technology, Gaithersburg, Maryland, USA

## A.1 Introduction

There are different scenarios for using radiometrically calibrated sensors in space. One scenario is that the sensor is launched into space or stationed in space to identify and track an object such as a missile in its midcourse travel in space. The sensor would see the distant object as a point source against the characteristic space background of a few Kelvin. Radiometric calibration of such a sensor in the laboratory has to be done for the flux levels expected in space using a well-known source such as a point-source blackbody in an environment simulating space. The calibration of such a point-source blackbody in order to use it as a standard for sensor calibration is discussed in this Appendix taking the data from Datla et al. [1] as an example.

## A.2 Calibration of a Cryogenic Point-Source Blackbody

In Section 1.5.1, the basic principles for the point-source blackbody calibration have been discussed. Figure 1.7 illustrates the basic experimental setup with an absolute cryogenic radiometer (ACR) as the detector. The ACR directly measures the radiant power irradiating its precision aperture in watts. The geometric data of the calibration setup for a blackbody calibration at National Institute of Standards and Technology (NIST) is given in Table A.1 with associated uncertainties. The values given in Column 2 are for the cryogenic temperatures of operation and are deduced from ambient measurements using known temperature dependence of expansion of materials.

Table A.2 shows the blackbody measurements [1]. Column 1 shows the nominal setting of the blackbody temperature. Column 2 shows the mean values of measurements of the platinum resistance thermometer mounted on the blackbody core. Column 3 shows the standard deviation for 1 s repetitions at each setting. The measured value of the radiant power $\Phi$ at the

*Contribution of the National Institute of Standards and Technology.



EXPERIMENTAL METHODS IN THE PHYSICAL SCIENCES, vol. 41
ISSN 1079-4042 DOI: 10.1016/S1079-4042(05)41011-5

TABLE A.1. Geometric Data for Cryogenic Blackbody Calibration Setup

| Quantity | Value | Uncertainty in measurement (1$\sigma$) (%) |
|---|---|---|
| 1. Blackbody aperture radius ($r_1$) at 20 K | 0.3244 mm | $\delta r_1/r_1 = 0.2$ |
| 2. Radius ($r_2$) of the ACR aperture at 2.2 K | 1.4971 cm | $\delta r_2/r_2 = 0.003$ |
| 3. Distance between the apertures ($R$) at 20 K | 30.77 cm | $\delta R/R = 0.136$ |

TABLE A.2. Blackbody Measurements, before and after Corrections, and Uncertainties

| (1) Nominal blackbody temp. (K) | (2) Blackbody sensor temp. (K) | (3) Std. dev. (K) | (4) Measured power (nW) | (5) Diffraction-corrected power (nW) | (6) Radiance temp. (K) | (7) Std. dev. $s_{ij}$ (K) | (8) Weight $w_{ij}$ |
|---|---|---|---|---|---|---|---|
| 200 | 199.874 | 0.002 | 72.0 | 73.29 | 201.713 | 1.420 | 0.4959 |
| | 199.897 | 0.005 | 71.9 | 73.19 | 201.643 | 3.500 | 0.0816 |
| | 199.947 | 0.002 | 70.6 | 71.87 | 200.725 | 2.130 | 0.2204 |
| 225 | 224.780 | 0.003 | 113.5 | 115.32 | 225.910 | 2.990 | 0.1119 |
| | 224.728 | 0.003 | 113.4 | 115.21 | 225.861 | 1.384 | 0.5221 |
| | 224.823 | 0.004 | 113.0 | 114.81 | 225.661 | 1.975 | 0.2564 |
| 250 | 249.662 | 0.004 | 171.3 | 173.70 | 250.272 | 2.100 | 0.2268 |
| | 249.639 | 0.007 | 172.6 | 175.02 | 250.750 | 1.166 | 0.7355 |
| | 249.733 | 0.005 | 172.2 | 174.61 | 250.600 | 1.865 | 0.2875 |
| 275 | 274.685 | 0.003 | 252.6 | 255.88 | 275.724 | 1.102 | 0.8234 |
| | 274.644 | 0.006 | 252.2 | 255.48 | 275.620 | 1.319 | 0.5748 |
| | 274.719 | 0.006 | 252.8 | 256.09 | 275.779 | 1.101 | 0.8249 |
| 300 | 299.531 | 0.007 | 358.4 | 362.70 | 300.851 | 0.715 | 1.9561 |
| | 299.521 | 0.003 | 357.3 | 361.59 | 300.620 | 1.135 | 0.7763 |
| | 299.588 | 0.005 | 358.4 | 362.70 | 300.851 | 0.887 | 1.2710 |
| 325 | 324.412 | 0.005 | 497.6 | 503.07 | 326.491 | 0.830 | 1.4516 |
| | 324.457 | 0.005 | 492.8 | 498.22 | 325.701 | 0.603 | 2.7502 |
| | 324.446 | 0.003 | 493.4 | 498.83 | 325.800 | 1.076 | 0.8637 |
| 350 | 349.306 | 0.003 | 665.8 | 672.46 | 351.059 | 0.644 | 2.4112 |
| | 349.337 | 0.008 | 663.7 | 670.34 | 350.782 | 0.997 | 1.0060 |
| | 349.445 | 0.006 | 664.8 | 671.45 | 350.927 | 0.869 | 1.3242 |
| 375 | 374.047 | 0.005 | 882.2 | 891.02 | 376.648 | 0.660 | 2.2957 |
| | 374.069 | 0.003 | 875.7 | 884.46 | 375.953 | 0.816 | 1.5018 |
| | 374.171 | 0.006 | 877.0 | 885.77 | 376.092 | 1.467 | 0.4647 |
| 400 | 399.026 | 0.005 | 1138.0 | 1148.24 | 401.303 | 0.872 | 1.3151 |
| | 399.056 | 0.004 | 1134.6 | 1144.81 | 401.003 | 1.266 | 0.6239 |
| | 399.131 | 0.003 | 1138.8 | 1149.10 | 401.373 | 1.217 | 0.6752 |

TABLE A.3. Calculated Diffraction Correction $(\Delta\Phi/\Phi)$ Given as a Percentage of the Measured Radiant Power ($\Phi$). Systematic Uncertainty (type B) in the Correction Due to All Approximations in the Calculations is $\pm 10\%$ of $\Delta\Phi/\Phi$

| (1) Nominal blackbody temp. (K) | (2) Diffraction correction, $\Delta\Phi/\Phi$ (%) | (3) Uncertainty ($1\sigma$) (%) |
|---|---|---|
| 200 | 1.8 | $\pm 0.18$ |
| 225 | 1.6 | $\pm 0.16$ |
| 250 | 1.4 | $\pm 0.14$ |
| 275 | 1.3 | $\pm 0.13$ |
| 300 | 1.2 | $\pm 0.12$ |
| 325 | 1.1 | $\pm 0.11$ |
| 350 | 1.0 | $\pm 0.10$ |
| 375 | 1.0 | $\pm 0.10$ |
| 400 | 0.9 | $\pm 0.09$ |

ACR aperture is shown in Column 4. Each of the measurements was repeated at 1.5 s intervals for 3 min. Three repeated measurements taken at different times to assess the reproducibility are shown. These measured values need to be corrected for diffraction effects in order to use geometrical optics and deduce the blackbody radiance temperature $T$. The diffraction effects are discussed in Section A.2.1, and the corrections are given in Table A.3. The values of radiant power corrected for diffraction, $\Phi_0$, are shown in Column 5 of Table A.2. The radiance temperature is then deduced by using the Stefan–Boltzmann formula,

$$T = \left( \frac{\Phi_0}{F_{12} A_1 \sigma} \right)^{1/4} \tag{A.1}$$

The value of the Stefan–Boltzmann constant is $\sigma = 5.6704 \times 10^{-8}$ W/m$^2$/K$^4$. The configuration factor $F_{12}$ is evaluated by using the data in Table A.1 and Eq. (1.13) in Section 1.3.3. The deduced radiance temperatures of the blackbody are shown in Table A.2, Column 6. Column 7 shows the standard deviations of the deduced temperatures, $s_{ij}$, using Eq. (A.1) for each of the 120 measurements of radiant flux at each setting.

### A.2.1 Diffraction Correction

As discussed in Chapter 9, the radiant power $\Phi$ from the blackbody incident at the ACR aperture is not solely determined according to geometrical optics because of diffraction effects at both limiting and non-limiting apertures in the beam path shown in Figure A.1. Therefore, diffraction losses at each one of the apertures in the beam path are estimated by using

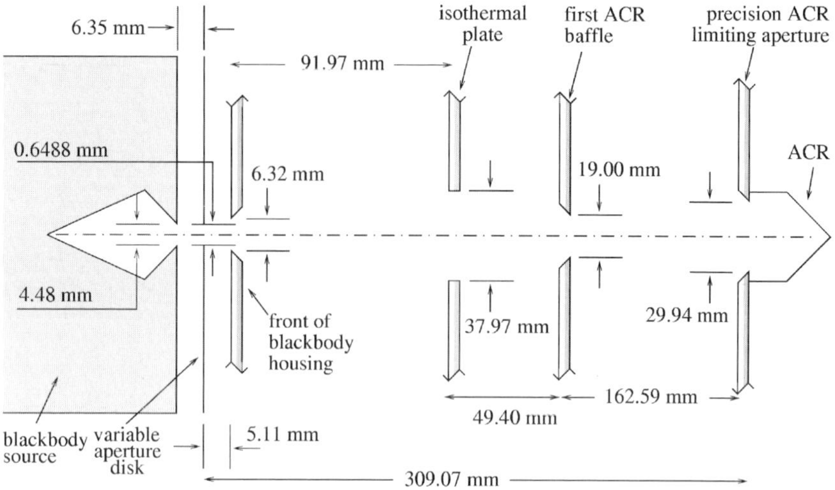


FIG. A.1. Optical setup for blackbody calibration.

the procedures described in Chapter 9. The measured radiant power values are corrected for diffraction effects at each aperture in the beam path. Table A.3 gives the calculated correction factors, $(\Phi_0 - \Phi)/\Phi = \Delta\Phi/\Phi$, as a percentage of the measured radiant power for various temperatures.

Figure A.1 illustrates the geometry of the optical setup. Items shown are not to relative scale. From left to right, they are (1) the blackbody cavity and its opening, (2) the defining aperture (diameter-0.6488 mm) on the variable aperture disk, (3) the baffle on the front of the blackbody housing, (4) an isothermal plate, (5) the first ACR baffle, and (6) the precision ACR limiting aperture. The blackbody cavity opening and its distance from the 0.6488-mm-diameter aperture define the filling angle of the 0.6488 mm aperture. This filling angle is small enough that blackbody radiation underfills the baffle on the front of the blackbody housing, which has minimal diffraction effects. Similarly, the first ACR baffle defines the field of view seen by the ACR, so that the isothermal plate also has minimal diffraction effects. This is fortunate because the perimeter of the latter is not a knife edge and it is hard to characterize related diffraction effects.

There are two main diffraction effects: a loss because of the 0.6488 mm defining aperture, and a gain because of the first ACR baffle. One can estimate the loss by treating the blackbody cavity opening as an extended source, treating the precision ACR limiting aperture as a detector, and considering diffraction effects of the 0.6488 mm defining aperture. One can estimate the gain by treating the 0.6488 mm defining aperture as a source, the precision ACR limiting aperture as a detector, and considering diffraction effects of the first ACR baffle. There are many options for treating diffraction effects, as discussed in greater detail in Chapter 9. For completeness, a few points are also mentioned here. Section 9.4.3.1 discusses how to estimate the loss and gain separately. Section 9.4.3.2 discusses how these effects may be approximately combined as described below.

Diffraction at the 0.6488 mm aperture affects the spectral power $\Phi_\lambda^{(1)}(\lambda)$ reaching the detector in the first hypothetical three-element setup by multiplying it by $F_1(\lambda) = \Phi_\lambda^{(1)}(\lambda)/\Phi_\lambda^{(1,0)}(\lambda)$, where $\Phi_\lambda^{(1,0)}(\lambda)$ is the corresponding ideal spectral power. Diffraction at the first ACR baffle affects the spectral power $\Phi_\lambda^{(2)}(\lambda)$ reaching the detector in the second hypothetical three-element setup by multiplying it by $F_2(\lambda) = \Phi_\lambda^{(2)}(\lambda)/\Phi_\lambda^{(2,0)}(\lambda)$, where $\Phi_\lambda^{(2,0)}(\lambda)$ is the corresponding ideal spectral power. The spectral power $\Phi_\lambda(\lambda)$ in the actual set is multiplied by a spectral diffraction factor $F(\lambda) = \Phi_\lambda(\lambda)/\Phi_\lambda^0(\lambda)$, where $\Phi_\lambda^0(\lambda)$ is the corresponding ideal spectral power. In this case one has $F(\lambda) \approx F_1(\lambda)F_2(\lambda)$.

The more detailed methods discussed in Section 9.5.1.2 can treat diffraction effects more precisely and account for the series of optics through which radiation passes. In the context of such a calculation, one can explicitly show that the diffraction effects of the front of the blackbody housing and isothermal plate are inconsequential. The dashed curve in Figure A.2 indicates $F(\lambda)$. For example, there is about a 2.5% loss of spectral power at 40 μm because of diffraction. In this and many other cases, the methods of

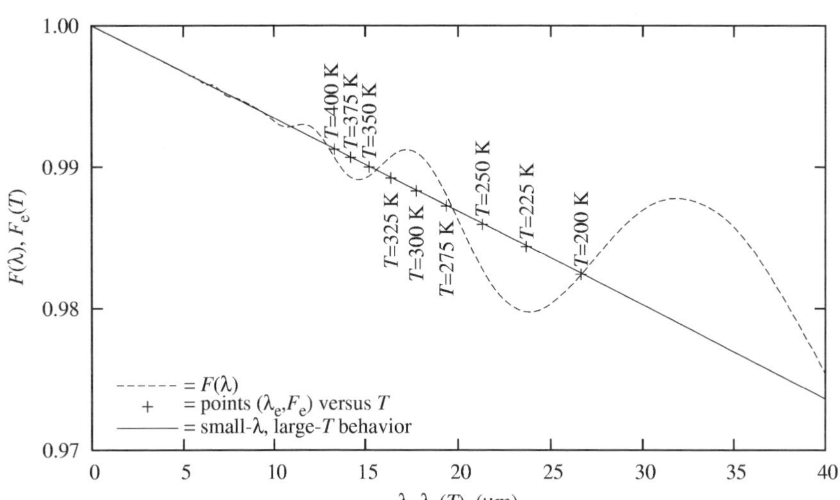


FIG. A.2. Diffraction effects on spectral and total power in setup in Figure A.1.

Sections 9.4.3.1, 9.4.3.2, 9.5.1.2 give very similar results. Diffraction effects on spectral power increase in size with wavelength, being almost linear at the shortest wavelengths, but with an increasingly larger oscillatory part at longer wavelengths.

For diffraction effects on total power $\Phi$, one may introduce an overall effective diffraction factor $F_e = \Phi/\Phi_0 = \int_0^\infty d\lambda\, F(\lambda)\Phi_\lambda^0(\lambda)/\int_0^\infty d\lambda\, \Phi_\lambda^0(\lambda)$, where $\Phi_0$ is the corresponding ideal total power, and one has $\Delta\Phi/\Phi = 1/F_e - 1$. The crosses in Figure A.2 indicate calculated values of $F_e$ at various temperatures. When plotting these results in such a way, the abscissa is $\lambda_e$, the effective wavelength (see Section 9.5.2.1). For a Planck source, one has

$$\lambda_e = \frac{\int_0^\infty d\lambda\, \lambda\Phi_\lambda^0(\lambda)}{\int_0^\infty d\lambda\, \Phi_\lambda^0(\lambda)} = \frac{\int_0^\infty d\lambda\, \lambda^{-4}\{\exp[c_2/(\lambda T)] - 1\}^{-1}}{\int_0^\infty d\lambda\, \lambda^{-5}\{\exp[c_2/(\lambda T)] - 1\}^{-1}} \approx \frac{5326.473\ \mu\text{m K}}{T}.$$

Here $\lambda_e$ is in μm, $T$ is in K, and $c_2 \approx 0.01437752(25)$ m K is the second radiation constant. (In the intermediate steps, there are also other, canceling factors in the numerator and denominator that are discussed in Chapter 1.) The solid curve in Figure A.2 extrapolates the small-$\lambda$ behavior of $F(\lambda)$ (see Section 9.4.2.1) or large-$T$ behavior of $F_e$ (see Section 9.5.2.2) to larger $\lambda$ or $\lambda_e$. The solid curve approximates the crosses very accurately, and much better than would using $F(\lambda_e)$, which would follow the dashed curve. This is a frequent trait of broadband radiation that is discussed further in Section 9.5.2.1.

### A.2.2 Radiance Temperature and Calibration Uncertainty

The final radiance temperatures for the calibration are obtained by the least-squares analysis of temperatures, $T_{ij}(i = 1, \ldots, n;\ j = 1, \ldots, 3)$, deduced from Eq. (A.1) as a function of blackbody sensor settings, $X_{ij}(i = 1, \ldots, n;\ j = 1, \ldots, 3)$. In order to evaluate a confidence band for the variability of the calibration curve, the following statistical procedure is followed. The calibration equation is assumed to follow the model

$$T_{ij} = a_0 + a_1 X_{ij} + a_2 X_{ij}^2 + a_3 X_{ij}^3 + \cdots + a_k X_{ij}^k + \varepsilon_{ij} \tag{A.2}$$

where $a_0, \ldots, a_k$ are to be estimated, and $\varepsilon_{ij}$, the random errors associated with the measurements $T_{ij}$, are assumed to be independent with heterogeneous variances $\sigma_{ij}^2$. The $\sigma_{ij}$ are estimated from the standard deviation $s_{ij}$ from approximately 120 data points for each temperature setting. Weighted least-squares analysis [2] accounts for the heterogeneity of variances where the weights, $w_{ij}$, are calculated from the empirical variances, $s_{ij}^2$, by

$$w_{ij} = 1/s_{ij}^2 \tag{A.3}$$

and are given in Table A.2, Column 8. The standard deviations associated with $T_{ij}$ are an order of magnitude larger than the standard deviations associated with $X_{ij}$ data. Therefore, random uncertainties associated with the measurements $X_{ij}$ are negligible for the purpose of least-squares analysis.

The degree $k$ for the polynomial in Eq. (A.2) is determined by a goodness-of-fit test that compares the agreement among three runs with the overall fit to the data. The lowest degree polynomial which satisfies the goodness-of-fit criterion is taken as the calibration curve. The least-squares analysis model is not an expression of physical law and it is only a statistical approximation for prediction of unknown values based on the given data. Therefore, one should use a minimal number of fitting coefficients [3]. For a blackbody, the ideal $T$ versus $X$ fit would be linear with unit slope and zero bias, i.e. the contact temperature sensor readout would be the same as the radiant temperature measured. If a linear fit is not satisfactory, one should critically examine why it fails before accepting a higher-order polynomial. A deviation from a linear fit in the current experiment would suggest that sources of radiation other than the blackbody are contributing to the radiometer output. These could be objects such as aperture holders and baffles that heat up radiatively due to the blackbody radiation and are in the field of view of the radiometer. In such an event, eliminating these possibilities by proper heat sinking and repeating the experiment would be appropriate. A slope other than unity also suggests a possible, uncorrected error in the presumed blackbody aperture radius $r_1$, radiometer aperture radius $r_2$, or distance between the apertures, $R$, or stray light from the blackbody. However, these errors can sometimes be calibrated out in the linear fit and the resulting expression.

In any case, for a polynomial fit, given a future blackbody sensor setting, $X_{\mathrm{h}}$, its calibrated radiance temperature value is given by

$$T_{\mathrm{h}} = a'_0 + a'_1 X_{\mathrm{h}} + a'_2 X_{\mathrm{h}}^2 + a'_3 X_{\mathrm{h}}^3 + \cdots + a'_k X_{\mathrm{h}}^k \qquad \text{(A.4)}$$

where $a'_0, \ldots, a'_k$ are least-squares estimates from calibration data. The random component of uncertainty $U_{\mathrm{r}}$ associated with the predicted value is computed as

$$U_{\mathrm{r}} = [(k+1)F(95, k+1, n-k-1)]^{1/2} s(T_{\mathrm{h}}) \qquad \text{(A.5)}$$

The constant $F(95, k+1, n-k-1)$ is the upper 95 percentile of Snedecor's $F$-distribution with $k+1$ degrees of freedom in the numerator and $n-k-1$ degrees of freedom in the denominator, and $s(T_{\mathrm{h}})$ is the standard deviation of the predicted value, $T_{\mathrm{h}}$. This uncertainty, based on the Working–Hoetelling confidence bands for the calibration curve, is valid for all future applications of the calibration curve as long as the model holds. The details

TABLE A.4. Predicted Radiance Temperatures (temp.) and Uncertainties (unc.) for Blackbody Nominal Temperature Settings

| (1) Nominal blackbody temp. (K) | (2) PRT sensor output (K) | (3) Predicted radiance temp. (K) | (4) 1σ random unc., $s(T_h)$ (K) | (5) ACR char. (K) | (6) Diffraction calculation unc. (K) | (7) Geometry measurement unc. (K) | (8) Total $b$ (K) | (9) Expanded unc. $U$ (K) |
|---|---|---|---|---|---|---|---|---|
| 200 | 199.92 | 200.48 | 0.11 | 0.06 | 0.09 | 0.24 | 0.26 | 0.75 (0.4%) |
| 225 | 224.78 | 225.52 | 0.09 | 0.07 | 0.09 | 0.27 | 0.29 | 0.80 (0.4%) |
| 250 | 249.68 | 250.60 | 0.07 | 0.08 | 0.09 | 0.30 | 0.32 | 0.87 (0.3%) |
| 275 | 274.68 | 275.78 | 0.06 | 0.09 | 0.09 | 0.33 | 0.35 | 0.94 (0.3%) |
| 300 | 299.55 | 300.83 | 0.04 | 0.09 | 0.09 | 0.36 | 0.38 | 1.01 (0.3%) |
| 325 | 324.45 | 325.91 | 0.04 | 0.10 | 0.09 | 0.39 | 0.41 | 1.09 (0.3%) |
| 350 | 349.36 | 351.01 | 0.04 | 0.11 | 0.09 | 0.42 | 0.44 | 1.17 (0.3%) |
| 375 | 374.12 | 375.94 | 0.06 | 0.11 | 0.09 | 0.45 | 0.47 | 1.25 (0.3%) |
| 400 | 399.07 | 401.07 | 0.07 | 0.12 | 0.09 | 0.48 | 0.50 | 1.33 (0.3%) |

of the statistical analysis can be found in References [4, 5]. Fitting is also discussed in Section 6.16.

In the least-squares analysis, the temperature values measured by the platinum resistance thermometer (PRT) sensor (Table A.2, Column 2) are used for the independent variable, $X$. Analysis shows that a linear function ($k = 1$) is sufficient for describing the data. The equation,

$$T_h = a'_0 + a'_1 X_h \tag{A.6}$$

gives the predicted radiance temperature $T_h$ for a blackbody temperature $X_h$ measured by the PRT sensor, as shown in Columns 3 and 2 of Table A.4, respectively. Column 1 shows the nominal temperature controller setting. The estimated coefficients and associated standard deviations are

$$\begin{aligned} a'_0 &= -0.89, \quad s(a'_0) = 0.26 \\ a'_1 &= +1.007, \quad s(a'_1) = 0.0008 \end{aligned} \tag{A.7}$$

Covariance between the fitted parameters is taken into account when propagating the uncertainties.

The percentage uncertainty in the deduced radiance temperature $\delta T/T$ is given by Eq. (A.1) and the theory of uncertainty propagation [6]. An approximation on a Taylor series expression gives the relationship between the variables as

$$\frac{\delta T}{T} \approx \frac{\delta \Phi}{4\Phi} + \frac{\delta r_1}{2r_1} + \frac{\delta r_2}{2r_2} + \frac{\delta R}{2R} \tag{A.8}$$

The first term on the right-hand side of Eq. (A.8) shows the relationship to the contribution from the uncertainty in the radiant power which has both type A and B components. The type A component of uncertainty is essentially folded into the least-squares analysis through weights given by Eq. (A.3) to each of the deduced temperature readings. So $s(T_h)$, obtained from the least-squares analysis which includes the covariance of the fitting parameters, gives the type A component of uncertainty and is shown in Column 4 of Table A.4. The type B component of the uncertainty due to systematic effects in the first term of Eq. (A.8) are listed separately in Columns 5 and 6. Column 5 gives the uncertainty in the power, $\delta\Phi/(4\Phi)$, due to the characterization of ACR as an absolute detector based on the principle of electrical substitution explained in detail in Chapter 2, and Column 6 gives the uncertainty in diffraction calculations. The other type B components are obtained by propagating the geometry uncertainties given in Table A.1 using the last three terms in Eq. (A.8). Their combined contribution to the temperature measurement uncertainty is shown in Column 7. Column 8 shows the total type B component, $b$, which is the square root of the sum of squares of all type B components given in Columns 5–7. The expanded uncertainty, $U$, as shown in Column 9, is obtained by expanding Eq. (A.5) as follows (see [1], p. 85):

$$U = [(k+1)F(95, k+1, n-k-1)]^{1/2}[s^2(T_h) + b^2]^{1/2} \tag{A.9}$$

The multiplying factor in Eq. (A.9) for $n = 27$, $k = 1$ for the data is 2.6. The value shown in parenthesis in Column 9 is the expanded uncertainty given in Kelvin and as a percentage of the measured temperature in parenthesis. Figure A.3 shows the 95% confidence band for the difference between the calibrated temperature and the PRT temperature. The points represent the same with measured radiance temperatures. The expanded uncertainty given in Table A.4 is within 0.4%. However, it can be seen from Figure A.3 that this model fits well in the middle range of temperatures and not at the extreme low and high temperatures.

The statistical procedure that leads to Eq. (A.9) did not take into account possible correlations between type B components because of the relationship shown in Eq. (A.8). However, calculations of expanded uncertainty, $U$, using the complete covariance matrix and least-squares fitting using appropriate weights have been carried out and the results are found to be the same as given by Eq. (A.9). The calibration constants in Eq. (A.7) are also found to be about the same as reported.

The blackbody calibration discussed above is for one aperture setting. In general, the point-source blackbodies are equipped with a filter wheel with multiple apertures of various sizes to be able to project different flux levels to the sensor. As the aperture size gets smaller, the uncertainty in its area measurement gets larger. It is possible to measure the effective aperture size at the cryogenic temperature of operation radiometrically to reduce the

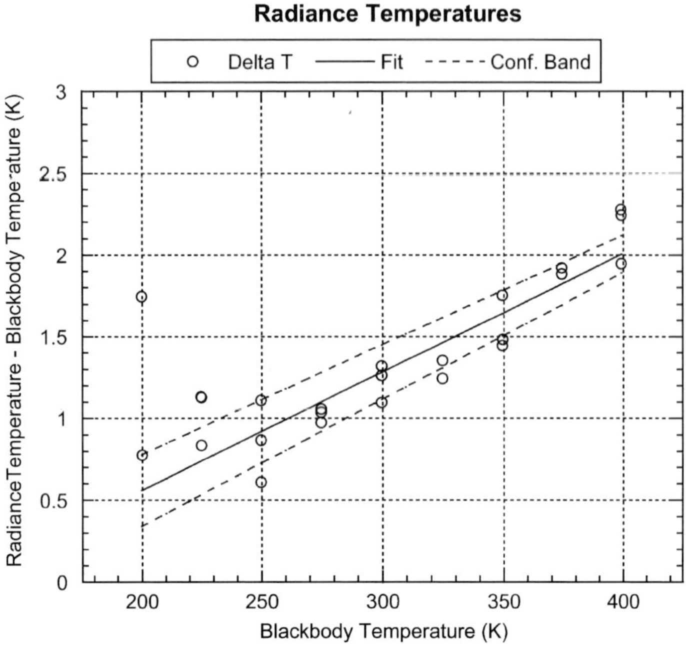


FIG. A.3. The solid line represent difference between the deduced radiance temperature and the temperature measured by the blackbody PRT sensor plotted as a function of the blackbody nominal temperature setting. The circles represent the same for measured radiance temperatures and are not of equal weights. The dashed lines represent upper and lower bounds of 95% confidence bands.

aperture measurement uncertainty. Reference [7] discusses this in detail. The method to follow is to determine the radiance of the blackbody core for its highest possible temperature setting and the largest aperture size to ensure that the diffraction correction and aperture area are characterized as well as possible. Under the assumption that the radiance does not change with aperture size, one can deduce the aperture size from the ACR radiant power measurements for other apertures, which probably involves diffraction corrections as well. In this way, the blackbody calibration described here can be carried out for each aperture using radiometrically determined aperture sizes.

### A.2.3 Example with Fewer Temperature Settings

In cases where measurements are done at only one or two temperature settings, least-squares analysis as described above is not applicable because of insufficient data. Then the ACR measurement of radiant power at each temperature setting is used to deduce the corresponding radiance temperature using Eq. (A.1). Again, the percentage propagated uncertainty in the deduced radiance temperature is given by Eq. (A.8). The type B components are evaluated as discussed in the previous Section A.2.2. The type A uncertainty in the radiant power measurement is analyzed by pooling the variances between groups of measurements. The radiant power data collected for 3 min leading to approximately 120 data points in each group is averaged to determine the average, $\bar{P}$, and the standard deviation, $s_{\mathrm{rpt}}$. It is a measure of the repeatability of the measurements in each set, and in general the ACR data show very good repeatability. At least at three different times, the data are taken to have three groups of data to estimate a standard deviation, $s_{\mathrm{rpr}}$, for reproducibility. The average power for three runs is calculated as

$$\bar{\bar{P}} = \frac{1}{3}\sum_{j=1}^{3} \bar{P}_j \tag{A.10}$$

and a standard deviation, $s_{\mathrm{t}}$, is calculated from

$$s_{\mathrm{t}}^2 = \frac{1}{2}\sum_{j=1}^{3} (\bar{P}_j - \bar{\bar{P}})^2 \tag{A.11}$$

The standard deviation $s_{\mathrm{t}}$ is a measure of both repeatability and reproducibility of the ACR power measurements [1]. The standard deviation of the mean, $s_{\mathrm{t}}/\sqrt{3}$, is the type A uncertainty in the power measurement. In Table A.5, the case for such analysis for measurements at 3 temperature settings are given based on the data of Table A.2. The average power after correcting for diffraction is given in Column 3 and the corresponding uncertainty evaluated using Eq. (A.11) is given in Column 4. Column 5 has the radiance temperature, and the propagated type A uncertainty due to uncertainty in radiant power measurement and total propagated type B uncertainty $b$,

TABLE A.5. Analysis of Uncertainty (unc.) in the Case of Fewer Temperature Settings

| (1) Nominal blackbody temp. (K) | (2) PRT sensor output (K) | (3) Measured diffraction-corrected power (nW) | (4) Std. dev., $s_{\mathrm{t}}$ (nW) | (5) Radiance temp. (K) | (6) Type A propagated unc. (K) | (7) Type B propagated unc. (K) | (8) Expanded unc., $U$ (K) |
|---|---|---|---|---|---|---|---|
| 300 | 299.55 | 361.2 | 0.4 | 300.77 | 0.08 | 0.38 | 0.78 (0.3%) |
| 350 | 349.36 | 669.8 | 0.8 | 350.93 | 0.10 | 0.44 | 0.90 (0.3%) |
| 400 | 399.07 | 1144.6 | 1.3 | 401.22 | 0.11 | 0.50 | 1.12 (0.3%) |

obtained as described in Table A.4, are given in Columns 6 and 7. The relative standard uncertainty, $u_{c,r}(T)$, which is the square root of the sum of squares of the propagated uncertainties, is determined. The expanded uncertainty, $U$, is obtained by multiplying $u_{c,r}(T)$ by the coverage factor 2, which will give less than 95% confidence band because of the small sample size [6]. In any case, the difference between the values predicted in Table A.4 and the deduced temperatures given in Table A.5 are within this expanded uncertainty.

## References


1. R. U. Datla, M. C. Croarkin, and A. C. Parr, Cryogenic blackbody calibrations at the National Institute of Standards and Technology low background infrared calibration facility, *J. Res. Natl. Inst. Stand. Technol.* 99, 77 (1994).
2. N. Draper and H. Smith, "Applied Regression Analysis." Wiley, New York, NY, 1981.
3. G. E. P. Box, W. G. Hunter, and J. S. Hunter, "Statistics for Experimenters." Wiley, New York, NY, 1978.
4. J. Neter, W. Wasserman, and M. H. Kutner, "Applied Linear Statistical Models," 2nd edition. Richard Irwin D. Irwin, Inc., Homewood, IL, 1985.
5. M. G. Natrella, "Experimental Statistics." Wiley, NY, 1966.
6. International Organization for Standardization, "Guide to the Expression of Uncertainty in Measurement." Geneva, Switzerland, 1993.
7. A. W. Smith, A. C. Carter, S. R. Lorentz, T. M. Jung, and R. U. Datla, Radiometrically deducing aperture sizes, *Metrologia* 40, S13 (2003).

# APPENDIX B. UNCERTAINTY EXAMPLE: SPECTRAL IRRADIANCE TRANSFER WITH ABSOLUTE CALIBRATION BY REFERENCE TO ILLUMINANCE

J. L. Gardner
National Measurement Institute, Lindfield, Australia

## B.1 Introduction

We have a lamp or blackbody whose relative spectral irradiance is known. We wish to transfer spectral irradiance values to another lamp by spectral comparison, and then determine the absolute spectral irradiance at a particular distance (in a given direction) from a measurement of illuminance (assuming that the full visible range is covered in the spectral irradiance transfer). What is the uncertainty of the resultant spectral irradiance values? Also, as these values may be combined in subsequent calculations, what are the correlations between the absolute spectral irradiance values?

This detailed example illustrates contribution of uncertainties (and their correlations) of the reference source, systematic and random effects during the spectral transfer, and handles the correlations between the individual spectral irradiance values and the illuminance integral.

Individual measurement effects during the transfer are handled generically as offsets, scaling or wavelength effects. We will consider calibration of a lamp with a distribution temperature of 3200 K against a 2300 K blackbody reference over the spectral range 250–1000 nm at 25 nm intervals. All values in this example are calculated with a value of $1.4388 \times 10^7$ nm K for Planck's second radiation constant $c_2$, and for wavelengths in air and no refractive index correction. All uncertainties will be quoted as standard uncertainties. Proper estimation of the illuminance from spectral data at this interval requires interpolation—any systematic effects due to interpolation are not considered here.

At each wavelength of the spectral comparison, we have the measurement equation for the relative spectral irradiance

$$S'_\lambda = t_\lambda S_{R,\lambda} \tag{B.1}$$

where $t_\lambda$ is the ratio of signals for the test and reference sources and $S_{R,\lambda}$ is the (relative) spectral irradiance of the reference source. The absolute value of spectral irradiance is given by normalising by a scaling constant,

$$S_\lambda = k_N S'_\lambda \tag{B.2}$$



EXPERIMENTAL METHODS IN THE PHYSICAL SCIENCES, vol. 41
ISSN 1079-4042 DOI: 10.1016/S1079-4042(05)41012-7

The illuminance $E_V$ of the source is an integral of the spectral irradiance over the response range of the eye (see Section 7.2.1),

$$E_V = K_m \Delta\lambda \sum_\lambda k_N S'_\lambda V_\lambda \tag{B.3}$$

where $V_\lambda$ is the spectral luminous efficiency function $V(\lambda)$, $K_m$ the maximum spectral luminous efficacy of 683 lm/W, and we have assumed that the data is recorded at a constant wavelength spacing $\Delta\lambda$. Hence

$$S_\lambda = \frac{E_V}{K_m \Delta\lambda \sum_\lambda S'_\lambda V_\lambda} S'_\lambda = \frac{E_V}{K_m \Delta\lambda} R_\lambda \tag{B.4}$$

where $E_V$ is the illuminance measured against photometric reference standards and the ratio

$$R_\lambda = \frac{S'_\lambda}{\sum_\lambda S'_\lambda V_\lambda} \tag{B.5}$$

is a useful form for propagating uncertainties, as we shall see below.

Uncertainty in the illuminance value transfers directly to the absolute spectral irradiance values. Strictly the spectral mismatch factor used to correct the illuminance value for the spectral power distribution of the lamp is correlated to the relative spectral irradiance values $S'_\lambda$, but this correction is small and the correlations are negligible.

Uncertainties are propagated by adding the variance and covariance of each contributing effect. The effects are considered below—here we formulate the required methods. As in Chapter 6, we use the notation $u(S)$ for uncertainty, ($u_s(S)$ for a signed uncertainty when dealing with systematic effects) and $u(S_1, S_2)$ for covariance, with $u^2(S_1) \equiv u(S_1, S_1)$. The subscript number indicates wavelength.

We require the uncertainties and correlations for values of the ratio $R_\lambda$. We first propagate uncertainties of $S'_\lambda$ to those of $\sum_\lambda S'_\lambda V_\lambda$.

(a) Random components of $S'_\lambda$ (Eq. (6.5)):

$$u^2\left(\sum_\lambda S'_\lambda V_\lambda\right) = \sum_\lambda V_\lambda^2 u^2(S'_\lambda) \tag{B.6}$$

(b) Systematic components of $S'_\lambda$ (Eq. (6.70)):

$$u_s\left(\sum_\lambda S'_\lambda V_\lambda\right) = \sum_\lambda V_\lambda u_s(S'_\lambda) \tag{B.7}$$

(Note the inclusion of the signed uncertainty—see Section 6.4.5.)

(c) Partially correlated components of $S'_\lambda$ (explicit form of Eq. (6.17)):

$$u^2\left(\sum_\lambda S'_\lambda V_\lambda\right) = \sum_{\lambda_1}\sum_{\lambda_2} V_{\lambda_1} V_{\lambda_2} u(S'_{\lambda_1}, S'_{\lambda_2}) \tag{B.8}$$

We then need to know the correlation between $S'_\lambda$ and $\sum_\lambda S'_\lambda V_\lambda$ (Eq. (6.19) in sum form).

(a) Random components of $S'_\lambda$: The two terms are correlated through only the component at wavelength $\lambda$. Hence their covariance is

$$u\left(S'_\lambda, \sum_\lambda S'_\lambda V_\lambda\right) = V_\lambda u^2(S'_\lambda) \tag{B.9}$$

(b) Systematic components of $S'_\lambda$: All terms in the sum are now completely correlated to $S'_\lambda$. Hence

$$u\left(S'_\lambda, \sum_\lambda S'_\lambda V_\lambda\right) = u_s(S'_\lambda)\sum_\lambda V_\lambda u_s(S'_\lambda) \tag{B.10}$$

or

$$u\left(S'_\lambda, \sum_\lambda S'_\lambda V_\lambda\right) = u_s(S'_\lambda) u_s\left(\sum_\lambda S'_\lambda V_\lambda\right) \tag{B.11}$$

i.e., $S'_\lambda$ and its weighted sum are also fully correlated.

(c) Partially correlated components of $S'_\lambda$:

$$u\left(S'_\lambda, \sum_\lambda S'_\lambda V_\lambda\right) = \sum_{\lambda_1} V_{\lambda_1} u(S'_\lambda, S'_{\lambda_1}) \tag{B.12}$$

Note that if the correlation coefficient $r$ between pairs of spectral irradiance values is constant, we can separately propagate variances and covariances using the random and systematic expressions, then add them multiplied by $(1 - r)$ and $r$, respectively.

Now we can calculate the uncertainty in the ratio $R_\lambda$ for each type of component. The spectral irradiance values will all be significant and so relative uncertainty is defined at all wavelengths. Hence the relative variance of the ratio in each case is

$$u^2_{\mathrm{rel}}(R_\lambda) = \frac{u^2(S'_\lambda)}{(S'_\lambda)^2} + \frac{u^2(\sum_\lambda V_\lambda S'_\lambda)}{(\sum_\lambda V_\lambda S'_\lambda)^2} - \frac{2u(S'_\lambda, \sum_\lambda V_\lambda S'_\lambda)}{S'_\lambda \sum_\lambda V_\lambda S'_\lambda} \tag{B.13}$$

For fully correlated components, this reduces to

$$u_{\mathrm{s,rel}}(R_\lambda) = \frac{u_{\mathrm{s}}(S'_\lambda)}{S'_\lambda} - \frac{u_{\mathrm{s}}\left(\sum_\lambda V_\lambda S'_\lambda\right)}{\sum_\lambda V_\lambda S'_\lambda} \tag{B.14}$$

This expression is zero for those fully correlated components that affect only the absolute values of the spectral irradiance.

The relative covariance between ratios at different wavelengths is given by

$$u_{\mathrm{rel}}(R_{\lambda_1}, R_{\lambda_2}) = \frac{u(S'_{\lambda_1}, S'_{\lambda_2})}{S'_{\lambda_1} S'_{\lambda_2}} + \frac{u^2\left(\sum_\lambda V_\lambda S'_\lambda\right)}{\left(\sum_\lambda V_\lambda S'_\lambda\right)^2} - \frac{u\left(S'_{\lambda_1}, \sum_\lambda V_\lambda S'_\lambda\right)}{S'_{\lambda_1} \sum_\lambda V_\lambda S'_\lambda} - \frac{u\left(S'_{\lambda_2}, \sum_\lambda V_\lambda S'_\lambda\right)}{S'_{\lambda_2} \sum_\lambda V_\lambda S'_\lambda} \tag{B.15}$$

This is most easily seen by use of Eq. (6.19) in sum form. Equation (B.13) follows from Eq. (B.15) for $\lambda_1 = \lambda_2$.

Relative variances and covariances for the absolute values of spectral irradiance (Eq. (B.4)) are then found by adding the relative variance of the illuminance measurement to those of the ratios. The first term of Eq. (B.15) is zero for random components. For fully correlated components, Eq. (B.15) reduces to

$$u_{\mathrm{rel}}(R_{\lambda_1}, R_{\lambda_2}) = u_{\mathrm{s,rel}}(R_{\lambda_1}) u_{\mathrm{s,rel}}(R_{\lambda_2}) \tag{B.16}$$

## B.2 Reference Lamp Uncertainties and Correlations

From Eq. (B.1) we have

$$u(S'_\lambda) = t_\lambda u(S_{\mathrm{R},\lambda}) = \frac{S'_\lambda}{S_{\mathrm{R},\lambda}} u(S_{\mathrm{R},\lambda}) \tag{B.17}$$

### B.2.1 Blackbody Reference

Suppose that we have calibrated the temperature $T$ of a blackbody reference to be 2300 K with an uncertainty $u(T)$ of 1 K. We assume here that the emissivity is constant throughout the wavelength range of interest. All spectral irradiance values are fully correlated (depending on a single variable), and the correlation is positive for all wavelength pairs. From Planck's law,

$$u(S_{\mathrm{R},\lambda}) = \frac{\partial L_{\mathrm{BB}}(\lambda, T)}{\partial T} u(T) \tag{B.18}$$

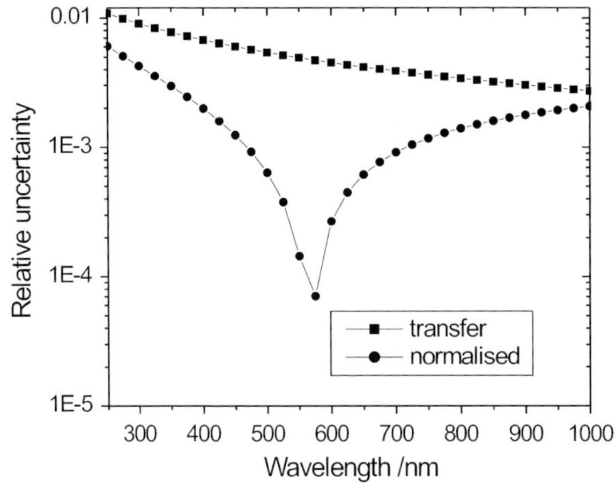


FIG. B.1. Relative uncertainty in the spectral irradiance values as directly transferred and after normalisation by the luminance integral, due to 1 K uncertainty in the reference blackbody temperature (at 2300 K).

The relative uncertainties in the transferred values of spectral irradiance $S'_\lambda$, and those after the normalisation of Eq. (B.4) to form the ratio $R_\lambda$, are shown in Figure B.1. All values are fully correlated. The relative uncertainty of the sum matches that of the transferred spectral irradiance values near the $V(\lambda)$ peak at 555 nm; hence the uncertainty contributed by that of the reference temperature passes through zero, and then at longer wavelengths tends to the constant value of that of the sum term as the relative uncertainty in the blackbody spectral irradiance falls. Normalisation significantly reduces the uncertainty.

## B.2.2 Spectral Irradiance Lamp Reference

Often the reference lamp is not a blackbody, but a spectral irradiance standard provided by a calibration laboratory, typically at a distribution temperature near 3200 K. The uncertainty information provided with the lamp should be given so that total uncertainty and sum of systematic uncertainties (i.e., those correlated between wavelengths) can be extracted. This information is usually given at a limited set of wavelengths rather than at all calibration wavelengths. These values can then be interpolated to all the wavelengths of the calibration. Then the correlation coefficient can be extracted in a manner similar to that given in Section 6.4.1, and the full variance–covariance matrix formed to propagate the partly correlated uncertainties from the reference calibration to the new lamp.

## B.3 Transfer Uncertainties

For transfer uncertainty components,

$$u(S'_\lambda) = S_{R,\lambda} u(t_\lambda) \tag{B.19}$$

The transfer ratio is strictly a ratio of signals, and hence affected by the efficiency of the monochromator. Here we take the signals to be the spectral values themselves. From

$$t_\lambda = \frac{S_\lambda}{S_{R,\lambda}} \tag{B.20}$$

for transfer uncertainty components in the test measurements we have

$$u_s(S'_\lambda) = u(S_\lambda) \tag{B.21}$$

and for those in the reference measurements

$$u_s(S'_\lambda) = -\frac{S_\lambda}{S_{R,\lambda}} u(S_{R,\lambda}) \tag{B.22}$$

(see Section 6.4.5 for a discussion of signed uncertainties).

Scaling factors affecting either signal path can be grouped together (in quadrature if independent, else a linear sum taking into account the sign), and Eq. (B.19) used directly.

### B.3.1 Scaling Factors

Systematic scaling factors (that is, those that multiply the absolute value of the transfer by a constant at all wavelengths such as amplifier gain settings, vignetting by baffles, different source distances) do not affect the uncertainty of the normalised ratio. A systematic drift of any scaling factors will affect the spectral shape, but here we assume that such effects have been corrected and only random effects (e.g., source noise from current fluctuations, amplifier gain fluctuations, vibration of a baffle affecting the beam transmittance, etc.) contribute to the uncertainty. Wavelength-independent random scaling factors can be grouped collectively (in quadrature). These uncertainties are usually given as relative, for which Eq. (B.19) reduces to

$$u(S'_\lambda) = S_\lambda u_{rel}(t_\lambda) \tag{B.23}$$

The lower two curves in Figure B.2 show the contribution from random scaling factors with a relative uncertainty of 0.0005. The $V(\lambda)$ weighting of the correlation between numerator and denominator of the ratio means that the relative uncertainty is summed for the two terms where the value of $V(\lambda)$ is small; the strong correlation near the peak of $V(\lambda)$ reduces the uncertainty of points at nearby wavelengths.

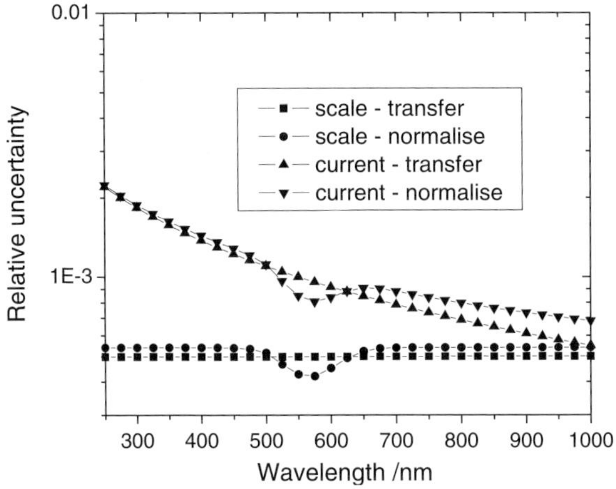


FIG. B.2. Random uncertainty components for source current noise (both lamp and blackbody reference are similar) and for transfer noise, before and after normalisation.

### B.3.2 Current Noise

Current noise is wavelength-dependent, equivalent to a change in temperature of the lamp. A relative uncertainty of 0.001 at a wavelength of 550 nm due to current fluctuations is equivalent to temperature fluctuations of 0.2 and 0.39 K at temperatures of 2300 and 3200 K, respectively. The contributions of current fluctuation in the reference blackbody and lamp cycles are similar, shown as the upper curves in Figure B.2. Similar structure to that of the previous section for a random scaling uncertainty is seen, but now varying with wavelength.

### B.3.3 Offset Factors

For offsets in the signals caused by background radiation, incorrect background subtraction or electronic offsets (all taken here as independent of wavelength), we have $u(S) = c$, with $c$ constant and specified as a fraction of the maximum signal $S$. Equations (B.21) and (B.22) are applied for test and reference signal offsets, respectively. These effects can have both systematic and random (noise) components. Figure B.3 shows the relative uncertainties in spectral irradiance before and after normalisation, for offsets of $10^{-4}$ of maximum signals. The rapid drop in spectral irradiance for the lower temperature blackbody reference at short wavelengths means that the effect is very large for the reference channel; much better background removal is required to use a 2300 K blackbody at 250 nm. Correlation effects

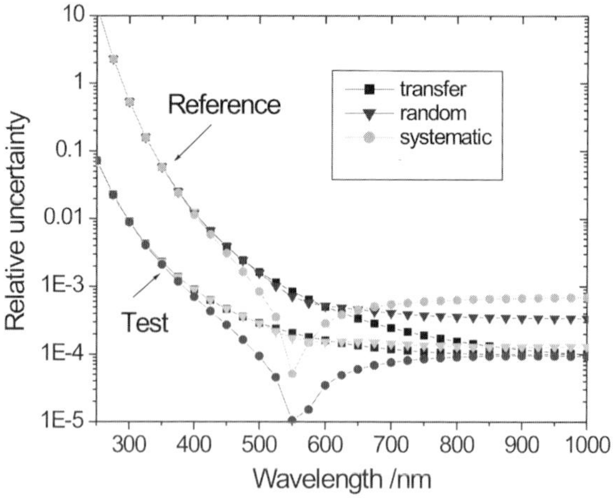


FIG. B.3. Uncertainty contributions from offsets at the level of $10^{-4}$ of the maximum signals in the sample (3200 K) and reference (2300 K) channels.

between individual values and the spectral integral are much larger for the systematic components. Here the systematic and random components have been taken at the same level; in practice, constant systematic offset components have a lower uncertainty than random ones, equivalent to the difference between standard deviation and standard deviation of the mean for a set of data.

### B.3.4 Wavelength Factors

Wavelength uncertainties can also occur as random (accuracy of the calibration of the positioning function) and systematic (offset common to all wavelengths). Assuming that the same wavelength setting is used for test and reference settings, we have

$$u_s(S_\lambda) = \frac{\partial S_\lambda}{\partial \lambda} u(\lambda), \quad u_s(S_{R,\lambda}) = \frac{\partial S_{R,\lambda}}{\partial \lambda} u(\lambda) \tag{B.24}$$

with the two values fully correlated at the one wavelength. Propagating these values to the transferred spectral irradiance using Eqs. (B.21) and (B.22) and then combining them yields

$$u_s(S') = \left( \frac{\partial S_\lambda}{\partial \lambda} - \frac{S_\lambda}{S_{R,\lambda}} \frac{\partial S_{R,\lambda}}{\partial \lambda} \right) u(\lambda) \tag{B.25}$$

Note that the use of signed uncertainties and correlations provides a much more straightforward derivation for wavelength uncertainties than that given in Section 6.5.3. Random setting accuracy can be high, by using high

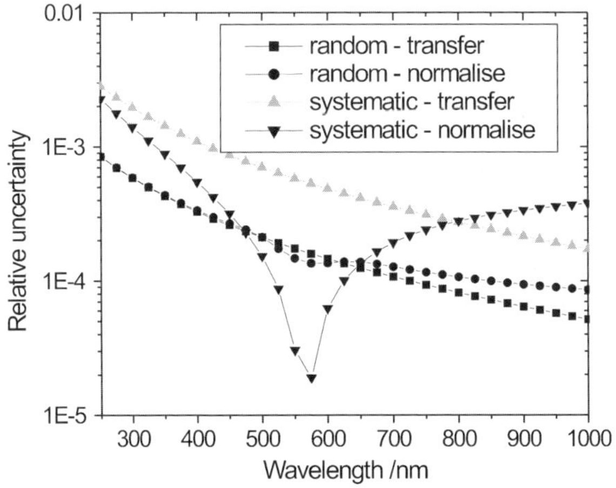


FIG. B.4. Uncertainties arising from random wavelength setting errors of 0.03 nm and systematic offsets of 0.1 nm.

resolution and many well-defined spectral lines during calibration of the wavelength scale. Offsets are more difficult to characterise, because they apply for the generally lower resolution used to increase signal levels during a measurement. Figure B.4 shows the effect of wavelength errors for random uncertainties of 0.03 nm and systematic offset uncertainties of 0.1 nm.

## B.4 Totals

The combined quadrature sum of all the independent effects described above is shown in Figure B.5. Correlations between the individual spectral irradiance values and their illuminance integral significantly reduce transferred uncertainties over most of the spectral range.

To these uncertainties must be added that of the illuminance measurement, fully correlated at all wavelengths. This component dominates in the middle of the spectral range and becomes insignificant at short wavelengths (given the uncertainty values assumed here for all other components). The correlation coefficient between pairs of different wavelengths varies significantly through the spectral range. Figure B.6 shows the surface of resultant correlation coefficients after adding a relative value of 0.3% (the dashed horizontal line shown in Fig. B.5). Correlations are strong for near-neighbours in the visible range. In the UV range, the dominant components (given the values chosen for the components in this example) come from the blackbody reference; Figure B.3 shows that random and systematic

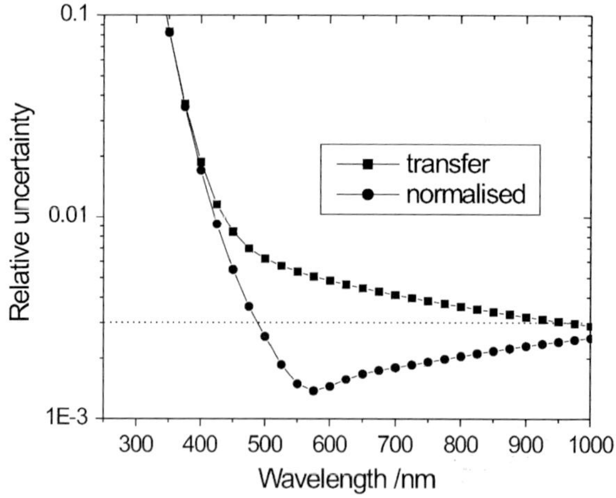


FIG. B.5. Total relative uncertainties of the transferred spectral irradiance values, before and after normalisation. The horizontal line represents the illuminance component to be added.

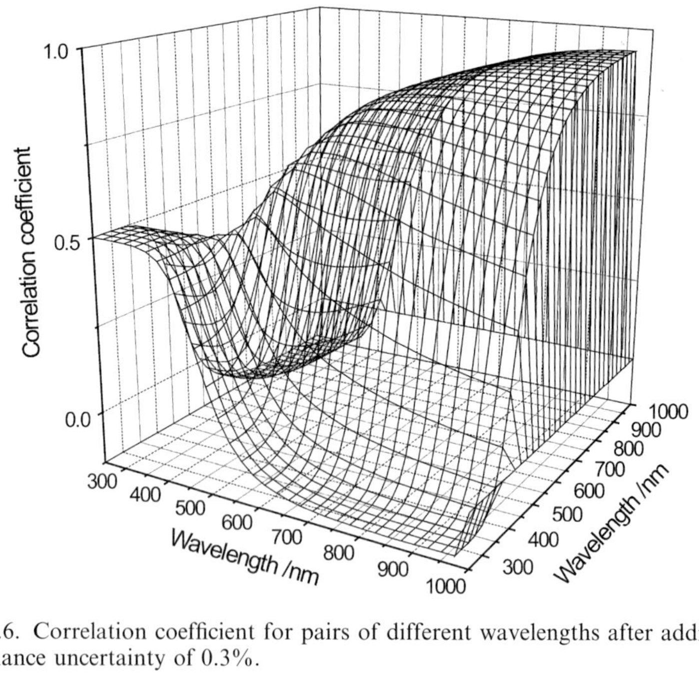


FIG. B.6. Correlation coefficient for pairs of different wavelengths after adding the illuminance uncertainty of 0.3%.

component contribute equally, and hence the correlation coefficient between near-neighbours is 0.5.

These correlations are both complex and important. If we calculate the illuminance of the lamp and propagate uncertainties including these correlations, the relative uncertainty of the illuminance value is 0.3%, as expected (and independent of the wavelength spacing of the transfer). The uncertainty in illuminance calculated from the transfer alone (including the partial correlations of the transfer process but not those of the normalisation) is 0.58%.

Ignoring the correlations leads to propagated uncertainties in illuminance of 0.14% if data are transferred at 25 nm intervals—the spacing used for the above calculations—and to a value of 0.06% if at 5 nm intervals. In practice, treating the complex correlations is difficult through subsequent uncertainty analyses when the spectral irradiance values at different wavelengths are combined. However, we note that for the uncertainty levels given above, systematic effects dominate; if we treat the normalised spectral irradiance values as fully correlated between wavelengths (at least in the visible range), the propagated uncertainty for the illuminance of the lamp becomes 0.34%, an acceptable increase over the true value of 0.30% given the simplicity of propagating the spectral irradiance uncertainty through subsequent combinations.

When calibrating filter radiometres, we combine the spectral irradiance values over a relatively narrow spectral range. Then the spectral irradiance values can be treated as having a constant correlation coefficient, that applying near the band centre; in the UV range this would be 0.5, as discussed above.

# INDEX